National Center for Construction Education and Research

Residential Electrical I

Annotated Instructor's Guide

Prentice Hall

Upper Saddle River, New Jersey
Columbus, Ohio

contren™ Learning Series

National Center for Construction Education and Research
President: Dan Bennet
Vice President of Training Operations and Program Development: Don Whyte
Director of Curriculum Revision and Development: Daniele Dixon
Electrical Project Manager: Daniele Dixon
Production Manager: Debie Ness
Editors: Tara Cohen and Lori Watson
Desktop Publisher: Jessica Martin
Cover Design Concept and Illustration: Buster O'Connor—www.eye4.com

The NCCER would like to acknowledge the contract service provider for this curriculum:
Topaz Publications, Liverpool, New York.

Pearson Education, Inc.

Editor in Chief: Stephen Helba
Product Manager: Lori Cowen
Production Editor: Stephen C. Robb
Design Coordinator: Karrie M. Converse-Jones
Text Design Concept: Rebecca Bobb
Copy Editor: Sheryl Rose
Scanning Coordinator: Karen L. Bretz
Scanning Technician: Janet Portisch
Production Manager: Pat Tonneman

This book was set in Palatino and Helvetica by Carlisle Communications, Ltd. It was printed and bound by Document Technology Resources. The cover was printed by Phoenix Color Corp.

Pearson Education Ltd.
Pearson Education Australia Pty. Limited
Pearson Education Singapore Pte. Ltd.
Pearson Education North Asia Ltd.
Pearson Education Canada, Ltd.
Pearson Educación de Mexico, S.A. de C.V.
Pearson Education—Japan
Pearson Education Malaysia Pte. Ltd.
Pearson Education, *Upper Saddle River, New Jersey*

10 9 8 7 6 5 4 3 2 1
ISBN 0-13-112235-5

Preface

THANK YOU

Thank you for selecting this textbook. We appreciate that you have chosen the Contren™ Learning Series and look forward to working with you as you prepare your students for a career in the construction industry.

PRENTICE HALL

Prentice Hall, the leading educational textbook publisher and part of the Pearson Education family, partnered with the National Center for Construction Education and Research (NCCER) in 1995 to become publisher of NCCER's Contren™ Learning Series. NCCER and Prentice Hall work closely to fulfill their joint goal of bringing together the best resources industry and education have to offer.

NATIONAL CENTER FOR CONSTRUCTION EDUCATION AND RESEARCH

The National Center for Construction Education and Research (NCCER) is a not-for-profit education foundation established by the nation's leading construction companies. NCCER was created to provide the industry with standardized construction education materials, the **Contren™ Learning Series,** and a system for tracking and recognizing students' training accomplishments—**NCCER's National Registry.**

CONTREN™ LEARNING SERIES

The Contren™ Learning Series is comprised of the complete line of NCCER's construction education materials. Contren™ includes curricula for nearly 40 construction crafts in residential, commercial, industrial, maintenance, highway/heavy, and pipeline construction; the *Construction Site Safety Program;* and management leadership programs ranging from *Crew Leader* to *Project Manager.*

The Contren™ Learning Series prepares individuals for every step along the construction career path. Every member of the construction workforce, from entry-level craftsperson to crew leader to project manager, benefits from Contren™. It is the complete construction education package.

Contren™ Development

What sets Contren™ apart from other construction curricula is the development process. Contren™ curricula are not written by one author, but by a team of the finest craft professionals in the country. For each craft, NCCER assembles a team of craft professionals referred to as Subject Matter Experts. The Subject Matter Experts meet frequently with an NCCER representative and a professional writer to develop the curriculum. The process of developing a complete new craft curriculum may take up to two years.

The experience and knowledge of the Subject Matter Experts, combined with NCCER's aggressive revision schedule, ensures exceptional up-to-date training programs that meet or exceed national industry standards.

Contren™ provides total training flexibility. The curricula are modular in format, allowing for program customization. Contren™ is competency-based and contains measurable objectives. Each Contren™ level meets or exceeds the Apprenticeship, Training, Employer, and Labor Services (ATELS) requirement of 144 classroom hours.

Contren™ Features and Supplements

We understand that the course material you choose may have a profound effect on your success. The best way we can help to ensure your success is to provide you with the finest tools we have available. That's why NCCER and Prentice Hall are releasing select titles as Annotated Instructor's Guides. Each Annotated Instructor's Guide is actually the Student Module enhanced with specific directions to the instructor, space for the instructor's notes, suggestions for session break-outs, a comprehensive materials and equipment list, and teaching tips to coincide with the performance examinations, laboratories, and demonstrations.

This Annotated Instructor's Guide is packaged with a separate test booklet that includes the performance tests, written exams, and answer keys. Also included is the instructor's version of the Student Workbook that contains practice exercises that can be assigned as homework or used to complement classroom instruction. Other supplements sold separately are test scrambling software, transparency masters, and PowerPoint® presentation slides with all the illustrations and photos from the Student Guide.

Special pedagogical features throughout the Student Guide augment the technical material to maintain the trainee's interest and foster a deeper appreciation of the trade.

Inside Track provides a head start for those entering the field by presenting tricks of the trade from master electricians.

Think About It uses "What If?" questions to help trainees apply theory to real-world experiences and put ideas into action.

Case History emphasizes the importance of safety by citing examples of the costly (and often devastating) consequences of ignoring safe work practices and OSHA regulations.

Profile in Success shares the apprenticeship and career experiences of various industry leaders, each of whom offers a unique perspective on how to succeed in industry.

Projects Workbook

We are also pleased to offer a hands-on, activities-based workbook designed to supplement the subjects presented in the residential student textbooks. In ***The Household Circuit: Class Projects for the Practice of Electrical Principles and Applications***, you'll find a wide variety of new construction projects; GFCI, duplex, and special receptacle installation projects; lighting installation and repair projects; communication systems projects; and circuit troubleshooting projects. Additionally, we've included a section on projects from SkillsUSA-VICA Championships to help your students to prepare for statewide and national competitions in their trade. SkillsUSA-VICA Championships showcase the best career and technical students in the nation. They involve industry in directly evaluating student performance and keep training relevant to the employer's needs. Each project features materials and tools needed, instructions with illustrations and photos where applicable, and the process or rationale behind the project. The projects also relate back to specific competencies in the tests for extra reinforcement.

If you'd like to order copies of the workbook for your students, please use **ISBN 0-13-102596-1.**

OTHER CONTREN™ TITLES

This textbook includes Contren™ training modules that relate to residential construction. However, NCCER publishes over 1,500 construction modules in nearly 40 crafts. Following is a complete list of crafts:

Boilermaking
Carpentry
Concrete Finishing
Construction Craft Laborer
Core Curriculum
Currículum Básico
Electrical
Electronic Systems Technician
Electrical Topics, Advanced
Heating, Ventilating, and Air Conditioning
Heavy Equipment Operations
Highway/Heavy Construction
Instrumentation
Insulating
Ironworking
Maintenance, Industrial
Masonry
Metal Building Assembly
Millwright
Mobile Crane Operations
Painting
Painting, Industrial
Pipefitting
Pipelayer
Pipeline Control Center Operations
Pipeline Corrosion Control
Pipeline Electrical and Instrumentation

Pipeline Field Operations, Gas
Pipeline Field Operations, Liquid
Pipeline Maintenance
Pipeline Mechanical
Plumbing
Scaffolding
Sheet Metal
Sprinkler Fitting
Welding (AWS)
Welding, Industrial

Modules from any of the above may be purchased separately and used to enhance your current program. Some modules that may be of interest to you include:

From NCCER's Contren™ Electrical Series

26202-02	Motors: Theory and Application
26204-02	Conduit Bending
26206-02	Conductor Installations
26207-02	Cable Tray
26211-02	Contactors and Relays
26301-02	Load Calculations—Branch and Feeder Circuits
26302-02	Conductor Selection and Calculations
26303-02	Overcurrent Protection
26306-02	Distribution Equipment
26307-02	Distribution System Transformers
26309-02	Motor Calculations
26310-02	Motor Maintenance
26311-02	Motor Controls
26401-03	Load Calculations—Feeder and Services
26402-03	Practical Applications of Lighting
26403-03	Standby and Emergency Systems
26404-03	Basic Electronic Theory
26405-03	Fire Alarm Systems
26406-03	Specialty Transformers
26407-03	Advanced Motor Controls
26408-03	HVAC Controls
26409-03	Heat Tracing and Freeze Protection
26410-03	Motor Maintenance Part Two
26411-03	High-Voltage Terminations/Splices

From NCCER's Contren™ Carpentry Series

27101-01	Orientation to the Trade
27102-01	Wood Building Materials, Fasteners, and Adhesives
27103-01	Hand and Power Tools
27203-01	Introduction to Concrete and Reinforcing Materials
27204-01	Foundations and Flatwork
27205-01	Concrete Forms
27104-01	Floor Systems
27105-01	Wall and Ceiling Framing
27106-01	Roof Framing
27107-01	Windows and Exterior Doors
27201-01	Reading Plans and Elevations
27202-01	Site Layout One: Distance Measurement and Leveling
27301-02	Exterior Finishing
27302-02	Roofing Applications
27303-02	Thermal and Moisture Protection
27304-02	Stairs
27305-02	Framing with Metal Studs
27308-02	Interior Finish One: Doors
27310-02	Interior Finish Three: Windows, Door, Floor, and Ceiling Trim
27311-02	Interior Finish Four: Cabinet Installation

From NCCER's Contren™ Plumbing Series

02101-00	Introduction to the Plumbing Trade
02102-00	Plumbing Tools
02103-00	Introduction to Plumbing Math
02106-00	Copper Pipe and Fittings
02107-00	Cast-Iron Pipe and Fittings
02105-00	Plastic Pipe and Fittings
02108-00	Carbon Steel Pipe and Fittings
02104-00	Introduction to Plumbing Drawings
02110-00	Introduction to Drain, Waste, and Vent (DWV) Systems
02111-00	Introduction to Water Distribution Systems
02209-01	Fuel Gas Systems
02208-01	Installing Water Heaters
02109-00	Fixtures and Faucets

Visit www.crafttraining.com for a complete list of Contren™ modules.

PROGRAM RESOURCES

NCCER'S National Registry

NCCER's system for tracking and recognizing students' training accomplishments is the National Registry. If your training facility is an NCCER Accredited Training Sponsor, an NCCER Accredited Training Unit, or an NCCER Training Unit, your students are eligible to receive the following industry credentials:

Transcript—Successful completion of the written and performance tests for a training module

Wallet Card *and* **Craft Level Certificate**—Successful completion of all the modules within an entire craft level

Craft Completion Certificate—Successful completion of all levels of a craft

These industry credentials can help your students in many ways, including finding employment. Many leading construction companies recognize NCCER credentials nationwide. The credentials may help them as they progress onto other NCCER training. Students can receive credit for the modules and levels that they have completed and can continue to build their education rather than repeat it.

NCCER credentials are important to your students. Make sure they receive them!

NCCER Partners

NCCER has over 30 national industry partners. These partnering associations can be a valuable resource to you and your students.

- American Fire Sprinkler Association
- American Petroleum Institute
- American Society for Training and Development
- American Welding Society
- Associated Builders and Contractors, Inc.
- Association for Career and Technical Education
- Associated General Contractors of America
- Carolinas AGC, Inc.
- Carolinas Electrical Contractors Association
- Citizens Democracy Corps
- Construction Industry Institute
- Construction Users Roundtable
- Design-Build Institute of America
- Merit Contractors Association of Canada
- Metal Building Manufacturers Association
- National Association of Minority Contractors
- National Association of State Supervisors for Trade and Industrial Education
- National Association of Women in Construction
- National Insulation Association
- National Ready Mixed Concrete Association
- National Utility Contractors Association
- National Vocational Technical Honor Society
- North American Crane Bureau
- Painting and Decorating Contractors of America
- Portland Cement Association
- SkillsUSA-VICA
- Steel Erectors Association of America
- Texas Gulf Coast Chapter ABC
- U.S. Army Corps of Engineers
- University of Florida
- Women Construction Owners and Executives, USA

NCCER Accredited Training Sponsors

Visit NCCER's Web site at www.nccer.org to obtain a current list of NCCER Accredited Training Sponsors. Contact sponsors in your area to establish the industry linkage needed for your program to succeed.

NCCER's Website

NCCER's Web site, www.nccer.org, is a valuable resource for your program. On NCCER's Web site you will find School-to-Career resources, all NCCER forms, NCCER's Accreditation Guidelines, Contren™ online catalog, and industry image materials, to name just a few items.

Acknowledgments

This curriculum was produced as a result of the farsightedness and leadership of the following sponsors:

Beacon Electric Company
Circle Electric, Inc.
Cypress Electric, an IES Company
Duck Creek Engineering
Encompass Electrical Technologies –Florida, LLC
Leading Edge Electrical Services
Pelican Chapter ABC
San Diego ABC
TIC, The Industrial Company
Washington CITC
Washington Rust Constructors Inc.
Wisconsin ABC
Zachry Construction

This curriculum would not exist were it not for the dedication and unselfish energy of those volunteers who served on the Authoring Team. A sincere thanks is extended to:

Clarence "Ed" Cockrell
Randy Cole
Gary Edgington
Tim Ely
E.L. Jarrell
Dan Lamphear
Leonard R. "Skip" Layne
L.J. LeBlanc
Daniel M. LePage
Neil Matthes
Jim Mitchem
Robert Mueller
Christine Porter
Mike Powers

Contents

Chapter 1

Electrical Safety

26101-02

MODULE OVERVIEW

This course introduces the electrical trainee to the safety rules and regulations for electricians, including the necessary precautions for avoiding various job-site hazards.

PREREQUISITES

Please refer to the Course Map in the Trainee Module. Prior to training with this module, it is recommended that the trainee shall have successfully completed the following modules:

Core Curriculum.

LEARNING OBJECTIVES

Upon completion of this module, the trainee will be able to:

1. Demonstrate safe working procedures in a construction environment.
2. Explain the purpose of OSHA and how it promotes safety on the job.
3. Identify electrical hazards and how to avoid or minimize them in the workplace.
4. Explain safety issues concerning lockout/tagout procedures, personal protection using assured grounding and isolation programs, confined space entry, respiratory protection, and fall protection systems.

PERFORMANCE OBJECTIVES

Under supervision of the instructor, the trainee should be able to:

1. Perform a visual inspection and an air test on rubber gloves.
2. Perform a hazard assessment of a job such as replacing the lights in your classroom.
   - Discuss the work to be performed and the hazards involved.
   - Locate the closest phone to the work site and ensure that the local emergency telephone numbers are either posted at the phone or known by you and your partner(s).
   - Plan an escape route from the location in the event of an accident.

NCCER STANDARDIZED CRAFT TRAINING PROGRAM

The National Center for Construction Education and Research (NCCER) provides a standardized national program of accredited craft training. Key features of the program include instructor certification, competency-based training, and performance testing. The program provides trainees, instructors, and companies with a standard form of recognition through a National Craft Training Registry. The program is described in full in the *Guidelines for Accreditation*, published by the NCCER. For more information on standardized craft training, contact the NCCER at P.O. Box 141104, Gainesville, FL 32614-1104, 352-334-0911, visit our Web site at www.nccer.org, or e-mail info@nccer.org.

HOW TO USE THIS ANNOTATED INSTRUCTOR'S GUIDE

Each page presents two sections of information. The larger section displays each page exactly as it appears in the Trainee Module. The narrow column ties suggested trainee and instructor actions to each page and provides icons to call your attention to material, safety, audiovisual, or testing requirements. The bottom of each page includes space for your notes.

If you see the Teaching Tip icon, that means there is a teaching tip associated with this section. Also refer to any suggested teaching tips at the end of the module.

SAFETY CONSIDERATIONS

Ensure that the trainees are equipped with appropriate personal protective equipment. Emphasize that the basic safety habits established in apprenticeship are crucial to a long and healthy career as an electrician.

PREPARATION

Before teaching this module, you should review the Module Outline, Learning and Performance Objectives, and the Materials and Equipment List. Be sure to allow ample time to prepare your own training or lesson plan and gather all required equipment and materials.

MATERIALS AND EQUIPMENT LIST

Materials:

Transparencies

Markers/chalk

Lockout/tagout devices and labels

Copy of the latest edition of the *National Electrical Code®*

OSHA Electrical Safety Guidelines (pocket guide)

Module Examinations*

Performance Profile Sheets*

Equipment:

Overhead projector and screen

Whiteboard/chalkboard

Various types of personal protective and safety equipment, including rubber gloves, insulating blankets, lockout/tagout devices, hot sticks, fuse pullers, shorting probes, safety glasses, and face shields

Access to an eye wash station

*Located in the Test Booklet packaged with this Annotated Instructor's Guide.

ADDITIONAL RESOURCES

This module is intended to present thorough resources for task training. The following reference works are suggested for both instructors and motivated trainees interested in further study. These are optional materials for continued education rather than for task training.

29 CFR Parts 1900 – 1910, Standards for General Industry. Occupational Safety and Health Administration, US Department of Labor.

29 CFR Part 1926, Standards for the Construction Industry. Occupational Safety and Health Administration, US Department of Labor.

National Electrical Code Handbook, Latest Edition. Quincy, MA: National Fire Protection Association.

National Electrical Safety Code, Latest Edition. Quincy, MA: National Fire Protection Association.

NOTES

The designations "National Electrical Code," "NE Code," and "NEC," where used in this document, refer to the *National Electrical Code®*, which is a registered trademark of the National Fire Protection Association, Quincy, MA. All National Electrical Code (NEC) references in this module refer to the 2002 edition of the NEC.

If you feel that additional math instruction would be helpful, Prentice Hall offers a basic math textbook entitled *Fundamentals of Electrical and Mechanical Mathematics.* It covers the basic math requirements for electrical trainees and may be ordered by contacting Prentice Hall Customer Service at 1-800-922-0579.

TEACHING TIME FOR THIS MODULE

An outline for use in developing your lesson plan is presented below. Note that each Roman numeral in the outline equates to one session of instruction. Each session has a suggested time period of 2½ hours. This includes 10 minutes at the beginning of each session for administrative tasks and one 10-minute break during the session. Approximately 12½ hours are suggested to cover *Electrical Safety*. You will need to adjust the time required for hands-on activity and testing based on your class size and resources.

Topic	**Planned Time**
Session I. Introduction to Electrical Hazards; Laboratory	
A. Electrical Shock	_______________
1. Body Resistance	_______________
2. Burns	_______________
B. Protective Equipment	_______________
1. Rubber Protective Equipment	_______________
2. Protective Apparel	_______________
3. Personal Clothing	_______________
4. Hot Sticks	_______________
5. Fuse Pullers	_______________
6. Shorting Probes	_______________
7. Eye and Face Protection	_______________
C. Laboratory	_______________
Under instructor supervision, have the trainees practice examining rubber gloves using the visual inspection and air tests.	_______________
D. Verifying Deenergized Circuits	_______________
E. Basic Safety Precautions	_______________
Session II. OSHA Standards	
A. Section-by-Section Review of OSHA Standards 1910 and 1926	_______________
1. Design Safety Standards for Electrical Systems	_______________
2. Electric Utilization Systems	_______________
3. General Requirements	_______________
4. Wiring Design and Protection	_______________
5. Wiring Methods, Components, and Equipment for General Use	_______________
6. Specific Purpose Equipment and Installations	_______________
7. Hazardous (Classified) Locations	_______________
8. Special Systems	_______________
9. Scope	_______________
10. Training	_______________
11. Selection and Use of Work Practices	_______________
12. Use of Equipment	_______________
13. Safeguards for Personnel Protection	_______________
B. OSHA Safety Topics	_______________
1. Safety Philosophy and General Safety Precautions	_______________
2. Electrical Regulations and Lockout/Tagout Rule	_______________
3. Other OSHA Regulations	_______________

Session III. Ladders and Scaffolds

A. Ladders ______________

1. Straight Ladders ______________
2. Extension Ladders ______________
3. Step Ladders ______________

B. Scaffolds ______________

Session IV. General Construction Safety Topics; Laboratory; Confined Space Entry; First Aid

A. Lifts, Hoists, and Cranes ______________

B. Lifting ______________

C. Laboratory ______________

Under instructor supervision, have the trainees practice proper lifting procedures. ______________

D. Basic Tool Safety ______________

E. Confined Space Entry Procedures ______________

1. General Guidelines ______________
2. Confined Space Hazard Review ______________
3. Entry and Work Procedures ______________

F. First Aid ______________

Session V. Hazardous Materials; Laboratory; Fall Protection; Module Examination and Performance Testing

A. Solvents and Toxic Vapors ______________

1. Precautions When Using Solvents ______________
2. Respiratory Protection ______________

B. Asbestos ______________

1. Monitoring ______________
2. Regulated Areas ______________
3. Methods of Compliance ______________

C. Batteries ______________

1. Acids ______________
2. Eye Wash Stations ______________

D. Laboratory ______________

Under instructor supervision, have the trainees practice using an eye wash station. ______________

E. PCBs ______________

F. Fall Protection ______________

1. Fall Protection Procedures ______________
2. Types of Fall Protection Systems ______________

G. Module Examination ______________

1. Trainees must score 70% or higher to receive recognition from the NCCER.
2. Record the testing results on Craft Training Report Form 200 and submit the results to the Training Program Sponsor.

H. Performance Testing ______________

1. Trainees must perform each task to the satisfaction of the instructor to receive recognition from the NCCER.
2. Record the testing results on Craft Training Report Form 200 and submit the results to the Training Program Sponsor.

Chapter 1

Module 26101-02

Electrical Safety

Instructor's Notes:

Course Map

This course map shows all of the modules in the first level of the Residential Electrical curriculum. The suggested training order begins at the bottom and proceeds up. Skill levels increase as you advance on the course map. The local Training Program Sponsor may adjust the training order.

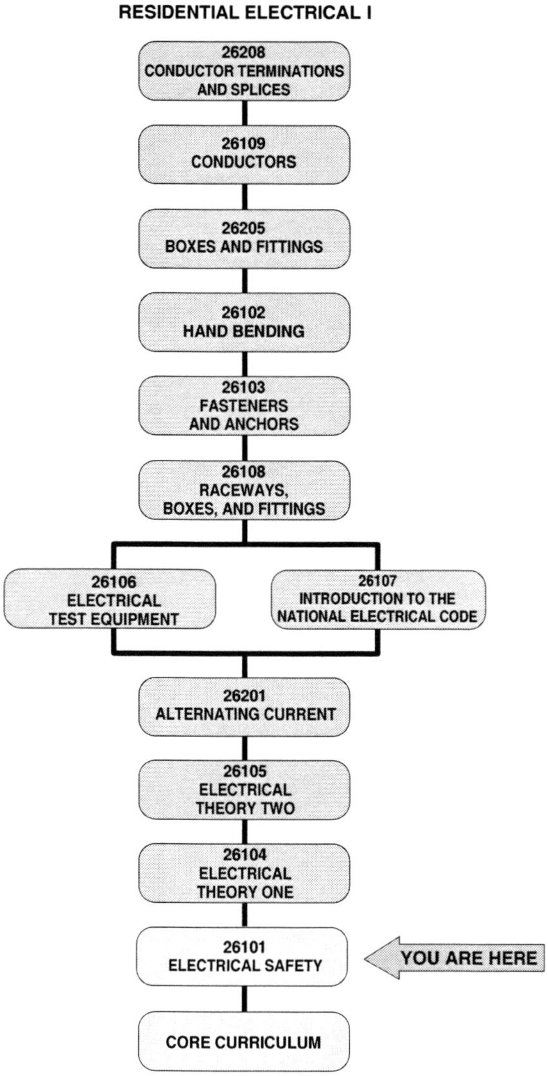

Homework

Assign reading of Module 26101.

MODULE 26101 CONTENTS

Instructor's Notes:

Figures

Tables

Chapter 1

MODULE 26101

Electrical Safety

Objectives

When you have completed this module, you will be able to do the following:

1. Demonstrate safe working procedures in a construction environment.
2. Explain the purpose of OSHA and how it promotes safety on the job.
3. Identify electrical hazards and how to avoid or minimize them in the workplace.
4. Explain safety issues concerning lockout/tagout procedures, personal protection using assured grounding and isolation programs, confined space entry, respiratory protection, and fall protection systems.

Prerequisites

Before you begin this module, it is recommended that you successfully complete the following: Core Curriculum.

Required Trainee Materials

1. Paper and pencil
2. Copy of the latest edition of the *National Electrical Code*
3. *OSHA 3075, Controlling Electrical Hazards*
4. Appropriate personal protective equipment

1.0.0 ◆ INTRODUCTION

As an electrician, you will be exposed to many potentially hazardous conditions that will exist on the job site. No training manual, set of rules and regulations, or listing of hazards can make working conditions completely safe. However, it is possible for an electrician to work a full career without serious accident or injury. To reach this goal, you need to be aware of potential hazards and stay constantly alert to these hazards. You must take the proper precautions and practice the basic rules of safety. You must be safety-conscious at all times. Safety should become a habit. Keeping a safe attitude on the job will go a long way in reducing the number and severity of accidents. Remember that your safety is up to you.

As an apprentice electrician, you need to be especially careful. You should only work under the direction of experienced personnel who are familiar with the various job site hazards and the means of avoiding them.

The most life-threatening hazards on a construction site are:

- Falls when you are working in high places
- Electrocution caused by coming into contact with live electrical circuits
- The possibility of being crushed by falling materials or equipment

Note: The designations "National Electrical Code," "NE Code," and "NEC," where used in this document, refer to the *National Electrical Code®*, which is a registered trademark of the National Fire Protection Association, Quincy, MA. *All National Electrical Code (NEC) references in this module refer to the 2002 edition of the NEC.*

Materials

Ensure you have everything required to teach the course. Check the Materials and Equipment List at the front of this Instructor's Guide.

Audiovisual

Show Transparency 1, Course Objectives.

Show Transparency 2, Performance Profile Tasks.

Classroom

Discuss the main types of hazards found on construction sites.

Electrical Safety in the Workplace

INSIDE TRACK

Each year in the U.S., there are approximately 20,000 electricity-related accidents at home and in the workplace. In a recent year, these accidents resulted in 700 deaths. Electrical accidents are the third leading cause of death in the workplace.

- The possibility of being struck by flying objects or moving equipment/vehicles such as trucks, forklifts, and construction equipment

Other hazards include cuts, burns, back sprains, and getting chemicals or objects in your eyes. Most injuries, both those that are life-threatening and those that are less severe, are preventable if the proper precautions are taken.

What's Wrong with This Picture?

101PO101.EPS

2.0.0 ◆ ELECTRICAL SHOCK

Electricity can be described as a potential that results in the movement of electrons in a conductor. This movement of electrons is called *electrical current*. Some substances, such as silver, copper, steel, and aluminum, are excellent conductors. The human body is also a conductor. The conductivity of the human body greatly increases when the skin is wet or moistened with perspiration.

Electrical current flows along the path of least resistance to return to its source. The source return point is called the *neutral* or *ground* of a circuit. If the human body contacts an electrically energized point and is also in contact with the ground or another point in the circuit, the human body becomes a path for the current to return to its source. *Table 1* shows the effects of current passing through the human body. One mA is one milliamp, or one one-thousandth of an ampere.

A primary cause of death from electrical shock is when the heart's rhythm is overcome by an electrical current. Normally, the heart's operation uses a very low-level electrical signal to cause the heart

Table 1 Current Level Effects on the Human Body

Current Value	Typical Effects
Less than 1mA	No sensation.
1 to 20mA	Sensation of shock, possibly painful. May lose some muscular control between 10 and 20mA.
20 to 50mA	Painful shock, severe muscular contractions, breathing difficulties.
50 to 200mA	Same symptoms as above, only more severe, up to 100mA. Between 100 and 200mA, ventricular fibrillation may occur. Typically results in almost immediate death unless special medical equipment and treatment are available.
Over 200mA	Severe burns and muscular contractions. The chest muscles contract and stop the heart for the duration of the shock.

Teaching Tip

Discuss "What's Wrong with This Picture?"

Classroom

Explain the effects of various current levels on the human body.

Instructor's Notes:

10 CHAPTER 1

Severity of Shock

THINK ABOUT IT

In *Table 1*, how many milliamps separate a mild shock from a potentially fatal one? What is the fractional equivalent of this in amps? How many amps are drawn by a 60W light bulb?

to contract and pump blood. When an abnormal electrical signal, such as current from an electrical shock, reaches the heart, the low-level heartbeat signals are overcome. The heart begins twitching in an irregular manner and goes out of rhythm with the pulse. This twitching is called **fibrillation.** Unless the normal heartbeat rhythm is restored using special defibrillation equipment (paddles), the individual will die. No known case of heart fibrillation has ever been corrected without the use of defibrillation equipment by a qualified medical practitioner. Other effects of electrical shock may include immediate heart stoppage and burns. In addition, the body's reaction to the shock can cause a fall or other accident. Delayed internal problems can also result.

2.1.0 The Effect of Current

The amount of current measured in amperes that passes through a body determines the outcome of an electrical shock. The higher the voltage, the greater the chance for a fatal shock. In a one-year study in California, the following results were observed by the State Division of Industry Safety:

- Thirty percent of all electrical accidents were caused by contact with conductors. Of these accidents, 66% involved low-voltage conductors (those carrying 600 volts [V] or less).

Note

Electric shocks or burns are a major cause of accidents in our industry. According to the Bureau of Labor Statistics, electrical shock is the leading cause of death in the electrical industry.

- Portable, electrically operated hand tools made up the second largest number of injuries (15%). Almost 70% of these injuries happened when the frame or case of the tool became energized. These injuries could have been prevented by following proper safety practices, using grounded or double-insulated tools, and using **ground fault circuit interrupter (GFCI)** protection.

In one ten-year study, investigators found 9,765 electrical injuries that occurred in accidents. Over 18% of these injuries involved contact with voltage levels of over 600 volts. A little more than 13% of these high-voltage injuries resulted in death. These high-voltage totals included limited-amperage contacts, which are often found on electronic equipment. When tools or equipment touch high-voltage overhead lines, the chance that a resulting injury will be fatal climbs to 28%. Of the low-voltage injuries, 1.4% were fatal.

CAUTION

High voltage, defined as 600 volts or more, is almost ten times as likely to kill as low voltage. However, on the job you spend most of your time working on or near lower voltages. Due to the frequency of contact, most electrocution deaths actually occur at low voltages. Attitude about the harmlessness of lower voltages undoubtedly contributes to this statistic.

These statistics have been included to help you gain respect for the environment where you work and to stress how important safe working habits really are.

Dangers of Electricity

INSIDE TRACK

Never underestimate the power of electricity. For example, the current through a 25W light bulb is more than enough to kill you.

Teaching Tip

Discuss the "Think About It."

Classroom

Explain that terms shown in bold (blue) are defined in the Glossary at the back of this module.

Safety

Emphasize that shock is the leading cause of death in the electrical industry.

Teaching Tip

Discuss the "Think About It."

Audiovisual

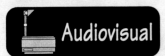

Show Transparency 3 (Figure 1).

Classroom

Explain the effect of electric shock on various parts of the body.

THINK ABOUT IT

Electrocution

Why can a bird perch safely on an electric wire? Squirrels are a common cause of shorts at substations; why does a squirrel get electrocuted when a bird does not?

2.1.1 Body Resistance

Electricity travels in closed circuits, and its normal route is through a conductor. Shock occurs when the body becomes part of the electric circuit (*Figure 1*). The current must enter the body at one point and leave at another. Shock normally occurs in one of three ways: the person must come in contact with both wires of the electric circuit; one wire of the electric circuit and the ground; or a metallic part that has become hot by being in contact with an energized wire while the person is also in contact with the ground.

To fully understand the harm done by electrical shock, we need to understand something about the physiology of certain body parts: the skin, the heart, and muscles.

- Skin covers the body and is made up of three layers. The most important layer, as far as electric shock is concerned, is the outer layer of dead cells referred to as the *horny layer.* This layer is composed mostly of a protein called *keratin,* and it is the keratin that provides the largest percentage of the body's electrical resistance. When it is dry, the outer layer of skin may have a resistance of several thousand ohms, but when it is moist, there is a radical drop in resistance, as is also the case if there is a cut or abrasion that pierces the horny layer. The amount of resistance provided by the skin will vary widely from individual to individual. A worker with a thick horny layer will have a much higher resistance than a child. The resistance will also vary widely at different parts of the body. For instance, the worker with high-resistance hands may have low-resistance skin on the back of his calf. The skin, like any insulator, has a breakdown voltage at which it ceases to act as a resistor and is simply punctured, leaving only the lower-resistance body tissue to impede the flow of current in the body. The breakdown voltage will vary with the individual, but is in the area of 600V. Since most industrial power distribution systems operate at 480V or higher, technicians working at these levels need to have special awareness of the shock potential.
- The heart is the pump that sends life-sustaining blood to all parts of the body. The blood flow is caused by the contractions of the heart muscle, which is controlled by electrical impulses. The electrical impulses are delivered by an intricate system of nerve tissue with built-in timing mechanisms, which make the chambers of the heart contract at exactly the right time. An outside electric current of as little as 75 milliamperes can upset the rhythmic, coordinated beating of the heart by disturbing the nerve impulses. When this happens, the heart is said to be in *fibrillation,* and the pumping action stops. Death will occur quickly if the normal beat is not restored. Remarkable as it may seem, what is needed to defibrillate the heart is a shock of an even higher intensity.
- The other muscles of the body are also controlled by electrical impulses delivered by nerves. Electric shock can cause loss of muscular control, resulting in the inability to let go of an electrical conductor. Electric shock can also cause injuries of an indirect nature in which involuntary muscle reaction from the electric shock can cause bruises, fractures, and even deaths resulting from collisions or falls.

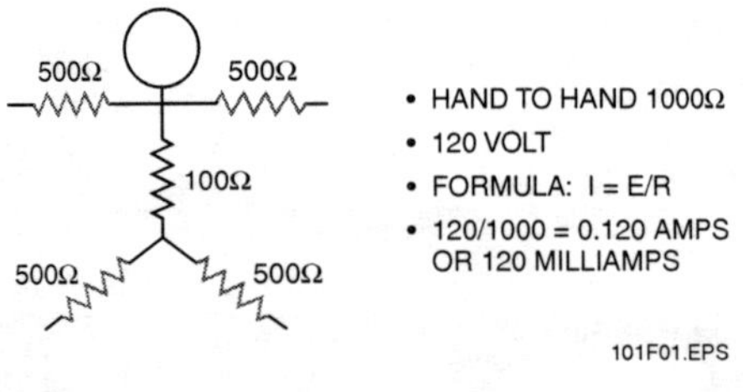

Figure 1 ◆ Body resistance.

Severity of shock—The severity of shock received when a person becomes a part of an electric circuit is affected by three primary factors: the amount of current flowing through the body (measured in amperes), the path of the current through the

Instructor's Notes:

12 CHAPTER 1

body, and the length of time the body is in the circuit. Other factors that may affect the severity of the shock are the frequency of the current, the phase of the heart cycle when shock occurs, and the general health of the person prior to the shock. Effects can range from a barely perceptible tingle to immediate cardiac arrest. Although there are no absolute limits, or even known values that show the exact injury at any given amperage range, *Table 1* lists the general effects of electric current on the body for different current levels. As this table illustrates, a difference of only 100 milliamperes exists between a current that is barely perceptible and one that can kill.

A severe shock can cause considerably more damage to the body than is visible. For example, a person may suffer internal hemorrhages and destruction of tissues, nerves, and muscle. In addition, shock is often only the beginning in a chain of events. The final injury may well be from a fall, cuts, burns, or broken bones.

2.1.2 Burns

The most common shock-related injury is a burn. Burns suffered in electrical accidents may be of three types: electrical burns, arc burns, and thermal contact burns.

- Electrical burns are the result of electric current flowing through the tissues or bones. Tissue damage is caused by the heat generated by the current flow through the body. An electrical burn is one of the most serious injuries you can receive, and should be given immediate attention. Since the most severe burning is likely to be internal, what may appear at first to be a small surface wound could, in fact, be an indication of severe internal burns.
- Arc burns make up a substantial portion of the injuries from electrical malfunctions. The electric arc between metals can be up to 35,000°F, which is about four times hotter than the surface of the sun. Workers several feet from the source of the arc can receive severe or fatal burns. Since most electrical safety guidelines recommend safe working distances based on shock considerations, workers can be following these guidelines and still be at risk from arc. Electric arcs can occur due to poor electrical contact or failed insulation. Electrical arcing is caused by the passage of substantial amounts of current through the vaporized terminal material (usually metal or carbon).

> **CAUTION**
> Since the heat of the arc is dependent on the short circuit current available at the arcing point, arcs generated on 480V systems can be just as dangerous as those generated at 13,000V.

- The third type of burn is a thermal contact burn. It is caused by contact with objects thrown during the blast associated with an electric arc. This blast comes from the pressure developed by the near-instantaneous heating of the air surrounding the arc, and from the expansion of the metal as it is vaporized. (Copper expands by a factor in excess of 65,000 times in boiling.) These pressures can be great enough to hurl people, switchgear, and cabinets considerable distances. Another hazard associated with the blast is the hurling of molten metal droplets, which can also cause thermal contact burns and associated damage. A possible beneficial side effect of the blast is that it could hurl a nearby person away from the arc, thereby reducing the effect of arc burns.

3.0.0 ◆ REDUCING YOUR RISK

There are many things that can be done to greatly reduce the chance of receiving an electrical shock. Always comply with your company's safety policy and all applicable rules and regulations, including job site rules. In addition, the Occupational Safety and Health Administration (OSHA) publishes the *Code of Federal Regulations (CFR).* CFR Part 1910

> **THINK ABOUT IT**
> ***Bodily Harm***
> What factors affect the amount of damage to the body during an electric shock?

Classroom

Discuss the effects of electrical, arc, and thermal contact burns.

Teaching Tip

Discuss the "Think About It."

Teaching Tip

Discuss the "Case History."

Classroom

Discuss approach distances.

Discuss the OSHA requirements for personal protective equipment.

CASE HISTORY

Arc Blast

An electrician in Louisville, KY was cleaning a high-voltage switch cabinet. He removed the padlock securing the switch enclosure and opened the door. He used a voltage meter to verify 0V on the three load phases at the rear of the cabinet. However, he did not test all potentially energized parts within the cabinet, and some components were still energized. He was wearing standard work boots and safety glasses, but was not wearing protective rubber gloves or an arc suit. As he used a paintbrush to clean the switch, an arc blast occurred, which lasted approximately one-sixth of a second. The electrician was knocked down by the blast. He suffered third-degree burns and required skin grafts on his arms and hands. The investigation determined that the blast was caused by debris, such as a cobweb, falling across the open switch.

The Bottom Line: Always wear the appropriate personal protective equipment and test all components for voltage before working in or around electrical devices.

covers the OSHA standards for general industry and CFR Part 1926 covers the OSHA standards for the construction industry.

Do not approach any electrical conductors closer than indicated in *Table 2* unless you are sure they are de-energized and your company has designated you as a qualified individual. Also, the values given in the table are *minimum* safe clearance distances; if you already have standard distances established, these are provided only as supplemental information. These distances are listed in CFR 1910.333/1926.416.

3.1.0 Protective Equipment

You should also become familiar with common personal protective equipment. In particular, know the voltage rating of each piece of equipment. Rubber gloves are used to prevent the skin from coming into contact with energized circuits. A separate leather cover protects the rubber glove from punctures and other damage (see *Figure 2*). OSHA addresses the use of protective equipment, apparel, and tools in CFR 1910.335(a). This article is divided into two sections: *Personal Protective Equipment* and *General Protective Equipment and Tools.*

The first section, *Personal Protective Equipment,* includes the following requirements:

- Employees working in areas where there are potential electrical hazards shall be provided with, and shall use, electrical protective equipment that is appropriate for the specific parts of the body to be protected and for the work to be performed.
- Protective equipment shall be maintained in a safe, reliable condition and shall be periodically

Table 2 Approach Distances for Qualified Employees—Alternating Current

Voltage Range (Phase-to-Phase)	Minimum Approach Distance
300V and less	Avoid contact
Over 300V, not over 750V	1 ft. 0 in. (30.5 cm)
Over 750V, not over 2kV	1 ft. 6 in. (46 cm)
Over 2kV, not over 15kV	2 ft. 0 in. (61 cm)
Over 15kV, not over 37kV	3 ft. 0 in. (91 cm)
Over 37kV, not over 87.5kV	3 ft. 6 in. (107 cm)
Over 87.5kV, not over 121kV	4 ft. 0 in. (122 cm)
Over 121kV, not over 140kV	4 ft. 6 in. (137 cm)

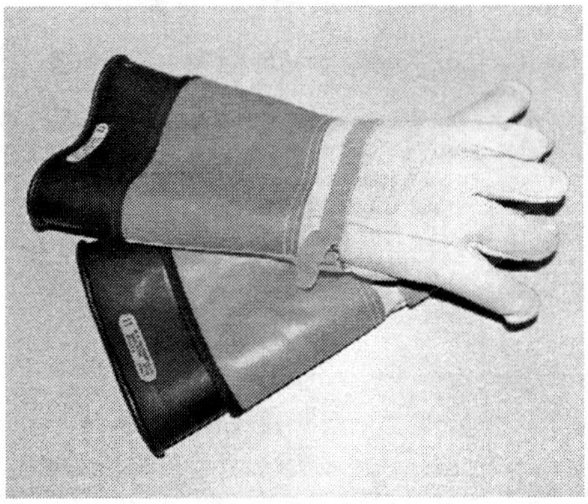

101F02.EPS

Figure 2 ◆ Rubber gloves and leather protectors.

Instructor's Notes:

inspected or tested, as required by CFR 1910.137/1926.95.

- If the insulating capability of protective equipment may be subject to damage during use, the insulating material shall be protected.
- Employees shall wear nonconductive head protection wherever there is a danger of head injury from electric shock or burns due to contact with exposed energized parts.
- Employees shall wear protective equipment for the eyes and face wherever there is danger of injury to the eyes or face from electric arcs or flashes or from flying objects resulting from an electrical explosion.

The second section, *General Protective Equipment and Tools,* includes the following requirements:

- When working near exposed energized conductors or circuit parts, each employee shall use insulated tools or handling equipment if the tools or handling equipment might make contact with such conductors or parts. If the insulating capability of insulated tools or handling equipment is subject to damage, the insulating material shall be protected.
- Fuse handling equipment, insulated for the circuit voltage, shall be used to remove or install fuses when the fuse terminals are energized.
- Ropes and handlines used near exposed energized parts shall be nonconductive.
- Protective shields, protective barriers, or insulating materials shall be used to protect each employee from shock, burns, or other electrically related injuries while that employee is working near exposed energized parts that might be accidentally contacted or where dangerous electric heating or arcing might occur. When normally enclosed live parts are exposed for maintenance or repair, they shall be guarded to protect unqualified persons from contact with the live parts.

The types of electrical safety equipment, protective apparel, and protective tools available for use are quite varied. We will discuss the most common types of safety equipment. These include:

- Rubber protective equipment, including gloves and blankets
- Protective apparel
- Personal clothing
- Hot sticks
- Fuse pullers
- Shorting probes
- Eye and face protection

3.1.1 1910.335(a)(1)/1926.951(a) Rubber Protective Equipment

At some point during the performance of their duties, all electrical workers will be exposed to energized circuits or equipment. Two of the most important articles of protection for electrical workers are insulated rubber gloves and rubber blankets, which must be matched to the voltage rating for the circuit or equipment. Rubber protective equipment is designed for the protection of the user. If it fails during use, a serious injury could occur.

Rubber protective equipment is available in two types. Type 1 designates rubber protective equipment that is manufactured of natural or synthetic rubber that is properly vulcanized, and Type 2 designates equipment that is ozone resistant, made from any elastomer or combination of elastomeric compounds. Ozone is a form of oxygen that is produced from electricity and is present in the air surrounding a conductor under high voltages. Normally, ozone is found at voltages of 10kV and higher, such as those found in electric utility transmission and distribution systems. Type 1 protective equipment can be damaged by *corona cutting,* which is the cutting action of ozone on natural rubber when it is under mechanical stress. Type 1 rubber protective equipment can also be damaged by ultraviolet rays. However, it is very important that the rubber protective equipment in use today be made of natural rubber or Type 1 equipment. Type 2 rubber protective equipment is very stiff and is not as easily worn as Type 1 equipment.

Various classes—The American National Standards Institute (ANSI) and the American Society for Testing of Materials (ASTM) have designated a specific classification system for rubber protective equipment. The voltage ratings are as follows:

- Class 0 1,000V
- Class 1 7,500V
- Class 2 17,500V
- Class 3 26,500V
- Class 4 36,000V

Inspection of protective equipment—Before rubber protective equipment can be worn by personnel in the field, all equipment must have a current test date stenciled on the equipment, and it must be inspected by the user. Insulating gloves must be tested each day by the user before they can be used. They must also be tested during the day if their insulating value is ever in question. Because rubber protective equipment is going to be used

Classroom

Discuss the various classes of rubber protective equipment.

Safety

Stress to trainees that work on energized equipment must be pre-approved by a supervisor or instructor.

Classroom

Discuss the proper use and maintenance of rubber gloves.

for personal protection and serious injury could result from its misuse or failure, it is important that an adequate safety factor be provided between the voltage on which it is to be used and the voltage at which it was tested.

All rubber protective equipment must be marked with the appropriate voltage rating and last inspection date. The markings that are required to be on rubber protective equipment must be applied in a manner that will not interfere with the protection that is provided by the equipment.

> **WARNING!**
> Never work on anything energized without direct instruction from your employer.

Gloves—Both high- and low-voltage rubber gloves are of the gauntlet type and are available in various sizes. To get the best possible protection and service life, here are a few general rules that apply whenever they are used in electrical work:

- Always wear leather protectors over your gloves. Any direct contact with sharp or pointed objects may cut, snag, or puncture the gloves and take away the protection you are depending on. Leather protectors are required by the National Fire Protection Association's Standard 70-E if the insulating capabilities of the gloves are subject to damage. (The standards of the National Fire Protection Association [NFPA] are incorporated into the OSHA standards.)
- Always wear rubber gloves right side out (serial number and size to the outside). Turning gloves inside out places a stress on the preformed rubber.
- Always keep the gauntlets up. Rolling them down sacrifices a valuable area of protection.
- Always inspect and field check gloves before using them. Always check the inside for any debris. The inspection of gloves is covered in more detail later in this section.
- Use light amounts of talcum powder or cotton liners with the rubber gloves. This gives the user more comfort, and it also helps to absorb some of the perspiration that can damage the gloves over years of use.
- Wash the rubber gloves in lukewarm, clean, fresh water after each use. Dry the gloves inside and out prior to returning to storage. Never use any type of cleaning solution on the gloves.
- Once the gloves have been properly cleaned, inspected, and tested, they must be properly stored. They should be stored in a cool, dry, dark place that is free from ozone, chemicals, oils, solvents, or other materials that could damage the gloves. Such storage should not be in the vicinity of hot pipes or direct sunlight. Both gloves and sleeves should be stored in their natural shape and kept in a bag or box inside their protectors. They should be stored undistorted, right side out, and unfolded.
- Gloves can be damaged by many different chemicals, especially petroleum-based products such as oils, gasoline, hydraulic fluid inhibitors, hand creams, pastes, and salves. If contact is made with these or other petroleum-based products, the contaminant should be wiped off immediately. If any signs of physical damage or chemical deterioration are found (e.g., swelling, softness, hardening, stickiness, ozone deterioration, or sun checking), the protective equipment must not be used.
- Never wear watches or rings while wearing rubber gloves; this can cause damage from the inside out and defeats the purpose of using rubber gloves. Never wear anything conductive.
- Rubber gloves must be tested every six months by a certified testing laboratory. Always check the inspection date before using gloves.
- Use rubber gloves only for their intended purpose, not for handling chemicals or other work. This also applies to the leather protectors.

Before rubber gloves are used, a visual inspection and an air test should be made. This should be done prior to use and as many times during the day as you feel necessary. To perform a visual inspection, stretch a small area of the glove, checking to see that no defects exist, such as:

- Embedded foreign material
- Deep scratches
- Pinholes or punctures
- Snags or cuts

Gloves and sleeves can be inspected by rolling the outside and inside of the protective equipment

Instructor's Notes:

16 CHAPTER 1

between the hands. This can be done by squeezing together the inside of the gloves or sleeves to bend the outside area and create enough stress to the inside surface to expose any cracks, cuts, or other defects. When the entire surface has been checked in this manner, the equipment is then turned inside out, and the procedure is repeated. It is very important not to leave the rubber protective equipment inside out.

Remember, any damage at all reduces the insulating ability of the rubber glove. Look for signs of deterioration from age, such as hardening and slight cracking. Also, if the glove has been exposed to petroleum products, it should be considered suspect because deterioration can be caused by such exposure. Gloves that are found to be defective must be turned in for disposal. Never leave a damaged glove lying around; someone may think it is a good glove and not perform an inspection prior to using it.

After visually inspecting the glove, other defects may be observed by applying the air test.

Step 1 Stretch the glove and look for any defects, as shown in *Figure 3.*

Step 2 Twirl the glove around quickly or roll it down from the glove gauntlet to trap air inside, as shown in *Figure 4.*

Step 3 Trap the air by squeezing the gauntlet with one hand. Use the other hand to squeeze the palm, fingers, and thumb to check for weaknesses and defects. See *Figures 5* and *6.*

Step 4 Hold the glove up to your ear to try to detect any escaping air.

Step 5 If the glove does not pass this inspection, it must be turned in for disposal.

CAUTION
Never use compressed gas for the air test as this can damage the glove.

Insulating blankets—An insulating blanket is a versatile cover-up device best suited for the protection of maintenance technicians against accidental contact with energized electrical equipment.

These blankets are designed and manufactured to provide insulating quality and flexibility for use in covering. Insulating blankets are designed only for covering equipment and should not be used on the floor. (Special rubber mats are available for floor use.) Use caution when installing on sharp edges or covering pointed objects.

Blankets must be inspected yearly and should be checked before each use. To check rubber blankets, place the blanket on a flat surface and roll the blanket from one corner to the opposite corner. If there are any irregularities in the rubber, this method will expose them. After the blanket has been rolled from each corner, it should then be turned over and the procedure repeated.

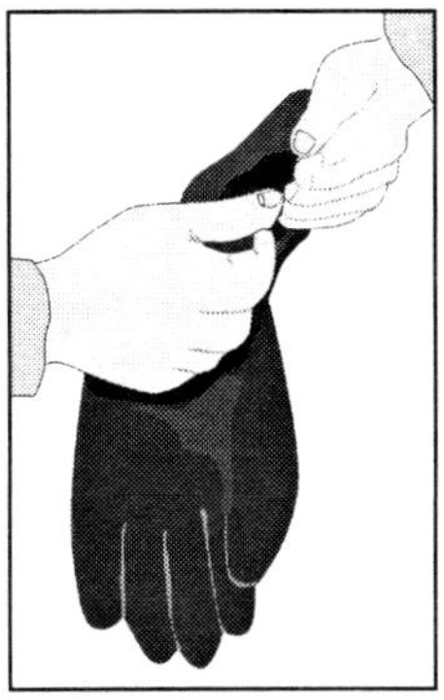
101F03.EPS

Figure 3 ◆ Inspection.

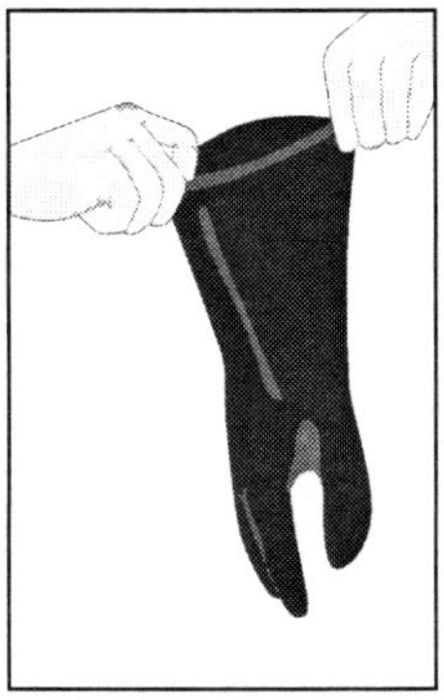
101F04.EPS

Figure 4 ◆ Trapping air.

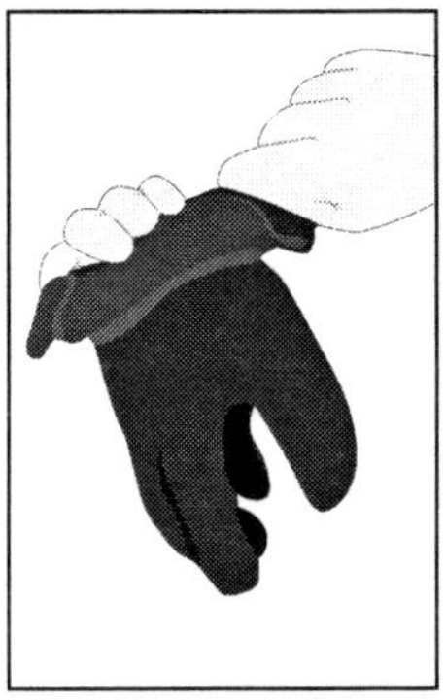
101F05.EPS

Figure 5 ◆ Inflated glove.

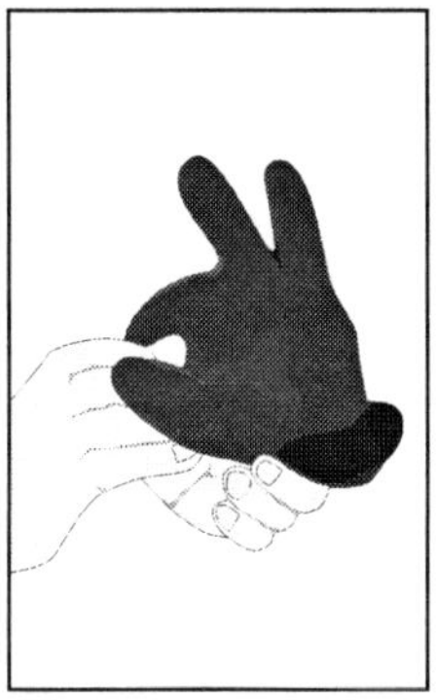
101F06.EPS

Figure 6 ◆ Inspecting glove.

Audiovisual

Show Transparencies 4 through 7 (Figures 3 through 6).

Demonstration

Demonstrate how to inspect rubber gloves using the visual inspection and air tests.

Laboratory

Have the trainees practice inspecting rubber gloves. Note the proficiency of each trainee.

Classroom

Discuss the use of insulating blankets.

Insulating blankets are cleaned in the same manner as rubber gloves. Once the protective equipment has been properly cleaned, inspected, and tested, it must be properly stored. It should be stored in a cool, dry, dark place that is free from ozone, chemicals, oils, solvents, or other materials that could damage the equipment. Such storage should not be in the vicinity of hot pipes or direct sunlight. Blankets may be stored rolled in containers that are designed for this use; the inside diameter of the roll should be at least two inches.

3.1.2 Protective Apparel

Besides rubber gloves, there are other types of special application protective apparel, such as fire suits, face shields, and rubber sleeves.

Manufacturing plants should have other types of special application protective equipment available for use, such as high-voltage sleeves, high-voltage boots, nonconductive protective helmets, nonconductive eyewear and face protection, and switchboard blankets.

All equipment should be inspected before use and during use, as necessary. The equipment used and the extent of the precautions taken depend on each individual situation; however, it is better to be overprotected than underprotected when you are trying to prevent electrocution.

When working with high voltages, flash suits may be required in some applications. Some plants require them to be worn for all switching and rack-in or rack-out operations.

Face shields should also be worn during all switching operations where arcs are a possibility. The thin plastic type of face shield should be avoided because it will melt when exposed to the extremely high temperatures of an electrical arc.

Rubber sleeves are another type of protective apparel that should be worn during switching operations and breaker racking. Sleeves must be inspected yearly.

3.1.3 Personal Clothing

Any individual who will perform work in an electrical environment or in plant substations should dress accordingly. Avoid wearing synthetic-fiber clothing; these types of materials will melt when exposed to high temperatures and will actually increase the severity of a burn. Wear cotton clothing, fiberglass-toe boots or shoes, and hard hats. Use hearing protection where needed.

3.1.4 Hot Sticks

Hot sticks are insulated tools designed for the manual operation of disconnecting switches, fuse removal and insertion, and the application and removal of temporary grounds.

A hot stick is made up of two parts, the head or *hood* and the insulating rod. The head can be made of metal or hardened plastic, while the insulating section may be wood, plastic, laminated wood, or other effective insulating materials. There are also telescoping sticks available.

Most plants have hot sticks available for different purposes. Select a stick of the correct type and size for the application.

INSIDE TRACK

Flash Protection Signs

NEC Section 110.16 requires that all switchboards, panelboards, industrial control panels, and motor control centers be clearly marked to warn qualified persons of potential electric arc flash hazards.

THINK ABOUT IT

Dressing for Safety

How could minor flaws in clothing cause harm? What about metal components, such as the rivets in jeans, or synthetic materials, such as polyester? How are these dangerous? How does protective apparel prevent accidents?

Demonstration

To the extent possible, demonstrate the use of, and allow the trainees to examine, various types of protective apparel.

Teaching Tip

Discuss the "Think About It."

Instructor's Notes:

Storage of hot sticks is important. They should be hung up vertically on a wall to prevent any damage. They should also be stored away from direct sunlight and prevented from being exposed to petroleum products. The preferred method of storage is to place the stick in a long section of capped pipe.

3.1.5 Fuse Pullers

Use the plastic or fiberglass style of fuse puller for removing and installing low-voltage cartridge fuses. All fuse pulling and replacement operations must be done using fuse pullers.

The best type of fuse puller is one that has a spread guard installed. This prevents the puller from opening if resistance is met when installing fuses.

3.1.6 Shorting Probes

Before working on de-energized circuits that have capacitors installed, you must discharge the capacitors using a safety shorting probe. When using a shorting probe, first connect the test clip to a good ground to make contact. If necessary, scrape the paint from the metal surface. Then, hold the shorting probe by the handle and touch the probe end of the shorting rod to the points to be shorted. The probe end can be hooked over the part or terminal to provide a constant connection to ground. Never touch any metal part of the shorting probe while grounding circuits or components. Whenever possible, especially when working on or near any de-energized high-voltage circuits, shorting probes should be connected and then left attached to the de-energized portion of any circuit for the duration of the work. This action serves as an extra safety precaution against any accidental application of voltage to the circuit.

3.1.7 Eye and Face Protection

NFPA 70-E requires that protective equipment for the eyes and face shall be used whenever there is danger of injury to the eyes or face from electrical arcs or flashes, or from flying or falling objects resulting from an electrical explosion.

INSIDE TRACK

Using a Shorting Probe

Using a shorting probe can be a very dangerous task. Always take the time to use the appropriate personal protective equipment. Remember, you only have one chance to protect yourself.

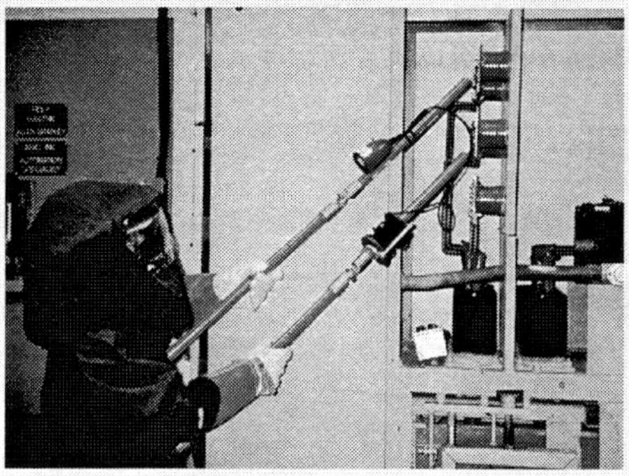

101PO102.EPS

Demonstration

To the extent possible, demonstrate the use of, and allow the trainees to examine, hot sticks, fuse pullers, shorting probes, safety glasses, and face shields.

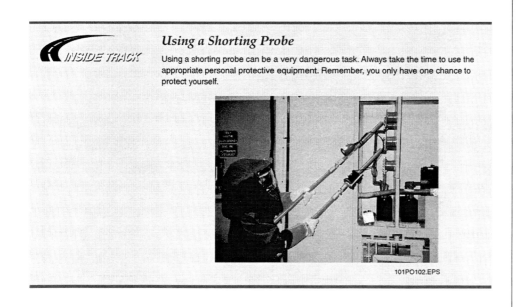

Classroom

Explain how to verify that a circuit has been deenergized.

Review the safety checklist.

3.2.0 Verify That Circuits Are De-energized

You should always assume that all the circuits are energized until you have verified that the circuit is de-energized. Follow these steps to verify that a circuit is de-energized:

Step 1 Ensure that the circuit is properly tagged and locked out (OSHA 1910.333/1926.417).

Step 2 Verify the test instrument operation on a known source.

Step 3 Using the test instrument, check the circuit to be de-energized. The voltage should be zero.

Step 4 Verify the test instrument operation, once again on a known power source.

3.3.0 Other Precautions

There are several other precautions you can take to help make your job safer. For example:

- Always remove all jewelry (e.g., rings, watches, bracelets, and necklaces) before working on electrical equipment. Most jewelry is made of conductive material and wearing it can result in a shock, as well as other injuries if the jewelry gets caught in moving components.
- When working on energized equipment, it is safer to work in pairs. In doing so, if one of the workers experiences a harmful electrical shock, the other worker can quickly deenergize the circuit and call for help.
- Plan each job before you begin it. Make sure you understand exactly what it is you are going to do. If you are not sure, ask your supervisor.
- You will need to look over the appropriate prints and drawings to locate isolation devices and potential hazards. Never defeat safety interlocks. Remember to plan your escape route before starting work. Know where the nearest phone is and the emergency number to dial for assistance.
- If you realize that the work will go beyond the scope of what was planned, stop and get instructions from your supervisor before continuing. Do not attempt to plan as you go.
- It is critical that you stay alert. Workplaces are dynamic, and situations relative to safety are always changing. If you leave the work area to pick up material, take a break, or have lunch, reevaluate your surroundings when you return. Remember, plan ahead.

4.0.0 ◆ OSHA

The purpose of the Occupational Safety and Health Administration (OSHA) is "to assure safe and healthful working conditions for working men and women." OSHA is authorized to enforce standards and assist and encourage the states in their efforts to ensure safe and healthful working conditions. OSHA assists states by providing for research, information, education, and training in the field of occupational safety and health.

The law that established OSHA specifies the duties of both the employer and employee with respect to safety. Some of the key requirements are outlined below. This list does not include everything, nor does it override the procedures called for by your employer.

- Employers shall provide a place of employment free from recognized hazards likely to cause death or serious injury.
- Employers shall comply with the standards of the act.
- Employers shall be subject to fines and other penalties for violation of those standards.

> **WARNING!**
> OSHA states that employees have a duty to follow the safety rules laid down by the employer. Additionally, some states can reduce the amount of benefits paid to an injured employee if that employee was not following known, established safety rules. Your company may also terminate you if you violate an established safety rule.

4.1.0 Safety Standards

The OSHA standards are split into several sections. As discussed earlier, the two that affect you the most are CFR 1926, construction specific, and CFR 1910, which is the standard for general industry. Either or both may apply depending on where you are working and what you are doing. If a job site condition is covered in the 1926 book, then that standard takes precedence. However, if a more stringent requirement is listed in the 1910 standard, it should also be met. An excellent example is the current difference in the two standards on confined spaces; if someone gets hurt or killed, the decision to use the less stringent 1926 standard could be called into question. OSHA's

Instructor's Notes:

General Duty Clause states that an employer should have known all recognized hazards and removed the hazard or protected the employee.

To protect workers from the occupational injuries and fatalities caused by electrical hazards, OSHA has issued a set of design safety standards for electrical utilization systems. These standards are 1926.400–449 and 1910.302–308. OSHA also recognizes the *National Electrical Code (NEC)* for certain installations.

> **Note**
>
> OSHA does *not* recognize the current edition of the NEC; it generally takes several years for that to occur.

The CFR 1910 standard must be followed whenever the construction standard CFR 1926 does not address an issue that is covered by CFR 1910, or for a pre-existing installation. If the CFR 1910 standard is more stringent than CFR 1926, then the more stringent standard should be followed. OSHA does not update their standards in a timely manner, and as such, there are often differences in similar sections of the two standards. Safety should always be first, and the more protective work rules should always be chosen.

4.1.1 1910.302–308/1926.402–408 Design Safety Standards for Electrical Systems

This section contains design safety regulations for all the electrical equipment and installations used to provide power and light to employee workplaces. The articles listed are outlined in the following sections.

4.1.2 1910.302/1926.402 Electric Utilization Systems

This article identifies the scope of the standard. Listings are included to show which electrical installations and equipment are covered under the standard, and which installations and equipment are not covered under the standard. Furthermore, certain sections of the standard apply only to utilization equipment installed after March 15, 1972, and some apply only to equipment installed after April 16, 1981. Article 1910.302 (1926.402) addresses these oddities and provides guidance to clarify them.

4.1.3 1910.303/1926.403 General Requirements

This article covers topics that mostly concern equipment installation clearances, identification,

INSIDE TRACK

Safety on the Job Site

101PO103.EPS

Uncovered openings present several hazards:

- Workers may trip over them.
- If they are large enough, workers may actually fall through them.
- If there is a work area below, tools or other objects may fall through them, causing serious injury to workers below.

Did you know that OSHA could cite your company for working around the open pipe in the photo shown here, even if you weren't responsible for creating it? If you are working near uncovered openings, you have only two options:

- Cover the opening properly (either you may do it or you may have the responsible contractor cover it).

or:

- Stop work and leave the job site until the opening has been covered.

Classroom

Show and have the trainees examine the NEC and OSHA books. Emphasize that the more stringent regulation should always be applied.

Review the applicable sections of *CFR 1910/1926.*

Emphasize a contractor's responsibility for maintaining a safe job site.

Classroom

Continue review of applicable sections of *CFR 1910/1926*.

and examination. Some of the major subjects addressed in this article are:

- Equipment installation examinations
- Splicing
- Marking
- Identification of disconnecting means
- Workspace around electrical equipment

4.1.4 1910.304/1926.404 Wiring Design and Protection

This article covers the application, identification, and protection requirements of grounding conductors, outside conductors, service conductors, and equipment enclosures. Some of the major topics discussed are:

- Grounded conductors
- Outside conductors
- Service conductors
- Overcurrent protection
- System grounding requirements

4.1.5 1910.305/1926.405 Wiring Methods, Components, and Equipment for General Use

In general, this article addresses the wiring method requirements of raceways, cable trays, pull and junction boxes, switches, and switchboards; the application requirements of temporary wiring installations; the equipment and conductor requirements for general wiring; and the protection requirements of motors, transformers, capacitors, and storage batteries. Some of the major topics are:

- Wiring methods
- Cabinets, boxes, and fittings
- Switches
- Switchboards and panelboards
- Enclosures for damp or wet locations
- Conductors for general wiring
- Flexible cords and cables
- Portable cables
- Equipment for general use

4.1.6 1910.306/1926.406 Specific Purpose Equipment and Installations

This article addresses the requirements of special equipment and installations not covered in other articles. Some of the major types of equipment and installations found in this article are:

- Electric signs and outline lighting
- Cranes and hoists
- Elevators, dumbwaiters, escalators, and moving walks
- Electric welders
- Data processing systems
- X-ray equipment
- Induction and dielectric heating equipment
- Electrolytic cells
- Electrically driven or controlled irrigation machines
- Swimming pools, fountains, and similar installations

4.1.7 1910.307/1926.407 Hazardous (Classified) Locations

This article covers the requirements for electric equipment and wiring in locations that are classified because they contain: (1) flammable vapors, liquids, and/or gases, or combustible dust or fibers; and (2) the likelihood that a flammable or combustible concentration or quantity is present. Some of the major topics covered in this article are:

- Scope
- Electrical installations in hazardous locations
- Conduit
- Equipment in Division 2 locations

4.1.8 1910.308/1926.408 Special Systems

This article covers the wiring methods, grounding, protection, identification, and other general requirements of special systems not covered in other articles. Some of the major subtopics found in this article are:

- Systems over 600 volts nominal
- Emergency power systems
- Class 1, 2, and 3 remote control, signaling, and power-limited circuits
- Fire-protective signaling systems
- Communications systems

4.1.9 1910.331/1926.416 Scope

This article serves as an overview of the following articles and also provides a summary of the installations that this standard allows qualified and unqualified persons to work on or near, as well as the installations that this standard does *not* cover.

Instructor's Notes:

4.1.10 1910.332 Training

The training requirements contained in this article apply to employees who face a risk of electric shock. Some of the topics that appear in this article are:

- Content of training
- Additional requirements for unqualified persons
- Additional requirements for qualified persons
- Type of training

4.1.11 1910.333/1926.416–417 Selection and Use of Work Practices

This article covers the implementation of safety-related work practices necessary to prevent electrical shock and other related injuries to the employee. Some of the major topics addressed in this article are listed below:

- General
- Working on or near exposed de-energized parts
- Working on or near exposed energized parts

4.1.12 1910.334/1926.431 Use of Equipment

This article was added to reinforce the regulations pertaining to portable electrical equipment, test equipment, and load break switches. Major topics include:

- Portable electric equipment
- Electric power and lighting circuits
- Test instruments and equipment

4.1.13 1910.335/1926.416 Safeguards for Personnel Protection

This article covers the personnel protection requirements for employees in the vicinity of electrical hazards. It addresses regulations that protect personnel working on equipment as well as personnel working nearby. Some of the major topics are:

- Use of protective equipment
- Alerting techniques

Now that background topics have been covered and an overview of the OSHA electrical safety standards has been provided, it is time to move on to topics related directly to safety. As we discuss these topics, we will continually refer to the OSHA standards to identify the requirements that govern them.

OSHA 1926 Subpart K also addresses electrical safety requirements that are necessary for the practical safeguarding of employees involved in construction work.

4.2.0 Safety Philosophy and General Safety Precautions

The most important piece of safety equipment required when performing work in an electrical environment is common sense. All areas of electrical safety precautions and practices draw upon common sense and attention to detail. One of the most dangerous conditions in an electrical work area is a poor attitude toward safety.

> **WARNING!**
> Only qualified individuals may work on electrical equipment. Your employer will determine who is qualified. Remember, your employer's safety rules must always be followed.

As stated in CFR 1910.333(a)/1926.403, safety-related work practices shall be employed to prevent electric shock or other injuries resulting from either direct or indirect electrical contact when work is performed near or on equipment or circuits that are or may be energized. The specific safety-related work practices shall be consistent with the nature and extent of the associated electrical hazards. The following are considered some of the basic and necessary attitudes and electrical safety precautions that lay the groundwork for a proper safety program. Before going on any electrical work assignment, these safety precautions should be reviewed and adhered to.

- *All work on electrical equipment should be done with circuits de-energized and cleared or grounded*—It is obvious that working on energized equipment is much more dangerous than working on equipment that is de-energized. Work on energized electrical equipment should be avoided if at all possible. CFR 1910.333(a)(1)/1926.403 states that live parts to which an employee may be exposed shall be de-energized before the employee works on or near them, unless the employer can demonstrate that de-energizing introduces additional or increased hazards or is not possible because of equipment design or operational limitations. Live parts that operate at less than 50 volts to ground need not be de-energized if there will be no increased exposure to electrical burns or to explosion due to electric arcs.

Safety

Define the term *qualified individual.* Explain that it involves having the proper training and experience to complete a certain task.

Classroom

Review OSHA general safety precautions.

Classroom

Continue review of OSHA general safety precautions.

- *All conductors, buses, and connections should be considered energized until proven otherwise*—As stated in 1910.333(b)(1)/1926.417, conductors and parts of electrical equipment that have not been locked out or tagged out in accordance with this section should be considered energized. Routine operation of the circuit breakers and disconnect switches contained in a power distribution system can be hazardous if not approached in the right manner. Several basic precautions that can be observed in switchgear operations are:
  –Wear proper clothing made of 100% cotton or fire-resistant fabric.
  –Eye, face, and head protection should be worn. Turn your head away whenever closing devices.
  –Whenever operating circuit breakers in low-voltage or medium-voltage systems, always stand off to the side of the unit.
  –Always try to operate disconnect switches and circuit breakers under a no-load condition.
  –Never intentionally force an interlock on a system or circuit breaker.
  –Always verify what you are closing a device into; you could violate a lockout or close into a hard fault.

Often, a circuit breaker or disconnect switch is used for providing lockout on an electrical system. To ensure that a lockout is not violated, perform the following procedures when using the device as a lockout point:

- Breakers must always be locked out and tagged as discussed previously whenever you are working on a circuit that is tied to an energized breaker. Breakers capable of being opened and racked out to the disconnected position should have this done. Afterward, approved safety locks must be installed. The breaker may be removed from its cubicle completely to prevent unexpected mishaps. Always follow the standard rack-out and removal procedures that were supplied with the switchgear. Once removed, a sign must be hung on the breaker identifying its use as a lockout point, and approved safety locks must be installed when the breaker is used for isolation. Breakers equipped with closing springs should be discharged to release all stored energy in the breaker mechanism.
- Some of the circuit breakers used are equipped with keyed interlocks for protection during operation. These locks are generally called *kirk-locks* and are relied upon to ensure proper sequence of operation only. These are not to be used for the purpose of locking out a circuit or system. Where disconnects are installed for use in isolation, they should never be opened under load. When opening a disconnect manually, it should be done quickly with a positive force. Again, lockouts should be used when the disconnects are open.
- Whenever performing switching or fuse replacements, always use the protective equipment necessary to ensure personnel safety. *Never* make the assumption that because things have gone fine the last 999 times, they will not go wrong this time. Always prepare yourself for the worst case accident when performing switching.
- Whenever reenergizing circuits following maintenance or removal of a faulted component, extreme care should be used. Always verify that the equipment is in a condition to be reenergized safely. All connections should be insulated and all covers should be installed. Have all personnel stand clear of the area for the initial reenergization. *Never* assume everything is in perfect condition. Verify the conditions!

The following procedure is provided as a guideline for ensuring that equipment and systems will not be damaged by reclosing low-voltage circuit breakers into faults. If a low-voltage circuit breaker has opened for no apparent reason, perform the following:

Step 1 Verify that the equipment being supplied is not physically damaged and shows no obvious signs of overheating or fire.

Step 2 Make all appropriate tests to locate any faults.

Step 3 Reclose the feeder breaker. Stand off to the side when closing the breaker.

Step 4 If the circuit breaker trips again, do not attempt to reclose the breaker. In a plant environment, Electrical Engineering should be notified, and the cause of the trip should be isolated and repaired.

The same general procedure should be followed for fuse replacement, with the exception of transformer fuses. If a transformer fuse blows, the transformer and feeder cabling should be inspected and tested before reenergizing. A blown fuse to a transformer is very significant because it normally indicates an internal fault. Transformer failures are catastrophic in nature and can be extremely dangerous. If applicable, contact the in-plant Electrical Engineering Department prior to commencing any effort to reenergize a transformer.

Power must always be removed from a circuit when removing and installing fuses. The air break

Instructor's Notes:

Working on Energized Systems

THINK ABOUT IT

Some electricians commonly work on energized systems because they think it's too much trouble to turn off the power. What practices have you seen around your home or workplace that could be deadly?

disconnects (or quick disconnects) provided on the upstream side of a large transformer must be opened prior to removing the transformer's fuses. Otherwise, severe arcing will occur as the fuse is removed. This arcing can result in personnel injury and equipment damage.

To replace fuses servicing circuits below 600 volts:

- Secure power to fuses or ensure all downstream loads have been disconnected.
- Always use a positive force to remove and install fuses.

When replacing fuses servicing systems above 600 volts:

- Open and lock out the disconnect switches.
- Unlock the fuse compartment.
- Verify that the fuses are de-energized.
- Attach the fuse removal hot stick to the fuse and remove it.

4.3.0 Electrical Regulations

OSHA has certain regulations that apply to job site electrical safety. These regulations include:

- All electrical work shall be in compliance with the latest NEC and OSHA standards.

Note

OSHA may not recognize the current edition of the NEC, which can sometimes cause problems; however, OSHA typically will *not* cite for any differences.

- The noncurrent-carrying metal parts of fixed, portable, and plug-connected equipment shall be grounded. It is best to choose **grounded tools.** However, portable tools and appliances protected by an approved system of double insulation need not be grounded. *Figure 7* shows an example of a **double-insulated/ ungrounded tool.**
- Extension cords shall be the three-wire type, shall be protected from damage, and shall not be fastened with staples or hung in a manner that could cause damage to the outer jacket or insulation.

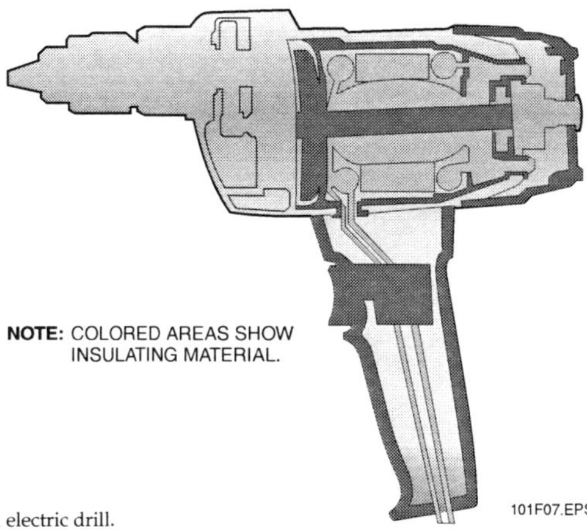
NOTE: COLORED AREAS SHOW INSULATING MATERIAL.
101F07.EPS

Figure 7 ◆ Double-insulated electric drill.

Teaching Tip

Discuss the "Think About It."

Audiovisual

Show Transparency 8 (Figure 7).

Teaching Tip

Discuss the "Case History."

Classroom

Ask the trainees whether they have ever put themselves at risk by working without GFCI protection. Continue to review the OSHA safety guidelines.

Review the OSHA lockout/tagout rule.

CASE HISTORY

Potential Hazards

A self-employed builder was using a metal cutting tool on a metal carport roof and was not using GFCI protection. The male and female plugs of his extension cord partially separated, and the active pin touched the metal roofing. When the builder grounded himself on the gutter of an adjacent roof, he received a fatal shock.

The Bottom Line: Always use GFCI protection and be on the lookout for potential hazards.

Never run an extension cord through a doorway or window that can pinch the cord. Also, never allow vehicles or equipment to drive over cords.

- Exposed lamps in temporary lights shall be guarded to prevent accidental contact, except where lamps are deeply recessed in the reflector. Temporary lights shall not be suspended, except in accordance with their listed labeling.
- Receptacles for attachment plugs shall be of an approved type and properly installed. Installation of the receptacle will be in accordance with the listing and labeling for each receptacle and shall be GFCI-protected if the setting is a temporarily wired construction site. If permanent receptacles are used with extension cords, then you must use GFCI protection.
- Each disconnecting means for motors and appliances and each service feeder or branch circuit at the point where it originates shall be legibly marked to indicate its purpose and voltage.
- Flexible cords shall be used in continuous lengths (no splices) and shall be of a type listed in ***NEC Table 400.4.***
- Ground fault protection is required when supplying temporary power to equipment used by personnel during any repair, remodel, maintenance, construction, and demolition activities. There are two methods for accomplishing this: an assured grounding program (limited to use in certain industrial applications only per *NEC Section 527.6*), or ground fault protection receptacles or breakers. Each employer will set the standard and method to be used. *Figure 8* shows a typical ground-fault circuit interrupter.

4.3.1 OSHA Lockout/Tagout Rule

OSHA released the 29 CFR 1926 lockout/tagout rule in December 1991. This rule covers the specific procedure to be followed for the "servicing and maintenance of machines and equipment in which the unexpected energization or startup of the machines or equipment, or releases of stored energy, could cause injury to employees. This standard establishes minimum performance requirements for the control of such hazardous energy."

The purpose of the OSHA procedure is to ensure that equipment is isolated from all potentially hazardous energy (for example, electrical, mechanical, hydraulic, chemical, or thermal), and tagged and locked out before employees perform any servicing or maintenance activities in which the unexpected energization, startup, or release of stored energy could cause injury. All employees shall be instructed in the lockout/tagout procedure.

CAUTION

Although 99% of your work may be electrical, be aware that you may also need to lock out mechanical equipment.

101F08.EPS

Figure 8 ◆ Typical GFCI.

Instructor's Notes:

GFCIs

THINK ABOUT IT

Explain how GFCIs protect people. Where should a GFCI be installed in the circuit to be most effective?

The following is an example of a lockout/tagout procedure. Make sure to use the procedure that is specific to your employer or job site.

WARNING!
This procedure is provided for your information only. The OSHA procedure provides only the minimum requirements for lockouts/tagouts. Consult the lockout/tagout procedure for your company and the plant or job site at which you are working. Remember that your life could depend on the lockout/tagout procedure. It is critical that you use the correct procedure for your site. The NEC requires that remote-mounted motor disconnects be permanently equipped with a lockout feature.

I. *Introduction*
   A. This lockout/tagout procedure has been established for the protection of personnel from potential exposure to hazardous energy sources during construction, installation, service, and maintenance of electrical energy systems.
   B. This procedure applies to and must be followed by all personnel who may be potentially exposed to the unexpected startup or release of hazardous energy (e.g., electrical, mechanical, pneumatic, hydraulic, chemical, or thermal).

Exception: This procedure does not apply to process and/or utility equipment or systems with cord and plug power supply systems when the cord and plug are the only source of hazardous energy, are removed from the source, and remain under the exclusive control of the authorized employee.

Exception: This procedure does not apply to troubleshooting (diagnostic) procedures and installation of electrical equipment and systems when the energy source cannot be de-energized because continuity of service is essential or shutdown of the system is impractical. Additional personal protective equipment for such work is required and the safe work practices identified for this work must be followed.

II. *Definitions*
- *Affected employee*—Any person working on or near equipment or machinery when maintenance or installation tasks are being performed by others during lockout/tagout conditions.
- *Appointed authorized employee*—Any person appointed by the job site supervisor to coordinate and maintain the security of a group lockout/tagout condition.
- *Authorized employee*—Any person authorized by the job site supervisor to use lockout/tagout procedures while working on electrical equipment.
- *Authorized supervisor*—The assigned job site supervisor who is in charge of coordination of procedures and maintenance of security of all lockout/tagout operations at the job site.
- *Energy isolation device*—An approved electrical disconnect switch capable of accepting approved lockout/tagout hardware for the purpose of isolating and securing a hazardous electrical source in an open or safe position.
- *Lockout/tagout hardware*—A combination of padlocks, danger tags, and other devices designed to attach to and secure electrical isolation devices.

III. *Training*
   A. Each authorized supervisor, authorized employee, and appointed authorized employee shall receive initial and as-needed user-level training in lockout/tagout procedures.
   B. Training is to include recognition of hazardous energy sources, the type and magnitude of energy sources in the workplace, and the procedures for energy isolation and control.
   C. Retraining will be conducted on an as-needed basis whenever lockout/tagout procedures are changed or there is evidence that procedures are not being followed properly.

Teaching Tip

Discuss the "Think About It."

Classroom

Continue review of lockout/tagout procedure.

Classroom

Continue review of lockout/tagout procedure.

Audiovisual

Show Transparency 9 (Figure 10).

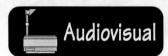

IV. Protective Equipment and Hardware

A. Lockout/tagout devices shall be used exclusively for controlling hazardous electrical energy sources.

B. All padlocks must be numbered and assigned to one employee only.

C. No duplicate or master keys will be made available to anyone except the site supervisor.

D. A current list with the lock number and authorized employee's name must be maintained by the site supervisor.

E. Danger tags must be of the standard white, red, and black *DANGER—DO NOT OPERATE* design and shall include the authorized employee's name, the date, and the appropriate network company (use permanent markers).

F. Danger tags must be used in conjunction with padlocks, as shown in *Figure 9*.

V. Procedures

A. Preparation for lockout/tagout:

1. Check the procedures to ensure that no changes have been made since you last used a lockout/tagout.
2. Identify all authorized and affected employees involved with the pending lockout/tagout.

B. Sequence for lockout/tagout:

1. Notify all authorized and affected personnel that a lockout/tagout is to be used and explain the reason why.
2. Shut down the equipment or system using the normal OFF or STOP procedures.
3. Lock out energy sources and test disconnects to be sure they cannot be moved to the ON position and open the control cutout switch. If there is no cutout switch, block the magnet in the switch open position before working on electrically operated equipment/apparatus such as motors, relays, etc. Remove the control wire.
4. Lock and tag the required switches in the open position. Each authorized employee must affix a separate lock and tag. An example is shown in *Figure 10*.

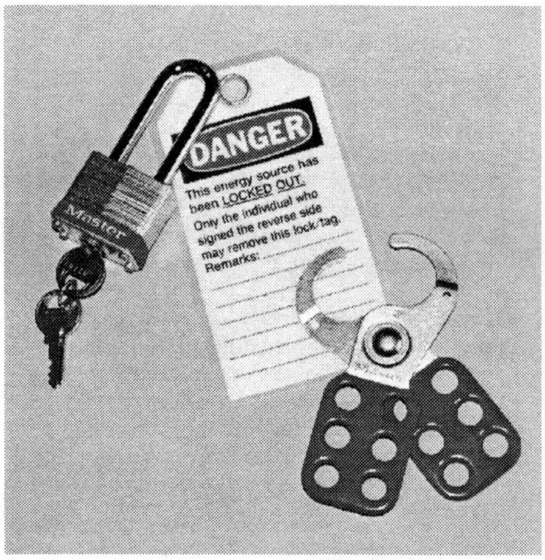

101F09.EPS

Figure 9 ◈ Lockout/tagout device.

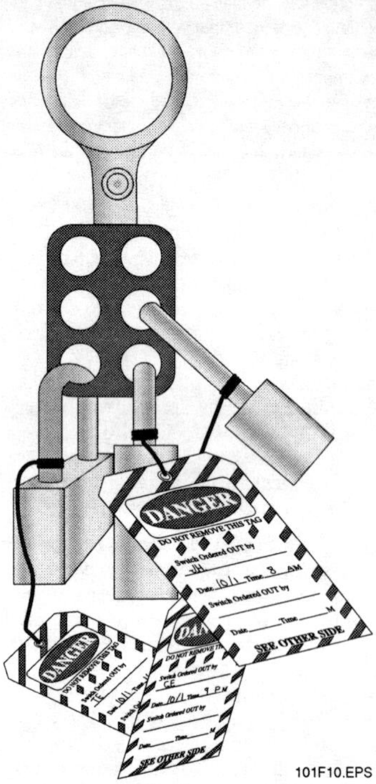

101F10.EPS

Figure 10 ◈ Multiple lockout/tagout device.

Instructor's Notes:

5. Dissipate any stored energy by attaching the equipment or system to ground.
6. Verify that the test equipment is functional via a known power source.
7. Confirm that all switches are in the open position and use test equipment to verify that all parts are de-energized.
8. If it is necessary to temporarily leave the area, upon returning, retest to ensure that the equipment or system is still de-energized.

C. Restoration of energy:

1. Confirm that all personnel and tools, including shorting probes, are accounted for and removed from the equipment or system.
2. Completely reassemble and secure the equipment or system.
3. Replace and/or reactivate all safety controls.
4. Remove locks and tags from isolation switches. Authorized employees must remove their own locks and tags.
5. Notify all affected personnel that the lockout/tagout has ended and the equipment or system is energized.
6. Operate or close isolation switches to restore energy.

VI. Emergency Removal Authorization

A. In the event a lockout/tagout device is left secured, and the authorized employee is absent, or the key is lost, the authorized supervisor can remove the lockout/tagout device.

B. The authorized employee must be informed that the lockout/tagout device has been removed.

C. Written verification of the action taken, including informing the authorized employee of the removal, must be recorded in the job journal.

What's Wrong with This Picture?

101PO104.EPS

Teaching Tip

Discuss "What's Wrong with This Picture?"

CASE HISTORY

Lockout/Tagout Dilemma

In Georgia, electricians found energized switches after the lockout of a circuit panel in an older system that had been upgraded several times. The existing wiring did not match the current site drawings. A subsequent investigation found many such situations in older facilities.

The Bottom Line: Never rely solely on drawings. It is mandatory that the circuit be tested after lockout to verify that it is de-energized.

Teaching Tip

Review the "Case History."

Teaching Tip

Discuss the "Think About It."

Safety

Emphasize the importance of assuming that all circuits are energized until proven otherwise.

Think About It

Lockout/Tagout—Who Does It and When?

What situations are likely to require lockout/tagout? Who is responsible for performing the lockout/tagout? When would more than one person be responsible?

4.4.0 Other OSHA Regulations

There are other OSHA regulations that you need to be aware of on the job site. For example:

- OSHA requires the posting of hard hat areas. Be alert to those areas and always wear your hard hat properly, with the bill in front. Hard hats should be worn whenever overhead hazards exist, or there is the risk of exposure to electric shock or burns.
- You should wear safety shoes on all job sites. Keep them in good condition.
- Do not wear clothing with exposed metal zippers, buttons, or other metal fasteners. Avoid wearing loose-fitting or torn clothing.
- Protect your eyes. Your eyesight is threatened by many activities on the job site. Always wear safety glasses with full side shields. In addition, the job may also require protective equipment such as face shields or goggles.

4.4.1 Testing for Voltage

OSHA also requires that you inspect or test existing conditions before beginning work on electrical equipment or lines. Usually, you will use a voltmeter/sensor or voltage tester to do this. You should assume that all electrical equipment and lines are energized until you have determined that they are not. Do not proceed to work on or near energized parts until the operating voltage is determined.

After the electrical equipment to be worked on has been locked and tagged out, the equipment must be verified as de-energized before work can proceed. This section sets the requirements that must be met before any circuits or equipment can be considered de-energized. First, and most importantly, only qualified persons may verify that a circuit or piece of equipment is de-energized. Before approaching the equipment to be worked on, the qualified person shall operate the equipment's normal operating controls to check that the proper energy sources have been disconnected.

Upon opening a control enclosure, the qualified person shall note the presence of any components that may store electrical energy. Initially, these components should be avoided.

To verify that the lockout was adequate and the equipment is indeed de-energized, a qualified person must use appropriate test equipment to check for power, paying particular attention to induced voltages and unrelated feedback voltage.

Ensure that your testing equipment is working properly by performing the *live-dead-live* check before each use. To perform this test, first check your voltmeter on a known live voltage source. This known source must be in the same range as the electrical equipment you will be working on. Next, without changing scales on your voltmeter, check for the presence of power in the equipment you have locked out. Finally, to ensure that your voltmeter did not malfunction, check it again on the known live source. Performing this test will assure you that your voltage testing equipment is reliable.

In accordance with OSHA section 1910.333(b)(2)(iv)/1926.417(d)(4)(ii), if the circuit to be tested normally operates at more than 600 volts, the live-dead-live check must be performed.

Once it has been verified that power is not present, stored electrical energy that might endanger personnel must be released. A qualified person must use the proper devices to release the stored energy, such as using a shorting probe to discharge a capacitor.

5.0.0 ◆ LADDERS AND SCAFFOLDS

Ladders and scaffolds account for about half of the injuries from workplace electrocutions. The involuntary recoil that can occur when a person is shocked can cause the person to be thrown from a ladder or high place.

5.1.0 Ladders

Many job site accidents involve the misuse of ladders. Make sure to follow these general rules every time you use any ladder. Following these rules can prevent serious injury or even death.

Instructor's Notes:

- Before using any ladder, inspect it. Look for loose or missing rungs, cleats, bolts, or screws, and check for cracked, broken, or badly worn rungs, cleats, or side rails.
- If you find a ladder in poor condition, do not use it. Report it and tag it for repair or disposal.
- Never modify a ladder by cutting it or weakening its parts.
- Do not set up ladders where they may be run into by others, such as in doorways or walkways. If it is absolutely necessary to set up a ladder in such a location, protect the ladder with barriers.
- Do not increase a ladder's reach by standing it on boxes, barrels, or anything other than a flat surface.
- Check your shoes for grease, oil, or mud before climbing a ladder. These materials could make you slip.
- Always face the ladder and hold on with both hands when climbing up or down.
- Never lean out from the ladder. Keep your belt buckle centered between the rails. If something is out of reach, get down and move the ladder.

WARNING!
When performing electrical work, always use ladders made of nonconductive material.

5.1.1 Straight and Extension Ladders

There are some specific rules to follow when working with straight and extension ladders:

- Always place a straight ladder at the proper angle. The distance from the ladder feet to the base of the wall or support should be about one-fourth the working height of the ladder (see *Figure 11*).
- Secure straight ladders to prevent slipping. Use ladder shoes or hooks at the top and bottom. Another method is to secure a board to the floor against the ladder feet. For brief jobs, someone can hold the straight ladder.
- Side rails should extend above the top support point by at least 36 inches (see *Figure 11*).
- It takes two people to safely extend and raise an extension ladder. Extend the ladder only after it has been raised to an upright position.
- Never carry an extended ladder.
- Never use two ladders spliced together.
- Ladders should not be painted because paint can hide defects.

What's Wrong with This Picture?

101PO105.EPS

5.1.2 Step Ladders

There are also a few specific rules to use with a step ladder:

- Always open the step ladder all the way and lock the spreaders to avoid collapsing the ladder accidentally.
- Use a step ladder that is high enough for the job so that you do not have to reach. Get someone to hold the ladder if it is more than 10 feet high.
- Never use a step ladder as a straight ladder.
- Never stand on or straddle the top two rungs of a step ladder.
- Ladders are not shelves.

WARNING!
Do not leave tools or materials on a step ladder.

Sometimes you will need to move or remove protective equipment, guards, or guardrails to

Teaching Tip

Discuss "What's Wrong with This Picture?"

Classroom

Review the safety rules for ladders.

Audiovisual

Show Transparency 10 (Figure 11).

ELECTRICAL SAFETY 31

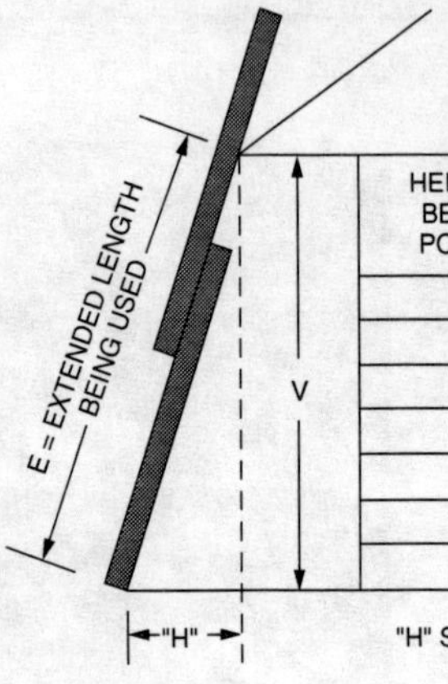

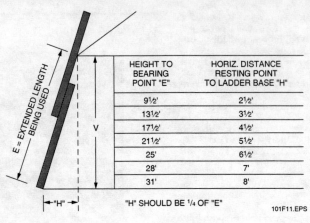

HEIGHT TO BEARING POINT "E"	HORIZ. DISTANCE RESTING POINT TO LADDER BASE "H"
9½'	2½'
13½'	3½'
17½'	4½'
21½'	5½'
25'	6½'
28'	7'
31'	8'

"H" SHOULD BE ¼ OF "E"

101F11.EPS

Figure 11 ◆ Straight ladder positioning.

complete a task using a ladder. Remember, always replace what you moved or removed before leaving the area.

Classroom

Review the safety guidelines for scaffolds.

5.2.0 Scaffolds

Working on scaffolds also involves being safe and alert to hazards. In general, keep scaffold platforms clear of unnecessary material or scrap. These can become deadly tripping hazards or falling objects. Carefully inspect each part of the scaffold as it is erected. Your life may depend on it! Makeshift scaffolds have caused many injuries and deaths on job sites. Use only scaffolding and planking materials designed and marked for their specific use. When working on a scaffold, follow the established specific requirements set by OSHA for the use of fall protection. When appropriate, wear an approved harness with a lanyard properly anchored to the structure.

> **Note**
> The following requirements represent a compilation of the more stringent requirements of both CFR 1910 and CFR 1926.

The following are some of the basic OSHA rules for working safely on scaffolds:

- Scaffolds must be erected on sound, rigid footing that can carry the maximum intended load.
- Guardrails and toe boards must be installed on the open sides and ends of platforms that are higher than six feet above the ground or floor.
- There must be a screen of ½-inch maximum openings between the toe board and the midrail where persons are required to work or pass under the scaffold.
- Scaffold planks must extend over their end supports not less than six inches nor more than 12 inches and must be properly blocked.
- If the scaffold does not have built-in ladders that meet the standard, then it must have an attached ladder access.
- All employees must be trained to erect, dismantle, and use scaffold(s).
- Unless it is impossible, fall protection must be worn while building or dismantling all scaffolding.
- Work platforms must be completely decked for use by employees.
- Your hard hat is the first line of protection from falling objects. Your hard hat, however, cannot protect your shoulders, arms, back, or feet from the danger of falling objects. The person working below depends on those working above. When you are working above the ground, be careful so that material, including your tools, cannot fall from your work site. Use trash containers or other similar means to keep debris from falling and never throw or sweep material from above.

Instructor's Notes:

32 CHAPTER 1

CASE HISTORY

Scaffolds and Electrical Hazards

Remember that scaffolds are excellent conductors of electricity. Recently, a maintenance crew needed to move a scaffold and although time was allocated in the work order to dismantle and rebuild the scaffold, the crew decided to push it instead. They did not follow OSHA recommendations for scaffold clearance and did not perform a job site survey. During the move, the five-tier scaffold contacted a 12,000V overhead power line. All four members of the crew were killed and the crew chief received serious injuries.

The Bottom Line: Never take shortcuts when it comes to your safety and the safety of others. Trained safety personnel should survey each job site prior to the start of work to assess potential hazards. Safe working distances should be maintained between scaffolding and power lines.

What's Wrong with This Picture?

101PO106.EPS

6.0.0 ◆ LIFTS, HOISTS, AND CRANES

On the job, you may be working in the operating area of lifts, hoists, or cranes. The following safety rules are for those who are working in the area with overhead equipment but are not directly involved in its operation.

- Stay alert and pay attention to the warning signals from operators.
- Never stand or walk under a load, regardless of whether it is moving or stationary.
- Always warn others of moving or approaching overhead loads.
- Never attempt to distract signal persons or operators of overhead equipment.
- Obey warning signs.
- Do not use equipment that you are not qualified to operate.

7.0.0 ◆ LIFTING

Back injuries cause many lost working hours every year. That is in addition to the misery felt by the person with the hurt back! Learn how to lift properly and size up the load. To lift, first stand close to the load. Then, squat down and keep your back straight. Get a firm grip on the load and keep the load close to your body. Lift by straightening your legs. Make sure that you lift with your legs and not your back. Do not be afraid to ask for help if you feel the load is too heavy. See *Figure 12* for an example of proper lifting.

8.0.0 ◆ BASIC TOOL SAFETY

When using any tools for the first time, read the operator's manual to learn the recommended safety precautions. If you are not certain about the operation of any tool, ask the advice of a more experienced worker. Before using a tool, you should know its function and how it works.

Always use the right tool for the job. Incorrectly using tools is one of the leading causes of job site injury. Using a hammer as a pry bar or a screwdriver as a chisel can cause damage to the tool and injure you in the process.

Teaching Tip

Review the "Case History."

Discuss "What's Wrong with This Picture?"

Audiovisual

Show Transparency 11 (Figure 12).

Demonstration

Demonstrate proper lifting procedures.

Laboratory

Under your supervision, have the trainees practice proper lifting procedures.

Classroom

Discuss the hazards involved with confined space entry.

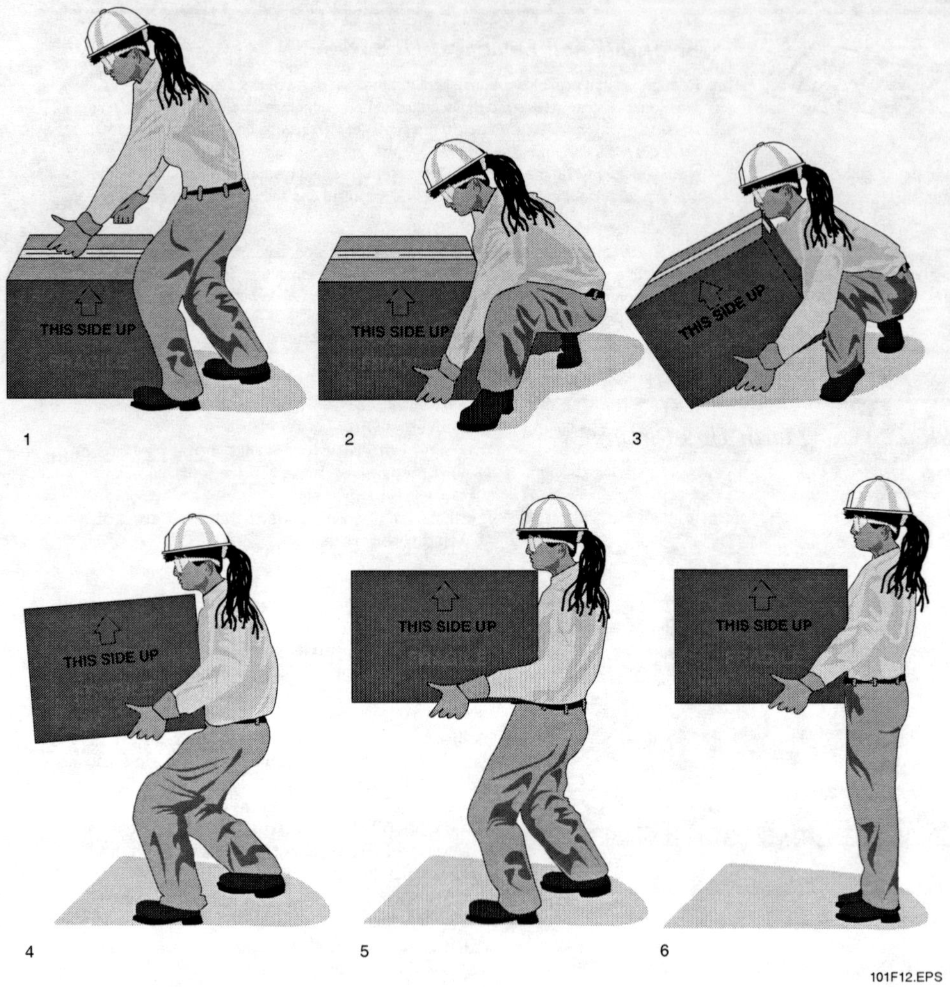

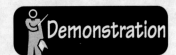

Figure 12 ◆ Proper lifting.

9.0.0 ◆ CONFINED SPACE ENTRY PROCEDURES

Occasionally, you may be required to do your work in a manhole or vault. If this is the case, there are some special safety considerations that you need to be aware of. For details on the subject of working in manholes and vaults, refer to 1910.146/1926.21(a)(6)(i) and (ii) and the *National Electrical Safety Code*. The general precautions are listed in the following paragraphs.

9.1.0 General Guidelines

A confined space includes (but is not limited to) any of the following: a manhole, boiler, tank,

Instructor's Notes:

INSIDE TRACK

Lifting

If you bend from the waist to pick up a 50-pound object, you are applying 10 times the amount of pressure (500 pounds) to your lower back. Lower back injuries are one of the most common workplace injuries because it's so easy to be careless about lifting, especially when you are in a hurry. Remember, it is much easier to ask for help than it is to nurse an injured back.

trench (four feet or deeper), tunnel, hopper, bin, sewer, vat, pipeline, vault, pit, air duct, or vessel. A confined space is identified as follows:

- It has limited entry and exit.
- It is not intended for continued human occupancy.
- It has poor ventilation.
- It has the potential for entrapment/engulfment.
- It has the potential for accumulating a dangerous atmosphere.
- Entry into a confined space occurs when any part of the body crosses the plane of entry. No employee shall enter a confined space unless the employee has been trained in confined space entry procedures.
- All hazards must be eliminated or controlled before a confined space entry is made.
- All appropriate personal protective equipment shall be worn at all times during confined space entry and work. The minimum required equipment includes a hard hat, safety glasses, full body harness, and life line.
- Ladders used for entry must be secured.
- A rescue retrieval system must be in use when entering confined spaces and while working in permit-required confined spaces (discussed later). Each employee must be capable of being rescued by the retrieval system.
- Only no-entry rescues will be performed by company personnel. Entry rescues will be performed by trained rescue personnel identified on the entry permit.
- The area outside the confined space must be properly barricaded, and appropriate warning signs must be posted.
- Entry permits can be issued and signed by job site supervisors only. Permits must be kept at the confined space while work is being conducted. At the end of the shift, the entry permits must be made part of the job journal and retained for one year.

9.2.0 Confined Space Hazard Review

Before determining the proper procedure for confined space entry, a hazard review shall be performed. The hazard review shall include, but not be limited to, the following conditions:

- The past and current uses of the confined space
- The physical characteristics of the space including size, shape, air circulation, etc.
- Proximity of the space to other hazards
- Existing or potential hazards in the confined space, such as:
  –Atmospheric conditions (oxygen levels, flammable/explosive levels, and/or toxic levels)
  –Presence/potential for liquids
  –Presence/potential for particulates
- Potential for mechanical/electrical hazards in the confined space (including work to be done)

Once the hazard review is completed, the supervisor, in consultation with the project managers and/or safety manager, shall classify the confined space as one of the following:

- A nonpermit confined space
- A permit-required confined space controlled by ventilation
- A permit-required confined space

THINK ABOUT IT

Working in Tight Spaces

We routinely work in cramped quarters, from closets to ceiling spaces, without concern. What hazards should we be aware of? What hazards do we typically ignore?

Classroom

Review the guidelines for conducting a confined space hazard review.

Teaching Tip

Discuss the "Think About It."

Classroom

Review the definitions of confined spaces.

Review entry permits and rescue plans.

Once the confined space has been properly classified, the appropriate entry and work procedures must be followed.

9.3.0 Entry and Work Procedures

Nonpermit spaces—A hazard review checklist must be completed before a confined space is designated as a *nonpermit space.* The checklist must be made part of the job journal, and a copy of the checklist must be sent to the safety office. A nonpermit confined space must meet the following criteria:

- There is no actual or potential atmospheric hazard.

> **Note**
> Using ventilation to clear the atmosphere does not meet this criterion.

- There are no actual or potential physical, electrical, or mechanical hazards capable of causing harm or death.

Documentation using the hazards checklist and entry permit forms, and verifying that the confined space is hazard-free, must be made available to employees and maintained at the confined space while work is conducted. If it is necessary to enter the space to verify that it is hazard-free or to eliminate hazards, entry must be made under the requirements of a permit-required space.

An employee may enter the confined space using the minimum fall protection of harness and anchored life line. Once in the space, the employee may disconnect the life line and reconnect it before exiting.

If the work being done creates a hazard, the space must be reclassified as a permit-required space. If any other atmospheric, physical, electrical, or mechanical hazards arise, the space is to be evacuated immediately and reclassified as a permit-required entry space.

Permit-required spaces controlled by ventilation—A hazard review checklist must be completed before a confined space is designated as a *permit-required space controlled by ventilation.* The checklist must be made part of the job journal, and a copy of the checklist must be sent to the safety office. A permit-required confined space controlled by ventilation must meet the following criteria:

- The only hazard in the confined space is an actual/potential atmospheric hazard.
- Continuous forced-air ventilation maintains a safe atmosphere (i.e., within the limits designated on the entry permit).
- Inspection and monitoring data are documented.
- No other physical, electrical, or mechanical hazard exists.

An entry permit must be issued and signed by the job site supervisor and be kept at the confined space while work is being conducted.

Atmospheric testing must be conducted before entry into the confined space and in the following order:

- Oxygen content
- Flammable gases and vapors
- Toxic contaminants

Unacceptable atmospheric conditions must be eliminated with forced-air ventilation. If continuous forced-air ventilation is required to maintain an acceptable atmosphere, employees may not enter until forced-air ventilation has eliminated any hazardous atmosphere. Periodic atmospheric testing must be conducted during the work shift to ensure that the atmosphere remains clear. Periodic monitoring must be documented on the entry permit. If atmospheric conditions change, employees must exit the confined space immediately, and atmospheric conditions must be re-evaluated. Continuous communication must be maintained with the employees working in the confined space.

If hot work is to be performed, a hot work permit is required, and the hazard analysis must document that the hot work does not create additional hazards that are not controlled by ventilation only. Hot work is defined as any work that produces arcs, sparks, flames, heat, or other sources of ignition.

A rescue plan using trained rescue personnel must be in place prior to the start of work in the confined space. All employees should be aware of the rescue plan and how to activate it.

Permit-required confined spaces—A hazard review checklist must be completed before a confined space is designated as a *permit-required confined space.* The checklist must be made part of the job journal, and a copy must be sent to the safety office. A permit-required space meets the following criteria:

- There are actual/potential hazards, other than a hazardous atmosphere.
- Ventilation alone does not eliminate atmospheric hazards.
- Conditions in and around the confined space must be continually monitored.

Instructor's Notes:

An entry permit must be issued and signed by the job site supervisor. The permit is to be kept at the confined space while work is being performed in the space.

Atmospheric testing must be conducted before entry into the confined space and in the following order:

- Oxygen content
- Flammable gases and contaminants
- Toxic contaminants

Unacceptable atmospheric conditions must be eliminated/controlled prior to employee entry. Methods of elimination may include isolation, purging, flushing, or ventilating. Continuous atmospheric monitoring must be conducted while employees are in the confined space. Triggering of a monitoring alarm means employees should evacuate the confined space immediately. Any other physical hazards must be eliminated or controlled by engineering and work practice controls before entry. Additional personal protective equipment should be used as a follow-up to the above methods. An attendant, whose job it is to monitor conditions in and around the confined space and to maintain contact with the employees in the space, must be stationed outside the confined space for the duration of entry operations.

If hot work is to be performed, a hot work permit is required, and the hazard analysis must document the additional hazards and precautions to be considered.

A rescue plan using trained rescue personnel must be in place before confined space entry. The attendant should be aware of the rescue plan and have the means to activate it.

10.0.0 ◆ FIRST AID

You should be prepared in case an accident does occur on the job site or anywhere else. First aid training that includes certification classes in CPR and artificial respiration could be the best insurance you and your fellow workers ever receive. Make sure that you know where first aid is available at your job site. Also, make sure you know the accident reporting procedure. Each job site should also have a first aid manual or booklet giving easy-to-find emergency treatment procedures for various types of injuries. Emergency first aid telephone numbers should be readily available to everyone on the job site. Refer to CFR 1910.151/1926.23 and 1926.50 for specific requirements.

11.0.0 ◆ SOLVENTS AND TOXIC VAPORS

The solvents that are used by electricians may give off vapors that are toxic enough to make people temporarily ill or even cause permanent injury. Many solvents are skin and eye irritants. Solvents can also be systemic poisons when they are swallowed or absorbed through the skin.

Solvents in spray or aerosol form are dangerous in another way. Small aerosol particles or solvent vapors mix with air to form a combustible mixture with oxygen. The slightest spark could cause an explosion in a confined area because the mix is perfect for fast ignition. There are procedures and methods for using, storing, and disposing of most solvents and chemicals. These procedures are normally found in the Material Safety Data Sheets (MSDSs) available at your facility.

An MSDS is required for all materials that could be hazardous to personnel or equipment. These sheets contain information on the material, such as the manufacturer and chemical makeup. As much information as possible is kept on the hazardous material to prevent a dangerous situation; or, in the event of a dangerous situation, the information is used to rectify the problem in as safe a manner as possible. See *Figure 13* for an example of procedures you may find on the job.

11.1.0 Precautions When Using Solvents

It is always best to use a nonflammable, nontoxic solvent whenever possible. However, any time solvents are used, it is essential that your work area be adequately ventilated and that you wear the appropriate personal protective equipment:

- A chemical face shield with chemical goggles should be used to protect the eyes and skin from sprays and splashes.
- A chemical apron should be worn to protect your body from sprays and splashes. Remember that some solvents are acid-based. If they come into contact with your clothes, solvents can eat through your clothes to your skin.
- A paper filter mask does not stop vapors; it is used only for nuisance dust. In situations where a paper mask does not supply adequate protection, chemical cartridge respirators might be needed. These respirators can stop many vapors if the correct cartridge is selected. In areas where ventilation is a serious problem, a self-contained breathing apparatus (SCBA) must be used.

Classroom

Review the hazards involved with the use of solvents.

Audiovisual

Show Transparency 12 (Figure 13).

Section VII — Precautions for Safe Handling and Use

Steps to Be Taken in Case Material is Released or Spilled

Isolate from oxidizers, heat, sparks, electric equipment, and open flames.

Waste Disposal Method

Recycle or incinerate observing local, state and federal health, safety and pollution laws.

Precautions to Be Taken in Handling and Storing

Store in a cool dry area. Observe label cautions and instructions.

Other Precautions

SEE ATTACHMENT PARA #3

Section VIII — Control Measures

Respiratory Protection *(Specify Type)*

Suitable for use with organic solvents

Ventilation	Local Exhaust: preferable	Special	none
	Mechanical *(General)*: acceptable	Other	none
Protective Gloves	recommended (must not dissolve in solvents)	Eye Protection	goggles

Other Protective Clothing or Equipment

none

Work/Hygenic Practices

Use with adequate ventilation. Observe label cautions.

Figure 13 ◈ Portion of an MSDS.

101F13.TIF

Classroom

List the various types of respiratory protection and explain the purpose of each type. Emphasize the importance of proper training and adherence to company safety policies.

- Make sure that you have been given a full medical evaluation and that you are properly trained in using respirators at your site.

11.2.0 Respiratory Protection

Protection against high concentrations of dust, mist, fumes, vapors, gases, and/or oxygen deficiency is provided by appropriate respirators.

Appropriate respiratory protective devices should be used for the hazardous material involved and the extent and nature of the work performed.

An air-purifying respirator is, as its name implies, a respirator that removes contaminants from air inhaled by the wearer. The respirators may be divided into the following types: particulate-removing (mechanical filter), gas- and vapor-removing (chemical filter), and a combination of particulate-removing and gas- and vapor-removing.

Particulate-removing respirators are designed to protect the wearer against the inhalation of particulate matter in the ambient atmosphere. They may be designed to protect against a single type of particulate, such as pneumoconiosis-producing and nuisance dust, toxic dust, metal fumes or mist, or against various combinations of these types.

Gas- and vapor-removing respirators are designed to protect the wearer against the inhalation of gases or vapors in the ambient atmosphere. They are designated as gas masks, chemical cartridge respirators (nonemergency gas respirators), and self-rescue respirators. They may be designed to protect against a single gas such as chlorine; a single type of gas, such as acid gases; or a combination of types of gases, such as acid gases and organic vapors.

If you are required to use a respiratory protective device, you must be evaluated by a physician to ensure that you are physically fit to use a respirator. You must then be fitted and thoroughly instructed in the respirator's use.

Any employee whose job entails having to wear a respirator must keep his face free of facial hair in the seal area.

Respiratory protective equipment must be inspected regularly and maintained in good condition. Respiratory equipment must be properly cleaned on a regular basis and stored in a sanitary, dustproof container.

Instructor's Notes:

Altered Respiratory Equipment

CASE HISTORY

A self-employed man applied a solvent-based coating to the inside of a tank. Instead of wearing the proper respirator, he used nonstandard air supply hoses and altered the face mask. All joints and the exhalation slots were sealed with tape. He collapsed and was not discovered for several hours.

The Bottom Line: Never alter or improvise safety equipment.

WARNING!
Do not use any respirator unless you have been fitted for it and thoroughly understand its use. As with all safety rules, follow your employer's respiratory program and policies.

12.0.0 ◆ ASBESTOS

Asbestos is a mineral-based material that is resistant to heat and corrosive chemicals. Depending on the chemical composition, asbestos fibers may range in texture from coarse to silky. The properties that make asbestos fibers so valuable to industry are its high tensile strength, flexibility, heat and chemical resistance, and good frictional properties.

Asbestos fibers enter the body by inhalation of airborne particles or by ingestion and can become embedded in the tissues of the respiratory or digestive systems. Years of exposure to asbestos can cause numerous disabling or fatal diseases. Among these diseases are asbestosis, an emphysema-like condition; lung cancer; mesothelioma, a cancerous tumor that spreads rapidly in the cells of membranes covering the lungs and body organs; and gastrointestinal cancer.

12.1.0 Monitoring

Employers who have a workplace or work operation covered by OSHA 3096 (*Asbestos Standard for the Construction Industry*) must perform initial monitoring to determine the airborne concentrations of asbestos to which employees may be exposed. If employers can demonstrate that employee exposures are below the action level and/or excursion limit by means of objective or historical data, initial monitoring is not required. If initial monitoring indicates that employee exposures are below the action level and/or excursion limit, then periodic monitoring is not required. Within regulated areas, the employer must conduct daily monitoring unless all workers are equipped with supplied-air respirators operated in the positive-pressure mode. If daily monitoring by statistically reliable measurements indicates that employee exposures are below the action level and/or excursion limit, then no further monitoring is required for those employees whose exposures are represented by such monitoring. Employees must be given the chance to observe monitoring, and affected employees must be notified as soon as possible following the employer's receipt of the results.

12.2.0 Regulated Areas

The employer must establish a regulated area where airborne concentrations of asbestos exceed or can reasonably be expected to exceed the locally determined exposure limit, or when certain types of construction work are performed, such as cutting asbestos-cement sheets and removing asbestos-containing floor tiles. Only authorized personnel may enter regulated areas. All persons entering a regulated area must be supplied with an appropriate respirator. No smoking, eating, drinking, or applying cosmetics is permitted in regulated areas. Warning signs must be displayed at each regulated area and must be posted at all approaches to regulated areas. These signs must bear the following information:

DANGER
ASBESTOS
CANCER AND LUNG DISEASE HAZARD
AUTHORIZED PERSONNEL ONLY
RESPIRATORS AND PROTECTIVE CLOTHING
ARE REQUIRED IN THIS AREA

Teaching Tip

Review the "Case History." Emphasize the dangers involved in altering any type of equipment.

Classroom

Discuss the hazards involved with handling asbestos.

Safety

Emphasize that even though contact with asbestos may not present any immediate discomfort, the long-term effects can be deadly.

Where feasible, the employer shall establish negative-pressure enclosures before commencing asbestos removal, demolition, and renovation operations. The setup and monitoring requirements for negative-pressure enclosures are as follows:

- A competent person shall be designated to set up the enclosure and ensure its integrity and supervise employee activity within the enclosure.
- Exemptions are given for small-scale, short-duration maintenance or renovation operations.
- The employer shall conduct daily monitoring of the exposure of each employee who is assigned to work within a regulated area. Short-term monitoring is required whenever asbestos concentrations will not be uniform throughout the workday and where high concentrations of asbestos may reasonably be expected to be released or created in excess of the local limit.

In addition, warning labels must be affixed on all asbestos products and to all containers of asbestos products, including waste containers, that may be in the workplace. The label must include the following information:

DANGER
CONTAINS ASBESTOS FIBERS
AVOID CREATING DUST
CANCER AND LUNG DISEASE HAZARD

12.3.0 Methods of Compliance

To the extent feasible, engineering and work practice controls must be used to reduce employee exposure to within the permissible exposure limit (PEL). The employer must use one or more of the following control methods to achieve compliance:

- Local exhaust ventilation equipped with high-efficiency particulate air (HEPA) filter dust collection systems
- General ventilation systems
- Vacuum cleaners equipped with HEPA filters
- Enclosure or isolation of asbestos dust-producing processes
- Use of wet methods, wetting agents, or removal encapsulants during asbestos handling, mixing, removal, cutting, application, and cleanup
- Prompt disposal of asbestos-containing wastes in leak-tight containers

Prohibited work practices include the following:

- The use of high-speed abrasive disc saws that are not equipped with appropriate engineering controls
- The use of compressed air to remove asbestos-containing materials, unless the compressed air is used in conjunction with an enclosed ventilation system

Where engineering and work practice controls have been instituted but are insufficient to reduce employee exposure to a level that is at or below the PEL, respiratory protection must be used to supplement these controls.

Classroom

Discuss the explosion and acid hazards involved when working with batteries.

13.0.0 ◆ BATTERIES

Working around wet cell batteries can be dangerous if the proper precautions are not taken. Batteries often give off hydrogen gas as a byproduct. When hydrogen mixes with air, the mixture can be explosive in the proper concentration. For this reason, smoking is strictly prohibited in battery rooms, and only insulated tools should be used. Proper ventilation also reduces the chance of explosion in battery areas. Follow your company's procedures for working near batteries. Also, ensure that your company's procedures are followed for lifting heavy batteries.

13.1.0 Acids

Batteries also contain acid, which will eat away human skin and many other materials. Personal protective equipment for battery work typically includes chemical aprons, sleeves, gloves, face shields, and goggles to prevent acid from contacting skin and eyes. Follow your site procedures for dealing with spills of these materials. Also, know the location of first aid when working with these chemicals.

Demonstration

If available, demonstrate the proper use of an eye wash station.

Laboratory

Have each trainee practice operating the wash station.

13.2.0 Wash Stations

Because of the chance that battery acid may contact someone's eyes or skin, wash stations are located near battery rooms. Do not connect or disconnect batteries without proper supervision. Everyone who works in the area should know where the nearest wash station is and how to use it. Battery acid should be flushed from the skin and eyes with large amounts of water or with a neutralizing solution.

Instructor's Notes:

14.0.0 ◆ PCBs

Polychlorinated biphenyls (PCBs) are chemicals that were marketed under various trade names as a liquid insulator/cooler in older transformers. In addition to being used in older transformers, PCBs are also found in some large capacitors and in the small ballast transformers used in street lighting and ordinary fluorescent light fixtures. Disposal of these materials is regulated by the EPA and must be done through a regulated disposal company; use extreme caution and follow your facility procedures.

> **CAUTION**
> If you come in contact with battery acid, report it immediately to your supervisor.

> **WARNING!**
> Do not come into contact with PCBs. They present a variety of serious health risks, including lung damage and cancer.

15.0.0 ◆ FALL PROTECTION

15.1.0 Fall Protection Procedures

Fall protection must be used when employees are on a walking or working surface that is six feet or more above a lower level and has an unprotected edge or side. The areas covered include, but are not limited to:

- Finished and unfinished floors or mezzanines
- Temporary or permanent walkways/ramps
- Finished or unfinished roof areas
- Elevator shafts and hoist-ways
- Floor, roof, or walkway holes
- Working six feet or more above dangerous equipment

Exception: If the dangerous equipment is unguarded, fall protection must be used at all heights regardless of the fall distance.

> **Note**
> Walking/working surfaces do not include ladders, scaffolds, vehicles, or trailers. Also, an unprotected edge or side is an edge/side where there is no guardrail system at least 39 inches high.

Fall protection is not required during inspection, investigation, or assessment of job site conditions before or after construction work.

These fall protection guidelines do not apply to the following areas. Fall protection for these areas is located in the subparts cited in parentheses.

- Cranes and derricks (1926 subpart N/1910 subpart N)
- Scaffolding (1926 subpart L/1910 subpart D)
- Electrical power transmission and distribution (1926 subpart V/1910 subpart R)
- Stairways and ladders (1926 subpart X/1910 subpart D)
- Excavations (1926 subpart P)

What's Wrong with This Picture?

101PO107.EPS

Classroom

Discuss PCB hazards.

Discuss fall protection systems.

Teaching Tip

Discuss "What's Wrong with This Picture?" Note the lack of hard hat and fall protection.

ELECTRICAL SAFETY 41

Teaching Tip

Discuss the "Think About It."

THINK ABOUT IT

Putting It All Together

This module has described a professional approach to electrical safety. How does this professional outlook differ from an everyday attitude? What do you think are the key features of a professional philosophy of safety?

Fall protection must be selected in order of preference as listed below. Selection of a lower-level system (e.g., safety nets) must be based only on feasibility of protection. The list includes, but is not limited to, the following:

- Guardrail systems and hole covers
- Personal fall arrest systems
- Safety nets

These fall protection procedures are designed to warn, isolate, restrict, or protect workers from a potential fall hazard.

15.2.0 Types of Fall Protection Systems

The type of system selected shall depend on the fall hazards associated with the work to be performed. First, a hazard analysis shall be conducted by the job site supervisor prior to the start of work. Based on the hazard analysis, the job site supervisor and project manager, in consultation with the safety manager, will select the appropriate fall protection system. All employees will be instructed in the use of the fall protection system before starting work.

Summary

Safety must be your concern at all times so that you do not become either the victim of an accident or the cause of one. Safety requirements and safe work practices are provided by OSHA and your employer. It is essential that you adhere to all safety requirements and follow your employer's safe work practices and procedures. Also, you must be able to identify the potential safety hazards of your job site. The consequences of unsafe job site conduct can often be expensive, painful, or even deadly. Report any unsafe act or condition immediately to your supervisor. You should also report all work-related accidents, injuries, and illnesses to your supervisor immediately. Remember, proper construction techniques, common sense, and a good safety attitude will help to prevent accidents, injuries, and fatalities.

Instructor's Notes:

Review Questions

1. The most life-threatening hazards on a construction site include all of the following, *except* _____ .
   a. falls
   b. electrocution
   c. being crushed or struck by falling or flying objects
   d. chemical burns
2. If a person's heart begins to fibrillate due to an electrical shock, the solution is to _____ .
   a. leave the person alone until the fibrillation stops
   b. administer heart massage
   c. use the Heimlich maneuver
   d. have a qualified person use emergency defibrillation equipment
3. The majority of injuries due to electrical shock are caused by _____.
   a. electrically operated hand tools
   b. contact with low-voltage conductors
   c. contact with high-voltage conductors
   d. lightning
4. Class 0 rubber gloves are used when working with voltages less than _____ .
   a. 1,000 volts
   b. 7,500 volts
   c. 17,500 volts
   d. 26,500 volts
5. An important use of a hot stick is to _____ .
   a. replace busbars
   b. test for voltage
   c. replace fuses
   d. test for continuity
6. Which of these statements correctly describes a double-insulated power tool?
   a. There is twice as much insulation on the power cord.
   b. It can safely be used in place of a grounded tool.
   c. It is made entirely of plastic or other nonconducting material.
   d. The entire tool is covered in rubber.
7. Which of the following applies in a lockout/tagout procedure?
   a. Only the supervisor can install lockout/tagout devices.
   b. If several employees are involved, the lockout/tagout equipment is applied only by the first employee to arrive at the disconnect.
   c. Lockout/tagout devices applied by one employee can be removed by another employee as long as it can be verified that the first employee has left for the day.
   d. Lockout/tagout devices are installed by every authorized employee involved in the work.
8. What is the proper distance from the feet of a straight ladder to the wall?
   a. one-fourth the working height of the ladder
   b. one-half the height of the ladder
   c. three feet
   d. one-fourth of the square root of the height of the ladder
9. What are the minimum and maximum distances (in inches) that a scaffold plank can extend beyond its end support?
   a. 4, 8
   b. 6, 10
   c. 6, 12
   d. 8, 12
10. Which of these conditions applies to a permit-required confined space, but not to a permit-required space controlled by ventilation?
    a. A hazard review checklist must be completed.
    b. An attendant, whose job is to monitor the space, must be stationed outside the space.
    c. Unacceptable atmospheric conditions must be eliminated.
    d. Atmospheric testing must be conducted.

Classroom

Have the trainees complete the Review Questions and go over the answers prior to administering the Module Examination.

Examination

Administer the Module Examination. Record the results on Craft Training Report Form 200 and submit the results to the Training Program Sponsor.

Performance Testing

Administer the Performance Test and fill out Performance Profile Sheets for each trainee. If desired, trainee proficiency noted during laboratory sessions may be used to complete the Performance Test. Be sure to record the results on Craft Training Report Form 200 and submit the results to the Training Program Sponsor.

PROFILE IN SUCCESS

Michael J. Powers, Encompass Electrical Technologies–Florida, LLC

How did you choose a career in the electrical field?
My father was an electrician and after I "burned out" with a career in fast-food management, I decided to choose a completely different field.

Tell us about your apprenticeship experience.
It was excellent! I worked under several very knowledgeable electricians and had a pretty good selection of teachers. Over my four-year apprenticeship, I was able to work on a variety of jobs, from Photomats to kennels to colleges.

What positions have you held and how did they help you to get where you are now?
I have been an electrical apprentice, a licensed electrician, a job site superintendent, a master electrician, and am currently a corporate safety and training director. The knowledge I acquired in electrical theory in apprenticeship school and preparing for my licensing exams, as well as the practical on-the-job experience in my twenty years in the electrical field, were wonderful training for my current position.

I also serve on the authoring team for NCCER's Electrical Curricula, which has provided me with not only the opportunity to share what I have learned, but is also a great way to keep current in other areas by meeting with electricians from a variety of disciplines (we have commercial, residential, and industrial electricians on the team, as well as instructors).

What would you say was the single greatest factor that contributed to your success?
A company that recognized and rewarded competent, hard workers and provided them with the support and guidance to allow them to develop and succeed in the industry.

What does your current job entail?
I am responsible for safe work practices and procedures through the job site management team at EET–FL, which currently has a workforce of 1,300. I also assist in developing, delivering, and administering the training program, from apprenticeship to in-house to outsourced training.

What advice do you have for trainees?
Training in all its aspects is the key to your success and advancement in the industry. Any time you are given a training opportunity, take it, even if it might not appear relevant at the time. Eventually, all knowledge can be applied to some situation. Most importantly, have fun! The construction industry is composed of good people. I firmly believe that construction workers, as a group, are much more honest and direct than any other comparable group. Wait—did I say comparable group? That's a misstatement—there is no comparable group. Construction workers build America!

44 CHAPTER 1

Instructor's Notes:

GLOSSARY

Trade Terms Introduced in This Module

Double-insulated/ungrounded tool: An electrical tool that is constructed so that the case is insulated from electrical energy. The case is made of a nonconductive material.

Fibrillation: Very rapid irregular contractions of the muscle fibers of the heart that result in the heartbeat and pulse going out of rhythm with each other.

Grounded tool: An electrical tool with a three-prong plug at the end of its power cord or some other means to ensure that stray current travels to ground without passing through the body of the user. The ground plug is bonded to the conductive frame of the tool.

Ground fault circuit interrupter (GFCI): A protective device that functions to deenergize a circuit or portion thereof within an established period of time when a current to ground exceeds some predetermined value that is less than that required to operate the overcurrent protective device of the supply circuit.

Polychlorinated biphenyls (PCBs): Toxic chemicals that may be contained in liquids used to cool certain types of large transformers and capacitors.

Additional Resources

This module is intended to present thorough resources for task training. The following reference works are suggested for further study. These are optional materials for continued education rather than for task training.

29 CFR Parts 1900–1910, Standards for General Industry. Occupational Safety and Health Administration, U.S. Department of Labor.

29 CFR Part 1926, Standards for the Construction Industry. Occupational Safety and Health Administration, U.S. Department of Labor.

National Electrical Code Handbook, Latest Edition. Quincy, MA: National Fire Protection Association.

National Electrical Safety Code, Latest Edition. Quincy, MA: National Fire Protection Association.

Instructor's Notes:

ANSWER KEY

Answers to Review Questions

Answer	Section
1. d	1.0.0
2. d	2.0.0
3. b	2.1.0
4. a	3.1.1
5. c	3.1.4
6. b	4.3.0
7. d	4.3.1
8. a	5.1.1
9. c	5.2.0
10. b	9.3.0

TEACHING TIPS

Section 1.0.0 *What's Wrong with This Picture?*

Have the trainees call out as many violations as they can find. These include improper lifting, reaching, ladder use, sloppy work area, etc. If desired, divide the class into teams and see which team comes up with the most hazards.

Section 2.0.0 *Think About It—Severity of Shock*

Note that less than 200mA ($\frac{1}{5}$A) separates a mild shock from a fatal shock. A 60W light bulb draws $\frac{1}{2}$A at 120V.

Section 2.1.0 *Think About It—Electrocution*

Explain that the bird is safe because it does not provide a path to ground. The squirrel does provide a path to ground when its body contacts both the pole and the wire as it attempts to get up onto the wire.

Section 2.1.1 *Think About It—Bodily Harm*

Explain that the amount of damage depends on the amount of current, length of contact, effectiveness of ground, path of the current through the body, and moisture present on the body, as well as the thickness of the skin and any broken areas on the skin.

Section 3.1.2 *Think About It—Dressing for Safety*

Explain that minor clothing flaws such as tears may become caught in machinery or expose the skin to toxic substances. Rivets are conductive, while polyester has a tendency to melt into the skin under intense heat. Protective apparel is designed to place a nonconductive shield between the body and electrical and/or chemical sources of harm.

Section 4.2.0 *Think About It—Working on Energized Systems*

Sample answers might include using metal utensils in a toaster in the home or using tools without GFCI protection in the home or on the job.

Section 4.3.1 *Think About It—GFCIs*

Explain that GFCIs respond to very small current imbalances, thereby protecting personnel from higher ground fault currents. GFCIs should be installed at the source.

Note: The NEC has specific requirements regarding the placement of GFCIs in areas such as kitchens and bathrooms. See *NEC Article 210.*

Section 4.3.1 *What's Wrong with This Picture?*

Explain how dangerous it is to trust your life to a piece of electrical tape. Emphasize that a proper lockout/tagout provides the electrician with the security of a physical lock to prevent unauthorized persons from reenergizing the circuit.

Section 4.3.1 *Case History—Lockout/Tagout Dilemma*

Ask the trainees to apply this scenario to their own homes. In how many cases does the existing wiring match what would have been shown on the original plans?

Section 4.3.1 ***Think About It—Lockout/Tagout—Who Does It and When?***

Note that whenever the power is removed, a lockout/tagout procedure should be followed. Each individual is responsible for protecting himself/herself, but the responsibility of coordination generally falls to the supervisor at the site.

Section 5.1.1 ***What's Wrong with This Picture?***

Note that the ladder is facing the wrong way for proper access and that it does not extend 36" beyond the top support. Ask the trainees to explain why this is dangerous.

Section 5.2.0 ***What's Wrong with This Picture?***

Note the dangerous construction of this scaffold. Have the trainees spot violations by comparing this photo with the scaffold guidelines on the previous page.

Section 9.2.0 ***Think About It—Working in Tight Spaces***

Ask the trainees to list some of the hazards of working in a small area, such as a crawl space. They should be concerned with the physical hazards, such as ventilation (particularly when exposed to toxic vapors or when working in hot weather); contact with protruding nails or fiberglass insulation; and proper work clearances. They should also be aware of potential electrical hazards, such as exposed wiring.

Section 15.1.0 ***Think About It—Putting It All Together***

Ask the trainees to identify some of the key points of this module. Emphasize that a professional approach to safety involves recognizing hazards, acquiring the proper training to minimize hazards, and accepting the responsibility for protecting both yourself and others.

CONTREN™ LEARNING SERIES — USER UPDATES

The NCCER makes every effort to keep these textbooks up-to-date and free of technical errors. We appreciate your help in this process. If you have an idea for improving this textbook, or if you find an error, a typographical mistake, or an inaccuracy in NCCER's Contren™ textbooks, please write us, using this form or a photocopy. Be sure to include the exact module number, page number, a detailed description, and the correction, if applicable. Your input will be brought to the attention of the Technical Review Committee. Thank you for your assistance.

Instructors – If you found that additional materials were necessary in order to teach this module effectively, please let us know so that we may include them in the Equipment/Materials list in the Instructor's Guide.

Write: Curriculum Revision and Development Department
National Center for Construction Education and Research
P.O. Box 141104, Gainesville, FL 32614-1104

Fax: 352-334-0932

E-mail: curriculum@nccer.org

Craft ______________ Module Name ______________

Copyright Date ______________ Module Number ______________ Page Number(s) ______________

Description

____

____

____

____

(Optional) Correction

____

____

____

(Optional) Your Name and Address

____

____

____

Chapter 2

Electrical Theory One

26104-02

MODULE OVERVIEW

This course introduces the electrical trainee to basic electrical theory.

PREREQUISITES

Please refer to the Course Map in the Trainee Module. Prior to training with this module, it is recommended that the trainee shall have successfully completed the following:

Core Curriculum; Residential Electrical I, Chapter 1

LEARNING OBJECTIVES

Upon completion of this module, the trainee will be able to:

1. Recognize what atoms are and how they are constructed.
2. Define voltage and identify the ways in which it can be produced.
3. Explain the difference between conductors and insulators.
4. Define the units of measurement that are used to measure the properties of electricity.
5. Explain how voltage, current, and resistance are related to each other.
6. Using the formula for Ohm's law, calculate an unknown value.
7. Explain the different types of meters used to measure voltage, current, and resistance.
8. Using the power formula, calculate the amount of power used by a circuit.

PERFORMANCE OBJECTIVES

Under supervision of the instructor, the trainee should be able to:

1. Use the formula for Ohm's law to calculate voltage, current, and resistance.
2. Given different resistors, identify the correct resistance value and tolerance using the color code.
3. Draw basic voltmeter and ohmmeter circuits and explain how they operate.
4. Use the power formula to calculate the amount of power used by a circuit.
5. Use a variation of the power formula to calculate the maximum current a resistor can carry based on the resistor's value and power rating.

NCCER STANDARDIZED CRAFT TRAINING PROGRAM

The National Center for Construction Education and Research (NCCER) provides a standardized national program of accredited craft training. Key features of the program include instructor certification, competency-based training, and performance testing. The program provides trainees, instructors, and companies with a standard form of recognition through a National Craft Training Registry. The program is described in full in the *Guidelines for Accreditation,* published by the NCCER. For more information on standardized craft training, contact the NCCER at P.O. Box 141104, Gainesville, FL 32614-1104, 352-334-0911, visit our Web site at www.nccer.org, or e-mail info@nccer.org.

HOW TO USE THIS ANNOTATED INSTRUCTOR'S GUIDE

Each page presents two sections of information. The larger section displays each page exactly as it appears in the Trainee Module. The narrow column ties suggested trainee and instructor actions to each page and provides icons to call your attention to material, safety, audiovisual, or testing requirements. The bottom of each page includes space for your notes.

Teaching Tip

If you see the Teaching Tip icon, that means there is a teaching tip associated with this section. Also refer to the suggested teaching tips at the end of the module.

SAFETY CONSIDERATIONS

Ensure that the trainees are equipped with appropriate personal protective equipment.

PREPARATION

Before teaching this module, you should review the Module Outline, Learning and Performance Objectives, and the Materials and Equipment List. Be sure to allow ample time to prepare your own training or lesson plan and gather all required equipment and materials.

MATERIALS AND EQUIPMENT LIST

Materials:

Transparencies

Markers/chalk

Copy of the latest edition of the *National Electrical Code®*

Various types of resistors

Module Examinations*

Performance Profile Sheets*

Equipment:

Overhead projector and screen

Whiteboard/chalkboard

Appropriate personal protective equipment

*Located in the Test Booklet packaged with this Annotated Instructor's Guide.

ADDITIONAL RESOURCES

This module is intended to present thorough resources for task training. The following reference works are suggested for both instructors and motivated trainees interested in further study. These are optional materials for continued education rather than for task training.

Electronics Fundamentals, Latest Edition. Upper Saddle River, NJ: Prentice Hall.

Principles of Electric Circuits, Latest Edition. Upper Saddle River, NJ: Prentice Hall.

NOTES

The designations "National Electrical Code," "NE Code," and "NEC," where used in this document, refer to the *National Electrical Code®*, which is a registered trademark of the National Fire Protection Association, Quincy, MA. All National Electrical Code (NEC) references in this module refer to the 2002 edition of the NEC.

If you feel that additional math instruction would be helpful, Prentice Hall offers a basic math textbook entitled *Fundamentals of Electrical and Mechanical Mathematics.* It covers the basic math requirements for electrical trainees and may be ordered by contacting Prentice Hall Customer Service at 1-800-922-0579.

TEACHING TIME FOR THIS MODULE

An outline for use in developing your lesson plan is presented below. Note that each Roman numeral in the outline equates to one session of instruction. Each session has a suggested time period of 2½ hours. This includes 10 minutes at the beginning of each session for administrative tasks and one 10-minute break during the session. Approximately 7½ hours are suggested to cover *Electrical Theory One*. You will need to adjust the time required for hands-on activity and testing based on your class size and resources.

Topic	Planned Time
Session I. Introduction to Electrical Theory; Conductors and Insulators; Electric Charge and Current	
A. Introduction to Electrical Theory	____________
B. Conductors and Insulators	____________
1. The Atom	____________
a. The Nucleus	____________
b. Electrical Charges	____________
2. Conductors and Insulators	____________
C. Electric Charge and Current	____________
1. Current Flow	____________
2. Voltage	____________
Session II. Resistance; Schematic Representation of Circuit Elements; Resistors; Measuring Voltage, Current, and Resistance	
A. Resistance	____________
1. Characteristics of Resistance	____________
2. Ohm's Law	____________
B. Schematic Representation of Circuit Elements	____________
C. Resistors	____________
1. Resistor Color Codes	____________
D. Measuring Voltage, Current, and Resistance	____________
1. Basic Meter Operation	____________
2. Voltmeter	____________
3. Ammeter	____________
4. Ohmmeter	____________
Session III. Electrical Power; Module Examination and Performance Testing	
A. Electrical Power	____________
1. Power Equation	____________
2. Power Rating of Resistors	____________
B. Module Examination	____________
1. Trainees must score 70% or higher to receive recognition from the NCCER.	
2. Record the testing results on Craft Training Report Form 200 and submit the results to the Training Program Sponsor.	
C. Performance Testing	____________
1. Trainees must perform each task to the satisfaction of the instructor to receive recognition from the NCCER.	
2. Record the testing results on Craft Training Report Form 200 and submit the results to the Training Program Sponsor.	

Chapter 2

Module 26104-02

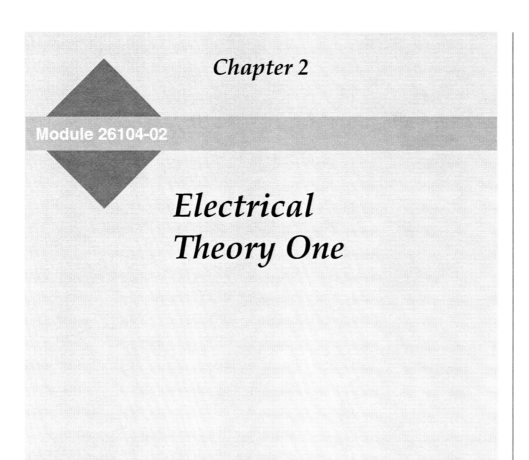

Electrical Theory One

Course Map

This course map shows all of the modules in the first level of the Residential Electrical curriculum. The suggested training order begins at the bottom and proceeds up. Skill levels increase as you advance on the course map. The local Training Program Sponsor may adjust the training order.

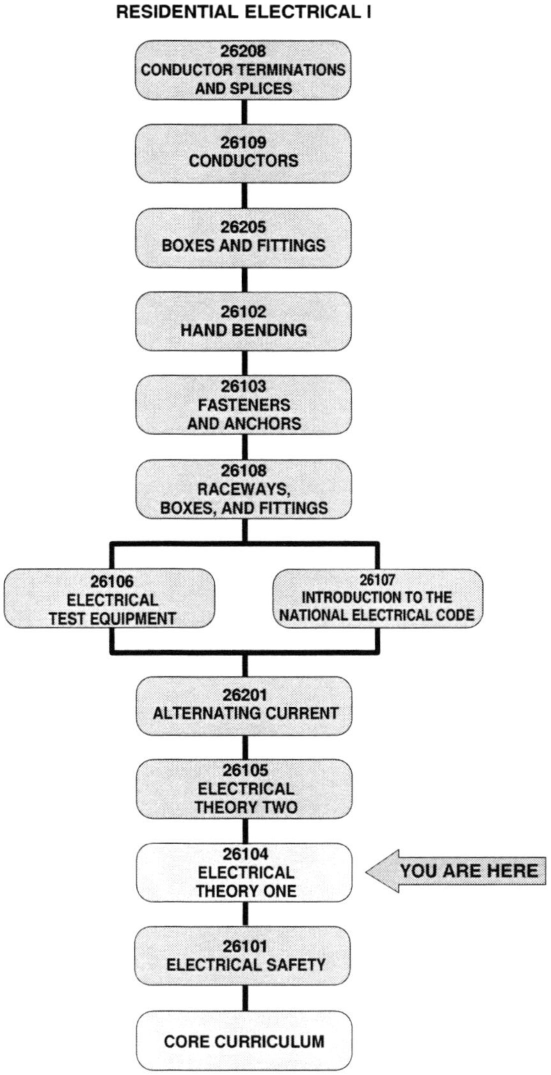

Homework

Assign reading of Module 26104.

MODULE 26104 CONTENTS

ELECTRICAL THEORY ONE 53

Figures

Tables

Instructor's Notes:

Chapter 2

MODULE 26104

Electrical Theory One

Objectives

When you have completed this module, you will be able to do the following:

1. Recognize what atoms are and how they are constructed.
2. Define voltage and identify the ways in which it can be produced.
3. Explain the difference between conductors and insulators.
4. Define the units of measurement that are used to measure the properties of electricity.
5. Explain how voltage, current, and resistance are related to each other.
6. Using the formula for Ohm's law, calculate an unknown value.
7. Explain the different types of meters used to measure voltage, current, and resistance.
8. Using the power formula, calculate the amount of power used by a circuit.

Prerequisites

Before you begin this module, it is recommended that you successfully complete the following: Core Curriculum; Residential Electrical I, Chapter 1.

Required Trainee Materials

1. Paper and pencil
2. Copy of the latest edition of the *National Electrical Code*
3. Appropriate personal protective equipment

1.0.0 ◆ INTRODUCTION TO ELECTRICAL THEORY

As an electrician, you must work with a force that cannot be seen. However, electricity is there on the job, every day of the year. It is necessary that you understand the forces of electricity so that you will be safe on the job. The first step is a basic understanding of the principles of electricity.

The relationships among **current, voltage, resistance,** and **power** in a basic direct current (DC) **series circuit** are common to all types of electrical **circuits.** This module provides a general introduction to the electrical concepts used in **Ohm's law.** It also presents the opportunity to practice applying these basic concepts to DC series circuits. In this way, you can prepare for further study in electrical and electronics theory and maintenance techniques. By practicing these techniques for all combinations of DC circuits, you will be prepared to work on any DC circuits you might encounter.

Note: The designations "National Electrical Code," "NE Code," and "NEC," where used in this document, refer to the *National Electrical Code®*, which is a registered trademark of the National Fire Protection Association, Quincy, MA. *All National Electrical Code (NEC) references in this module refer to the 2002 edition of the NEC.*

Materials

Ensure you have everything required to teach the course. Check the Materials and Equipment List at the front of this Instructor's Guide.

Audiovisual

Show Transparency 1, Course Objectives.

Show Transparency 2, Performance Profile Tasks.

Classroom

Explain that terms shown in bold (blue) are defined in the Glossary at the back of this module.

ELECTRICAL THEORY ONE 55

Teaching Tip

Discuss the "Think About It." Emphasize that true professionals take pride in knowing as much as possible about their chosen field.

Classroom

Review basic atomic theory.

Audiovisual

Show Transparency 3 (Figure 1).

THINK ABOUT IT

Why Bother Learning Theory?

Many trainees wonder why they need to bother learning the theory behind how things operate. They figure, why should I learn how it works as long as I know how to install it? The answer is, if you only know how to install something (e.g., run wire, connect switches, etc.), that's all you are ever going to be able to do. For example, if you don't know how your car operates, how can you troubleshoot it? The answer is, you can't. You can only keep changing out the parts until you finally hit on what is causing the problem. (How many times have you seen people do this?) Remember, unless you understand not only how things work but why they work, you'll only ever be a parts changer. With theory behind you, there is no limit to what you can do.

2.0.0 ◆ CONDUCTORS AND INSULATORS

2.1.0 The Atom

The **atom** is the smallest part of an element that enters into a chemical change, but it does so in the form of a charged particle. These charged particles are called *ions,* and are of two types—positive and negative. A positive ion may be defined as an atom that has become positively charged. A negative ion may be defined as an atom that has become negatively charged. One of the properties of charged ions is that ions of the same charge tend to repel one another, whereas ions of unlike charge will attract one another. The term *charge* can be taken to mean a quantity of electricity that is either positive or negative.

The structure of an atom is best explained by a detailed analysis of the simplest of all atoms, that of the element hydrogen. The hydrogen atom in *Figure 1* is composed of a **nucleus** containing one **proton** and a single orbiting **electron.** As the electron revolves around the nucleus, it is held in this orbit by two counteracting forces. One of these forces is called *centrifugal force,* which is the force that tends to cause the electron to fly outward as it travels around its circular orbit. The second force acting on the electron is *electrostatic force.* This force tends to pull the electron in toward the nucleus and is provided by the mutual attraction between the positive nucleus and the negative electron. At some given radius, the two forces will balance each other, providing a stable path for the electron.

- A proton (+) repels another proton (+).
- An electron (−) repels another electron (−).
- A proton (+) attracts an electron (−).

Basically, an atom contains three types of subatomic particles that are of interest in electricity: electrons, protons, and **neutrons.**

The protons and neutrons are located in the center, or nucleus, of the atom, and the electrons travel about the nucleus in orbits.

Because protons are relatively heavy, the repulsive force they exert on one another in the nucleus of an atom has little effect.

The attracting and repelling forces on charged materials occur because of the electrostatic lines of force that exist around the charged materials. In a negatively charged object, the lines of force of the excess electrons add to produce an electrostatic field that has lines of force coming into the object from all directions. In a positively charged object, the lines of force of the excess protons add to produce an electrostatic field that has lines of force going out of the object in all directions. The electrostatic fields either aid or oppose each other to attract or repel.

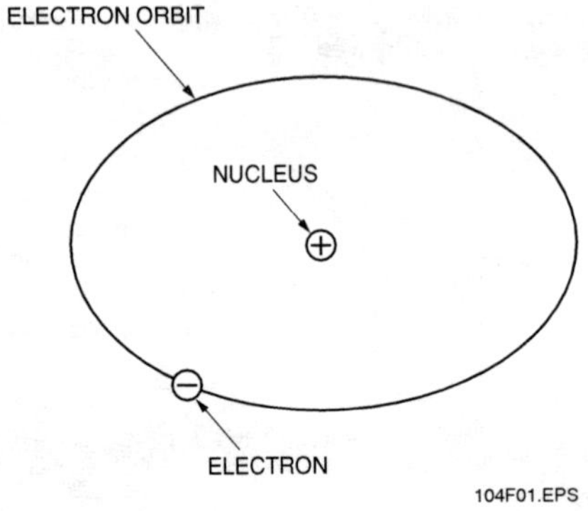

Figure 1 ◆ Hydrogen atom.

Instructor's Notes:

THINK ABOUT IT

Electrical Charges

Think about the things you come in contact with every day. Where do you see or find examples of electrostatic attraction?

2.1.1 The Nucleus

The nucleus is the central part of the atom. It is made up of heavy particles called protons and neutrons. The proton is a charged particle containing the smallest known unit of positive electricity. The neutron has no electrical charge. The number of protons in the nucleus determines how the atom of one element differs from the atom of another element.

Although a neutron is actually a particle by itself, it is generally thought of as an electron and proton combined and is electrically neutral. Since neutrons are electrically neutral, they are not considered important to the electrical nature of atoms.

2.1.2 Electrical Charges

The negative charge of an electron is equal but opposite to the positive charge of a proton. The charges of an electron and a proton are called *electrostatic charges.* The lines of force associated with each particle produce electrostatic fields. Because of the way these fields act together, charged particles can attract or repel one another. The Law of Electrical Charges states that particles with like charges repel each other and those with unlike charges attract each other. This is shown in *Figure 2.*

2.2.0 Conductors and Insulators

The difference between atoms, with respect to chemical activity and stability, depends on the number and position of the electrons included within the atom. In general, the electrons reside in groups of orbits called *shells.* The shells are arranged in steps that correspond to fixed energy levels.

The number of electrons in the outermost shell determines the valence of an atom. For this reason, the outer shell of an atom is called the **valence shell,** and the electrons contained in this shell are called *valence electrons* (*Figure 3*). The valence of an atom determines its ability to gain or lose an electron, which in turn determines the chemical and electrical properties of the atom. An atom that is lacking only one or two electrons from its outer shell will easily gain electrons to complete its shell, but a large amount of energy is required to free any of its electrons. An atom having a relatively small number of electrons in its outer shell in comparison to the number of electrons required to fill the shell will easily lose these valence electrons.

It is the valence electrons that we are most concerned with in electricity. These are the electrons that are easiest to break loose from their parent atom. Normally, a **conductor** has three or less valence electrons, an **insulator** has five or more

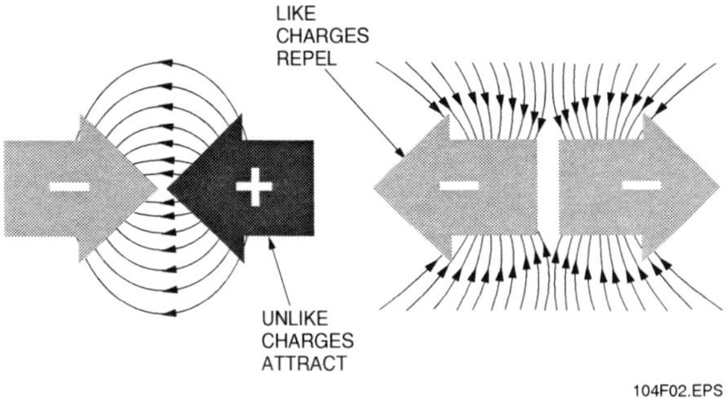

LIKE CHARGES REPEL
UNLIKE CHARGES ATTRACT
104F02.EPS

Figure 2 ◆ Law of electrical charges.

Teaching Tip

Discuss the "Think About It." Answers will vary from static cling to shocks when touching metal objects.

Audiovisual

Show Transparency 4 (Figure 2).

Classroom

Discuss the differences between conductors and insulators.

Audiovisual

Show Transparency 5 (Figure 3).

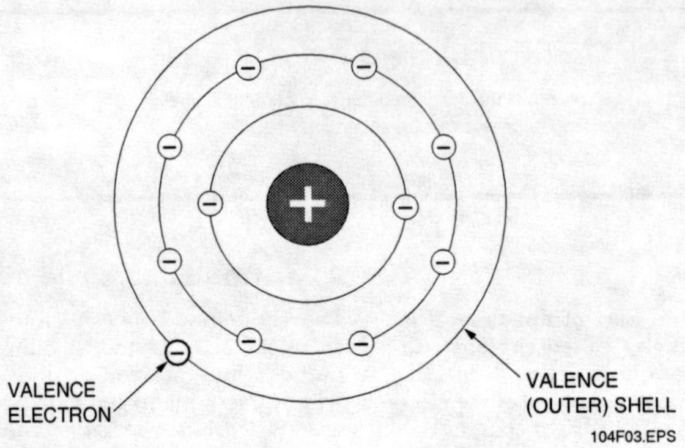

Figure 3 ◆ Valence shell and electrons.

Teaching Tip

Discuss the "Think About It."

THINK ABOUT IT

Conductors

Why do some substances conduct? What happens inside a conductor? What makes a good conductor?

INSIDE TRACK

Insulative Materials Become Conductive When Wet

While pure (distilled) water is an insulator, even trace levels of minerals make it into a conductor.

Classroom

Provide examples of conductors and insulators.

valence electrons, and semiconductors usually have four valence electrons.

All the elements of which **matter** is made may be placed into one of three categories: conductors, insulators, and semiconductors.

Conductors, for example, are elements such as copper and silver that will conduct a flow of electricity very readily. Because of their good conducting abilities, they are formed into wire and used whenever it is desired to transfer electrical energy from one point to another.

Insulators, on the other hand, do not conduct electricity to any great degree and are used when it is desirable to prevent the flow of electricity. Compounds such as porcelain and plastic are good insulators.

Materials such as germanium and silicon are not good conductors but cannot be used as insulators either, since their electrical characteristics fall between those of conductors and those of insulators. These in-between materials are classified as *semiconductors.* As you will learn later in your training, semiconductors play a crucial role in electronic circuits.

3.0.0 ◆ ELECTRIC CHARGE AND CURRENT

An electric charge has the ability to do the work of moving another charge by attraction or repulsion. The ability of a charge to do work is called its *potential.* When one charge is different from another, there must be a difference in potential between them. The sum of the difference of potential of all the charges in the electrostatic field is referred to as *electromotive force (emf)* or *voltage.* Voltage is frequently represented by the letter *E.*

Instructor's Notes:

58 CHAPTER 2

INSIDE TRACK

Units of Electricity and Volta

A disagreement with a fellow scientist over the twitching of a frog's leg eventually led 18th-century physicist Alessandro Volta to theorize that when certain objects and chemicals come into contact with each other, they produce an electric current. Believing that electricity came from contact between metals only, Volta coined the term *metallic electricity.* To demonstrate his theory, Volta placed two discs, one of silver and the other of zinc, into a weak acidic solution. When he linked the discs together with wire, electricity flowed through the wire. Thus, Volta introduced the world to the battery, also known as the *Voltaic pile.* Now Volta needed a term to measure the strength of the electric push or the flowing charge; the volt is that measure.

Electric charge is measured in **coulombs.** An electron has 1.6×10^{-19} coulombs of charge. Therefore, it takes 6.25×10^{18} electrons to make up one coulomb of charge, as shown below.

$$\frac{1}{1.6 \times 10^{-19}} = 6.25 \times 10^{18} \text{ electrons}$$

If two particles, one having charge Q_1 and the other charge Q_2, are a distance (d) apart, then the force between them is given by Coulomb's law, which states that the force is directly proportional to the product of the two charges and inversely proportional to the square of the distance between them:

$$\text{Force} = \frac{k \times Q_1 \times Q_2}{d^2}$$

If Q_1 and Q_2 are both positive or both negative, then the force is positive; it is repulsive. If Q_1 and Q_2 are of opposite charges, then the force is negative; it is attractive. k equals a constant with a value of 10^9.

3.1.0 Current Flow

The movement of the flow of electrons is called *current.* To produce current, the electrons are moved by a potential difference. Current is represented by the letter *I.* The basic unit in which current is measured is the **ampere,** also called the *amp.* The symbol for the ampere is *A.* One ampere of current is defined as the movement of one coulomb past any point of a conductor during one second of time. One coulomb is equal to 6.25×10^{18} electrons; therefore, one ampere is equal to 6.25×10^{18} electrons moving past any point of a conductor during one second of time.

The definition of current can be expressed as an equation:

$$I = \frac{Q}{T}$$

Where:

I = current (amperes)
Q = charge (coulombs)
T = time (seconds)

Charge differs from current in that Q is an accumulation of charge, while I measures the intensity of moving charges.

In a conductor, such as copper wire, the free electrons are charges that can be forced to move with relative ease by a potential difference. If a potential difference is connected across two ends of a copper wire, as shown in *Figure 4,* the applied

INSIDE TRACK

Law of Electrical Force and de Coulomb

In the 18th century, a French physicist named Charles de Coulomb was concerned with how electric charges behaved. He watched the repelling forces electric charges exerted by measuring the twist in a wire. An object's weight acted as a turning force to twist the wire, and the amount of twist was proportional to the object's weight. After many experiments with opposing forces, de Coulomb proposed the Inverse Square Law, later known as the *Law of Electrical Force.*

Classroom

Discuss Coulomb's law and current flow.

Audiovisual

Show Transparency 6 (Figure 4).

ELECTRICAL THEORY ONE 59

Teaching Tip

Discuss both of the "Think About It" features.

Classroom

Explain how voltage may be generated using a battery.

Teaching Tip

Discuss the "Think About It."

THINK ABOUT IT

The Magic of Electricity

The flow of electrons occurs at close to the speed of light, about 186,000 miles per second. How long does it take the light from the end of a flashlight to reach the floor? If you ran a light circuit from Maine to California and flipped the switch, how long would it take for the light to come on?

THINK ABOUT IT

Current Flow

Why do you need two wires to use electrical devices? Why can't current simply move to a lamp and be released as light energy?

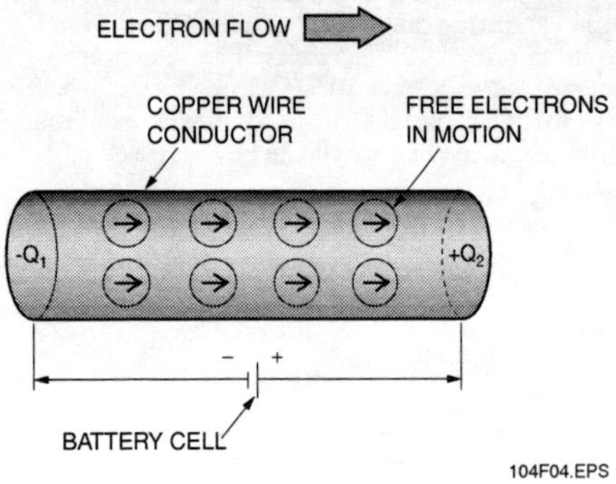

Figure 4 ◆ Potential difference causing electric current.

voltage forces the free electrons to move. This current is a flow of electrons from the point of negative charge (−) at one end of the wire, moving through the wire to the positive charge (+), at the other end. The direction of the electron flow is from the negative side of the battery, through the wire, and back to the positive side of the battery. The direction of current flow is therefore from a point of negative potential to a point of positive potential.

3.2.0 Voltage

The force that causes electrons to move is called *voltage, potential difference,* or *electromotive force (emf).* One **volt (V)** is the potential difference between two points for which one coulomb of electricity will do one **joule (J)** of work. A **battery** is one of several means of creating voltage. It chemically creates a large reserve of free electrons at the negative (−) terminal. The positive (+) terminal has electrons chemically removed and will therefore accept them if an external path is provided from the negative (−) terminal. When a battery is no longer able to chemically deposit electrons at the negative (−) terminal, it is said to be dead, or in need of recharging. Batteries are normally rated in volts. Large batteries are also rated in ampere-hours, where one ampere-hour is a current of one amp supplied for one hour.

THINK ABOUT IT

Voltage

Why is voltage called *electrical potential?* Why are we interested in potential charges—why not simply measure the actual flow of current? Can there be very high potential and no current, or a very small current? What happens when lightning strikes in terms of electrical potential?

Instructor's Notes:

60 CHAPTER 2

INSIDE TRACK

Joule's Law

While other scientists of the 19th century were experimenting with batteries, cells, and circuits, James Joule was theorizing about the relationship between heat and energy. He discovered, contrary to popular belief, that work did not just move heat from one place to another; work, in fact, generated heat. Furthermore, he demonstrated that over time, a relationship existed between the temperature of water and electric current. These ideas formed the basis for the concept of energy. In his honor, the modern unit of energy was named the *joule.*

INSIDE TRACK

The Visual Language of Electricity

Learning to read circuit diagrams is like learning to read a book—first you learn to read the letters, then you learn to read the words, and before you know it, you are reading without paying attention to the individual letters anymore. Circuits are the same way—you will struggle at first with the individual pieces, but before you know it, you will be reading a circuit without even thinking about it. Studying the table below will help you to understand the fundamental language of electricity.

What's Measured	Unit of Measurement	Symbol	Ohm's Law Symbol
Amount of current	Amp	A	I
Electrical power	Watt	W	P
Force of current	Volt	V	E
Resistance to current	Ohm	Ω	R

4.0.0 ◆ RESISTANCE

4.1.0 Characteristics of Resistance

Resistance is directly related to the ability of a material to conduct electricity. Conductors have very low resistance; insulators have very high resistance.

Resistance can be defined as the opposition to current flow. To add resistance to a circuit, electrical components called **resistors** are used. A resistor is a device whose resistance to current flow is a known, specified value. Resistance is measured in ohms and is represented by the symbol *R* in equations. One **ohm** is defined as the amount of resistance that will limit the current in a conductor to one ampere when the voltage applied to the conductor is one volt. The symbol for an ohm is Ω.

The resistance of a wire is proportional to the length of the wire, inversely proportional to the cross-sectional area of the wire, and dependent upon the kind of material of which the wire is made. The relationship for finding the resistance of a wire is:

$$R = \rho\frac{L}{A}$$

Where:

R = resistance (ohms)
L = length of wire (feet)
A = area of wire (circular mils, CM, or cm^2)
ρ = specific resistance (ohm-CM/ft. or microhm-CM)

A *mil* equals 0.001 inch; a circular mil is the cross-sectional area of a wire one mil in diameter.

The specific resistance is a constant that depends on the material of which the wire is made. *Table 1* shows the properties of various wire conductors.

Table 1 shows that at 75°F, a one-mil diameter, pure annealed copper wire that is one foot long has a resistance of 10.351 ohms; while a one-mil diameter, one-foot-long aluminum wire has a resistance of 16.758 ohms. Temperature is important in determining the resistance of a wire. The hotter a wire, the greater its resistance.

Classroom

Demonstrate how to calculate resistance.

Have the trainees practice resistance calculations.

Classroom

Demonstrate how to find voltage, current, and resistance using Ohm's law.

Have the trainees practice using Ohm's law.

Audiovisual

Show Transparency 7 (Figure 5).

Teaching Tip

Discuss both of the "Think About It" features.

4.2.0 Ohm's Law

Ohm's law defines the relationship between current, voltage, and resistance. There are three ways to express Ohm's law mathematically.

- The current in a circuit is equal to the voltage applied to the circuit divided by the resistance of the circuit:

$$I = \frac{E}{R}$$

- The resistance of a circuit is equal to the voltage applied to the circuit divided by the current in the circuit:

$$R = \frac{E}{I}$$

Table 1 Conductor Properties

Metal	Specific Resistance (Resistance of 1 CM/ft. in ohms)	
	32°F or 0°C	75°F or 23.8°C
Silver, pure annealed	8.831	9.674
Copper, pure annealed	9.39	10.351
Copper, annealed	9.59	10.505
Copper, hard-drawn	9.81	10.745
Gold	13.216	14.404
Aluminum	15.219	16.758
Zinc	34.595	37.957
Iron	54.529	62.643

- The applied voltage to a circuit is equal to the product of the current and the resistance of the circuit:

$$E = I \times R = IR$$

Where:

I = current (amperes)
R = resistance (ohms)
E = voltage or emf (volts)

If any two of the quantities E, I, or R are known, the third can be calculated.

The Ohm's law equations can be memorized and practiced effectively by using an Ohm's law circle, as shown in *Figure 5.* To find the equation for E, I, or R when two quantities are known, cover the unknown third quantity. The other two quantities in the circle will indicate how the covered quantity may be found.

Example 1:
Find I when E = 120V and R = 30Ω.

$$I = \frac{E}{R}$$

$$I = \frac{120V}{30\Omega}$$

$$I = 4A$$

This formula shows that in a DC circuit, current (I) is directly proportional to voltage (E) and inversely proportional to resistance (R).

Example 2:
Find R when E = 240V and I = 20A.

THINK ABOUT IT

Voltage Matters

Standard household voltage is different the world over, from 100V in Japan to 600V in Bombay, India. Many countries have no standard voltage; for example, France varies from 110V to 360V. If you were to plug a 120V hair dryer into England's 240V, you would burn out the dryer. Use basic electric theory to explain exactly what would happen to destroy the hair dryer.

THINK ABOUT IT

Using Ohm's Law

Study the resistance values in *Table 1.* Assuming the same length and area of wire, approximately how much will the current decrease from copper annealed wire to aluminum wire to iron wire?

Instructor's Notes:

62 CHAPTER 2

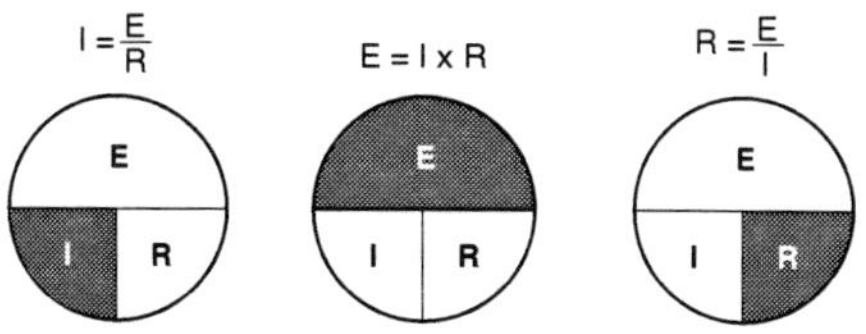

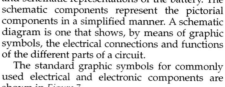

	LETTER SYMBOL	UNIT OF MEASUREMENT
CURRENT	I	AMPERES (A)
RESISTANCE	R	OHMS (Ω)
VOLTAGE	E	VOLTS (V)

104F05.EPS

Figure 5 ◈ Ohm's law circle.

$$R = \frac{E}{I}$$

$$R = \frac{240V}{20A}$$

$$R = 12\Omega$$

Example 3:
Find E when I = 15A and R = 8Ω.

$$E = I \times R$$

$$E = 15A \times 8\Omega$$

$$E = 120V$$

5.0.0 ◈ SCHEMATIC REPRESENTATION OF CIRCUIT ELEMENTS

A simple electric circuit is shown in both pictorial and **schematic** forms in *Figure 6*. The schematic diagram is a shorthand way to draw an electric circuit, and circuits are usually represented in this way. In addition to the connecting wire, three components are shown symbolically: the battery, the switch, and the lamp. Note the positive (+) and negative (−) markings in both the pictorial and schematic representations of the battery. The schematic components represent the pictorial components in a simplified manner. A schematic diagram is one that shows, by means of graphic symbols, the electrical connections and functions of the different parts of a circuit.

The standard graphic symbols for commonly used electrical and electronic components are shown in *Figure 7*.

6.0.0 ◈ RESISTORS

The function of a resistor is to offer a particular resistance to current flow. For a given current and known resistance, the change in voltage across the component, or **voltage drop**, can be predicted using Ohm's law. Voltage drop refers to a specific amount of voltage used, or developed, by that component. An example is a very basic circuit of a 10V battery and a single resistor in a series circuit. The voltage drop across that resistor is 10V because it is the only component in the circuit and all voltage must be dropped across that resistor. Similarly, for a given applied voltage, the current that flows may be predetermined by selection of the

Classroom

Introduce schematic diagrams.

Audiovisual

Show Transparencies 8 and 9 (Figures 6 and 7).

Classroom

Discuss resistors.

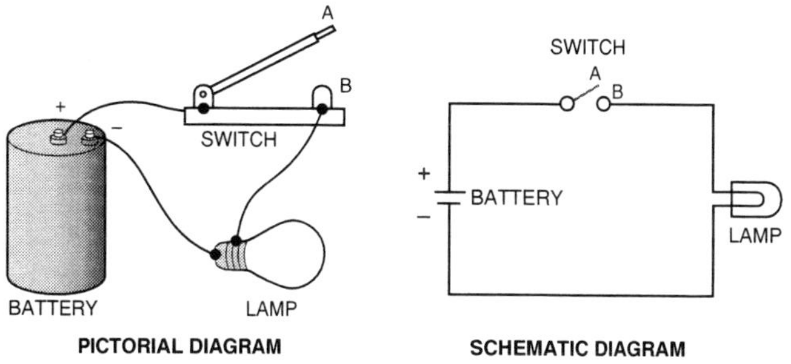

104F06.EPS

Figure 6 ◈ Simple electrical symbols.

ELECTRICAL THEORY ONE 63

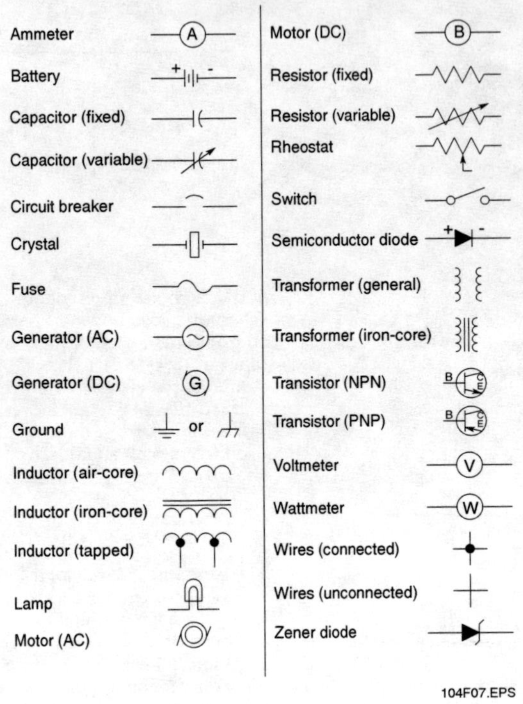

Figure 7 ◆ Standard schematic symbols.

Classroom

Discuss the "Think About It." Draw a simple schematic on the whiteboard/chalkboard.

Audiovisual

Show Transparency 10 (Figure 8A).

THINK ABOUT IT

Drawing a Schematic

Draw a schematic diagram showing a voltage source, switch, motor, and fuse.

resistor value. The required power dissipation largely dictates the construction and physical size of a resistor.

The two most common types of electronic resistors are *wire-wound* and *carbon composition construction.* A typical wire-wound resistor consists of a length of nickel wire wound on a ceramic tube and covered with porcelain. Low-resistance connecting wires are provided, and the resistance value is usually printed on the side of the component. *Figure 8* illustrates the construction of typical resistors. Carbon composition resistors are constructed by molding mixtures of powdered carbon and insulating materials into a cylindrical shape. An outer sheath of insulating material affords mechanical and electrical protection, and copper connecting wires are provided at each end. Carbon composition resistors are smaller and less expensive than the wire-wound type. However, the wire-wound type is the more rugged of the two and is able to survive much larger power dissipations than the carbon composition type.

Instructor's Notes:

Using Your Intuition

INSIDE TRACK

Learning the meanings of various electrical symbols may seem overwhelming, but if you take a moment to study *Figure 7*, you will see that most of them are intuitive—that is, they are shaped (in a symbolic way) to represent the actual object. For example, the battery shows + and −, just like an actual battery. The motor has two arms that suggest a spinning rotor. The transformer shows two coils. The resistor has a jagged edge to suggest pulling or resistance. Connected wires have a black dot that reminds you of solder. Unconnected wires simply cross. The fuse stretches out in both directions as though to provide extra slack in the line. The circuit breaker shows a line with a break in it. The capacitor shows a gap. The variable resistor has an arrow like a swinging compass needle. As you learn to read schematics, take the time to make mental connections between the symbol and the object it represents.

Teaching Tip

Use *Figure 7* to create schematic flashcards. Divide the class into teams and test them on symbol recognition.

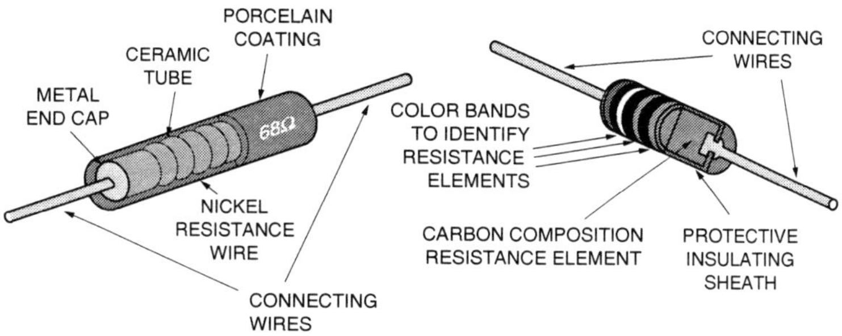

104F08A.EPS

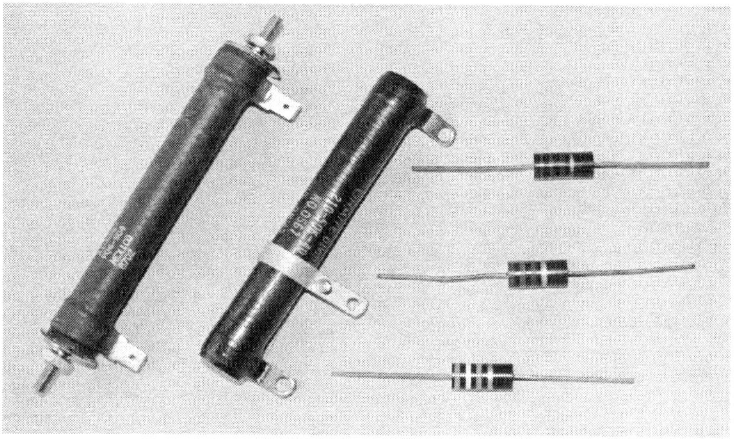

104F08B.EPS

Figure 8 ◈ Common resistors.

Classroom

Pass around resistors for the trainees to examine.

ELECTRICAL THEORY ONE 65

Audiovisual

Show Transparency 11 (Figure 9).

Classroom

Discuss resistor color codes.

Audiovisual

Show Transparencies 12 and 13 (Figures 10 and 11).

Most resistors have standard fixed values, so they can be termed *fixed resistors*. Variable resistors, also known as *adjustable resistors,* are used a great deal in electronics. Two common symbols for a variable resistor are shown in *Figure 9*.

A variable resistor consists of a coil of closely wound insulated resistance wire formed into a partial circle. The coil has a low-resistance terminal at each end, and a third terminal is connected to a movable contact with a shaft adjustment facility. The movable contact can be set to any point on a connecting track that extends over one (uninsulated) edge of the coil.

Using the adjustable contact, the resistance from either end terminal to the center terminal may be adjusted from zero to the maximum coil resistance.

Another type of variable resistor is known as a *decade resistance box.* This is a laboratory component that contains precise values of switched series-connected resistors.

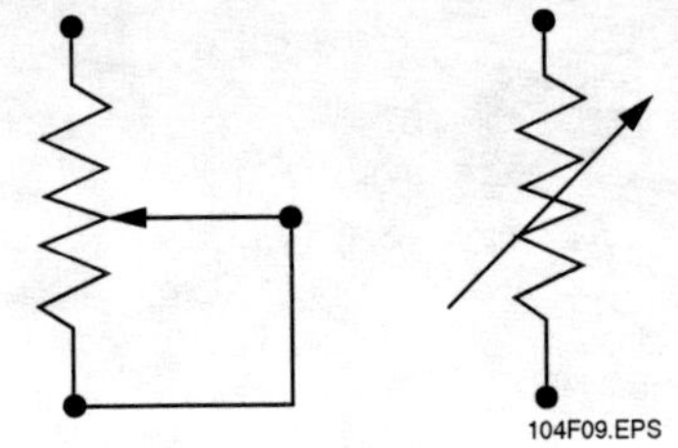

Figure 9 ◈ Symbols used for variable resistors.

6.1.0 Resistor Color Codes

Because carbon composition resistors are physically small (some are less than 1 cm in length), it is not convenient to print the resistance value on the side. Instead, a color code in the form of colored bands is employed to identify the resistance value and tolerance. The color code is illustrated in *Figure 10.* Starting from one end of the resistor, the first two bands identify the first and second digits of the resistance value, and the third band indicates the number of zeros. An exception to this is when the third band is either silver or gold, which indicates a 0.01 or 0.1 multiplier, respectively. The fourth band is always either silver or gold, and in this position, silver indicates a ± 10% tolerance and gold indicates a ± 5% tolerance. Where no fourth band is present, the resistor tolerance is ± 20%.

We can put this information to practical use by determining the range of values for the carbon resistor in *Figure 11.*

The color code for this resistor is as follows:

- Brown = 1, black = 0, red = 2, gold = a tolerance of ± 5%
- First digit = 1, second digit = 0, number of zeros (2) = 1,000Ω

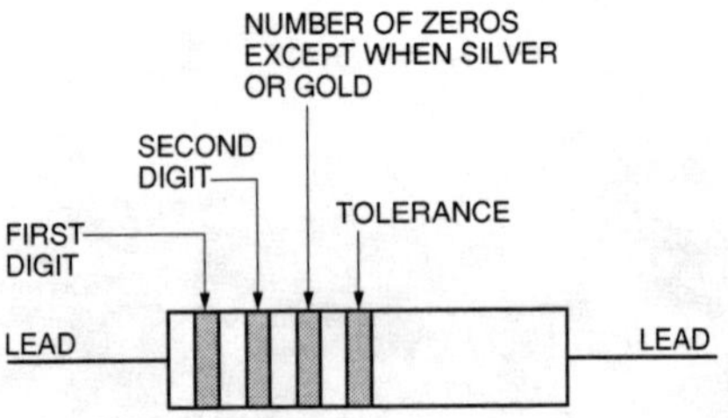

0	BLACK	7	VIOLET
1	BROWN	8	GREY
2	RED	9	WHITE
3	ORANGE	0.1	GOLD
4	YELLOW	0.01	SILVER
5	GREEN	5%	GOLD – TOLERANCE
6	BLUE	10%	SILVER – TOLERANCE

104F10.EPS

Figure 10 ◈ Resistor color codes.

Instructor's Notes:

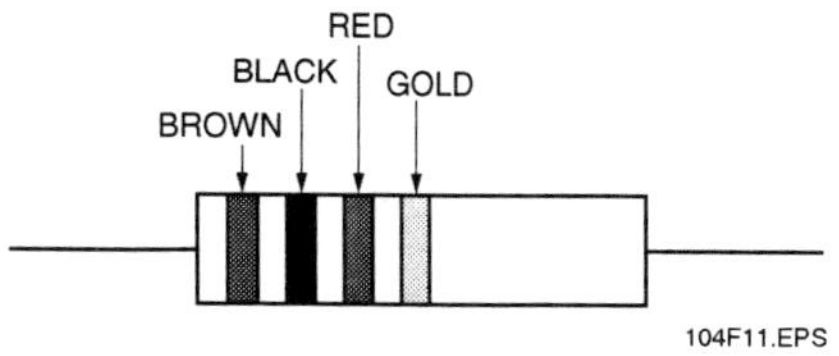

Figure 11 ◆ Sample color codes on a fixed resistor.

Since this resistor has a value of 1,000Ω ± 5%, the resistor can range in value from 950Ω to 1,050Ω.

7.0.0 ◆ MEASURING VOLTAGE, CURRENT, AND RESISTANCE

Working with electricity requires making accurate measurements. This section will discuss the basic meters used to measure voltage, current, and resistance: the **voltmeter, ammeter,** and **ohmmeter.**

> **WARNING!**
> Only qualified individuals may use these meters. Consult your company's safety policy for applicable rules.

7.1.0 Basic Meter Operation

When troubleshooting or testing equipment, you will need various meters to check for proper circuit voltages, currents, and resistances and to determine if the wiring is defective. Meters are used in repairing, maintaining, and troubleshooting electrical circuits and equipment. The best and most expensive measuring instrument is of no use to you unless you know what you are measuring and what each reading indicates. Remember that the purpose of a meter is to measure quantities existing within a circuit. For this reason, when the meter is connected to a circuit, it must not change the condition of the circuit.

The three basic electrical quantities discussed in this section are current, voltage, and resistance. Actually, it is really current that causes the meter to respond even when voltage or resistance is being measured. In a basic meter, the measurement of current can be calibrated to indicate almost any electrical quantity based on the principle of Ohm's law. The amount of current that flows through a meter is determined by the voltage applied to the meter and the resistance of the meter, as stated by I = E/R.

For a given meter resistance, different values of applied voltage will cause specific values of current to flow. Although the meter actually measures current, the meter scale can be calibrated in units of voltage. Similarly, for a given applied voltage, different values of resistance will cause specific values of current to flow; therefore, the meter scale can also be calibrated in units of resistance rather than current. The same holds true for power, since power is proportional to current, as stated by P = EI. It is on this principle that the meter was developed and its construction allows for the measurement of various parameters by actually measuring current.

You must understand the purpose and function of each individual piece of test equipment and any limitations associated with it. It is also extremely important that you understand how to safely use each piece of equipment. If you understand the capabilities of the test equipment, you can better use the equipment, better understand the indications on the equipment, and know what substitute or backup meters can be used.

7.2.0 Voltmeter

A simple voltmeter consists of the meter movement in series with the internal resistance of the voltmeter itself. For example, a meter with a 50-microamp (μA) meter movement and a 1,000Ω internal resistance can be used to directly measure voltages up to 0.05V, as shown in *Figure 12.* (The prefix **micro** means one-millionth.) When the meter is placed across the voltage source, a current determined by the internal resistance of the meter flows through the meter movement. A voltmeter's internal resistance is typically high to minimize meter loading effects on the source.

To measure larger voltages, a multiplier resistor is used. This increased series resistance limits the current that can flow through the meter movement, thus extending the range of the meter.

To avoid damage to the meter movement, the following precautions should be observed when using a voltmeter:

- Always set the full-scale voltage of the meter to be larger than the expected voltage to be measured.
- Always ensure that the internal resistance of the voltmeter is much greater than the resistance of the component to be measured. This means that the current it takes to drive the voltmeter (about 50 microamps) should be a negligible fraction of

Classroom

Introduce basic meter operation.

Review the operation of a basic voltmeter.

Audiovisual

Show Transparency 14 (Figure 12).

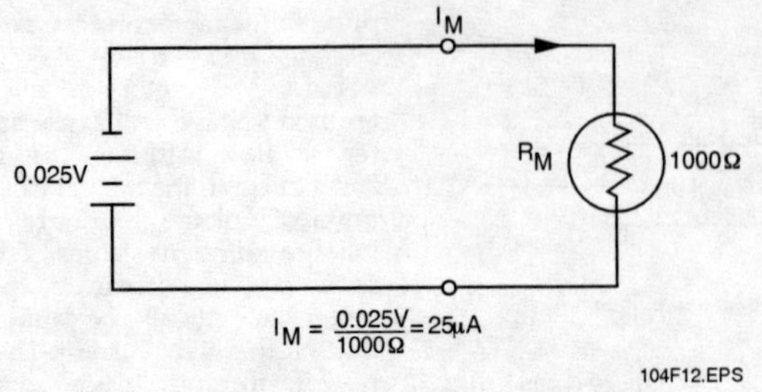

Figure 12 ◆ Simple voltmeter.

the current flowing through the circuit element being measured.

- If you are unsure of the level of the voltage to be measured, take a reading at the highest range of the voltmeter and progressively (step-by-step) lower the range until the reading is obtained.

In most commercial voltmeters, the internal resistance is expressed by the ohms-per-volt rating of the meter. A typical meter has a rating of 20,000 ohms-per-volt with a 50-microamp movement. This quantity tells what the internal resistance of the meter is on any particular full-scale setting. In general, the meter's internal resistance is the ohms-per-volt rating multiplied by the full-scale voltage. The higher the ohms-per-volt rating, the higher the internal resistance of the meter, and the smaller the effect of the meter on the circuit.

Classroom

Review the operation of a basic ammeter.

7.3.0 Ammeter

A current meter, usually called an *ammeter,* is used by placing the meter in series with the wire through which the current is flowing. This method of connection is shown in *Figure 13.* Notice how the magnitude of load current will flow through the ammeter. Because of this, an ammeter's internal resistance must be low to minimize the circuit-loading effects as seen by the source. Also, high current magnitudes flowing through an ammeter can damage it. For this reason, ammeter shunts are employed to reduce the ammeter circuit current to a fraction of the current flowing through the load.

Audiovisual

Show Transparency 15 (Figure 13).

To avoid damage to the meter movement, the following precautions should be observed when taking current measurements with an ammeter:

- Always check the polarity of the ammeter. Make certain that the meter is connected to the circuit so that electrons flow into the negative lead and out of the positive lead. It is easy to tell which is the positive lead because it is normally red. The negative lead is usually black.

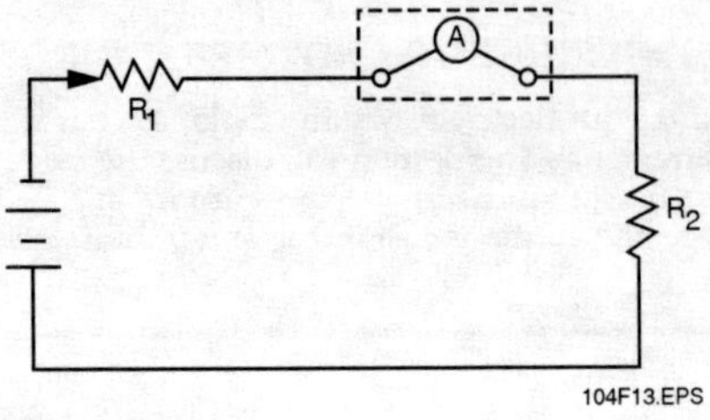

Figure 13 ◆ Ammeter connection.

- Always set the full-scale deflection of the meter to be larger than the expected current. To be safe, set the full-scale current several times larger than the expected current, and then slowly increase the meter sensitivity to the appropriate scale.
- Always connect the ammeter in series with the circuit element through which the current to be measured is flowing. Never connect the ammeter in parallel. When an ammeter is connected across a constant-potential source of appreciable voltage, the low internal resistance of the meter bypasses the circuit resistance. This results in the application of the source voltage directly to the meter terminals. The resulting excess current will burn out the meter coil.

Classroom

Review the operation of a basic ohmmeter.

7.4.0 Ohmmeter

An ohmmeter is used to measure resistance and check continuity. The deflection of the pointer of an ohmmeter is controlled by the amount of battery current passing through the coil. Current flow

Instructor's Notes:

68 CHAPTER 2

depends on the applied voltage and the circuit resistance. By applying a constant source voltage to the circuit under test, the resultant current flow depends only on circuit resistance. This magnitude of current will create meter movement. By knowing the relationship between current and resistance, an ohmmeter's scale can be calibrated to indicate circuit resistance based on the magnitude of current for a constant source voltage. Refer to *Figure 14,* a simple ohmmeter circuit.

8.0.0 ◆ ELECTRICAL POWER

Power is defined as the rate of doing work. This is equivalent to the rate at which energy is used or dissipated. Electrons passing through a resistance dissipate energy in the form of heat. In electrical circuits, power is measured in units called **watts (W).** The power in watts equals the rate of energy conversion. One watt of power equals the work done in one second by one volt of potential difference in moving one coulomb of charge. One coulomb per second is an ampere; therefore, power in watts equals the product of amperes times volts.

The work done in an electrical circuit can be useful work or it can be wasted work. In both cases, the rate at which the work is done is still measured in power. The turning of an electric motor is useful work. On the other hand, the heating

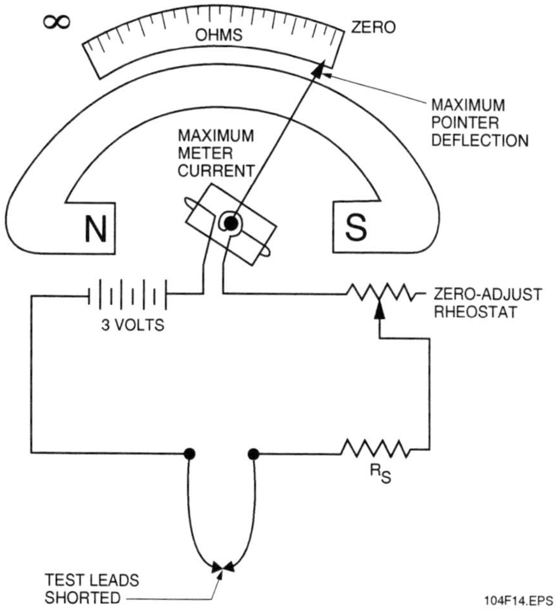

Figure 14 ◆ Simple ohmmeter circuit.

> **INSIDE TRACK**
>
> ### *Using an Ohmmeter*
>
> An ohmmeter has its own battery to test the resistance or continuity of a circuit. Therefore, the circuit must be deenergized because the ohmmeter is calibrated for its own power source.

Audiovisual

Show Transparency 16 (Figure 14).

Classroom

Discuss power and watts.

Teaching Tip

Discuss the first "Think About It." The hair dryer (heating element) and light bulb (filament) are resistors.

Discuss the second "Think About It." Encourage the widest possible range of answers, from cars to calculators.

Audiovisual

Show Transparency 17 (Figure 15).

Teaching Tip

Discuss the third "Think About It." Utility companies are concerned with the power measurements made by electric meters.

THINK ABOUT IT

Resistors

Which of the following items are resistors?

- Hair dryer
- Incandescent light bulb
- Switch
- Receptacle
- Circuit breaker

THINK ABOUT IT

Power

We take electrical power for granted, never stopping to think how surprising it is that a flow of submicroscopic electrons can pump thousands of gallons of water or illuminate a skyscraper. Our lives now constantly rely on the ability of the electron to do work. Think about your day up to this moment. How has electrical power shaped your experience?

of wires or resistors in a circuit is wasted work, since no useful function is performed by the heat.

The unit of electrical work is the joule. This is the amount of work done by one coulomb flowing through a potential difference of one volt. Thus, if five coulombs flow through a potential difference of one volt, five joules of work are done. The time it takes these coulombs to flow through the potential difference has no bearing on the amount of work done.

It is more convenient when working with circuits to think of amperes of current rather than coulombs. As previously discussed, one ampere equals one coulomb passing a point in one second. Using amperes, one joule of work is done in one second when one ampere moves through a potential difference of one volt. This rate of one joule of work in one second is the basic unit of power, and is called a *watt*. Therefore, a watt is the power used when one ampere of current flows through a potential difference of one volt, as shown in *Figure 15*.

Mechanical power is usually measured in units of horsepower (hp). To convert from horsepower to watts, multiply the number of horsepower by 746. To convert from watts to horsepower, divide the number of watts by 746. Conversions for common units of power are given in *Table 2*.

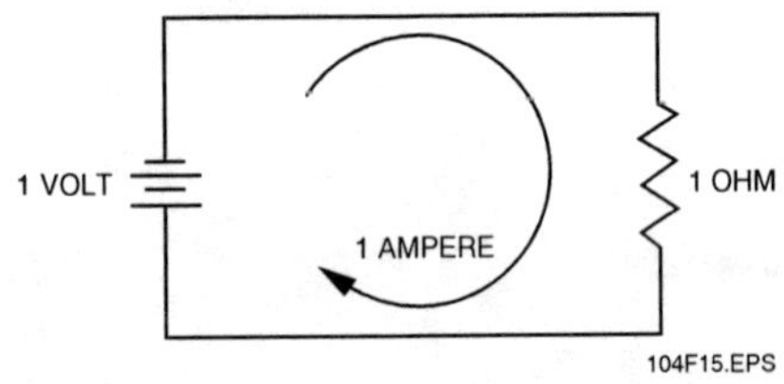

Figure 15 ◆ One watt.

THINK ABOUT IT

Measuring Watts

Electricians are less interested in measuring watts than in measuring amperage and resistance. Who would be most interested in power measurements? What is a common example of a device used to measure watts?

Instructor's Notes:

70 CHAPTER 2

Table 2 Conversion Table

1,000 watts (W)	= 1 kilowatt (kW)
1,000,000 watts (W)	= 1 megawatt (MW)
1,000 kilowatts (kW)	= 1 megawatt (MW)
1 watt (W)	= 0.00134 horsepower (hp)
1 horsepower (hp)	= 746 watts (W)

The kilowatt-hour (kWh) is commonly used for large amounts of electrical work or energy. (The prefix **kilo** means one thousand.) The amount is calculated simply as the product of the power in kilowatts multiplied by the time in hours during which the power is used. If a light bulb uses 300W or 0.3kW for 4 hours, the amount of energy is 0.3×4, which equals 1.2kWh.

Very large amounts of electrical work or energy are measured in megawatts (MW). (The prefix **mega** means one million.)

Electricity usage is figured in kilowatt-hours of energy. The power line voltage is fairly constant at 120V. Suppose the total load current in the main line equals 20A. Then the power in watts from the 120V line is:

$$P = 120V \times 20A$$
$$P = 2{,}400W \text{ or } 2.4kW$$

If this power is used for five hours, then the energy of work supplied equals:

$$2.4 \times 5 = 12kWh$$

8.1.0 Power Equation

When one ampere flows through a difference of two volts, two watts must be used. In other words, the number of watts used is equal to the number of amperes of current times the potential difference. This is expressed in equation form as:

$$P = I \times E \text{ or } P = IE$$

Where:

P = power used in watts
I = current in amperes
E = potential difference in volts

The equation is sometimes called *Ohm's law for power,* because it is similar to *Ohm's law.* This equation is used to find the power consumed in a circuit or load when the values of current and voltage are known. The second form of the equation is used to find the voltage when the power and current are known:

$$E = \frac{P}{I}$$

The third form of the equation is used to find the current when the power and voltage are known:

$$I = \frac{P}{E}$$

Using these three equations, the power, voltage, or current in a circuit can be calculated whenever any two of the values are already known.

Example 1:
Calculate the power in a circuit where the source of 100V produces 2A in a 50Ω resistance.

$$P = IE$$
$$P = 2 \times 100$$
$$P = 200W$$

This means the source generates 200W of power while the resistance dissipates 200W in the form of heat.

Example 2:
Calculate the source voltage in a circuit that consumes 1,200W at a current of 5A.

$$E = \frac{P}{I}$$
$$E = \frac{1{,}200}{5}$$
$$E = 240V$$

Example 3:
Calculate the current in a circuit that consumes 600W with a source voltage of 120V.

$$I = \frac{P}{E}$$
$$I = \frac{600}{120}$$
$$I = 5A$$

Components that use the power dissipated in their resistance are generally rated in terms of power. The power is rated at normal operating voltage, which is usually 120V. For instance, an appliance that draws 5A at 120V would dissipate 600W. The rating for the appliance would then be 600W/120V.

Classroom

Demonstrate how to use the power equation.

Have the trainees practice using the power equation.

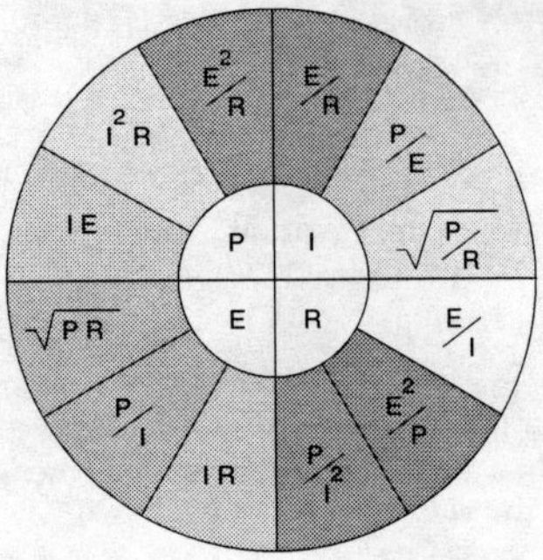

Figure 16 ◆ Expanded Ohm's law circle.

Classroom

Explain that the power formula can be used to derive power, voltage, current, or resistance.

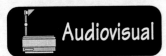

Audiovisual

Show Transparency 18 (Figure 16).

Classroom

Discuss resistor power ratings.

Teaching Tip

Discuss the "Think About It." Answers will vary depending upon the building devices.

To calculate I or R for components rated in terms of power at a specified voltage, it may be convenient to use the power formula in different forms. There are actually three basic power formulas, but each can be rearranged into three other forms for a total of nine combinations:

$$P = IE \qquad P = I^2R \qquad P = \frac{E^2}{R}$$

$$I = \frac{P}{E} \qquad R = \frac{P}{I^2} \qquad R = \frac{E^2}{P}$$

$$E = \frac{P}{I} \qquad I = \sqrt{\frac{P}{R}} \qquad E = \sqrt{PR}$$

Note that all of these formulas are based on Ohm's law (E = IR) and the power formula (P = I × E). *Figure 16* shows all of the applicable power, voltage, resistance, and current equations.

8.2.0 Power Rating of Resistors

If too much current flows through a resistor, the heat caused by the current will damage or destroy the resistor. This heat is caused by I^2R heating, which is power loss expressed in watts. Therefore, every resistor is given a wattage, or power rating, to show how much I^2R heating it can take before it burns out. This means that a resistor with a power rating of one watt will burn out if it is used in a circuit where the current causes it to dissipate heat at a rate greater than one watt.

If the power rating of a resistor is known, the maximum current it can carry is found by using an equation derived from $P = I^2R$:

$$P = I^2R \text{ becomes } I^2 = P/R,$$
$$\text{which becomes } I = \sqrt{P/R}$$

Using this equation, find the maximum current that can be carried by a 1Ω resistor with a power rating of 4W:

$$I = \sqrt{P/R} = \sqrt{4/1} = 2 \text{ amperes}$$

If such a resistor conducts more than 2 amperes, it will dissipate more than its rated power and burn out.

Power ratings assigned by resistor manufacturers are usually based on the resistors being mounted in an open location where there is free air circulation, and where the temperature is not higher than 104°F (40°C). Therefore, if a resistor is mounted in a small, crowded, enclosed space, or where the temperature is higher than 104°F, there is a good chance it will burn out even before its power rating is exceeded. Also, some resistors are designed to be attached to a chassis or frame that will carry away the heat.

Summary

The relationships among current, voltage, resistance, and power are consistent for all types of DC circuits and can be calculated using Ohm's law and Ohm's law for power. Understanding and being able to apply these concepts is necessary for effective circuit analysis and troubleshooting.

> **THINK ABOUT IT**
>
> ***Putting It All Together***
>
> Notice the common electrical devices in the building you're in. What is their wattage rating? How much current do they draw? How would you test their voltage or amperage?

Instructor's Notes:

Review Questions

1. A type of subatomic particle with a negative charge is a(n) _____.
   a. proton
   b. neutron
   c. electron
   d. nucleus

2. A type of subatomic particle with a positive charge is a(n) _____.
   a. proton
   b. neutron
   c. electron
   d. nucleus

3. Like charges _____ each other.
   a. attract
   b. repel
   c. have no effect on
   d. complement

4. An electron has _____ coulombs of charge.
   a. 1.6×10^{-9}
   b. 1.6×10^{9}
   c. 1.6×10^{19}
   d. 1.6×10^{-19}

5. The quantity that Ohm's law does not express a relationship for in an electrical circuit is _____.
   a. charge
   b. resistance
   c. voltage
   d. current

6. The color band that represents tolerance on a resistor is the _____.
   a. 4th band
   b. 3rd band
   c. 2nd band
   d. 1st band

7. A resistor with a color code of red/red/orange indicates a value of _____.
   a. 22,000 ohms
   b. 66 ohms
   c. 223 ohms
   d. 220 ohms

8. An ammeter is placed in _____ with the circuit being tested.
   a. parallel
   b. series

9. The basic unit of power is the _____ .
   a. volt
   b. ampere
   c. coulomb
   d. watt

10. The power in a circuit with 120 volts and 5 amps is _____.
    a. 24 watts
    b. 600 watts
    c. 6,000 watts
    d. $\frac{1}{24}$ watt

Classroom

Have the trainees complete the Review Questions and go over the answers prior to administering the Module Examination.

Examination

Administer the Module Examination. Record the results on Craft Training Report Form 200 and submit the results to the Training Program Sponsor.

Performance Testing

Administer the Performance Test and fill out Performance Profile Sheets for each trainee. If desired, trainee proficiency noted during laboratory sessions may be used to complete the Performance Test. Be sure to record the results on Craft Training Report Form 200 and submit the results to the Training Program Sponsor.

PROFILE IN SUCCESS

Charles Michael (Mike) Holt, Mike Holt Enterprises

Mike Holt's story is one of a high school dropout who became a nationally respected and well-educated master electrician, electrical contractor, author, speaker, and educator. He is recognized as an expert on the NEC and is sought after by major corporations and government agencies as a consultant and lecturer. Along the way, Mike fathered seven children, was twice crowned National Barefoot Waterskiing Champion, and set five national waterskiing records.

How did you choose a career in the electrical field?
I was always fascinated by electricity. I read books about Thomas Edison when I was young and I could just envision myself doing electrical experiments. I tried getting into the high school electricity program, but the class was full. I had to leave high school in the eleventh grade, and after a couple of years I got a job as an electrician's helper.

Tell us about your apprenticeship experience.
I wasn't able to get into the electrician's union because none of my family belonged, so I took a job as a helper in a non-union shop. There was no formal training program, and I really envied the electricians who had gone to apprenticeship school. I didn't let it stop me, though. I bought every book I could afford and studied on my own, and eventually was able to get my journeyman's license.

What positions have you held in the industry?
Since I became a journeyman, I've been a foreman, project manager, estimator, electrical inspector, and instructor. I had my own electrical contracting business for five years, but gave that up to devote full time to developing training materials. Now I spend my time developing new training programs, writing books and articles, consulting, and conducting seminars.

What factor or factors contributed most to your success?
I would have to say my personal commitment and sense of responsibility. I wanted to be an electrician and was proud when I got my journeyman's license through self-study. I did not have a lot of people encouraging me early in my career, so I had to do it on my own. Some of my co-workers even put me down because I studied for my journeyman's license before and after work and during breaks, but I didn't let that stop me.

A short time after I left high school, I came to realize that success depends on education. I got my GED certificate and continued my education, eventually enrolling in the University of Miami's MBA program.

Instructor's Notes:

GLOSSARY

Trade Terms Introduced in This Module

Ammeter: An instrument for measuring electrical current.

Ampere (A): A unit of electrical current. For example, one volt across one ohm of resistance causes a current flow of one ampere.

Atom: The smallest particle to which an element may be divided and still retain the properties of the element.

Battery: A DC voltage source consisting of two or more cells that convert chemical energy into electrical energy.

Circuit: A complete path for current flow.

Conductor: A material that offers very little resistance to current flow.

Coulomb: An electrical charge equal to 6.25×10^{18} electrons or 6,250,000,000,000,000,000 electrons. A coulomb is the common unit of quantity used for specifying the size of a given charge.

Current: The movement, or flow, of electrons in a circuit. Current (I) is measured in amperes.

Electron: A negatively charged particle that orbits the nucleus of an atom.

Insulator: A material that offers resistance to current flow.

Joule (J): A unit of measurement that represents one newton-meter (Nm), which is a unit of measure for doing work.

Kilo: A prefix used to indicate one thousand; for example, one kilowatt is equal to one thousand watts.

Matter: Any substance that has mass and occupies space.

Mega: A prefix used to indicate one million; for example, one megawatt is equal to one million watts.

Micro: A prefix used to indicate one-millionth; for example, one microwatt is equal to one-millionth of a watt.

Neutrons: Electrically neutral particles (neither positive nor negative) that have the same mass as a proton and are found in the nucleus of an atom.

Nucleus: The center of an atom. It contains the protons and neutrons of the atom.

Ohm (Ω): The basic unit of measurement for resistance.

Ohmmeter: An instrument used for measuring resistance.

Ohm's law: A statement of the relationships among current, voltage, and resistance in an electrical circuit: current (I) equals voltage (E) divided by resistance (R). Generally expressed as a mathematical formula: $I = E/R$.

Power: The rate of doing work or the rate at which energy is used or dissipated. Electrical power is the rate of doing electrical work. Electrical power is measured in watts.

Protons: The smallest positively charged particles of an atom. Protons are contained in the nucleus of an atom.

Resistance: An electrical property that opposes the flow of current through a circuit. Resistance (R) is measured in ohms.

Resistor: Any device in a circuit that resists the flow of electrons.

Schematic: A type of drawing in which symbols are used to represent the components in a system.

Series circuit: A circuit with only one path for current flow.

Valence shell: The outermost ring of electrons that orbit about the nucleus of an atom.

Volt (V): The unit of measurement for voltage (electromotive force). One volt is equivalent to the force required to produce a current of one ampere through a resistance of one ohm.

Voltage: The driving force that makes current flow in a circuit. Voltage (E) is also referred to as *electromotive force* or *potential.*

Voltage drop: The change in voltage across a component that is caused by the current flowing through it and the amount of resistance opposing it.

Voltmeter: An instrument for measuring voltage. The resistance of the voltmeter is fixed. When the voltmeter is connected to a circuit, the current passing through the meter will be directly proportional to the voltage at the connection points.

Watt (W): The basic unit of measurement for electrical power.

Instructor's Notes:

Additional Resources

This module is intended to present thorough resources for task training. The following reference works are suggested for further study. These are optional materials for continued education rather than for task training.

Electronics Fundamentals, Latest Edition. New York: Prentice Hall.
Principles of Electric Circuits, Latest Edition. New York: Prentice Hall.

ANSWER KEY

Answers to Review Questions

Answer	Section
1. c	Terms/2.1.0
2. a	Terms/2.1.0
3. b	2.1.0
4. d	3.0.0
5. a	4.2.0
6. a	6.1.0
7. a	6.1.0/Figure 10
8. b	7.3.0
9. d	8.0.0
10. b	8.1.0

TEACHING TIPS

Section 2.2.0 *Think About It—Conductors*

Conductors are materials in which current can be established with relative ease. Conductors have a large number of free electrons that are capable of moving from one atom to another in the material. It is these free electrons that make current flow possible.

Section 3.1.0 *Think About It—The Magic of Electricity*

It takes only a fraction of a second for the light from a flashlight to strike the floor. The distance from Maine to California is approximately 2,500 miles. Since electrons travel at about 186,000 miles per second, it would take about 0.013 second (2,500 ÷ 186,000) for the light to come on in California.

Section 3.1.0 *Think About It—Current Flow*

Two wires are needed to provide a closed path for current flow through an electrical device such as a lamp. These wires, one connected to the input of the device and the other to its output, must be connected to a battery or other source of potential difference in order for current to flow in the circuit. Light produced by an incandescent lamp depends upon the current flow through the lamp filament. The current flow causes the filament element to heat and incandesce, thus converting electrical energy to light.

Section 3.2.0 *Think About It—Voltage*

Voltage is called *electrical potential* because it has the capability or potential for doing work. An electrical potential can exist without current flow. For example, an electrical potential of 12V exists between the negative and positive terminals of a fully charged 12V battery when it is disconnected from a circuit. However, when connected to an external circuit, the battery has the potential difference needed to cause current flow through the circuit. Capacitors connected to electric high-voltage power transmission lines provide an example of a device where high electrical potential exists with no or little current as a result of the electrical energy being stored by the capacitors.

The ionization of atoms within turbulent clouds, or between clouds and the earth, causes cloud centers to become oppositely charged, creating an electrical potential between them. Lightning arcs occur when the buildup of this electrical potential exceeds the natural electrical insulator quality of the surrounding air. The result is a high-current electrical discharge between the charged bodies as they act to equalize their respective charges, thereby dissipating the electrical potential between them.

Section 4.2.0 *Think About It—Voltage Matters*

Doubling the voltage applied to the hair dryer will double the amount of current that flows through the fixed resistance of the hair dryer's heating element. This will most likely cause the heating element to burn out.

Section 4.2.0 *Think About It—Ohm's Law*

As shown in *Table 1*, the resistance of copper annealed wire is somewhat lower than that of aluminum wire, and greatly lower than that of iron wire. The current flow through each type of wire is inversely proportional to the wire's resistance. The higher the resistance, the lower the current flow. Therefore, the copper annealed wire will carry the most current, followed by the aluminum wire, and then the iron wire.

CONTREN™ LEARNING SERIES — USER UPDATES

The NCCER makes every effort to keep these textbooks up-to-date and free of technical errors. We appreciate your help in this process. If you have an idea for improving this textbook, or if you find an error, a typographical mistake, or an inaccuracy in NCCER's Contren™ textbooks, please write us, using this form or a photocopy. Be sure to include the exact module number, page number, a detailed description, and the correction, if applicable. Your input will be brought to the attention of the Technical Review Committee. Thank you for your assistance.

Instructors – If you found that additional materials were necessary in order to teach this module effectively, please let us know so that we may include them in the Equipment/Materials list in the Instructor's Guide.

Write: Curriculum Revision and Development Department
National Center for Construction Education and Research
P.O. Box 141104, Gainesville, FL 32614-1104

Fax: 352-334-0932

E-mail: curriculum@nccer.org

Craft ____________________ Module Name ____________________

Copyright Date __________ Module Number __________ Page Number(s) __________

Description

____

____

____

____

(Optional) Correction

____

____

____

(Optional) Your Name and Address

____

____

____

Chapter 3

Electrical Theory Two

26105-02

Electrical Theory Two
Annotated Instructor's Guide

Module 26105

MODULE OVERVIEW

This course introduces the electrical trainee to circuit calculations involving the application of Ohm's and Kirchhoff's laws.

PREREQUISITES

Please refer to the Course Map in the Trainee Module. Prior to training with this module, it is recommended that the trainee shall have successfully completed the following:

Core Curriculum; Residential Electrical I, Chapters 1 and 2

LEARNING OBJECTIVES

Upon completion of this module, the trainee will be able to:

1. Explain the basic characteristics of a series circuit.
2. Explain the basic characteristics of a parallel circuit.
3. Explain the basic characteristics of a series-parallel circuit.
4. Calculate, using Kirchhoff's voltage law, the voltage drop in series, parallel, and series-parallel circuits.
5. Calculate, using Kirchhoff's current law, the total current in parallel and series-parallel circuits.
6. Find the total amount of resistance in a series circuit.
7. Find the total amount of resistance in a parallel circuit.
8. Find the total amount of resistance in a series-parallel circuit.

PERFORMANCE OBJECTIVES

Under supervision of the instructor, the trainee should be able to:

1. Calculate the total resistance for selected series, parallel, and series-parallel circuits.
2. Use Kirchhoff's current law to calculate the total and unknown currents in parallel and series-parallel circuits.
3. Use Kirchhoff's voltage law to calculate voltage drops in series, parallel, and series-parallel circuits.

NCCER STANDARDIZED CRAFT TRAINING PROGRAM

The National Center for Construction Education and Research (NCCER) provides a standardized national program of accredited craft training. Key features of the program include instructor certification, competency-based training, and performance testing. The program provides trainees, instructors, and companies with a standard form of recognition through a National Craft Training Registry. The program is described in full in the *Guidelines for Accreditation,* published by the NCCER. For more information on standardized craft training, contact the NCCER at P.O. Box 141104, Gainesville, FL 32614-1104, 352-334-0911, visit our Web site at www.nccer.org, or e-mail info@nccer.org.

HOW TO USE THIS ANNOTATED INSTRUCTOR'S GUIDE

Each page presents two sections of information. The larger section displays each page exactly as it appears in the Trainee Module. The narrow column ties suggested trainee and instructor actions to each page and provides icons to call your attention to material, safety, audiovisual, or testing requirements. The bottom of each page includes space for your notes.

If you see the Teaching Tip icon, that means there is a teaching tip associated with this section. Also refer to the suggested teaching tips at the end of the module.

SAFETY CONSIDERATIONS

Ensure that the trainees are equipped with appropriate personal protective equipment.

PREPARATION

Before teaching this module, you should review the Module Outline, Learning and Performance Objectives, and the Materials and Equipment List. Be sure to allow ample time to prepare your own training or lesson plan and gather all required equipment and materials.

MATERIALS AND EQUIPMENT LIST

Materials:

Transparencies

Markers/chalk

Copy of the latest edition of the *National Electrical Code®*

Module Examinations*

Performance Profile Sheets*

Equipment:

Overhead projector and screen

Whiteboard/chalkboard

Appropriate personal protective equipment

*Located in the Test Booklet packaged with this Annotated Instructor's Guide.

ADDITIONAL RESOURCES

This module is intended to present thorough resources for task training. The following reference works are suggested for both instructors and motivated trainees interested in further study. These are optional materials for continued education rather than for task training.

Electronics Fundamentals, Latest Edition. Upper Saddle River, NJ: Prentice Hall.

Principles of Electric Circuits, Latest Edition. Upper Saddle River, NJ: Prentice Hall.

NOTES

The designations "National Electrical Code," "NE Code," and "NEC," where used in this document, refer to the *National Electrical Code®*, which is a registered trademark of the National Fire Protection Association, Quincy, MA. All National Electrical Code (NEC) references in this module refer to the 2002 edition of the NEC.

If you feel that additional math instruction would be helpful, Prentice Hall offers a basic math textbook entitled *Fundamentals of Electrical and Mechanical Mathematics.* It covers the basic math requirements for electrical trainees and may be ordered by contacting Prentice Hall Customer Service at 1-800-922-0579.

TEACHING TIME FOR THIS MODULE

An outline for use in developing your lesson plan is presented below. Note that each Roman numeral in the outline equates to one session of instruction. Each session has a suggested time period of 2½ hours. This includes 10 minutes at the beginning of each session for administrative tasks and one 10-minute break during the session. Approximately 7½ hours are suggested to cover *Electrical Theory Two*. You will need to adjust the time required for hands-on activity and testing based on your class size and resources.

Topic	Planned Time
Session I. Resistive Circuits	
A. Resistive Circuits	________
1. Resistances in Series	________
2. Resistances in Parallel	________
a. Simplified Formulas	________
3. Series-Parallel Circuits	________
a. Reducing Series-Parallel Circuits	________
Session II. Applying Ohm's Law to Resistive Circuits	
A. Voltage and Current in Series Circuits	________
B. Voltage and Current in Parallel Circuits	________
C. Voltage and Current in Series-Parallel Circuits	________
Session III. Kirchhoff's Laws; Module Examination and Performance Testing	
A. Kirchhoff's Laws	________
1. Kirchhoff's Current Law	________
2. Kirchhoff's Voltage Law	________
3. Loop Equations	________
B. Module Examination	________
1. Trainees must score 70% or higher to receive recognition from the NCCER.	
2. Record the testing results on Craft Training Report Form 200 and submit the results to the Training Program Sponsor.	
C. Performance Testing	________
1. Trainees must perform each task to the satisfaction of the instructor to receive recognition from the NCCER.	
2. Record the testing results on Craft Training Report Form 200 and submit the results to the Training Program Sponsor.	

Chapter 3

Module 26105-02

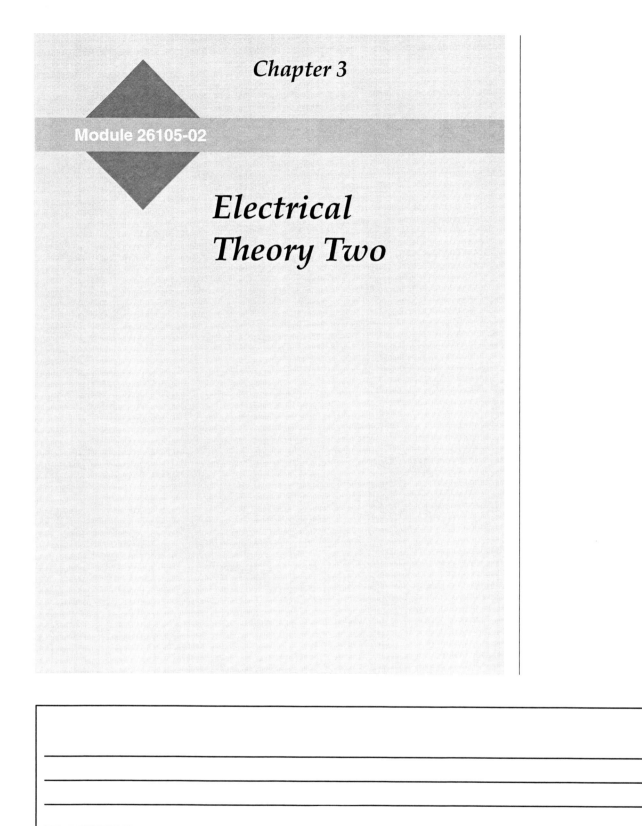

Electrical Theory Two

Course Map

This course map shows all of the modules in the first level of the Residential Electrical curriculum. The suggested training order begins at the bottom and proceeds up. Skill levels increase as you advance on the course map. The local Training Program Sponsor may adjust the training order.

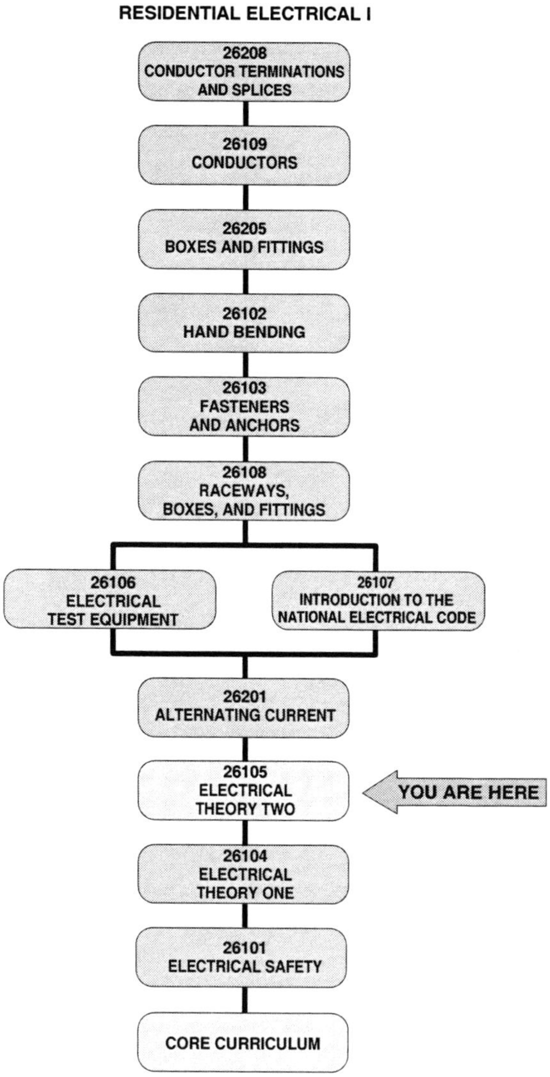

Homework

Assign reading of Module 26105.

MODULE 26105 CONTENTS

ELECTRICAL THEORY TWO 85

Figures

Instructor's Notes:

Chapter 3

MODULE 26105

Electrical Theory Two

Objectives

When you have completed this module, you will be able to do the following:

1. Explain the basic characteristics of a series circuit.
2. Explain the basic characteristics of a parallel circuit.
3. Explain the basic characteristics of a series-parallel circuit.
4. Calculate, using Kirchhoff's voltage law, the voltage drop in series, parallel, and series-parallel circuits.
5. Calculate, using Kirchhoff's current law, the total current in parallel and series-parallel circuits.
6. Find the total amount of resistance in a series circuit.
7. Find the total amount of resistance in a parallel circuit.
8. Find the total amount of resistance in a series-parallel circuit.

Prerequisites

Before you begin this module, it is recommended that you successfully complete the following: Core Curriculum; Residential Electrical I, Chapters 1 and 2.

Required Trainee Materials

1. Paper and pencil
2. Copy of the latest edition of the *National Electrical Code*
3. Appropriate personal protective equipment

1.0.0 ◆ INTRODUCTION

Ohm's law was explained in the module *Electrical Theory One.* This fundamental concept is now going to be used to analyze more complex **series circuits, parallel circuits,** and **series-parallel circuits.** This module will explain how to calculate resistance, current, and voltage in these complex circuits. Ohm's law will be used to develop a new law for voltage and current determination. This law, called *Kirchhoff's law,* will become the new foundation for analyzing circuits.

2.0.0 ◆ RESISTIVE CIRCUITS

2.1.0 Resistances in Series

A series circuit is a circuit in which there is only one path for current flow. Resistance is measured in ohms (Ω). In the series circuit shown in *Figure 1,* the current (I) is the same in all parts of the circuit. This means that the current flowing through R_1 is the same as the current flowing through R_2 and R_3, and it is also the same as the current supplied by the battery.

When resistances are connected in series as in this example, the total resistance in the circuit is

Note: The designations "National Electrical Code," "NE Code," and "NEC," where used in this document, refer to the *National Electrical Code®*, which is a registered trademark of the National Fire Protection Association, Quincy, MA. *All National Electrical Code (NEC) references in this module refer to the 2002 edition of the NEC.*

Materials

Ensure you have everything required to teach the course. Check the Materials and Equipment List at the front of this Instructor's Guide.

Audiovisual

Show Transparency 1, Course Objectives.

Show Transparency 2, Performance Profile Tasks.

Classroom

Explain that terms shown in bold (blue) are defined in the Glossary at the back of this module.

Discuss basic series circuits.

Audiovisual

Show Transparency 3 (Figure 1).

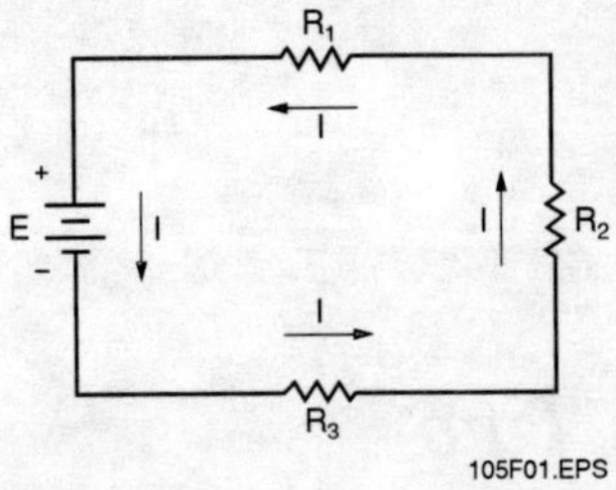

Figure 1 ◆ Series circuit.

equal to the sum of the resistances of all the parts of the circuit:

$$R_T = R_1 + R_2 + R_3$$

Where:

R_T = total resistance

$R_1 + R_2 + R_3$ = resistances in series

Example 1:

The circuit shown in *Figure 2(A)* has 50Ω, 75Ω, and 100Ω resistors in series. Find the total resistance of the circuit.

Add the values of the three resistors in series:

$$R_T = R_1 + R_2 + R_3 = 50 + 75 + 100 = 225\Omega$$

Example 2:

The circuit shown in *Figure 2(B)* has three lamps connected in series with the resistances shown. Find the total resistance of the circuit.

Add the values of the three lamp resistances in series:

$$R_T = R_1 + R_2 + R_3 = 20 + 40 + 60 = 120\Omega$$

2.2.0 Resistances in Parallel

The total resistance in a parallel resistive circuit is given by the formula:

$$R_T = \cfrac{1}{\cfrac{1}{R_1} + \cfrac{1}{R_2} + \cfrac{1}{R_3} + \cfrac{1}{R_n}}$$

Where:

R_T = total resistance in parallel

R_1, R_2, R_3, and R_n = branch resistances

Example 1:

Find the total resistance of the 2Ω, 4Ω, and 8Ω resistors in parallel shown in *Figure 3.*

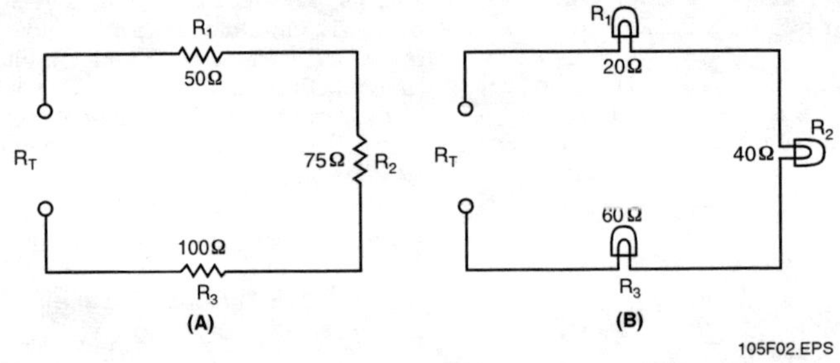

Figure 2 ◆ Total resistance.

> ### *Series Circuits*
> THINK ABOUT IT
>
> Simple series circuits are seldom encountered in practical wiring. The only simple series circuit you may recognize is older strands of Christmas lights, in which the entire string went dead when one lamp burned out. Think about what the actual wiring of a series circuit would look like in household receptacles. How would the circuit physically be wired? What kind of illumination would you get if you wired your household receptacles in series and plugged half a dozen lamps into those receptacles?

Classroom

Explain how to calculate total resistance in a basic series circuit.

Audiovisual

Show Transparency 4 (Figure 2).

Classroom

Have the trainees practice calculating the total resistance in a basic series circuit.

Teaching Tip

Discuss the "Think About It."

Instructor's Notes:

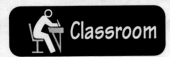

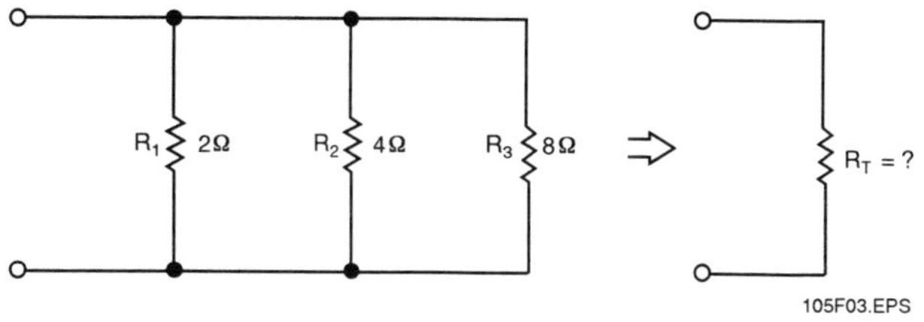

Figure 3 ◆ Parallel branch.

Write the formula for the three resistances in parallel:

$$R_T = \frac{1}{\frac{1}{R_1} + \frac{1}{R_2} + \frac{1}{R_3}}$$

Substitute the resistance values:

$$R_T = \frac{1}{\frac{1}{2} + \frac{1}{4} + \frac{1}{8}}$$

$$R_T = \frac{1}{0.5 + 0.25 + 0.125}$$

$$R_T = \frac{1}{0.875}$$

$$R_T = 1.14\Omega$$

Note that when resistances are connected in parallel, the total resistance is always less than the resistance of any single branch.

In this case:

$R_T = 1.14\Omega < R_1 = 2\Omega$, $R_2 = 4\Omega$, and $R_3 = 8\Omega$

Example 2:

Add a fourth parallel resistor of 2Ω to the circuit in *Figure 3*. What is the new total resistance, and what is the net effect of adding another resistance in parallel?

Write the formula for four resistances in parallel:

$$R_T = \frac{1}{\frac{1}{R_1} + \frac{1}{R_2} + \frac{1}{R_3} + \frac{1}{R_4}}$$

Substitute values:

$$R_T = \frac{1}{\frac{1}{2} + \frac{1}{4} + \frac{1}{8} + \frac{1}{2}}$$

$$R_T = \frac{1}{0.5 + 0.25 + 0.125 + 0.5}$$

$$R_T = \frac{1}{1.375}$$

$$R_T = 0.73\Omega$$

The net effect of adding another resistance in parallel is a reduction of the total resistance from 1.14Ω to 0.73Ω.

2.2.1 Simplified Formulas

The total resistance of *equal* resistors in parallel is equal to the resistance of one resistor divided by the number of resistors:

$$R_T = \frac{R}{N}$$

Where:

R_T = total resistance of equal resistors in parallel
R = resistance of one of the equal resistors
N = number of equal resistors

If two resistors with the same resistance are connected in parallel, the equivalent resistance is half of that value, as shown in *Figure 4*.

The two 200Ω resistors in parallel are the equivalent of one 100Ω resistor; the two 100Ω resistors are the equivalent of one 50Ω resistor; and the two 50Ω resistors are the equivalent of one 25Ω resistor.

When any two *unequal* resistors are in parallel, it is often easier to calculate the total resistance by multiplying the two resistances and then dividing the product by the sum of the resistances:

$$R_T = \frac{R_1 \times R_2}{R_1 + R_2}$$

Where:

R_T = total resistance of unequal resistors in parallel
R_1, R_2 = two unequal resistors in parallel

Classroom

Discuss basic parallel circuits.

Audiovisual

Show Transparency 5 (Figure 3).

Classroom

Explain how to calculate the total resistance in a basic parallel circuit.

Discuss simplified resistance formulas and have the trainees practice using them.

Audiovisual

Show Transparency 6 (Figure 4).

ELECTRICAL THEORY TWO 89

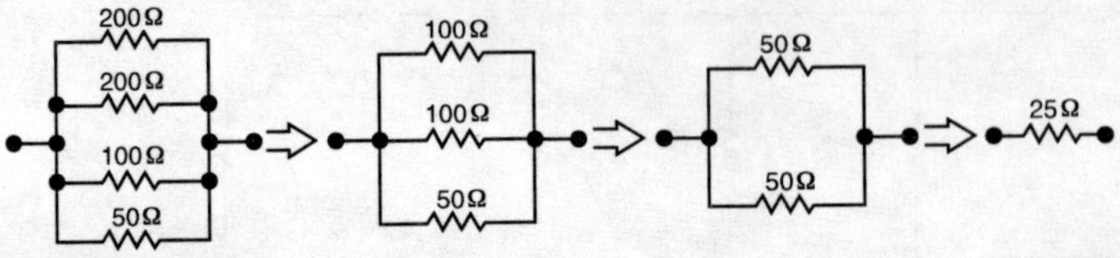

Figure 4 ◆ Equal resistances in a parallel circuit.

INSIDE TRACK

Parallel Circuits

Most practical circuits are wired in parallel, like the pole lights shown here.

105PO501.EPS

Teaching Tip

Discuss the "Think About It."

THINK ABOUT IT

Parallel Circuits

An interesting fact about circuits is the drop in resistance in a parallel circuit as more resistors are added. But this fact does not mean that you can add an endless number of devices, such as lamps, in a parallel circuit. Why not?

Instructor's Notes:

Example 1:
Find the total resistance of a 6Ω (R_1) resistor and an 18Ω (R_2) resistor in parallel:

$$R_T = \frac{R_1 \times R_2}{R_1 + R_2} = \frac{6 \times 18}{6 + 18} = \frac{108}{24} = 4.5\Omega$$

Example 2:
Find the total resistance of a 100Ω (R_1) resistor and a 150Ω (R_2) resistor in parallel:

$$R_T = \frac{R_1 \times R_2}{R_1 + R_2} = \frac{100 \times 150}{100 + 150} = \frac{15{,}000}{250} = 60\Omega$$

2.3.0 Series-Parallel Circuits

To find current, voltage, and resistance in series circuits and parallel circuits is fairly easy. When working with either type, use only the rules that apply to that type. In a series-parallel circuit, some parts of the circuit are series connected and other parts are parallel connected. Thus, in some parts the rules for series circuits apply, and in other parts, the rules for parallel circuits apply. To analyze or solve a problem involving a series-parallel circuit, it is necessary to recognize which parts of the circuit are series connected and which parts are parallel connected. This is obvious if the circuit is simple. Many times, however, the circuit must be redrawn, putting it into a form that is easier to recognize.

In a series circuit, the current is the same at all points. In a parallel circuit, there are one or more points where the current divides and flows in separate branches. In a series-parallel circuit, there are both separate branches and series loads. The easiest way to find out whether a circuit is a series, parallel, or series-parallel circuit is to start at the negative terminal of the power source and trace the path of current through the circuit back to the positive terminal of the power source. If the current does not divide anywhere, it is a series circuit. If the current divides into separate branches, but there are no series loads, it is a parallel circuit. If the current divides into separate branches and there are also series loads, it is a series-parallel circuit. *Figure 5* shows electric lamps connected in series, parallel, and series-parallel circuits.

After determining that a circuit is series-parallel, redraw the circuit so that the branches and the series loads are more easily recognized. This is especially helpful when computing the total resistance of the circuit. *Figure 6* shows resistors connected in a series-parallel circuit and the equivalent circuit redrawn to simplify it.

2.3.1 Reducing Series-Parallel Circuits

Very often, all that is known about a series-parallel circuit is the applied voltage and the values of the individual resistances. To find the voltage drop across any of the loads or the current in any of the branches, the total circuit current must usually be known. But to find the total current, the total resistance of the circuit must be known. To find the total

Classroom

Discuss series-parallel circuits and explain how they can be redrawn to simplify calculations.

Audiovisual

Show Transparencies 7 and 8 (Figures 5 and 6).

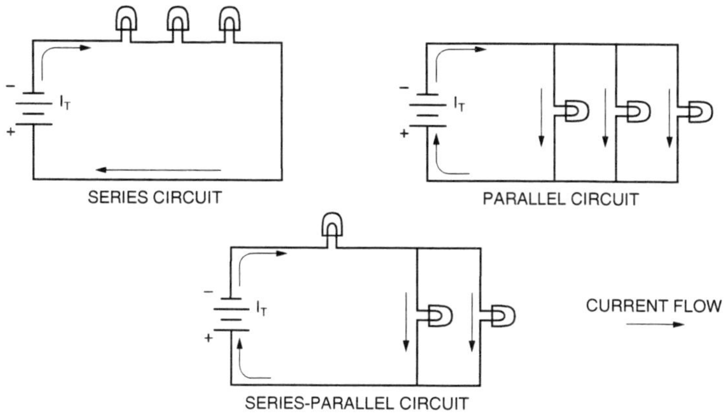

105F05.EPS

Figure 5 ◈ Series, parallel, and series-parallel circuits.

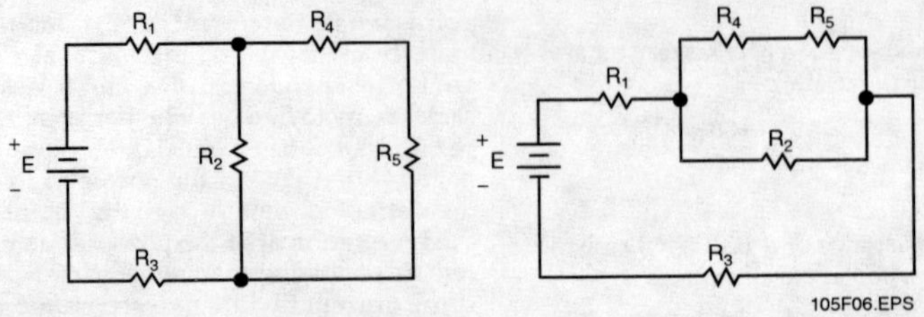

Figure 6 ◆ Redrawing a series-parallel circuit.

Teaching Tip

Discuss the "Think About It."

Series-Parallel Circuits

THINK ABOUT IT

Explain *Figure 6.* Which resistors are in series and which are in parallel?

Audiovisual

Show Transparency 9 (Figure 7).

Classroom

Explain how to calculate the total resistance in a series-parallel circuit.

Have the trainees practice calculating the total resistance in a series-parallel circuit.

resistance, reduce the circuit to its simplest form, which is usually one resistance that forms a series circuit with the voltage source. This simple series circuit has the equivalent resistance of the series-parallel circuit it was derived from, and also has the same total current. There are four basic steps in reducing a series-parallel circuit:

- If necessary, redraw the circuit so that all parallel combinations of resistances and series resistances are easily recognized.
- For each parallel combination of resistances, calculate its effective resistance.
- Replace each of the parallel combinations with one resistance whose value is equal to the effective resistance of that combination. This provides a circuit with all series loads.
- Find the total resistance of this circuit by adding the resistances of all the series loads.

Examine the series-parallel circuit shown in *Figure 7* and reduce it to an equivalent series circuit.

In this circuit, resistors R_2 and R_3 are connected in parallel, but resistor R_1 is in series with both the battery and the parallel combination of R_2 and R_3. The current I_T leaving the negative terminal of the voltage source travels through resistor R_1 before it is divided at the junction of resistors R_1, R_2, and R_3 (Point A) to go through the two branches formed by resistors R_2 and R_3.

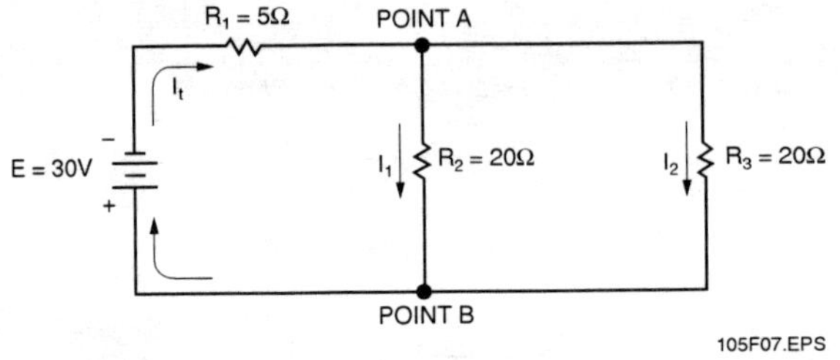

Figure 7 ◆ Reducing a series-parallel circuit.

Instructor's Notes:

Given the information in *Figure 7,* calculate the resistance of R_2 and R_3 in parallel and the total resistance of the circuit, R_T.

The total resistance of the circuit is the sum of R_1 and the equivalent resistance of R_2 and R_3 in parallel. To find R_T, first find the resistance of R_2 and R_3 in parallel. Because the two resistances have the same value of 20Ω, the resulting equivalent resistance is 10Ω. Therefore, the total resistance (R_T) is 15Ω (5Ω + 10Ω).

2.4.0 Applying Ohm's Law

2.4.1 Voltage and Current in Series Circuits

In resistive circuits, unknown circuit parameters can be found by using Ohm's law and the techniques for determining equivalent resistance. Ohm's law may be applied to an entire series circuit or to the individual parts of the circuit. When it is used on a particular part of a circuit, the voltage across that part is equal to the current in that part multiplied by the resistance of that part.

For example, given the information in *Figure 8,* calculate the total resistance (R_T) and the total current (I_T).

To find R_T:

$$R_T = R_1 + R_2 + R_3$$
$$R_T = 20 + 50 + 120$$
$$R_T = 190\Omega$$

To find I_T using Ohm's law:

$$I_T = \frac{E_T}{R_T}$$
$$I_T = \frac{95}{190}$$
$$I_T = 0.5\text{A}$$

Find the voltage across each resistor. In a series circuit, the current is the same; that is, I = 0.5A through each resistor:

$$E_1 = IR_1 = 0.5(20) = 10\text{V}$$
$$E_2 = IR_2 = 0.5(50) = 25\text{V}$$
$$E_3 = IR_3 = 0.5(120) = 60\text{V}$$

The voltages E_1, E_2, and E_3 found for *Figure 8* are known as *voltage drops* or *IR drops.* Their effect is to reduce the voltage that is available to be applied across the rest of the components in the circuit. The sum of the voltage drops in any series circuit is always equal to the voltage that is applied to the circuit. The total voltage (E_T) is the same as the applied voltage and can be verified in this example (E_T = 10 + 25 + 60 or 95V).

2.4.2 Voltage and Current in Parallel Circuits

A parallel circuit is a circuit in which two or more components are connected across the same voltage source, as illustrated in *Figure 9.* The resistors R_1, R_2, and R_3 are in parallel with each other and with the battery. Each parallel path is then a

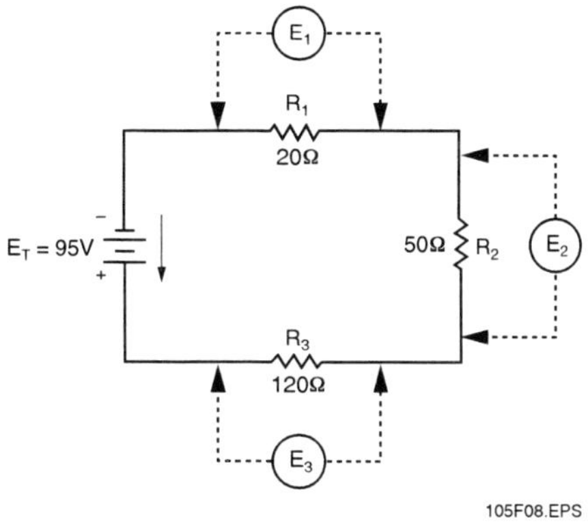

Figure 8 ◆ Calculating voltage drops.

THINK ABOUT IT

Voltage Drops

Calculating voltage drops is not just a schoolroom exercise. It is important to know the voltage drop when sizing circuit components. What would happen if you sized a component without accounting for a substantial voltage drop in the circuit?

Classroom

Demonstrate how to use Ohm's law to solve for unknown values in series circuits.

Audiovisual

Show Transparency 10 (Figure 8).

Classroom

Demonstrate how to use Ohm's law to solve for unknown values in parallel circuits.

Audiovisual

Show Transparency 11 (Figure 9).

Teaching Tip

Discuss the "Think About It."

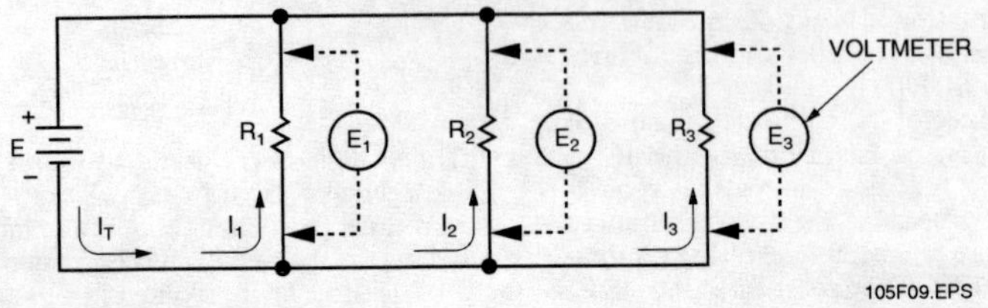

Figure 9 ◆ Parallel circuit.

Classroom

Show examples of how to solve for an unknown branch current.

Audiovisual

Show Transparency 12 (Figure 10).

branch with its own individual current. When the total current I_T leaves the voltage source E, part I_1 of the current I_T will flow through R_1, part I_2 will flow through R_2, and the remainder I_3 will flow through R_3. The branch currents I_1, I_2, and I_3 can be different. However, if a voltmeter is connected across R_1, R_2, and R_3, the respective voltages E_1, E_2, and E_3 will be equal to the source voltage E.

The total current I_T is equal to the sum of all branch currents.

This formula applies for any number of parallel branches whether the resistances are equal or unequal.

Using Ohm's law, each branch current equals the applied voltage divided by the resistance between the two points where the voltage is applied. Hence, for each branch in *Figure 9* we have the following equations:

$$\text{Branch 1: } I_1 = \frac{E_1}{R_1} = \frac{E}{R_1}$$

$$\text{Branch 2: } I_2 = \frac{E_2}{R_2} = \frac{E}{R_2}$$

$$\text{Branch 3: } I_3 = \frac{E_3}{R_3} = \frac{E}{R_3}$$

With the same applied voltage, any branch that has less resistance allows more current through it than a branch with higher resistance.

Example 1:

The two branches R_1 and R_2, shown in *Figure 10(A)*, across a 110V power line draw a total line current of 20A. Branch R_1 takes 12A. What is the current I_2 in branch R_2?

Transpose to find I_2 and then substitute given values:

$$I_T = I_1 + I_2$$
$$I_2 = I_T - I_1$$
$$I_2 = 20 - 12 = 8A$$

Example 2:

As shown in *Figure 10(B)*, the two branches R_1 and R_2 across a 240V power line draw a total line current of 35A. Branch R_2 takes 20A. What is the current I_1 in branch R_1?

Transpose to find I_1 and then substitute given values:

$$I_T = I_1 + I_2$$
$$I_1 = I_T - I_2$$
$$I_1 = 35 - 20 = 15A$$

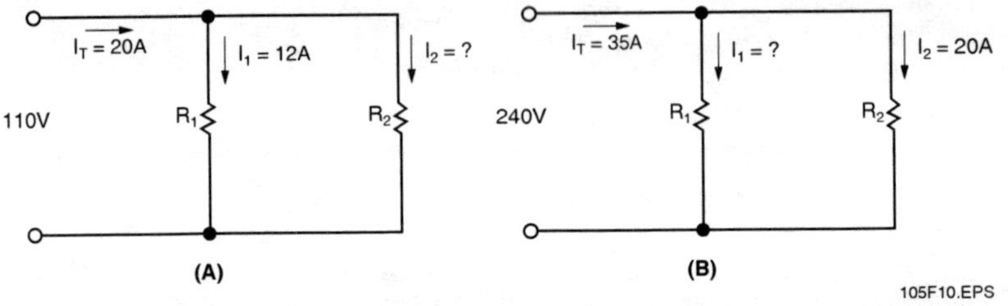

Figure 10 ◆ Solving for an unknown current.

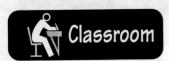

Instructor's Notes:

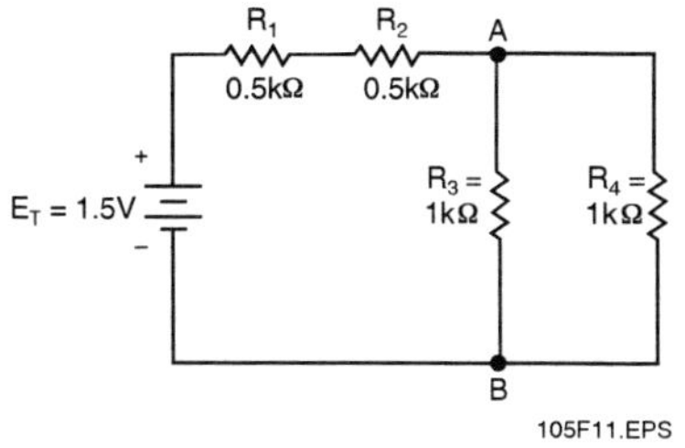

Figure 11 ◆ Series-parallel circuit.

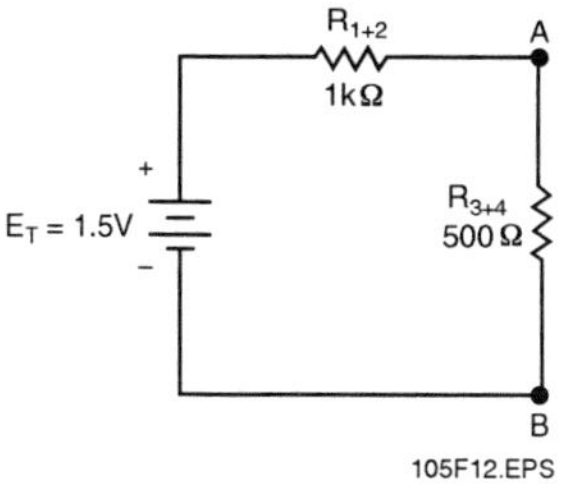

Figure 12 ◆ Simplified series-parallel circuit.

2.4.3 Voltage and Current in Series-Parallel Circuits

Series-parallel circuits combine the elements and characteristics of both the series and parallel configurations. By properly applying the equations and methods previously discussed, the values of individual components of the circuit can be determined. *Figure 11* shows a simple series-parallel circuit with a 1.5V battery.

The current and voltage associated with each component can be determined by first simplifying the circuit to find the total current, and then working across the individual components.

This circuit can be broken into two components: the series resistances R_1 and R_2, and the parallel resistances R_3 and R_4.

R_1 and R_2 can be added together to form the equivalent series resistance R_{1+2}:

$$R_{1+2} = R_1 + R_2$$
$$R_{1+2} = 0.5k\Omega + 0.5k\Omega$$
$$R_{1+2} = 1k\Omega$$

R_3 and R_4 can be totaled using either the general reciprocal formula or, since there are two resistances in parallel, the product over sum method. Both methods are shown below.

$$R_{3+4} = \frac{1}{\frac{1}{R_3} + \frac{1}{R_4}} = \frac{1}{\frac{1}{1k\Omega} + \frac{1}{1k\Omega}}$$
$$= \frac{1}{\frac{2}{1{,}000\Omega}} = \frac{1}{.002} = 500\Omega$$
$$R_{3+4} = \frac{R_3 \times R_4}{R_3 + R_4} = \frac{1k\Omega \times 1k\Omega}{1k\Omega + 1k\Omega}$$
$$= \frac{1{,}000{,}000\Omega}{2{,}000\Omega} = 500\Omega$$

The equivalent circuit containing the R_{1+2} resistance of 1kΩ and the R_{3+4} resistance of 500Ω is shown in *Figure 12*.

Using the Ohm's law relationship that total current equals voltage divided by circuit resistance, the circuit current can be determined. First, however, total circuit resistance must be found. Since the simplified circuit consists of two resistances in series, they are simply added together to obtain total resistance.

$$R_T = R_{1+2} + R_{3+4}$$
$$R_T = 1k\Omega + 500\Omega$$
$$R_T = 1.5k\Omega$$

Applying this to the current/voltage equation:

$$I_T = \frac{E_T}{R_T}$$
$$I_T = \frac{1.5V}{1.5k\Omega}$$
$$I_T = 1mA \text{ or } 0.001A$$

Now that the total current is known, voltage drops across individual components can be determined:

$$E_{R1} = I_T R_1 = 1mA \times 0.5k\Omega = 0.5V$$
$$E_{R2} = I_T R_2 = 1mA \times 0.5k\Omega = 0.5V$$

Since the total voltage equals the sum of all voltage drops, the voltage drop from A to B can be determined by subtraction:

$$E_T = E_{R1} + E_{R2} + E_{A+B}$$
$$E_T - E_{R1} - E_{R2} = E_{A+B}$$
$$1.5V - 0.5V - 0.5V = E_{A+B} = 0.5V$$

Since R_3 and R_4 are in parallel, some of the total current must pass through each resistor. R_3 and R_4

Classroom

Demonstrate how to use Ohm's law to solve for unknown values in series-parallel circuits.

Audiovisual

Show Transparencies 13 and 14 (Figures 11 and 12).

Classroom

Have the trainees practice series-parallel circuit calculations.

ELECTRICAL THEORY TWO 95

are equal, so the same current should flow through each branch. Using the relationship:

$$I = \frac{E}{R}$$

$$I_{R3} = \frac{E_{R3}}{R_3} \qquad I_{R4} = \frac{E_{R4}}{R_4}$$

$$I_{R3} = \frac{0.5V}{1k\Omega} \qquad I_{R4} = \frac{0.5V}{1k\Omega}$$

$$I_{R3} = 0.5mA \qquad I_{R4} = 0.5mA$$

$$0.5mA + 0.5mA = 1mA$$

Therefore, the total current for the circuit passes through R_1 and R_2 and is evenly divided between R_3 and R_4.

3.0.0 ◆ KIRCHHOFF'S LAWS

Kirchhoff's laws provide a simple, practical method of solving for unknown parameters in a circuit.

3.1.0 Kirchhoff's Current Law

In its most general form, **Kirchhoff's current law,** which is also called *Kirchhoff's first law,* states that at any point in a circuit, the total current entering that point must equal the total current leaving that point. For parallel circuits, this implies that the current in a parallel circuit is equal to the sum of the currents in each branch.

When using Kirchhoff's laws to solve circuits, it is necessary to adopt conventions that determine the algebraic signs for current and voltage terms. A convenient system for current is to consider all current flowing into a branch point as positive, and all current directed away from that point as negative.

As an example, in *Figure 13,* the currents can be written as:

$$I_A + I_B - I_C = 0$$

or

$$5A + 3A - 8A = 0$$

Currents I_A and I_B are positive terms because these currents flow into P, but I_C, directed out of P, is negative.

For a circuit application, refer to Point C at the top of the diagram in *Figure 14.* The 6A I_T into Point C divides into the 2A I_3 and 4A I_{4+5}, both directed out. Note that I_{4+5} is the current through R_4 and R_5. The algebraic equation is:

$$I_T - I_3 - I_{4+5} = 0$$

Substituting the values for each current:

$$6A - 2A - 4A = 0$$

For the opposite direction, refer to Point D at the bottom of *Figure 14.* Here, the branch currents into Point D combine to equal the mainline current I_T returning to the voltage source. Now, I_T is directed out from Point D, with I_3 and I_{4+5} directed in. The algebraic equation is:

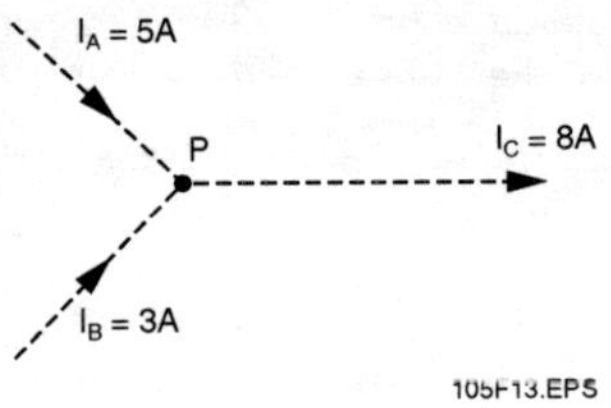

Figure 13 ◆ Kirchhoff's current law.

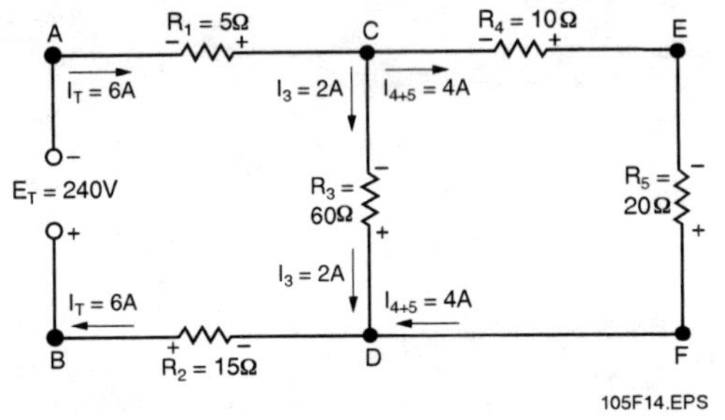

Figure 14 ◆ Application of Kirchhoff's current law.

Classroom

Demonstrate how to use Kirchhoff's current law to solve for unknown circuit values.

Audiovisual

Show Transparencies 15 and 16 (Figures 13 and 14).

Classroom

Have the trainees practice using Kirchhoff's current law to solve for unknown circuit values.

Instructor's Notes:

$$I_3 + I_{4+5} - I_T = 0$$
$$2A + 4A - 6A = 0$$

Note that at either Point C or Point D, the sum of the 2A and 4A branch currents must equal the 6A total line current. Therefore, Kirchhoff's current law can also be stated as:

$$I_{IN} = I_{OUT}$$

For *Figure 14,* the equations for current can be written as shown below.

At Point C:

$$6A = 2A + 4A$$

At Point D:

$$2A + 4A = 6A$$

Kirchhoff's current law is really the basis for the practical rule in parallel circuits that the total line current must equal the sum of the branch currents.

3.2.0 Kirchhoff's Voltage Law

Kirchhoff's voltage law states that the algebraic sum of the voltages around any closed path is zero.

Referring to *Figure 15,* the sum of the voltage drops around the circuit must equal the voltage applied to the circuit:

$$E_A = E_1 + E_2 + E_3$$

Where:

E_A = voltage applied to the circuit

E_1, E_2, and E_3 = voltage drops in the circuit

Another way of stating this law is that the algebraic sum of the voltage rises and voltage drops must be equal to zero. A voltage source is considered a voltage rise; a voltage across a resistor is a voltage drop. (For convenience in labeling, letter subscripts are shown for voltage sources and numerical subscripts are used for voltage drops.) This form of the law can be written by transposing the right members to the left side:

$$\text{Voltage applied} - \text{sum of voltage drops} = 0$$

Substitute letters:

$$E_A - E_1 - E_2 - E_3 = 0$$
$$E_A - (E_1 + E_2 + E_3) = 0$$

3.3.0 Loop Equations

Any closed path is called a *loop.* A loop equation specifies the voltages around the loop. Refer to *Figure 16.*

Consider the inside loop A, C, D, B, A, including the voltage drops E_1, E_3, and E_2, and the source E_T.

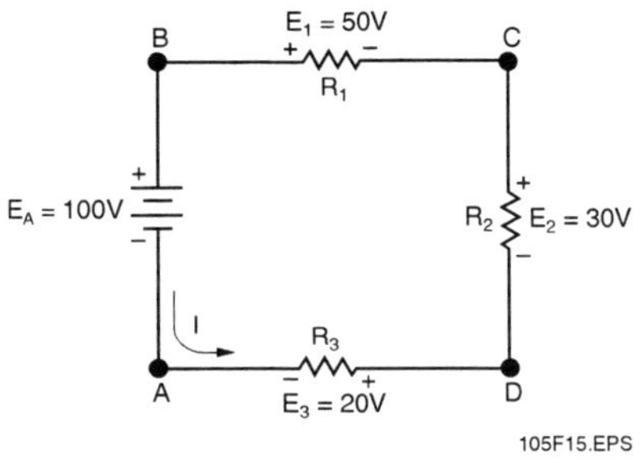

Figure 15 ◈ Kirchhoff's voltage law.

Demonstrate how to use Kirchhoff's voltage law to solve for unknown circuit values.

Show Transparency 17 (Figure 15).

Have the trainees practice using Kirchhoff's voltage law to solve for unknown circuit values.

Discuss loop equations.

Show Transparency 18 (Figure 16).

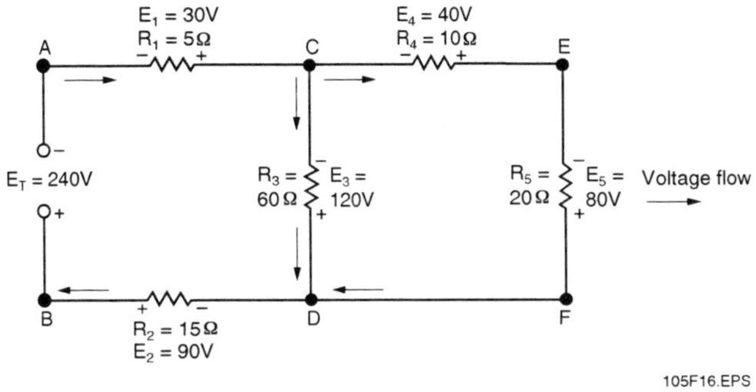

Figure 16 ◈ Loop equation.

ELECTRICAL THEORY TWO 97

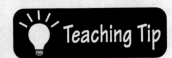

Teaching Tip

Discuss the "Think About It."

Putting It All Together

THINK ABOUT IT

Draw four 60W lamps in parallel with a 120V power source. What is the amperage in the circuit? What would happen to the amperage if we doubled the voltage?

In a clockwise direction, starting at Point A, the algebraic sum of the voltages is:

$$-E_1 - E_3 - E_2 + E_T = 0$$

or

$$-30V - 120V - 90V + 240V = 0$$

Voltages E_1, E_3, and E_2 have a negative value, because there is a decrease in voltage seen across each of the resistors in a clockwise direction. However, the source E_T is a positive term because an increase in voltage is seen in that same direction.

For the opposite direction, going counterclockwise in the same loop from Point A, E_T is negative while E_1, E_2, and E_3 have positive values. Therefore:

$$-E_T + E_2 + E_3 + E_1 = 0$$

or

$$-240V + 90V + 120V + 30V = 0$$

When the negative term is transposed, the equation becomes:

$$240V = 90V + 120V + 30V$$

In this form, the loop equation shows that Kirchhoff's voltage law is really the basis for the practical rule in series circuits that the sum of the voltage drops must equal the applied voltage.

Audiovisual

Show Transparency 19 (Figure 17).

For example, determine the voltage E_B for the circuit shown in *Figure 17*. The direction of the current flow is shown by the arrow. First mark the polarity of the voltage drops across the resistors and trace the circuit in the direction of the current flow starting at Point A. Then write the voltage equation around the circuit:

$$-E_3 - E_B - E_2 - E_1 + E_A = 0$$

Solve for E_B:

$$E_B = E_A - E_3 - E_2 - E_1$$
$$E_B = 15V - 2V - 6V - 3V$$
$$E_B = 4V$$

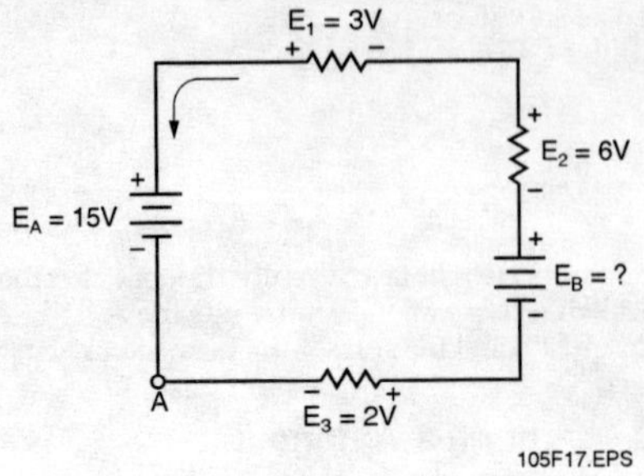

Figure 17 ◆ Applying Kirchhoff's voltage law.

Since E_B was found to be positive, the assumed direction of current is in fact the actual direction of current.

In its most general form, Kirchhoff's voltage law (also called *Kirchhoff's second law*) states that the algebraic sum of all the potential differences in a closed loop is equal to zero. A closed loop means any completely closed path consisting of wire, resistors, batteries, or other components. For series circuits, this implies that the sum of the voltage drops around the circuit is equal to the applied voltage. For parallel circuits, this implies that the voltage drops across all branches are equal.

Summary

The relationships among current, voltage, resistance, and power in Ohm's law are the same for both DC series and DC parallel circuits. Understanding and being able to apply these concepts is necessary for effective circuit analysis and troubleshooting. DC series-parallel circuits also have these fundamental relationships. Since DC series-parallel circuits are a combination of simple series and parallel circuits, Kirchhoff's voltage and cur-

Instructor's Notes:

rent laws will apply. Calculating I, E, R, and P for series-parallel circuits is no more difficult than calculating these values for simple series or parallel circuits. However, for series-parallel circuits, these calculations require more careful circuit analysis in order to use Ohm's law correctly.

Review Questions

1. The formula for calculating the total resistance in a series circuit with three resistors is _____.
   a. $R_T = R_1 + R_2 + R_3$
   b. $R_T = R_1 - R_2 - R_3$
   c. $R_T = R_1 \times R_2 \times R_3$
   d. $R_T = \dfrac{1}{\frac{1}{R_1} + \frac{1}{R_2} + \frac{1}{R_3}}$
2. The formula for calculating the total resistance in a parallel circuit with three resistors is _____.
   a. $R_T = R_1 + R_2 + R_3$
   b. $R_T = R_1 - R_2 - R_3$
   c. $R_T = R_1 \times R_2 \times R_3$
   d. $R_T = \dfrac{1}{\frac{1}{R_1} + \frac{1}{R_2} + \frac{1}{R_3}}$
3. The total resistance in *Figure 18* is _____.
   a. 1,035Ω
   b. 129Ω
   c. 100Ω
   d. 157Ω
4. Find the total resistance in a series circuit with three resistances of 10Ω, 20Ω, and 30Ω.
   a. 15Ω
   b. 1Ω
   c. 20Ω
   d. 60Ω
5. In a parallel circuit, the voltage across each path is equal to the _____.
   a. total circuit resistance times path current
   b. source voltage minus path voltage
   c. path resistance times total current
   d. source voltage
6. The value for total current in *Figure 19* is _____ amps.
   a. 1.25
   b. 2.50
   c. 5
   d. 10
7. A resistor of 32Ω is in parallel with a resistor of 36Ω, and a 54Ω resistor is in series with the pair. When 350V is applied to the combination, the current through the 54Ω resistor is _____ amps.
   a. 2.87
   b. 3.26
   c. 5.86
   d. 4.93

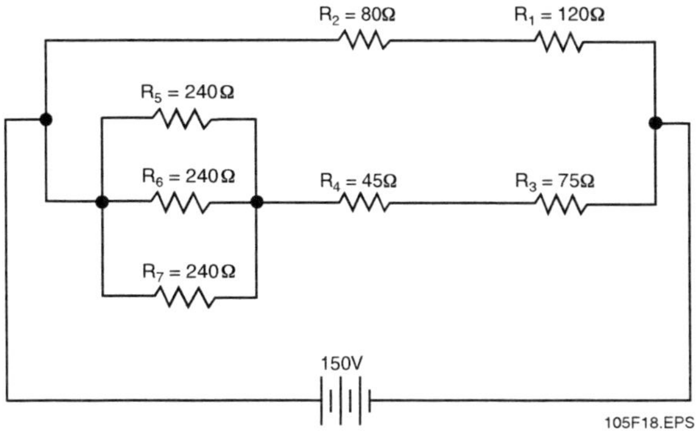

Figure 18 ◈ Series-parallel circuit.

Classroom

Have the trainees complete the Review Questions and go over the answers prior to administering the Module Examination.

Audiovisual

Show Transparencies 20 and 21 (Figures 18 and 19).

Examination

Administer the Module Examination. Record the results on Craft Training Report Form 200 and submit the results to the Training Program Sponsor.

Performance Testing

Administer the Performance Test and fill out Performance Profile Sheets for each trainee. If desired, trainee proficiency noted during laboratory sessions may be used to complete the Performance Test. Be sure to record the results on Craft Training Report Form 200 and submit the results to the Training Program Sponsor.

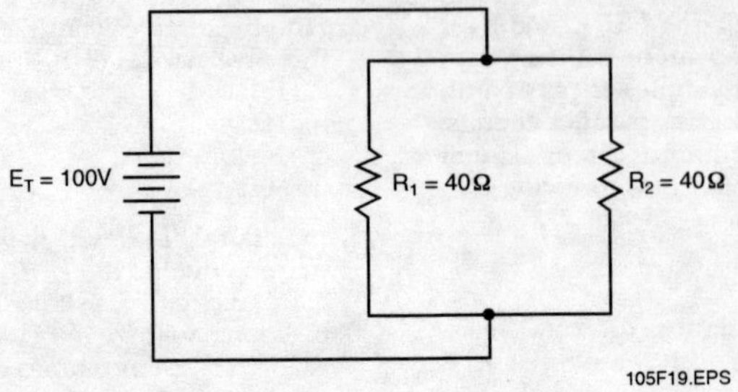

Figure 19 ◆ Parallel circuit.

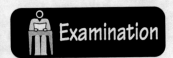

8. A 242Ω resistor is in parallel with a 180Ω resistor, and a 420Ω resistor is in series with the combination. A current of 22mA flows through the 242Ω resistor. The current through the 180Ω resistor is _____ mA.
a. 29.6
b. 40.2
c. 19.8
d. 36.4

9. Two 24Ω resistors are in parallel, and a 42Ω resistor is in series with the combination. When 78V is applied to the three resistors, the voltage drop across the 42Ω resistor is about _____ volts.
a. 55.8
b. 60.5
c. 65.3
d. 49.8

10. Kirchhoff's voltage law states that the algebraic sum of the voltages around any closed path is _____.
a. infinity
b. zero
c. twice the current
d. always less than the individual voltages due to voltage drop

Instructor's Notes:

PROFILE IN SUCCESS

James Mitchem, TIC—The Industrial Company

Jim Mitchem serves as a troubleshooter for a large electrical contractor. During his career in the electrical industry, he worked his way up from apprentice to technical services manager.

How did you become an electrician?
Quite by accident. A couple of years after college, I was working as a relief operator in a plant when the lead electrician retired, creating a vacancy. I liked the idea that electricians were expected to use their knowledge and initiative to keep the place running. I applied and was accepted as a trainee.

How did you get your training?
I took an electrical apprenticeship course by correspondence, and I was fortunate enough to work with good people who helped me along. I worked in an environment that exposed me to a variety of equipment and applications, and just about everyone I've ever worked with has taught me something. Now I'm passing my knowledge on to others.

What kinds of work have you done in your career?
I've worked as an apprentice, journeyman, instrument and controls technician, instrument fitter, foreman, general foreman, superintendent, and startup engineer. Each of these positions required that I learn new skills, both technical and managerial. My experience in many disciplines and types of projects has given me a high level of credibility with my employer and our clients.

Now I act as a technical resource and troubleshooter for job sites and in-house functions such as safety, quality assurance, and training. I visit job sites to help solve problems and help out with commissioning and startup.

What factor or factors have contributed the most to your success?
There are several factors. Two very important ones have been a desire to learn and a willingness to do whatever is asked of me. I also keep an eye on the big picture. When I'm on a job, I'm not just pulling wire, I'm building a power plant or whatever the project is. I also think it has helped me to remain with the same employer for 18 years.

Any advice for apprentices just beginning their careers?
Keep learning! And don't depend on others to train you. Take the initiative to buy or borrow books and trade journals. Take licensing tests and do whatever is necessary to keep your licenses current. Finally, make sure you know your own personal and professional values and work with a company that shares those values.

ELECTRICAL THEORY TWO 101

GLOSSARY

Trade Terms Introduced in This Module

Kirchhoff's current law (KCL): The statement that the total amount of current flowing through a parallel circuit is equal to the sum of the amounts of current flowing through each current path.

Kirchhoff's voltage law (KVL): The statement that the sum of all the voltage drops in a circuit is equal to the source voltage of the circuit.

Parallel circuits: Circuits containing two or more parallel paths through which current can flow.

Series circuits: Circuits with only one path for current flow.

Series-parallel circuits: Circuits that contain both series and parallel current paths.

Instructor's Notes:

Additional Resources

This module is intended to present thorough resources for task training. The following reference works are suggested for further study. These are optional materials for continued education rather than for task training.

Electronics Fundamentals, Latest Edition. New York: Prentice Hall.
Principles of Electric Circuits, Latest Edition. New York: Prentice Hall.

ANSWER KEY

Answers to Review Questions

Answer	Section
1. a	2.1.0
2. d	2.2.0
3. c	2.3.1
4. d	2.1.0
5. d	2.4.2
6. c	2.4.2
7. d	2.4.3
8. a	2.4.3
9. b	3.3.0
10. b	3.2.0

Section 2.2.0 *Think About It—Series Circuits*

Receptacles wired in series would have the output terminal of one receptacle wired to the input terminal of the next receptacle, and so on. Assuming the source voltage remains the same, the amount of light produced by each lamp in a series circuit depends on the total number of lamps installed in the circuit. This is because each lamp added to the circuit increases the total circuit resistance. A higher circuit resistance decreases the total current flow through the series circuit and therefore the current through each lamp, causing the amount of light produced by each lamp to decrease.

Section 2.2.1 *Think About It—Parallel Circuits*

Adding additional lamps in parallel increases the total current in a parallel circuit. The total number of devices or lamps that can be connected in parallel is limited by the ampacity of the conductors and the rating of the overcurrent device (fuse or circuit breaker) protecting the circuit. For example, if each lamp has a resistance of 240Ω, you can connect a maximum of 30 lamps in parallel in a 120V, 15A circuit. Thirty 240Ω lamps connected in parallel have a total resistance of 8Ω (240V ÷ 30). The total current in this circuit is equal to 15A (120V ÷ 8Ω).

Section 2.3.1 *Think About It—Series-Parallel Circuits*

Resistors R_4 and R_5 are in series with each other and in parallel with R_2. This combination (R_4, R_5, and R_2) is in series with R_1 and R_3.

Section 2.4.2 *Think About It—Voltage Drops*

Sizing a component without accounting for a substantial voltage drop across another component in a series circuit can result in specifying a component with too much resistance and/or the capability of handling a higher voltage than necessary. In a parallel circuit, the voltage drop across all branches is equal so there is less chance of error.

Section 3.3.0 *Think About It—Putting It All Together*

A 60W lamp across 120V draws 0.5A of current (I = 60W ÷ 120V). With four 60W lamps connected in parallel across a 120V source, the total circuit current is 2A. If the source voltage is raised from 120V to 240V, the total circuit current would decrease to 1A (0.25A per lamp).

CONTREN™ LEARNING SERIES — USER UPDATES

The NCCER makes every effort to keep these textbooks up-to-date and free of technical errors. We appreciate your help in this process. If you have an idea for improving this textbook, or if you find an error, a typographical mistake, or an inaccuracy in NCCER's Contren™ textbooks, please write us, using this form or a photocopy. Be sure to include the exact module number, page number, a detailed description, and the correction, if applicable. Your input will be brought to the attention of the Technical Review Committee. Thank you for your assistance.

Instructors – If you found that additional materials were necessary in order to teach this module effectively, please let us know so that we may include them in the Equipment/Materials list in the Instructor's Guide.

Write: Curriculum Revision and Development Department
National Center for Construction Education and Research
P.O. Box 141104, Gainesville, FL 32614-1104

Fax: 352-334-0932

E-mail: curriculum@nccer.org

Craft ____________ Module Name ____________

Copyright Date ____________ Module Number ____________ Page Number(s) ____________

Description

____

____

____

____

(Optional) Correction

____

____

____

(Optional) Your Name and Address

____

____

____

Chapter 4

Alternating Current

26201-03

MODULE OVERVIEW

This course introduces the electrical trainee to the principles of alternating current.

PREREQUISITES

Please refer to the Course Map in the Trainee Module. Prior to training with this module, it is recommended that the trainee shall have successfully completed the following modules:

Core Curriculum; Residential Electrical I, Chapters 1 through 3

OBJECTIVES

Upon completion of this module, the trainee will be able to:

1. Calculate the peak and effective voltage or current values for an AC waveform.
2. Calculate the phase relationship between two AC waveforms.
3. Describe the voltage and current phase relationship in a resistive AC circuit.
4. Describe the voltage and current transients that occur in an inductive circuit.
5. Define *inductive reactance* and state how it is affected by frequency.
6. Describe the voltage and current transients that occur in a capacitive circuit.
7. Define *capacitive reactance* and state how it is affected by frequency.
8. Explain the relationship between voltage and current in the following types of AC circuits:
   - RL circuit
   - RC circuit
   - LC circuit
   - RLC circuit
9. Describe the effect that resonant frequency has on impedance and current flow in a series or parallel resonant circuit.
10. Define *bandwidth* and describe how it is affected by resistance in a series or parallel resonant circuit.
11. Explain the following terms as they relate to AC circuits:
    - True power
    - Apparent power
    - Reactive power
    - Power factor
12. Explain basic transformer action.

PERFORMANCE TASKS

This is a knowledge-based module; there are no performance tasks.

NCCER STANDARDIZED CRAFT TRAINING PROGRAM

The National Center for Construction Education and Research (NCCER) provides a standardized national program of accredited craft training. Key features of the program include instructor certification, competency-based training, and performance testing. The program provides trainees, instructors, and companies with a standard form of recognition through a National Craft Training Registry. The program is described in full in the *Guidelines for Accreditation*, published by the NCCER. For more information on standardized craft training, contact the NCCER by writing us at P.O. Box 141104, Gainesville, FL 32614-1104, calling 352-334-0911, or emailing info@nccer.org. More information can be found at our Web site, www.nccer.org.

HOW TO USE THIS ANNOTATED INSTRUCTOR'S GUIDE

Each page presents two sections of information. The larger section displays each page exactly as it appears in the Trainee Module. The narrow column ties suggested trainee and instructor actions to each page and provides icons to call your attention to material, safety, audiovisual, or testing requirements. The bottom of each page includes space for your notes.

If you see the Teaching Tip icon, that means there is a teaching tip associated with this section. Also refer to any suggested teaching tips at the end of the module.

SAFETY CONSIDERATIONS

Ensure that the trainees are equipped with appropriate personal protective equipment. Stress the importance of using the proper safety equipment, precautions, and procedures when working with alternating current.

PREPARATION

Before teaching this module, you should review the Module Outline, Objectives, Performance Tasks, and the Materials and Equipment List. Be sure to allow ample time to prepare your own training or lesson plan and gather all required equipment and materials.

MATERIALS AND EQUIPMENT LIST

Materials:

Transparencies

Markers/chalk

Copy of the latest edition of the *National Electrical Code*®

Module Examinations*

Equipment:

Overhead projector and screen

Whiteboard/chalkboard

Appropriate personal protective equipment

Scientific calculator or trigonometric tables

Examples of capacitors

*Located in the Test Booklet packaged with this Annotated Instructor's Guide.

ADDITIONAL RESOURCES

This module is intended to present thorough resources for task training. The following reference works are suggested for both instructors and motivated trainees interested in further study. These are optional materials for continued education rather than for task training.

Introduction to Electric Circuits, Latest Edition. New York, NY: Prentice Hall.

Principles of Electric Circuits, Latest Edition. New York, NY: Prentice Hall.

NOTES

The designations "National Electrical Code," "NE Code," and "NEC," where used in this document, refer to the *National Electrical Code*®, which is a registered trademark of the National Fire Protection Association, Quincy, MA. All National Electrical Code (NEC) references in this module refer to the 2002 edition of the NEC.

If you feel that additional math instruction would be helpful, Prentice Hall offers a basic math textbook entitled *Fundamentals of Electrical and Mechanical Mathematics*. It covers the basic math requirements for electrical trainees and may be ordered by contacting Prentice Hall Customer Service at 1-800-922-0579.

TEACHING TIME FOR THIS MODULE

An outline for use in developing your lesson plan is presented below. Note that each Roman numeral in the outline equates to one session of instruction. Each session has a suggested time period of 2½ hours. This includes 10 minutes at the beginning of each session for administrative tasks and one 10-minute break during the session. Approximately 15 hours are suggested to cover *Alternating Current*. You will need to adjust the time required for hands-on activity and testing based on your class size and resources.

Topic	Planned Time
Session I. Introduction; Sine Wave Generation; Sine Wave Terminology	
A. Introduction	______
B. Sine Wave Generation	______
C. Sine Wave Terminology	______
1. Frequency	______
a. Period	______
2. Wavelength	______
3. Peak Value	______
4. Average Value	______
5. Root-Mean-Square or Effective Value	______
Session II. AC Phase Relationships; Nonsinusoidal Waveforms	
A. AC Phase Relationships	______
1. Phase Angle	______
2. Phase Angle Diagrams	______
B. Nonsinusoidal Waveforms	______
Session III. Resistance in AC Circuits; Inductance in AC Circuits	
A. Resistance in AC Circuits	______
B. Inductance in AC Circuits	______
1. Factors Affecting Inductance	______
2. Voltage and Current in an Inductive AC Circuit	______
3. Inductive Reactance	______
Session IV. Capacitance	
A. Capacitance	______
1. Factors Affecting Capacitance	______
2. Calculating Equivalent Capacitance	______
3. Capacitor Specifications	______
a. Voltage Rating	______
b. Leak Resistance	______
4. Voltage and Current in a Capacitive AC Circuit	______
5. Capacitive Reactance	______
Session V. LC and RLC Circuits; Power in AC Circuits	
A. LC and RLC Circuits	______
1. RL Circuits	______
a. Series RL Circuit	______
b. Parallel RL Circuit	______
2. RC Circuits	______
a. Series RC Circuit	______
b. Parallel RC Circuit	______

3. LC Circuits ______
   a. Series LC Circuit ______
   b. Parallel LC Circuit ______
4. RLC Circuits ______
   a. Series RLC Circuit ______
   b. Series Resonance ______
   c. Parallel RLC Circuit ______
   d. Parallel Resonance ______

B. Power in AC Circuits ______
1. True Power ______
2. Apparent Power ______
3. Reactive Power ______
4. Power Factor ______
5. Power Triangle ______

Session VI. Transformers; Module Examination

A. Transformers ______
1. Transformer Construction ______
   a. Core Characteristics ______
   b. Transformer Windings ______
2. Operating Characteristics ______
   a. Energized with No Load ______
   b. Phase Relationship ______
3. Turns and Voltage Ratios ______
4. Types of Transformers ______
   a. Isolation Transformer ______
   b. Autotransformer ______
   c. Current Transformer ______
   d. Potential Transformer ______

B. Review ______

C. Module Examination ______
1. Trainees must score 70% or higher to receive recognition from the NCCER.
2. Record the testing results on Craft Training Report Form 200 and submit the results to the Training Program Sponsor.

Chapter 4

Module 26201-03

Alternating Current

Instructor's Notes:

Course Map

This course map shows all of the modules in the first level of the Residential Electrical curriculum. The suggested training order begins at the bottom and proceeds up. Skill levels increase as you advance on the course map. The local Training Program Sponsor may adjust the training order.

RESIDENTIAL ELECTRICAL I

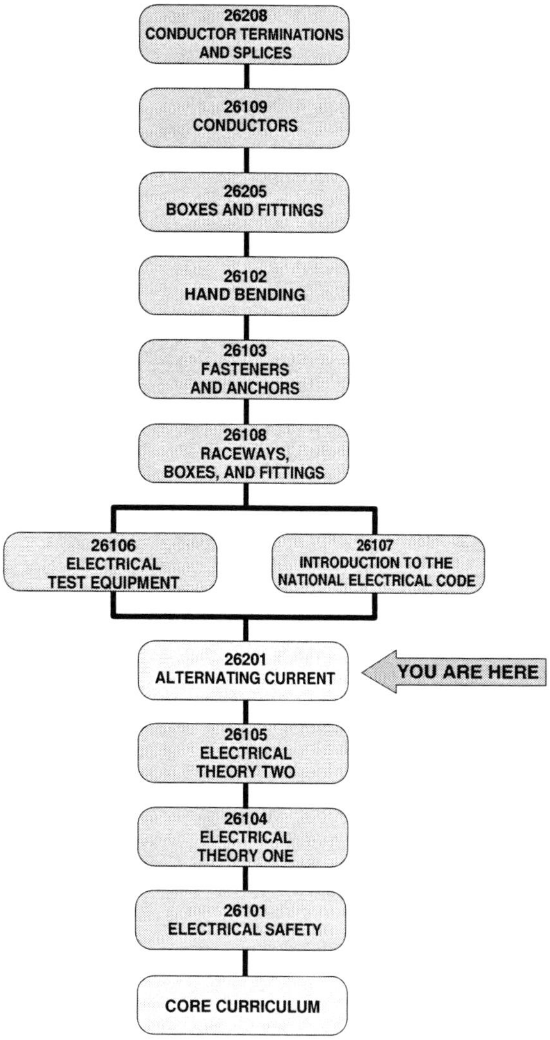

Homework

Assign reading of Module 26201.

MODULE 26201 CONTENTS

Instructor's Notes:

Figures

ALTERNATING CURRENT 113

Tables

Instructor's Notes:

MODULE 26201

Alternating Current

Objectives

When you have completed this module, you will be able to do the following:

1. Calculate the peak and effective voltage or current values for an AC waveform.
2. Calculate the phase relationship between two AC waveforms.
3. Describe the voltage and current phase relationship in a resistive AC circuit.
4. Describe the voltage and current transients that occur in an inductive circuit.
5. Define *inductive reactance* and state how it is affected by frequency.
6. Describe the voltage and current transients that occur in a capacitive circuit.
7. Define *capacitive reactance* and state how it is affected by frequency.
8. Explain the relationship between voltage and current in the following types of AC circuits:
   - RL circuit
   - RC circuit
   - LC circuit
   - RLC circuit
9. Describe the effect that resonant frequency has on impedance and current flow in a series or parallel resonant circuit.
10. Define *bandwidth* and describe how it is affected by resistance in a series or parallel resonant circuit.
11. Explain the following terms as they relate to AC circuits:
    - True power
    - Apparent power
    - Reactive power
    - Power factor
12. Explain basic transformer action.

Prerequisites

Before you begin this module, it is recommended that you successfully complete the following: Core Curriculum; Residential Electrical I, Chapters 1 through 3.

Required Trainee Materials

1. Pencil and paper
2. Appropriate personal protective equipment
3. Copy of the latest edition of the *National Electrical Code®*

1.0.0 ◆ INTRODUCTION

Alternating current (AC) and its associated voltage reverses between positive and negative polarities and varies in amplitude with time. One complete waveform or cycle includes a complete set of variations, with two alternations in polarity. Many sources of voltage change direction with time and produce a resultant waveform. The most common AC waveform is the *sine wave*.

Note: The designations "National Electrical Code," "NE Code," and "NEC," where used in this document, refer to the *National Electrical Code®*, which is a registered trademark of the National Fire Protection Association, Quincy, MA. *All National Electrical Code (NEC) references in this module refer to the 2002 edition of the NEC.*

Materials

Ensure that you have everything required to teach the course. Check the Materials and Equipment List at the front of this Instructor's Guide.

Classroom

Review the goals of the course.

Explain that the terms shown in bold (blue) are defined in the Glossary at the back of this module.

Audiovisual

Show Transparency 1, Course Objectives.

Classroom

Discuss alternating current and polarity.

Classroom

Review the basic concepts of electromagnetism.

Audiovisual

Show Transparency 2 (Figure 1).

INSIDE TRACK

Why Do Power Companies Generate and Distribute AC Power Instead of DC Power?

The transformer is the key. Power plants generate and distribute AC power because it permits the use of transformers, which makes power delivery more economical. Transformers used at generation plants step the AC voltage up, which decreases the current. Decreased current allows smaller-sized wires to be used for the power transmission lines. Smaller wire is less expensive and easier to support over the long distances that the power must travel from the generation plant to remotely located substations. At the substations, transformers are again used to step AC voltages back down to a level suitable for distribution to homes and businesses.

There is no such thing as a DC transformer. This means DC power would have to be transmitted at low voltages and high currents over very large-sized wires, making the process very uneconomical. When DC is required for special applications, the AC voltage may be converted to DC voltage by using rectifiers, which make the change electrically, or by using AC motor-DC generator sets, which make the change mechanically.

2.0.0 ◆ SINE WAVE GENERATION

To understand how the alternating current sine wave is generated, some of the basic principles learned in magnetism should be reviewed. Two principles form the basis of all electromagnetic phenomena:

- An electric current in a conductor creates a magnetic field that surrounds the conductor.
- Relative motion between a conductor and a magnetic field, when at least one component of that relative motion is in a direction that is perpendicular to the direction of the field, creates a voltage in the conductor.

Figure 1 shows how these principles are applied to generate an AC waveform in a simple one-loop rotary generator. The conductor loop rotates through the magnetic field to generate the induced AC voltage across its open terminals. The magnetic flux shown here is vertical.

There are several factors affecting the magnitude of voltage developed by a conductor through a magnetic field. They are the strength of the magnetic field, the length of the conductor, and the rate at which the conductor cuts directly across or perpendicular to the magnetic field.

Assuming that the strength of the magnetic field and the length of the conductor making the

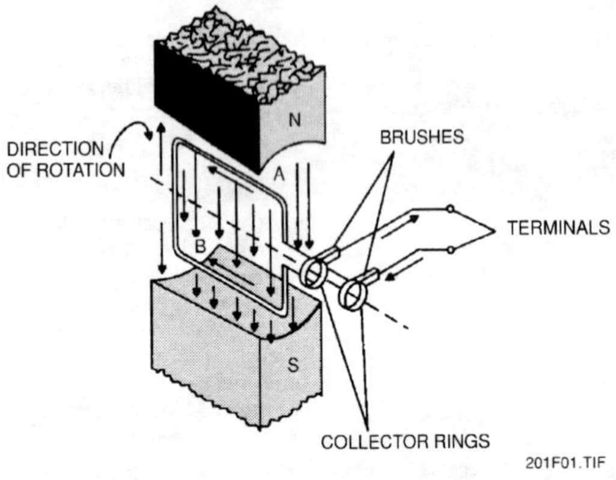

Figure 1 ◆ Conductor moving across a magnetic field.

Instructor's Notes:

loop are both constant, the voltage produced will vary depending on the rate at which the loop cuts directly across the magnetic field.

The rate at which the conductor cuts the magnetic field depends on two things: the speed of the generator in revolutions per minute (rpm) and the angle at which the conductor is traveling through the field. If the generator is operated at a constant rpm, the voltage produced at any moment will depend on the angle at which the conductor is cutting the field at that instant.

In *Figure 2*, the magnetic field is shown as parallel lines called *lines of flux*. These lines always go from the north to south poles in a generator. The motion of the conductor is shown by the large arrow.

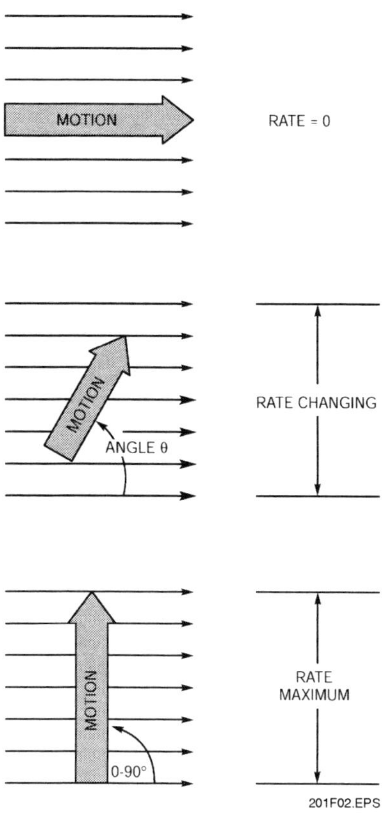

Figure 2 ◆ Angle versus rate of cutting lines of flux.

Assuming the speed of the conductor is constant, as the angle between the flux and the conductor motion increases, the number of flux lines cut in a given time (the rate) increases. When the conductor is moving parallel to the lines of flux (angle of 0°), it is not cutting any of them, and the voltage will be zero.

The angle between the lines of flux and the motion of the conductor is called θ (theta). The magnitude of the voltage produced will be proportional to the sine of the angle. Sine is a trigonometric function. Each angle has a sine value that never changes.

The sine of 0° is 0. It increases to a maximum of 1 at 90°. From 90° to 180°, the sine decreases back to 0. From 180° to 270°, the sine decreases to −1. Then from 270° to 360° (back to 0°), the sine increases to its original 0.

Because voltage is proportional to the sine of the angle, as the loop goes 360° around the circle the voltage will increase from 0 to its maximum at 90°, back to 0 at 180°, down to its maximum negative value at 270°, and back up to 0 at 360°, as shown in *Figure 3.*

Notice that at 180° the polarity reverses. This is because the conductor has turned completely around and is now cutting the lines of flux in the opposite direction. This can be shown using the left-hand rule for generators. The curve shown in *Figure 3* is called a *sine wave* because its shape is generated by the trigonometric function sine. The value of voltage at any point along the sine wave can be calculated if the angle and the maximum obtainable voltage (E_{max}) are known.

The formula used is:

$$E = E_{max} \sin \theta$$

Where:

E = voltage induced
E_{max} = maximum induced voltage
θ = angle at which the voltage is induced

Using the above formula, the values of voltage anywhere along the sine wave in *Figure 3* can be calculated. Sine values can be found using either a scientific calculator or trigonometric tables. With an E_{max} of 10 volts (V), the following values are calculated as examples:

$\theta = 0°$, sine = 0.0	$\theta = 45°$, sine = 0.707
$E = E_{max}\sin\theta$	$E = E_{max}\sin\theta$
$E = (10V)(0)$	$E = (10V)(0.707)$
$E = 0V$	$E = 7.07V$

Classroom

Explain the relationship between conductor motion and voltage production in a generator.

Audiovisual

Show Transparencies 3 and 4 (Figures 2 and 3).

Classroom

Demonstrate how to calculate induced voltage using the sine function.

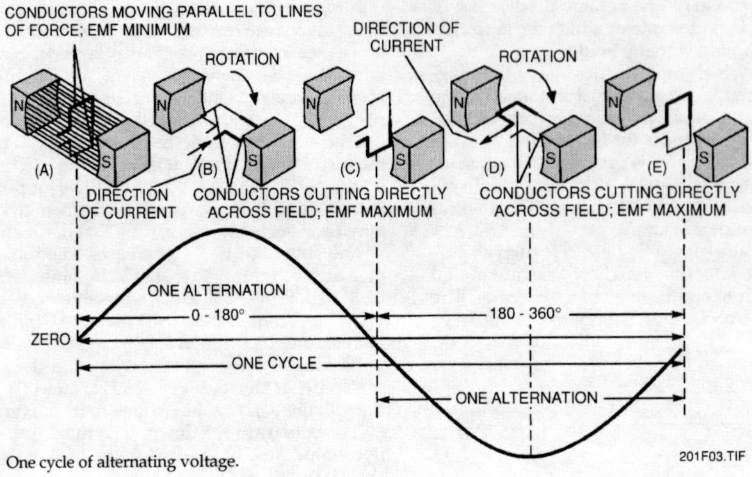

Figure 3 ◆ One cycle of alternating voltage.

$\theta = 90°$, sine = 1.0
$E = E_{max}\text{sine } \theta$
$E = (10V)(1.0)$
$E = 10V$

$\theta = 135°$, sine = 0.707
$E = E_{max}\text{sine } \theta$
$E = (10V)(0.707)$
$E = 7.07V$

$\theta = 180°$, sine = 0
$E = E_{max}\text{sine } \theta$
$E = (10V)(0)$
$E = 0V$

$\theta = 225°$, sine = −0.707
$E = E_{max}\text{sine } \theta$
$E = (10V)(-0.707)$
$E = -7.07V$

$\theta = 270°$, sine = −1.0
$E = E_{max}\text{sine } \theta$
$E = (10V)(-1.0)$
$E = -10V$

$\theta = 315°$, sine = −0.707
$E = E_{max}\text{sine } \theta$
$E = (10V)(-0.707)$
$E = -7.07V$

Classroom

Explain how to calculate frequency and period.

3.0.0 ◆ SINE WAVE TERMINOLOGY

3.1.0 Frequency

The **frequency** of a waveform is the number of times per second an identical pattern repeats itself. Each time the waveform changes from zero to a peak value and back to zero is called an *alternation.* Two alternations form one cycle. The number of cycles per second is the frequency. The unit of frequency is **hertz (Hz).** One hertz equals one cycle per second (cps).

Audiovisual

Show Transparency 5 (Figure 4).

For example, let us determine the frequency of the waveform shown in *Figure 4.*

In one-half second, the basic sine wave is repeated five times. Therefore, the frequency (f) is:

$$f = \frac{5 \text{ cycles}}{0.5 \text{ second}} = 10 \text{ cycles per second (Hz)}$$

3.1.1 Period

The period of a waveform is the time (t) required to complete one cycle. The period is the inverse of frequency:

$$t = \frac{1}{f}$$

Where:

t = period (seconds)
f = frequency (Hz or cps)

For example, let us determine the period of the waveform in *Figure 4.* If there are five cycles in one-half second, then the frequency for one cycle is 10 cps (0.5 ÷ 5 = 10). Therefore, the period is:

$$t = \frac{1}{\text{cps}}$$

$$t = \frac{1}{10} = 0.1 \text{ second}$$

3.2.0 Wavelength

The wavelength or λ (lambda) is the distance traveled by a waveform during one period. Since electricity travels at the speed of light (186,000 miles/second or 300,000,000 meters/second), the wave-

Instructor's Notes:

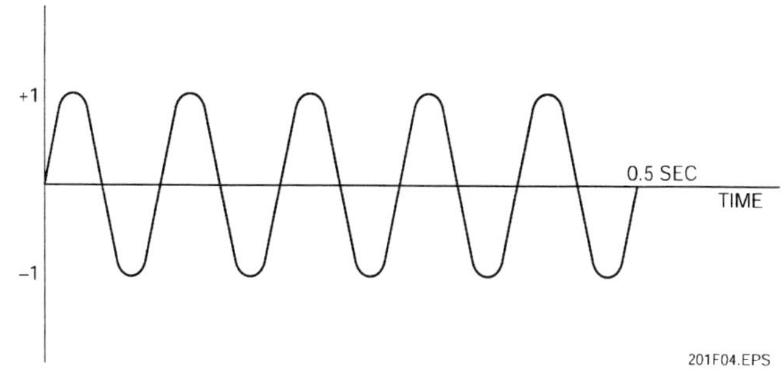

Figure 4 ◈ Frequency measurement.

> ### *Frequency*
> THINK ABOUT IT
>
> The frequency of the utility power generated in the United States is normally 60Hz. In some European countries and elsewhere, utility power is often generated at a frequency of 50Hz. Which of these frequencies (60Hz or 50Hz) has the shortest period?

length of electrical waveforms equals the product of the period and the speed of light (c):

$$\lambda = tc$$

or:

$$\lambda = \frac{c}{f}$$

Where:

λ = wavelength (meters)
t = period (seconds)
c = speed of light (meters/second)
f = frequency (Hz or cps)

3.3.0 Peak Value

The peak value is the maximum value of voltage (V_M) or current (I_M). For example, specifying that a sine wave has a **peak voltage** of 170V applies to either the positive or the negative peak. To include both peak amplitudes, the peak-to-peak (p–p) value may be specified. In the above example, the peak-to-peak value is 340V, double the peak value of 170V, because the positive and negative peaks are symmetrical. However, the two opposite peak values cannot occur at the same time. Furthermore, in some waveforms the two peaks are not equal. The positive peak value and peak-to-peak value of a sine wave are shown in *Figure 5.*

3.4.0 Average Value

The average value is calculated from all the values in a sine wave for one alternation or half cycle. The half cycle is used for the average because over a full cycle the average value is zero, which is useless for comparison purposes. If the sine values for all angles up to 180° in one alternation are added and then divided by the number of values, this average equals 0.637.

Since the peak value of the sine is 1 and the average equals 0.637, the average value can be calculated as follows:

$$\text{Average value} = 0.637 \times \text{peak value}$$

For example, with a peak of 170V, the average value is 0.637 × 170V, which equals approximately 108V. *Figure 5* shows where the average value would fall on a sine wave.

3.5.0 Root-Mean-Square or Effective Value

Meters used in AC circuits indicate a value called the *effective value.* The effective value is the value

Classroom

Discuss the "Think About It." 60Hz has the shortest period (1 ÷ 60 = .017 second). 50Hz has a period of .020 second (1 ÷ 50).

Explain how to calculate the wavelength of a waveform.

Define various voltage values:
- **Peak**
- **Average**
- **Root-mean-square (rms)**

Audiovisual

Show Transparency 6 (Figure 5).

Demonstrate the left-hand rule for generators.

INSIDE TRACK

Left-Hand Rule for Generators

Hand rules for generators and motors give direction to the basic principles of induction. For a generator, if you move a conductor through a magnetic field made up of flux lines, you will induce an EMF, which drives current through a conductor. The left-hand rule for generators will help you determine which direction the current will flow in the conductor. It states that if you hold the thumb, first, and middle fingers of the left hand at right angles to one another with the first finger pointing in the flux direction (from the north pole to the south pole), and the thumb pointing in the direction of motion of the conductor, the middle finger will point in the direction of the induced voltage (EMF). The polarity of the EMF determines the direction in which current will flow as a result of this induced EMF. The left-hand rule for generators is also called *Fleming's first rule.*

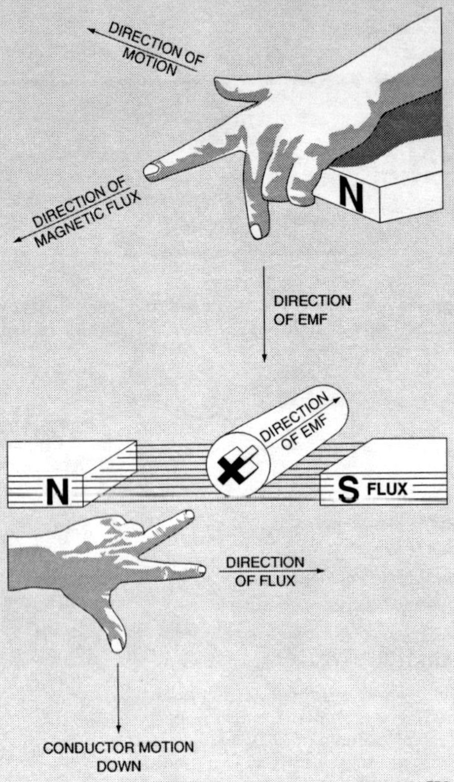

201UA0101.EPS

Instructor's Notes:

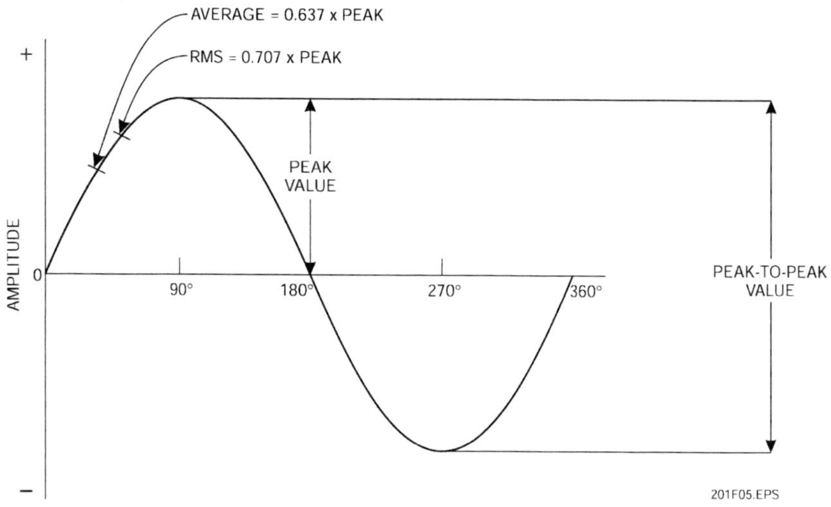

Figure 5 ◈ Amplitude values for a sine wave.

of the AC current or voltage wave that indicates the same energy transfer as an equivalent direct current (DC) or voltage.

The direct comparison between DC and AC is in the heating effect of the two currents. Heat produced by current is a function of current amplitude only and is independent of current direction. Thus, heat is produced by both alternations of the AC wave, although the current changes direction during each alternation.

In a DC circuit, the current maintains a steady amplitude. Therefore, the heat produced is steady and is equal to I^2R. In an AC circuit, the current is continuously changing; periodically high, periodically low, and periodically zero. To produce the same amount of heat from AC as from an equivalent amount of DC, the instantaneous value of the AC must at times exceed the DC value.

By averaging the heating effects of all the instantaneous values during one cycle of alternating current, it is possible to find the average heat produced by the AC current during the cycle. The amount of DC required to produce that heat will be equal to the effective value of the AC.

The most common method of specifying the amount of a sine wave of voltage or current is by stating its value at 45°, which is 70.7% of the peak. This is its **root-mean-square (rms)** value. Therefore:

$$\text{Value of rms} = 0.707 \times \text{peak value}$$

For example, with a peak of 170V, the rms value is 0.707 × 170, or approximately 120V. This is the voltage of the commercial AC power line, which is always given in rms value.

4.0.0 ◈ AC PHASE RELATIONSHIPS

In AC systems, phase is involved in two ways: the location of a point on a voltage or current wave with respect to the starting point of the wave or with respect to some corresponding point on the same wave. In the case of two waves of the same frequency, it is the time at which an event of one takes place with respect to a similar event of the other.

Often, the event is the starting of the waves at zero or the points at which the waves reach their maximum values. When two waves are compared in this manner, there is a phase lead or lag of one with respect to the other unless they are alternating in unison, in which case they are said to be in phase.

4.1.0 Phase Angle

Suppose that a generator started its cycle at 90° where maximum voltage output is produced instead of starting at the point of zero output. The two output voltage waves are shown in *Figure 6*.

Classroom

Explain how to calculate the rms value.

Discuss phase relationships and phase angles.

ALTERNATING CURRENT 121

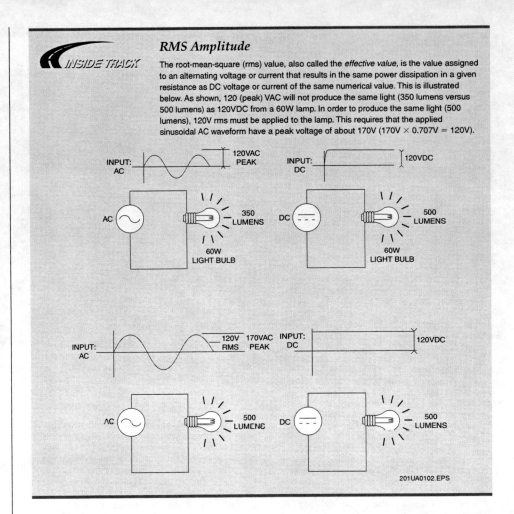

RMS Amplitude

INSIDE TRACK

The root-mean-square (rms) value, also called the *effective value*, is the value assigned to an alternating voltage or current that results in the same power dissipation in a given resistance as DC voltage or current of the same numerical value. This is illustrated below. As shown, 120 (peak) VAC will not produce the same light (350 lumens versus 500 lumens) as 120VDC from a 60W lamp. In order to produce the same light (500 lumens), 120V rms must be applied to the lamp. This requires that the applied sinusoidal AC waveform have a peak voltage of about 170V (170V × 0.707V = 120V).

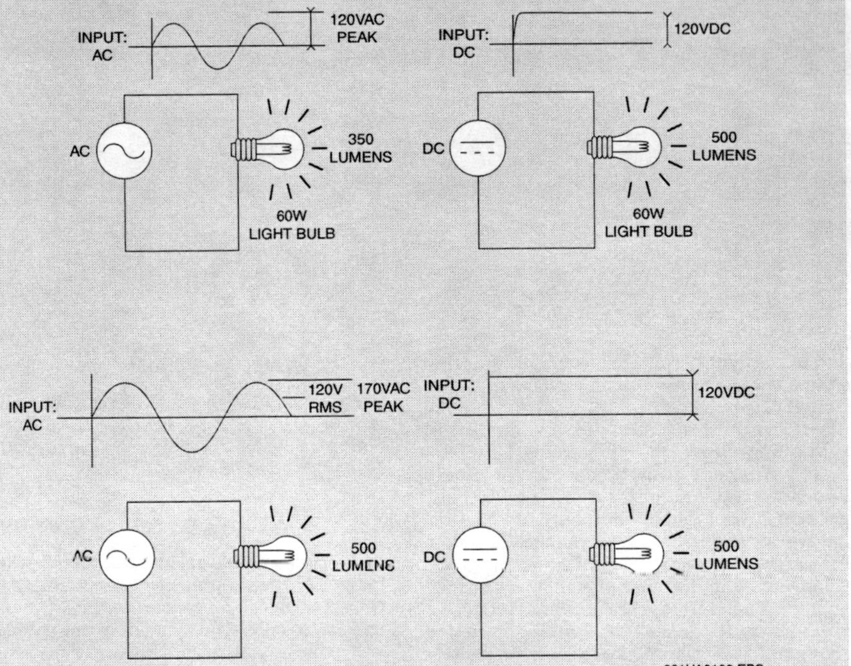

201UA0102.EPS

Audiovisual

Show Transparency 7 (Figure 6).

Each is the same waveform of alternating voltage, but wave B starts at the maximum value while wave A starts at zero. The complete cycle of wave B through 360° takes it back to the maximum value from which it started.

Wave A starts and finishes its cycle at zero. With respect to time, wave B is ahead of wave A in its values of generated voltage. The amount it leads in time equals one quarter revolution, which is 90°. This angular difference is the phase angle between waves B and A. Wave B leads wave A by the phase angle of 90°.

The 90° phase angle between waves B and A is maintained throughout the complete cycle and in all successive cycles as long as they both have the same frequency. At any instant in time, wave B has the value that A will have 90° later. For instance, at 180°, wave A is at zero, but B is already at its negative maximum value, the point where wave A will be later at 270°.

Instructor's Notes:

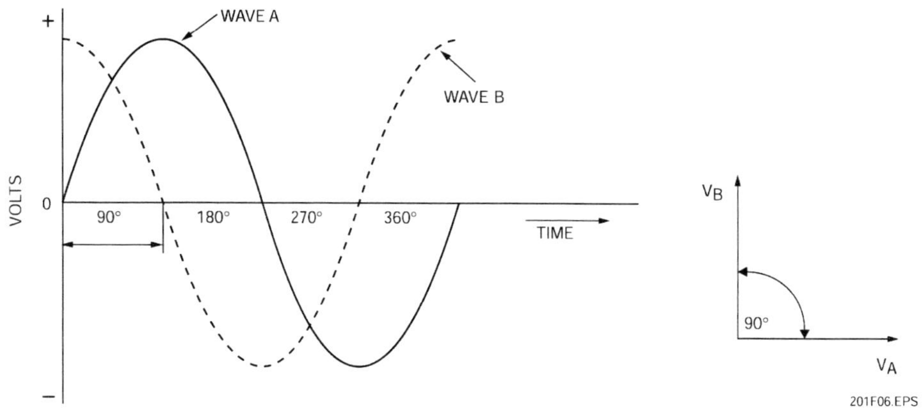

Figure 6 ◆ Voltage waveforms 90° out of phase.

To compare the phase angle between two waves, both waves must have the same frequency. Otherwise, the relative phase keeps changing. Both waves must also have sine wave variations, because this is the only kind of waveform that is measured in angular units of time. The amplitudes can be different for the two waves. The phases of two voltages, two currents, or a current with a voltage can be compared.

4.2.0 Phase Angle Diagrams

To compare AC phases, it is much more convenient to use vector diagrams corresponding to the voltage and current waveforms, as shown in *Figure 6*. V_A and V_B represent the vector quantities corresponding to the generator voltage.

A vector is a quantity that has magnitude and direction. The length of the arrow indicates the magnitude of the alternating voltage in rms, peak, or any AC value as long as the same measure is used for all the vectors. The angle of the arrow with respect to the horizontal axis indicates the phase angle.

In *Figure 6,* the vector V_A represents the voltage wave A, with a phase angle of 0°. This angle can be considered as the plane of the loop in the rotary generator where it starts with zero output voltage. The vector V_B is vertical to show the phase angle of 90° for this voltage wave, corresponding to the vertical generator loop at the start of its cycle. The angle between the two vectors is the phase angle.

The symbol for a phase angle is θ (theta). In *Figure 7,* $\theta = 0°$. *Figure 7* shows the waveforms and phasor diagram of two waves that are in phase but have different amplitudes.

5.0.0 ◆ NONSINUSOIDAL WAVEFORMS

The sine wave is the basic waveform for AC variations for several reasons. This waveform is produced by a rotary generator, as the output is proportional to the angle of rotation. Because of its derivation from circular motion, any sine wave can be analyzed in angular measure, either in degrees from 0° to 360° or in **radians** from 0 to 2π radians.

THINK ABOUT IT

Phase Angles

Why is the phase angle 90° in *Figure 6* and 0° in *Figure 7?* Why is the vector diagram in *Figure 7* shown as a straight line?

Classroom

Discuss phase angle diagrams.

Audiovisual

Show Transparency 8 (Figure 7).

Classroom

Discuss nonsinusoidal waveforms.

Teaching Tip

Discuss the "Think About It." See the Teaching Tip at the end of this module.

ALTERNATING CURRENT 123

Classroom

Compare sinusoidal vs. nonsinusoidal waveforms.

Audiovisual

Show Transparency 9 (Figure 8).

Classroom

Discuss resistance-only AC circuits and their associated waveforms.

Audiovisual

Show Transparency 10 (Figure 9).

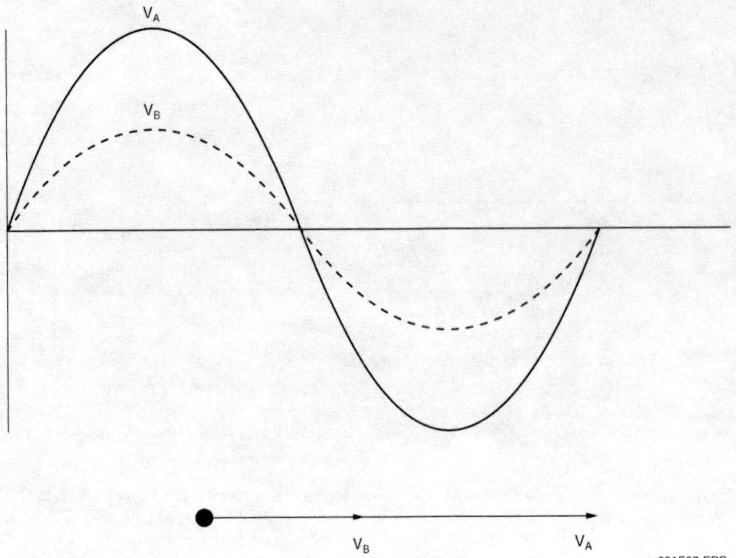

201F07.EPS

Figure 7 ◆ Waves in phase.

In many electronic applications, however, other waveshapes are important. Any waveform that is not a sine (or cosine) wave is a nonsinusoidal waveform. Common examples are the square wave and sawtooth wave in *Figure 8.*

With nonsinusoidal waveforms for either voltage or current, there are important differences and similarities to consider. Note the following comparisons with sine waves:

- In all cases, the cycle is measured between two points having the same amplitude and varying in the same direction. The period is the time for one cycle.
- Peak amplitude is measured from the zero axis to the maximum positive or negative value. However, peak-to-peak amplitude is better for measuring nonsinusoidal waveshapes because they can have asymmetrical peaks, as with the rectangular wave in *Figure 8.*
- The rms value 0.707 of peak applies only to sine waves, as this factor is derived from the sine values in the angular measure used only for the sine waveform.
- Phase angles apply only to sine waves, as angular measure is used only for sine waves. Note that the phase angle is indicated only on the sine wave of *Figure 8.*

6.0.0 ◆ RESISTANCE IN AC CIRCUITS

An AC circuit has an AC voltage source. Note the circular symbol with the sine wave inside it shown in *Figure 9.* It is used for any source of sine wave alternating voltage. This voltage connected across an external load resistance produces alternating current of the same waveform, frequency, and phase as the applied voltage.

According to Ohm's law, current (I) equals voltage (E) divided by resistance (R). When E is an rms value, I is also an rms value. For any instantaneous value of E during the cycle, the value of I is for the corresponding instant of time.

In an AC circuit with only resistance, the current variations are in phase with the applied voltage, as shown in *Figure 9.* This in-phase relationship between E and I means that such an AC circuit can be analyzed by the same methods used for DC circuits since there is not a phase angle to consider. Components that have only resistance include resistors, the filaments for incandescent light bulbs, and vacuum tube heaters.

Instructor's Notes:

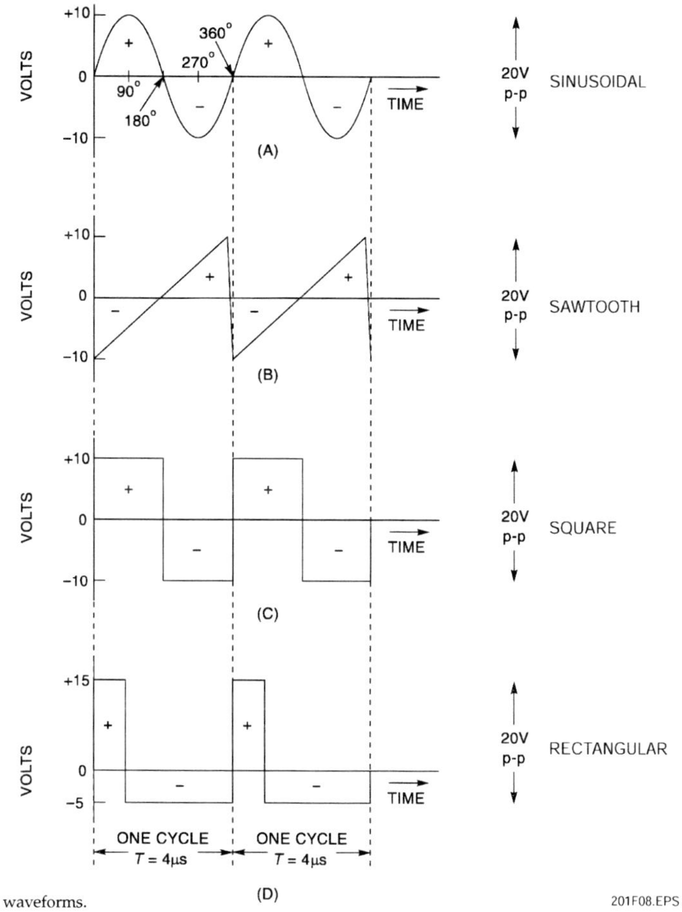

Figure 8 ◈ AC waveforms.

201F08.EPS

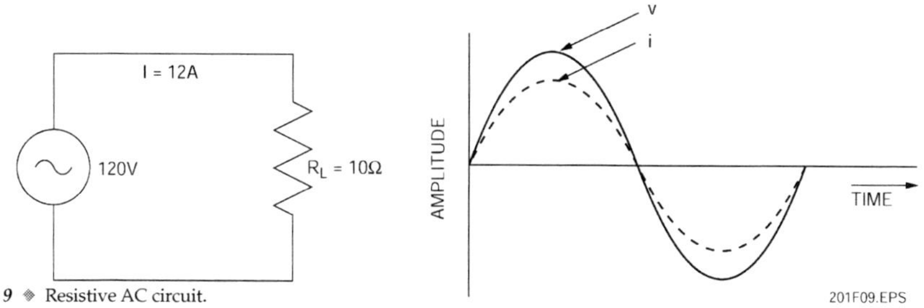

Figure 9 ◈ Resistive AC circuit.

201F09.EPS

ALTERNATING CURRENT 125

Discuss "What's wrong with this picture?" Explain that phase angles apply only to sine waves, not rectangular waves as shown here.

Explain the application of Ohm's law to purely resistive AC circuits.

Show Transparency 11 (Figure 10).

Define *inductance.*

What's wrong with this picture?

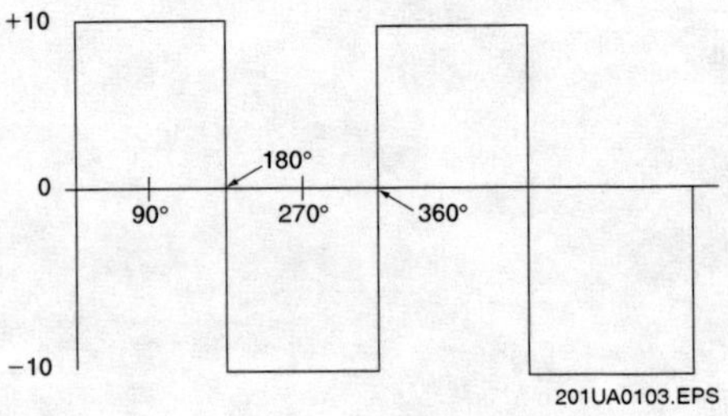

In purely resistive AC circuits, the voltage, current, and resistance are related by Ohm's law because the voltage and current are in phase.

$$I = \frac{E}{R}$$

Unless otherwise noted, the calculations in AC circuits are generally in rms values. For example, in *Figure 9,* the 120V applied across the 10Ω resistance R_L produces an rms current of 12A. This is determined as follows:

$$I = \frac{E}{R_L} = \frac{120V}{10\Omega} = 12A$$

Furthermore, the rms power (true power) dissipation is I^2R or:

$$P = (12A)^2 \times 10\Omega = 1{,}440W$$

Figure 10 shows the relationship between voltage and current in purely resistive AC circuits. The voltage and current are in phase, their cycles begin and end at the same time, and their peaks occur at the same time.

The value of the voltage shown in *Figure 10* depends on the applied voltage to the circuit. The value of the current depends on the applied voltage and the amount of resistance. If resistance is changed, it will affect only the magnitude of the current.

The total resistance in any AC circuit, whether it is a series, parallel, or series-parallel circuit, is calculated using the same rules that were learned and applied to DC circuits with resistance.

Power computations are discussed later in this module.

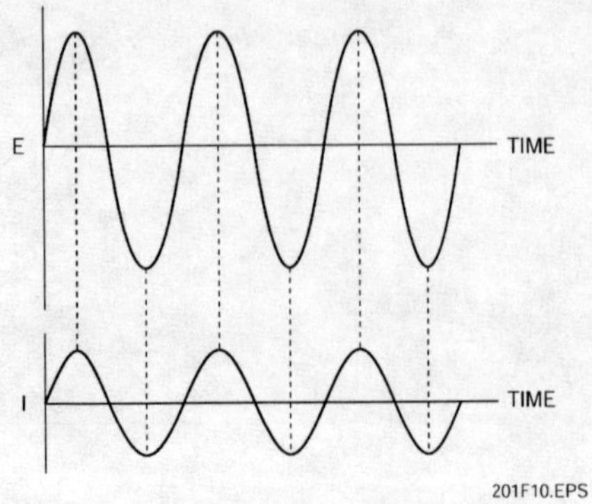

Figure 10 ◆ Voltage and current in a resistive AC circuit.

7.0.0 ◆ INDUCTANCE IN AC CIRCUITS

Inductance is the characteristic of an electrical circuit that opposes the change of current flow. It is the result of the expanding and collapsing field caused by the changing current. This moving flux cuts across the conductor that is providing the current, producing induced voltage in the wire itself. Furthermore, any other conductor in the field, whether carrying current or not, is also cut by the varying flux and has induced voltage. This induced current opposes the current flow that generated it.

In DC circuits, a change must be initiated in the circuit to cause inductance. The current must change to provide motion of the flux. A steady DC of 10A cannot produce any induced voltage as long as the current value is constant. A current of 1A changing to 2A does induce voltage. Also, the faster the current changes, the higher the induced voltage becomes, because when the flux moves at a higher speed it can induce more voltage.

However, in an AC circuit the current is continuously changing and producing induced voltage. Lower frequencies of AC require more inductance to produce the same amount of induced voltage as a higher frequency current. The current can have any waveform as long as the amplitude is changing.

The ability of a conductor to induce voltage in itself when the current changes is its **self-inductance** or simply inductance. The symbol for inductance is L and its unit is the henry (H). One henry is the amount of inductance that allows one volt to be induced when the current changes at the rate of one ampere per second.

Instructor's Notes:

7.1.0 Factors Affecting Inductance

An inductor is a coil of wire that may be wound on a core of metal or paper, or it may be self-supporting. It may consist of turns of wire placed side by side to form a layer of wire over the core or coil form. The inductance of a coil or inductor depends on its physical construction. Some of the factors affecting inductance are:

- *Number of turns*—The greater the number of turns, the greater the inductance. In addition, the spacing of the turns on a coil also affects inductance. A coil that has widely-spaced turns has a lower inductance than one which has the same number of more closely-spaced turns. The reason for this higher inductance is that the closely-wound turns produce a more concentrated magnetic field, causing the coil to exhibit a greater inductance.
- *Coil diameter*—The inductance increases directly as the cross-sectional area of the coil increases.
- *Length of the coil*—When the length of the coil is decreased, the turn spacing is decreased, increasing the inductance of the coil.
- *Core material*—The core of the coil can be either a magnetic material (such as iron) or a non-magnetic material (such as paper or air). Coils wound on a magnetic core produce a stronger magnetic field than those with non-magnetic cores, giving them higher values of inductance. Air-core coils are used where small values of inductance are required.
- *Winding the coil in layers*—The more layers used to form a coil, the greater the effect the magnetic field has on the conductor. Layering a coil can increase the inductance.

Factors affecting the inductance of a coil can be seen in *Figure 11.*

7.2.0 Voltage and Current in an Inductive AC Circuit

The self-induced voltage across an inductance L is produced by a change in current with respect to time (Δi/Δt) and can be stated as:

$$V_L = L\frac{\Delta i}{\Delta t}$$

Where:

Δ = change
V_L = volts
L = henrys
Δi/Δt = amperes per second

This gives the voltage in terms of how much magnetic flux is cut per second. When the magnetic flux associated with the current varies the same as I, this formula gives the same results for calculating induced voltage. Remember that the

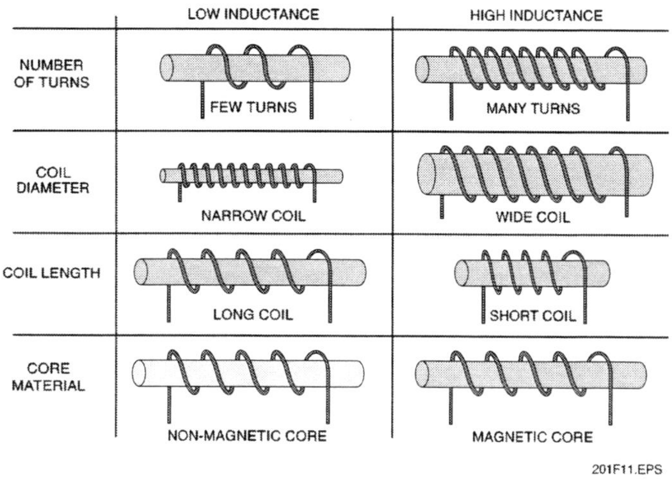

Figure 11 ◈ Factors affecting the inductance of a coil.

Classroom

Discuss the factors that affect inductance.

Audiovisual

Show Transparency 12 (Figure 11).

Classroom

Explain how to calculate the induced voltage in an AC circuit.

ALTERNATING CURRENT 127

Teaching Tip

Discuss the "Think About It." Answers include motors, transformers, and contactor/relay coils.

Classroom

Discuss the relationship between inductor voltage and current.

Audiovisual

Show Transparency 13 (Figure 12).

THINK ABOUT IT

Inductance in an AC Circuit

Can you name three commonly used electrical devices that insert inductance into an AC circuit?

induced voltage across the coil is actually the result of inducing electrons to move in the conductor, so there is also an induced current.

For example, what is the self-induced voltage V_L across a 4h inductance produced by a current change of 12A per second?

$$V_L = L\frac{\Delta i}{\Delta t}$$

$$V_L = 4h \times \frac{12A}{1}$$

$$V_L = 4 \times 12$$

$$V_L = 48V$$

The current through a 200 microhenry (μh) inductor changes from 0 to 200 milliamps (mA) in 2 microseconds (μsec). (The prefix **micro** means one-millionth). What is the V_L?

$$V_L = L\frac{\Delta i}{\Delta t}$$

$$V_L = (200 \times 10^{-6})\frac{200 \times 10^{-3}}{2 \times 10^{-6}}$$

$$V_L = 20V$$

The induced voltage is an actual voltage that can be measured, although V_L is produced only while the current is changing. When $\Delta i/\Delta t$ is present for only a short time, V_L is in the form of a voltage pulse. With a sine wave current that is always changing, V_L is a sinusoidal voltage that is 90° out of phase with I_L.

The current that flows in an inductor is induced by the changing magnetic field that surrounds the inductor. This changing magnetic field is produced by an AC voltage source that is applied to the inductor. The magnitude and polarity of the induced current depend on the field strength, direction, and rate at which the field cuts the inductor windings. The overall effect is that the current is out of phase and lags the applied voltage by 90°.

At 270° in *Figure 12,* the applied electromotive force (EMF) is zero, but it is increasing in the positive direction at its greatest rate of change. Likewise, electron flow due to the applied EMF is also increasing at its greatest rate. As the electron flow increases, it produces a magnetic field that is building with it. The lines of flux cut the conductor as they move outward from it with the expanding field.

As the lines of flux cut the conductor, they induce a current into it. The induced current is at its maximum value because the lines of flux are expanding outward through the conductor at their greatest rate. The direction of the induced current is in opposition to the force that generated it. Therefore, at 270° the applied voltage is zero and is increasing to a positive value, while the current is at its maximum negative value.

At 0° in *Figure 12,* the applied voltage is at its maximum positive value, but its rate of change is zero. Therefore, the field it produces is no longer expanding and is not cutting the conductor. Because there is no relative motion between the field and conductor, no current is induced. Therefore, at 0° voltage is at its maximum positive value, while current is zero.

At 90° in *Figure 12,* voltage is once again zero, but this time it is decreasing toward negative at its greatest rate of change. Because the applied voltage is decreasing, the magnetic field is collapsing inward on the conductor. This has the effect of reversing the direction of motion between the field and conductor that existed at 0°.

Therefore, the current will flow in a direction opposite of what it was at 0°. Also, because the applied voltage is decreasing at its greatest rate, the field is collapsing at its greatest rate. This causes the flux to cut the conductor at the greatest rate, causing the induced current magnitude to be maximum. At 90°, the applied voltage is zero decreasing toward negative, while the current is maximum positive.

At 180° in *Figure 12,* the applied voltage is at its maximum negative value, but just as at 0°, its rate of change is zero. At 180°, therefore, current will be zero. This explanation shows that the voltage peaks positive first, then 90° later the current peaks positive. Current thus lags the applied voltage in an inductor by 90°. This can easily be remembered using the phrase *ELI the ICE man.* ELI represents voltage (E), inductance (L), and current (I). In an inductor, the voltage leads the

Instructor's Notes:

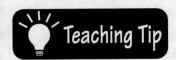

current just like the letter E leads or comes before the letter I. The word ICE will be explained in the section on **capacitance.**

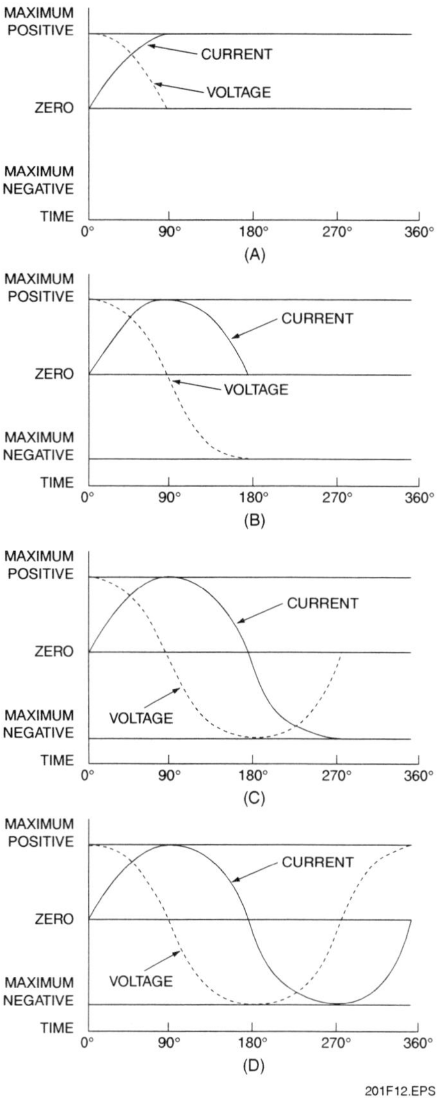

Figure 12 ◈ Inductor voltage and current relationship.

7.3.0 Inductive Reactance

The opposing force that an inductor presents to the flow of alternating current cannot be called resistance since it is not the result of friction within a conductor. The name given to this force is inductive **reactance** because it is the reaction of the inductor to alternating current. Inductive reactance is measured in ohms and its symbol is X_L.

Remember that the induced voltage in a conductor is proportional to the rate at which magnetic lines of force cut the conductor. The greater the rate or higher the frequency, the greater the counter-electromotive force (CEMF). Also, the induced voltage increases with an increase in inductance; the more turns, the greater the CEMF. Reactance then increases with an increase of frequency and with an increase in inductance. The formula for inductive reactance is as follows:

$$X_L = 2\pi fL$$

Where:

X_L = inductive reactance in ohms
2π = a constant in which the Greek letter pi (π) represents 3.14 and 2 × pi = 6.28
f = frequency of the alternating current in hertz
L = inductance in henrys

For example, if f is equal to 60Hz and L is equal to 20h, find X_L:

$$X_L = 2\pi fL$$
$$X_L = 6.28 \times 60\text{Hz} \times 20\text{h}$$
$$X_L = 7{,}536\Omega$$

Once calculated, the value of X_L is used like resistance in a form of Ohm's law:

$$I = \frac{E}{X_L}$$

Where:

I = effective current (amps)
E = effective voltage (volts)
X_L = inductive reactance (ohms)

Unlike a resistor, there is no power dissipation in an ideal inductor. An inductor limits current, but it uses no net energy since the energy required to build up the field in the inductor is given back to the circuit when the field collapses.

Classroom

Define *reactance.*

Explain how to calculate inductive reactance in an AC circuit.

ALTERNATING CURRENT 129

Classroom

Define *capacitance*.

Discuss capacitors and their applications. Pass around examples for the trainees to examine.

Audiovisual

Show Transparencies 14 and 15 (Figures 13 A/B and 14).

INSIDE TRACK

ELI in ELI the ICE Man

Remembering the phrase "ELI" as in "ELI the ICE man" is an easy way to remember the phase relationships that always exist between voltage and current in an inductive circuit. An inductive circuit is a circuit where there is more inductive reactance than capacitive reactance. The L in ELI indicates inductance. The E (voltage) is stated before the I (current) in ELI, meaning that the voltage leads the current in an inductive circuit.

8.0.0 ◆ CAPACITANCE

A capacitor is a device that stores an electric charge in a dielectric material. Capacitance is the ability to store a charge. In storing a charge, a capacitor opposes a change in voltage. *Figure 13* shows a simple capacitor in a circuit, schematic representations of two types of capacitors, and a photo of common capacitors.

Figure 14(A) shows a capacitor in a DC circuit. When voltage is applied, the capacitor begins to charge, as shown in *Figure 14(B)*. The charging continues until the potential difference across the capacitor is equal to the applied voltage. This charging current is transient or temporary since it flows only until the capacitor is charged to the applied voltage. Then there is no current in the circuit. *Figure 14(C)* shows this with the voltage across the capacitor equal to the battery voltage or 10V.

The capacitor can be discharged by connecting a conducting path across the dielectric. The stored charge across the dielectric provides the potential difference to produce a discharge current, as shown in *Figure 14(D)*. Once the capacitor is completely discharged, the voltage across it equals zero, and there is no discharge current.

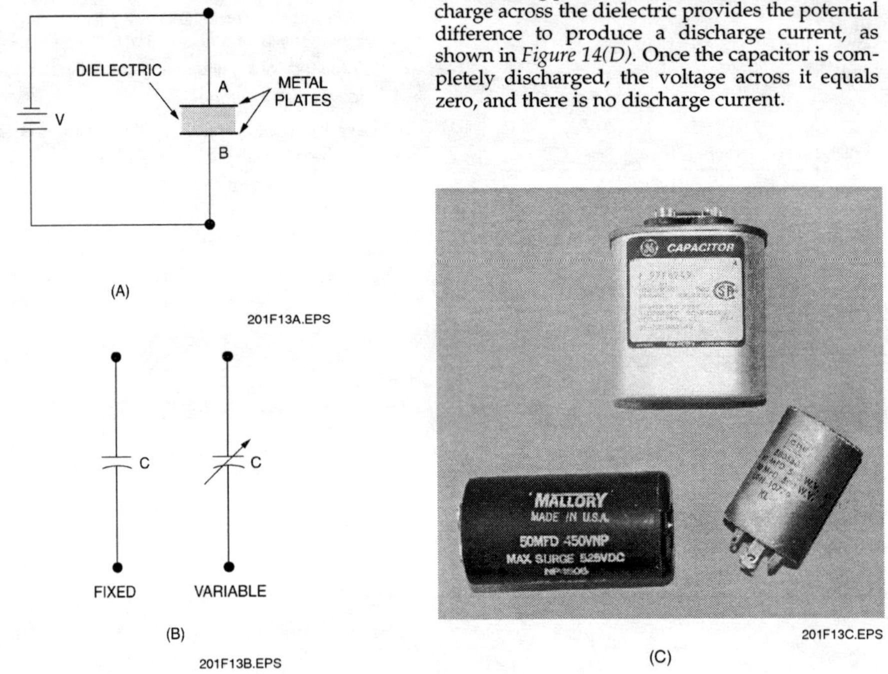

Figure 13 ◆ Capacitors.

Instructor's Notes:

130 CHAPTER 4

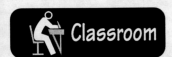

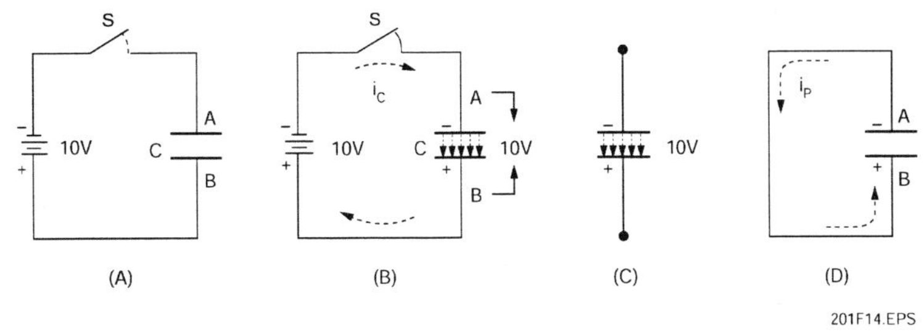

Figure 14 ◈ Charging and discharging a capacitor.

In a capacitive circuit, the charge and discharge current must always be in opposite directions. Current flows in one direction to charge the capacitor and in the opposite direction when the capacitor is allowed to discharge.

Current will flow in a capacitive circuit with AC voltage applied because of the capacitor charge and discharge current. There is no current through the dielectric, which is an insulator. While the capacitor is being charged by increasing applied voltage, the charging current flows in one direction to the plates. While the capacitor is discharging as the applied voltage decreases, the discharge current flows in the reverse direction. With alternating voltage applied, the capacitor alternately charges and discharges.

First, the capacitor is charged in one polarity, and then it discharges; next, the capacitor is charged in the opposite polarity, and then it discharges again. The cycles of charge and discharge current provide alternating current in the circuit at the same frequency as the applied voltage. The amount of capacitance in the circuit will determine how much current is allowed to flow.

Capacitance is measured in farads (F), where one farad is the capacitance when one coulomb is stored in the dielectric with a potential difference of one volt. Smaller values are measured in microfarads (μF). A small capacitance will allow less charge and discharge current to flow than a larger capacitance. The smaller capacitor has more opposition to alternating current, because less current flows with the same applied voltage.

In summary, capacitance exhibits the following characteristics:

- DC is blocked by a capacitor. Once charged, no current will flow in the circuit.
- AC flows in a capacitive circuit with AC voltage applied.
- A smaller capacitance allows less current.

8.1.0 Factors Affecting Capacitance

A capacitor consists of two conductors separated by an insulating material called a *dielectric.* There are many types and sizes of capacitors with different dielectric materials. The capacitance of a capacitor is determined by three factors:

- *Area of the plates*—The initial charge displacement on a set of capacitor plates is related to the number of free electrons in each plate. Larger plates will produce a greater capacitance than smaller ones. Therefore, the capacitance of a capacitor varies directly with the area of the plates. For example, if the area of the plates is doubled, the capacitance is doubled. If the size of the plates is reduced by 50%, the capacitance would also be reduced by 50%.
- *Distance between plates*—As two capacitor plates are brought closer together, more electrons will move away from the positively charged plate and move into the negatively charged plate. This is because the mutual attraction between the opposite charges on the plates increases as we move the plates closer together. This added movement of charge is an increase in the capacitance of the capacitor. In a capacitor composed of two plates of equal area, the capacitance varies inversely with the distance between the plates. For example, if the distance between the plates is decreased by one-half, the capacitance will be doubled. If the distance between the plates is doubled, the capacitance would be one-half as great.
- *Dielectric permittivity*—Another factor that determines the value of capacitance is the permittivity of the dielectric. The dielectric is the material between the capacitor plates in which the electric field appears. Relative permittivity expresses the ratio of the electric field strength in a dielectric to that in a vacuum. Permittivity has nothing to do

Classroom

Discuss the characteristics of capacitance.

Discuss the factors that affect capacitance:
- **Area of plates**
- **Distance between plates**
- **Dielectric permittivity**

ALTERNATING CURRENT 131

INSIDE TRACK

Capacitance

The concept of capacitance, like many electrical quantities, is often hard to visualize or understand. A comparison with a balloon may help to make this concept clearer. Electrical capacitance has a charging effect similar to blowing up a balloon and holding it closed. The expansion *capacity* of the balloon can be changed by changing the thickness of the balloon walls. A balloon with thick walls will expand less (have less capacity) than one with thin walls. This is like a small 10μF capacitor that has less capacity and will charge less than a larger 100μF capacitor.

with the dielectric strength of the medium or the breakdown voltage. An insulating material that will withstand a higher applied voltage than some other substance does not always have a higher dielectric permittivity. Many insulating materials have a greater dielectric permittivity than air. For a given applied voltage, a greater attraction exists between the opposite charges on the capacitor plates, and an electric field can be set up more easily than when the dielectric is air. The capacitance of the capacitor is increased when the permittivity of the dielectric is increased if all the other parameters remain unchanged.

8.2.0 Calculating Equivalent Capacitance

Classroom

Explain how to calculate equivalent capacitance:
- **In parallel**
- **In series**

Connecting capacitors in parallel is equivalent to adding the plate areas. Therefore, the total capacitance is the sum of the individual capacitances, as illustrated in *Figure 15*.

A 10μF capacitor in parallel with a 5μF capacitor, for example, provides a 15μF capacitance for the parallel combination. The voltage is the same across the parallel capacitors. Note that adding parallel capacitance is opposite to the case of inductances in parallel and resistances in parallel.

Connecting capacitances in series is equivalent to increasing the thickness of the dielectric. Therefore, the combined capacitance is less than the smallest individual value. The combined equivalent capacitance is calculated by the reciprocal formula, as shown in *Figure 16.*

Audiovisual

Show Transparencies 16 and 17 (Figures 15 and 16).

Capacitors connected in series are combined like resistors in parallel. Any of the shortcut calculations for the reciprocal formula apply. For example, the combined capacitance of two equal capacitances of 10μF in series is 5μF.

Capacitors are used in series to provide a higher voltage breakdown rating for the combination. For instance, each of three equal capacitances in series has one-third the applied voltage.

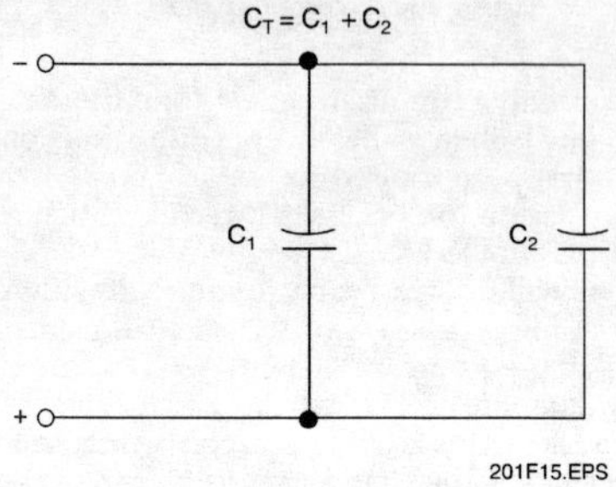

Figure 15 ◈ Capacitors in parallel.

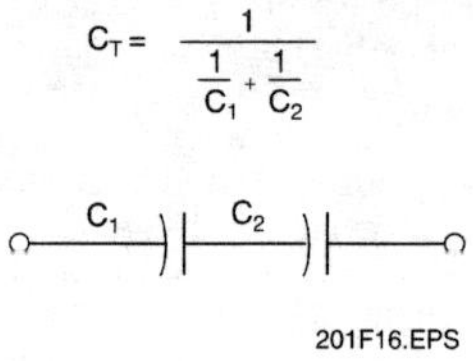

Figure 16 ◈ Capacitors in series.

In series, the voltage across each capacitor is inversely proportional to its capacitance. The smaller capacitance has the larger proportion of the applied voltage. The reason is that the series capacitances all have the same charge because they are in one current path. With equal charge, a smaller capacitance has a greater potential difference.

8.3.0 Capacitor Specifications

This specifies the maximum potential difference that can be applied across the plates without puncturing the dielectric.

Instructor's Notes:

THINK ABOUT IT

Capacitance

Suppose you had a motor with a bad 30µF starting capacitor and no 30µF direct replacement capacitor was available. As a temporary measure, you are authorized to substitute two equal-value capacitors in its place. What size capacitors (µF) should be used if you are connecting them in parallel?

8.3.1 Voltage Rating

Usually, the voltage rating is for temperatures up to about 60°C. High temperatures result in a lower voltage rating. Voltage ratings for general-purpose paper, mica, and ceramic capacitors are typically 200V to 500V. Ceramic capacitors with ratings of 1 to 5kV are also available.

Electrolytic capacitors are commonly used in 25V, 150V, and 450V ratings. In addition, 6V and 10V electrolytic capacitors are often used in transistor circuits. For applications where a lower voltage rating is permissible, more capacitance can be obtained in a smaller physical size.

The potential difference across the capacitor depends on the applied voltage and is not necessarily equal to the voltage rating. A voltage rating higher than the potential difference applied across the capacitor provides a safety factor for long life in service. With electrolytic capacitors, however, the actual capacitor voltage should be close to the rated voltage to produce the oxide film that provides the specified capacitance.

The voltage ratings are for applied DC voltage. The breakdown rating is lower for AC voltage because of the internal heat produced by continuous charge and discharge.

8.3.2 Leak Resistance

Consider a capacitor charged by a DC voltage source. After the charging voltage is removed, a perfect capacitor would keep its charge indefinitely. After a long period of time, however, the charge will be neutralized by a small leakage current through the dielectric and across the insulated case between terminals, because there is no perfect insulator. For paper, ceramic, and mica capacitors, the leakage current is very slight, or inversely, the leakage resistance is very high. For paper, ceramic, or mica capacitors, R_l is 100MΩ or more. However, electrolytic capacitors may have a leakage resistance of 0.5MΩ or less.

8.4.0 Voltage and Current in a Capacitive AC Circuit

In a capacitive circuit driven by an AC voltage source, the voltage is continuously changing. Thus, the charge on the capacitor is also continuously changing. The four parts of *Figure 17* show the variation of the alternating voltage and current in a capacitive circuit for each quarter of one cycle.

The solid line represents the voltage across the capacitor, and the dotted line represents the current. The line running through the center is the zero or reference point for both the voltage and the current. The bottom line marks off the time of the cycle in terms of electric degrees. Assume that the AC voltage has been acting on the capacitor for some time before the time represented by the starting point of the sine wave.

At the beginning of the first quarter-cycle (0° to 90°), the voltage has just passed through zero and is increasing in the positive direction. Since the zero point is the steepest part of the sine wave, the voltage is changing at its greatest rate.

The charge on a capacitor varies directly with the voltage; therefore, the charge on the capacitor is also changing at its greatest rate at the beginning of the first quarter-cycle. In other words, the greatest number of electrons are moving off one plate and onto the other plate. Thus, the capacitor current is at its maximum value.

As the voltage proceeds toward maximum at 90°, its rate of change becomes lower and lower, making the current decrease toward zero. At 90°, the voltage across the capacitor is maximum, the capacitor is fully charged, and there is no further movement of electrons from plate to plate. That is why the current at 90° is zero.

At the end of the first quarter-cycle, the alternating voltage stops increasing in the positive direction and starts to decrease. It is still a positive voltage; but to the capacitor, the decrease in voltage means that the plate that has just accumulated an excess of electrons must lose some electrons.

Teaching Tip

Discuss the "Think About It." 15µF capacitors should be used.

Classroom

Discuss capacitor specifications:
- **Voltage rating**
- **Leak resistance**

Discuss the relationship between voltage and current in a capacitive AC circuit.

Audiovisual

Show Transparency 18 (Figure 17).

ALTERNATING CURRENT 133

The current flow must reverse its direction. The second part of the figure shows the current curve to be below the zero line (negative current direction) during the second quarter-cycle (90° to 180°).

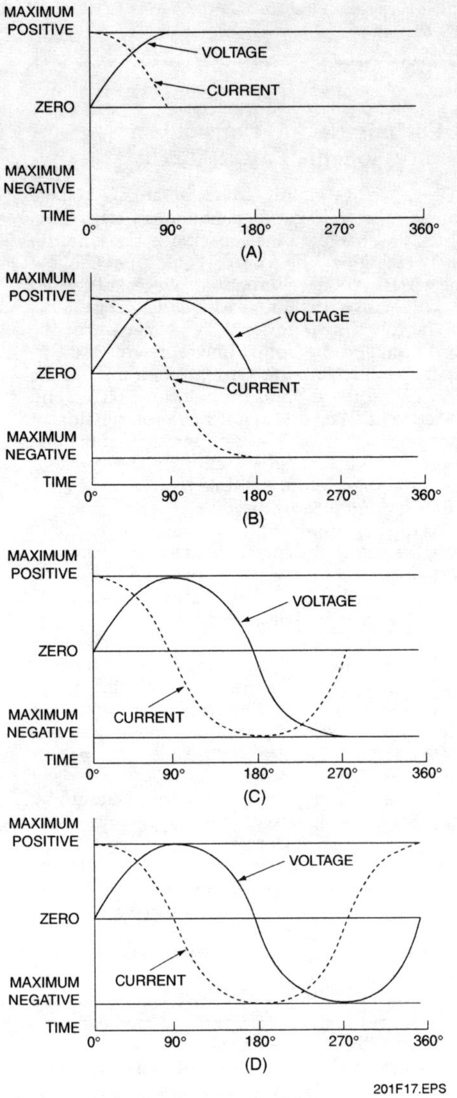

Figure 17 ◆ Voltage and current in a capacitive AC circuit.

At 180°, the voltage has dropped to zero. This means that, for a brief instant, the electrons are equally distributed between the two plates; the current is maximum because the rate of change of voltage is maximum.

Just after 180°, the voltage has reversed polarity and starts building to its maximum negative peak, which is reached at the end of the third quarter-cycle (180° to 270°). During the third quarter-cycle, the rate of voltage change gradually decreases as the charge builds to a maximum at 270°. At this point, the capacitor is fully charged and carries the full impressed voltage. Because the capacitor is fully charged, there is no further exchange of electrons and the current flow is zero at this point. The conditions are exactly the same as at the end of the first quarter-cycle (90°), but the polarity is reversed.

Just after 270°, the impressed voltage once again starts to decrease, and the capacitor must lose electrons from the negative plate. It must discharge, starting at a minimum rate of flow and rising to a maximum. This discharging action continues through the last quarter-cycle (270° to 360°) until the impressed voltage has reached zero. The beginning of the entire cycle is 360°, and everything starts over again.

In *Figure 17,* note that the current always arrives at a certain point in the cycle 90° ahead of the voltage because of the charging and discharging action. This voltage-current phase relationship in a capacitive circuit is exactly opposite to that in an inductive circuit. The current through a capacitor leads the voltage across the capacitor by 90°. A convenient way to remember this is the phrase *ELI the ICE man* (ELI refers to inductors, as previously explained). ICE pertains to capacitors as follows:

I = current
C = capacitor
E = voltage

In capacitors (C), current (I) leads voltage (E) by 90°.

It is important to realize that the current and voltage are both going through their individual cycles at the same time during the period the AC voltage is impressed. The current does not go through part of its cycle (charging or discharging) and then stop and wait for the voltage to catch up. The amplitude and polarity of the voltage and the amplitude and direction of the current are continually changing.

Their positions, with respect to each other and to the zero line at any electrical instant or any degree between 0° and 360°, can be seen by reading upward from the time-degree line. The current swing from

Classroom

Contrast the voltage-current relationships in inductive vs. capacitive circuits.

Instructor's Notes:

INSIDE TRACK

ICE in ELI the ICE Man

Remembering the phrase "ICE" as in "ELI the ICE man" is an easy way to remember the phase relationships that always exist between voltage and current in a capacitive circuit. A capacitive circuit is a circuit in which there is more capacitive reactance than inductive reactance. This is indicated by the C in ICE. The I (current) is stated before the E (voltage) in ICE, meaning that the current leads the voltage in a capacitive circuit.

the positive peak at 0° to the negative peak at 180° is not a measure of the number of electrons or the charge on the plates. It is a picture of the direction and strength of the current in relation to the polarity and strength of the voltage appearing across the plates.

8.5.0 Capacitive Reactance

Capacitors offer a very real opposition to current flow. This opposition arises from the fact that, at a given voltage and frequency, the number of electrons that go back and forth from plate to plate is limited by the storage ability or the capacitance of the capacitor. As the capacitance is increased, a greater number of electrons changes plates every cycle. Since current is a measure of the number of electrons passing a given point in a given time, the current is increased.

Increasing the frequency will also decrease the opposition offered by a capacitor. This occurs because the number of electrons that the capacitor is capable of handling at a given voltage will change plates more often. As a result, more electrons will pass a given point in a given time (greater current flow). The opposition that a capacitor offers to AC is therefore inversely proportional to frequency and capacitance. This opposition is called *capacitive reactance.* Capacitive reactance decreases with increasing frequency or, for a given frequency, the capacitive reactance decreases with increasing capacitance. The symbol for capacitive reactance is X_C. The formula is:

$$X_C = \frac{1}{2\pi fC}$$

Where:

X_C = capacitive reactance in ohms
f = frequency in hertz
C = capacitance in farads
$2\pi = 6.28\ (2 \times 3.14)$

For example, what is the capacitive reactance of a 0.05μF capacitor in a circuit whose frequency is 1 megahertz?

$$X_C = \frac{1}{2\pi fC} = \frac{1}{(6.28)(10^6 \text{ hertz})(5 \times 10^{-8} \text{ farads})}$$

$$X_C = \frac{1}{3.14 \times 10^{-1}} = \frac{1}{0.314} = 3.18 \text{ ohms}$$

The capacitive reactance of a 0.05μF capacitor operated at a frequency of 1 megahertz is 3.18 ohms. Suppose this same capacitor is operated at a lower frequency of 1,500 hertz instead of 1 megahertz. What is the capacitive reactance now? Substituting where $1{,}500 = 1.5 \times 10^3$ hertz:

$$X_C = \frac{1}{2\pi fC} = \frac{1}{(6.28)(1.5 \times 10^3 \text{ hertz})(5 \times 10^{-8} \text{ farads})}$$

$$X_C = \frac{1}{4.71 \times 10^{-4}} = 2{,}123 \text{ ohms}$$

Note a very interesting point from these two examples. As frequency is decreased from 1 megahertz to 1,500 hertz, the capacitive reactance increases from 3.18 ohms to 2,123 ohms. Capacitive reactance increases as the frequency decreases.

THINK ABOUT IT

Frequency and Capacitive Reactance

A variable capacitor is used in the tuner of an AM radio to tune the radio to the desired station. Will its capacitive reactance value be higher or lower when it is tuned to the low end of the frequency band (550kHz) than it would be when tuned to the high end of the band (1,440kHz)?

Classroom

Discuss capacitive reactance and explain how it is calculated.

Teaching Tip

Discuss the "Think About It." Explain that it would be higher because capacitive reactance increases as frequency decreases.

ALTERNATING CURRENT 135

Classroom

Review the voltage-current relationships in resistive, inductive, and capacitive circuits.

Audiovisual

Show Transparency 19 (Figure 18).

Classroom

Discuss impedance and explain how it is calculated.

Demonstrate how to find unknown values in series RL circuits.

Audiovisual

Show Transparency 20 (Figure 19).

9.0.0 ◆ LC AND RLC CIRCUITS

AC circuits often contain inductors, capacitors, and/or resistors connected in series or parallel combinations. When this is done, it is important to determine the resulting phase relationship between the applied voltage and the current in the circuit. The simplest method of combining factors that have different phase relationships is vector addition with the trigonometric functions. Each quantity is represented as a vector, and the resultant vector and phase angle are then calculated.

In purely resistive circuits, the voltage and current are in phase. In inductive circuits, the voltage leads the current by 90°. In capacitive circuits, the current leads the voltage by 90°. *Figure 18* shows the phase relationships of these components used in AC circuits. Recall that these characteristics are summarized by the phrase *ELI the ICE man.*

ELI = *E* Leads *I* (inductive)

ICE = *I* Capacitive (leads) *E*

The **impedance** Z of a circuit is defined as the total opposition to current flow. The magnitude of the impedance Z is given by the following equation in a series circuit:

$$Z = \sqrt{R^2 + X^2}$$

Where:

Z = impedance (ohms)
R = resistance (ohms)
X = net reactance (ohms)

The current through a resistance is always in phase with the voltage applied to it; thus resistance is shown along the 0° axis. The voltage across an inductor leads the current by 90°; thus inductive reactance is shown along the 90° axis. The voltage across a capacitor lags the current by 90°; thus capacitive reactance is shown along the −90° axis. The net reactance is the difference between the inductive reactance and the capacitive reactance:

X = net reactance (ohms)
X_L = inductive reactance (ohms)
X_C = capacitive reactance (ohms)

The impedance Z is the vector sum of the resistance R and the net reactance X. The angle, called the *phase angle,* gives the phase relationship between the applied voltage and current.

9.1.0 RL Circuits

RL circuits combine resistors and inductors in a series, parallel, or series-parallel configuration. In a pure inductive circuit, the current lags the voltage by an angle of 90°. In a circuit containing both resistance and inductance, the current will lag the voltage by some angle between zero and 90°.

9.1.1 Series RL Circuit

Figure 19 shows a series RL circuit. Since it is a series circuit, the current is the same in all portions of the loop. Using the values shown, the circuit will be analyzed for unknown values such as X_L, Z, I, E_L, and E_R.

The solution would be worked as follows:

Step 1 Compute the value of X_L.

$$X_L = 2\pi fL$$
$$X_L = 6.28 \times 100 \times 4 = 2{,}512 \text{ ohms}$$

Step 2 Draw vectors R and X_L as shown in *Figure 19*. R is drawn horizontally because the circuit current and voltage across R are in phase. It therefore becomes the reference line from which other angles are measured. X_L is drawn upward at 90° from R because voltage across X_L leads circuit current through R.

Step 3 Compute the value of circuit impedance Z, which is equal to the vector sum of X_L and R.

$$\tan = \frac{X_L}{R} = \frac{2{,}512}{1{,}500} = 1.67$$

$$\arctan 1.67 = 59.1°$$

$$\cos 59.1° = .5135$$

Find Z using the cosine function:

$$\cos = \frac{R}{Z}$$

$$Z = \frac{R}{\cos}$$

$$Z = \frac{1{,}500}{.5135} = 2{,}921 \text{ ohms}$$

Step 4 Compute the circuit current using Ohm's law for AC circuits.

$$I = \frac{E}{Z} = \frac{100V}{2.921\Omega} = 0.034A$$

Instructor's Notes:

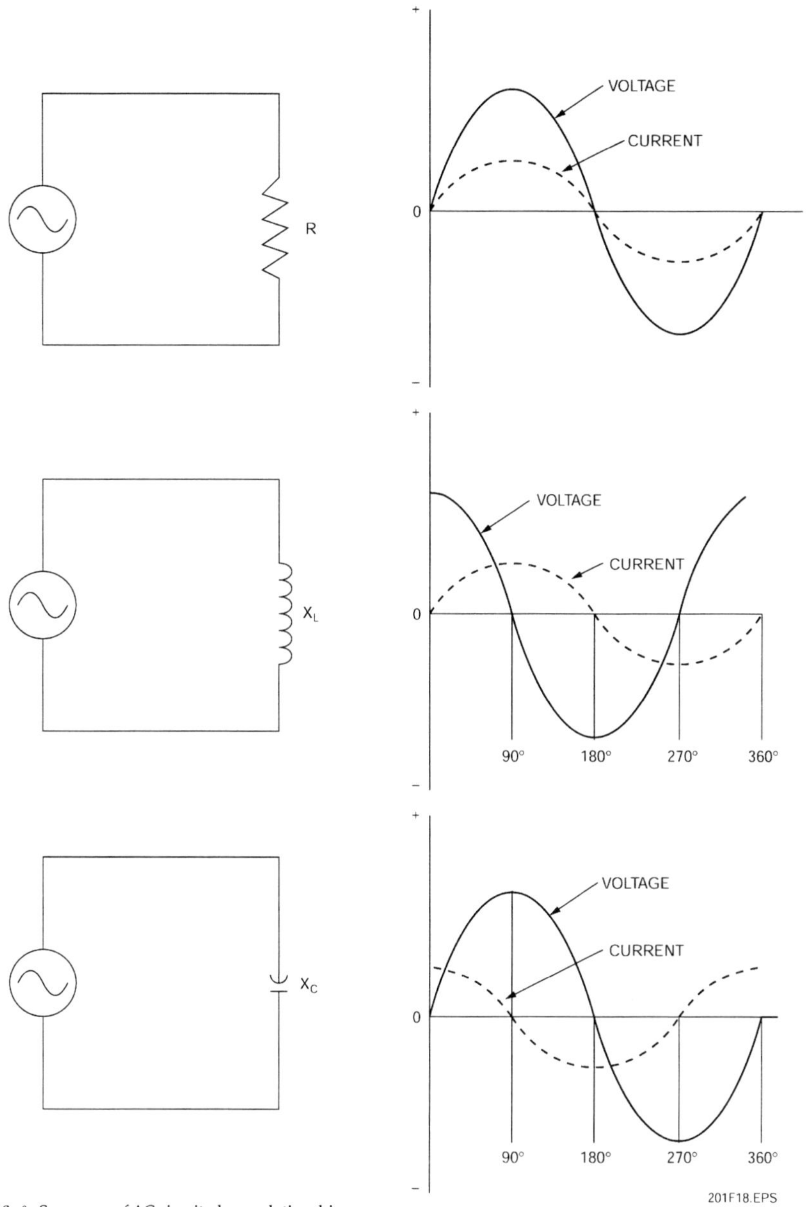

Figure 18 ◈ Summary of AC circuit phase relationships.

ALTERNATING CURRENT 137

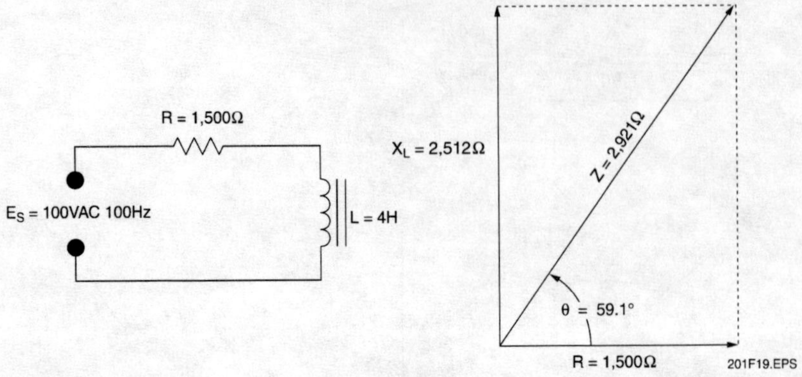

Figure 19 ◆ Series RL circuit and vector diagram.

Step 5 Compute voltage drops in the circuit.

$$E_L = IX_L = .034 \times 2{,}512 = 85 \text{ volts}$$

$$E_R = IR = .034 \times 1{,}500 = 51 \text{ volts}$$

Note that the voltage drops across the resistor and inductor do not equal the supply voltage because they must be added vectorially. This would be done as follows (because of rounding, numbers are not exact):

$$\tan = \frac{E_L}{E_R} = \frac{85}{51} = 1.67$$

$$\arctan 1.67 = 59.1°$$

$$\cos = \frac{E_R}{E_Z} \qquad E_Z = \frac{E_R}{\cos}$$

$$E_Z = \frac{51}{.5135} = \text{approx. } 100\text{V} = E_S$$

In this inductive circuit, the current lags the applied voltage by an angle equal to 59.1°.

Figure 20 shows another series RL circuit, its associated waveforms, and vector diagrams. This circuit is used to summarize the characteristics of a series RL circuit:

- The current I flows through all the series components.
- The voltage across X_L, labeled V_L, can be considered an IX_L voltage drop, just as V_R is used for an IR voltage drop.
- The current I through X_L must lag V_L by 90°, as this is the angle between current through an inductance and its self-induced voltage.
- The current I through R and its IR voltage drop have the same phase. There is no reactance to sine wave current in any resistance. Therefore, I and IR have the same phase, or this phase angle is 0°.
- V_T is the vector sum of the two out-of-phase voltages V_R and V_L.

Note
Since we are dealing with right triangles, we could also use the Pythagorean theorem (discussed later) to find this answer.

Classroom

Review the characteristics of series RL circuits.

Audiovisual

Show Transparency 21 (Figure 20).

INSIDE TRACK

Vector Analysis

When using vector analysis, the horizontal line is the in-phase value and the vertical line pointing up represents the leading value. The vertical line pointing down represents the lagging value.

Instructor's Notes:

138 CHAPTER 4

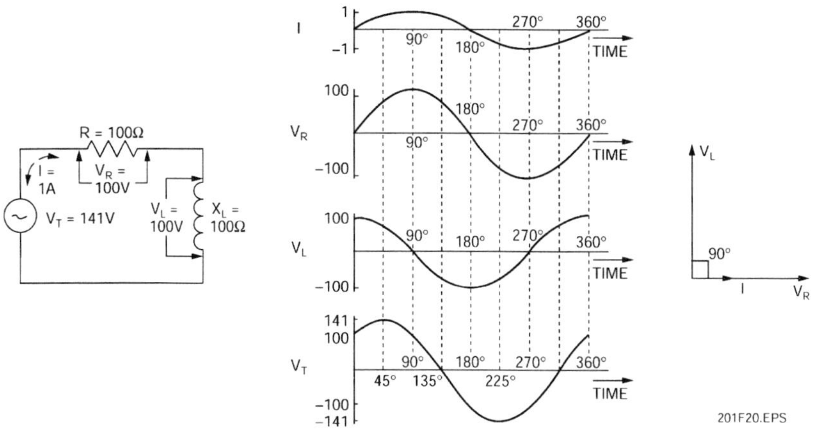

Figure 20 ◆ Series RL circuit with waveforms and vector diagram.

- Circuit current I lags V_T by the phase angle.
- Circuit impedance is the vector sum of R and XL.

In a series circuit, the higher the value of X_L compared with R, the more inductive the circuit is. This means there is more voltage drop across the inductive reactance, and the phase angle increases toward 90°. The series current lags the applied generator voltage.

Several combinations of X_L and R in series are listed in *Table 1* with their resultant impedance and phase angle. Note that a ratio of 10:1 or more for X_L/R means that the circuit is practically all inductive. The phase angle of 84.3° is only slightly less than 90° for the ratio of 10:1, and the total impedance Z is approximately equal to X_L. The voltage drop across X_L in the series circuit will be equal to the applied voltage, with almost none across R.

At the opposite extreme, when R is 10 times as large as X_L, the series circuit is mainly resistive. The phase angle of 5.7° means the current has almost the same phase as the applied voltage, the total impedance Z is approximately equal to R, and the voltage drop across R is practically equal to the applied voltage, with almost none across X_L.

Table 1 Series R and X_L Combinations

R (Ω)	X_L (Ω)	Z (Ω) (Approx.)	Phase Angle (θ) (°)
1	10	$\sqrt{101} = 10$	84.3°
10	10	$\sqrt{200} = 14$	45°
10	1	$\sqrt{101} = 10$	5.7°

9.1.2 Parallel RL Circuit

In a parallel RL circuit, the resistance and inductance are connected in parallel across a voltage source. Such a circuit thus has a resistive branch and an inductive branch.

The 90° phase angle must be considered for each of the branch currents, instead of voltage drops in a series circuit. Remember that any series circuit has different voltage drops, but one common current. A parallel circuit has different branch currents, but one common voltage.

In the parallel circuit in *Figure 21*, the applied voltage V_A is the same across X_L, R, and the generator, since they are all in parallel. There cannot be any phase difference between these voltages. Each branch, however, has its individual current. For the resistive branch $I_R = V_A/R$; in the inductive branch $I_L = V_A/X_L$.

The resistive branch current I_R has the same phase as the generator voltage V_A. The inductive branch current I_L lags V_A, however, because the current in an inductance lags the voltage across it by 90°.

The total line current, therefore, consists of I_R and I_L, which are 90° out of phase with each other. The phasor sum of I_R and I_L equals the total line

Classroom

Discuss the characteristics of parallel RL circuits.

Audiovisual

Show Transparency 22 (Figure 21).

ALTERNATING CURRENT 139

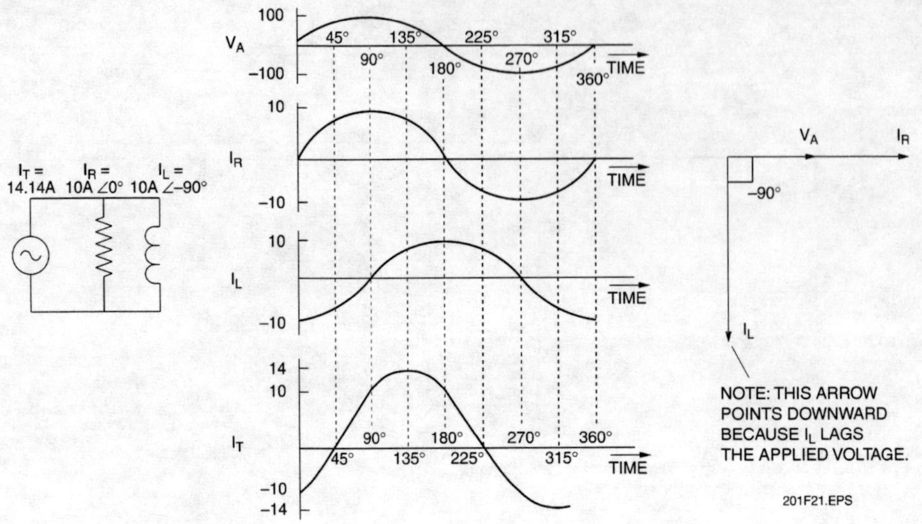

Figure 21 ◆ Parallel RL circuit with waveforms and vector diagram.

Classroom

Demonstrate how impedance is calculated in parallel RL circuits.

current I_T. These phase relations are shown by the waveforms and vectors in *Figure 21.* I_T will lag V_A by some phase angle that results from the vector addition of I_R and I_L.

The impedance of a parallel RL circuit is the total opposition to current flow by the R of the resistive branch and the X_L of the inductive branch. Since X_L and R are vector quantities, they must be added vectorially.

If the line current and the applied voltage are known, Z can also be calculated by the equation:

$$Z = \frac{V_A}{I_{Line}}$$

The Z of a parallel RL circuit is always less than the R or X_L of any one branch. The branch of a parallel RL circuit that offers the most opposition to current flow has the lesser effect on the phase angle of the current.

Several combinations of X_L and R in parallel are listed in *Table 2.* When X_L is 10 times R, the parallel circuit is practically resistive because there is little inductive current in the line. The small value of I_L results from the high X_L. The total impedance of the parallel circuit is approximately equal to the resistance then, since the high value of X_L in a parallel branch has little effect. The phase angle of −5.7° is practically 0° because almost all the line current is resistive.

As X_L becomes smaller, it provides more inductive current in the main line. When X_L is ⅒R, practically all the line current is the I_L component. Then, the parallel circuit is practically all inductive, with a total impedance practically equal to X_L. The phase angle of −84.3° is almost −90° because the line current is mostly inductive. Note that these conditions are opposite from the case of X_L and R in series.

Table 2 Parallel R and X_L Combinations

R (Ω)	X_L (Ω)	I_R (A)	I_L (A)	I_T (A) (Approx.)	$Z_T = V_A/I_T$ (Ω)	Phase Angle (θ) (°)
1	10	10	1	$\sqrt{101} = 10$	1	−5.7°
10	10	1	1	$\sqrt{2} = 1.4$	7.07	−45°
10	1	1	10	$\sqrt{101} = 10$	1	−84.3°

Instructor's Notes:

140 CHAPTER 4

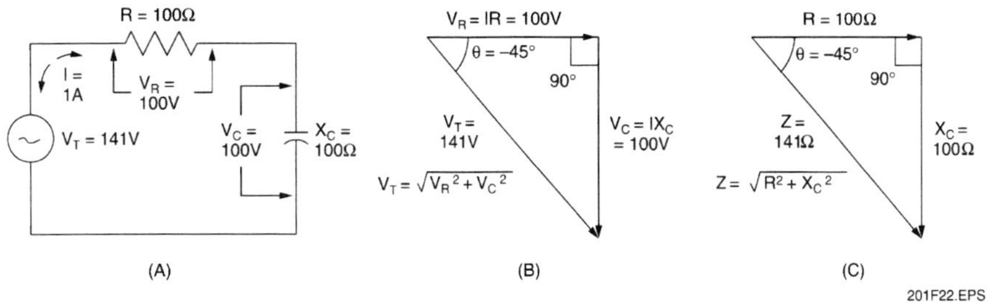

Figure 22 ◆ Series RC circuit with vector diagrams.

9.2.0 RC Circuits

In a circuit containing resistance only, the current and voltage are in phase. In a circuit of pure capacitance, the current leads the voltage by an angle of 90°. In a circuit that has both resistance and capacitance, the current will lead the voltage by some angle between 0° and 90°.

9.2.1 Series RC Circuit

Figure 22(A) shows a series RC circuit with resistance R in series with capacitive reactance X_C. Current I is the same in X_C and R since they are in series. Each has its own series voltage drop, equal to IR for the resistance and IX_C for the reactance.

In *Figure 22(B)*, the current phasor is shown horizontal as the reference phase, because I is the same throughout the series circuit. The resistive voltage drop IR has the same phase as I. The capacitor voltage IX_C must be 90° clockwise from I and IR, as the capacitive voltage lags. Note that the IX_C phasor is downward, exactly opposite from an IX_L phasor, because of the opposite phase angle.

If the capacitive reactance alone is considered, its voltage drop lags the series current I by 90°. The IR voltage has the same phase as I, however, because resistance provides no phase shift. Therefore, R and X_C combined in series must be added by vectors because they are 90° out of phase with each other, as shown in *Figure 22(C)*.

As with inductive reactance, θ (theta) is the phase angle between the generator voltage and its series current. As shown in *Figure 22(B)* and *Figure 22(C)*, θ can be calculated from the voltage or impedance triangle.

With series X_C the phase angle is negative, clockwise from the zero reference angle of I because the X_C voltage lags its current. To indicate the negative phase angle, this 90° phasor points downward from the horizontal reference, instead of upward as with the series inductive reactance.

In series, the higher the X_C compared with R, the more capacitive the circuit. There is more voltage drop across the capacitive reactance, and the phase angle increases toward −90°. The series X_C always makes the current lead the applied voltage. With all X_C and no R, the entire applied voltage is across X_C and equals −90°. Several combinations of X_C and R in series are listed in *Table 3*.

9.2.2 Parallel RC Circuit

In a parallel RC circuit, as shown in *Figure 23(A)*, a capacitive branch as well as a resistive branch are connected across a voltage source. The current that leaves the voltage source divides among the branches, so there are different currents in each branch. The current is therefore not a common quantity, as it is in the series RC circuit.

In a parallel RC circuit, the applied voltage is directly across each branch. Therefore, the branch voltages are equal in value to the applied voltage and all voltages are in phase. Since the voltage is common throughout the parallel RC circuit, it serves as the common quantity in any vector representation of parallel RC circuits. This means the reference vector will have the same phase relationship or direction as the circuit voltage. Note in *Figure 23(B)* that V_A and I_R are both shown as the 0° reference.

Table 3 Series R and X_C Combinations

R (Ω)	X_C (Ω)	Z (Ω) (Approx.)	Phase Angle $(\theta)_Z$ (°)
1	10	$\sqrt{101} = 10$	84.3°
10	10	$\sqrt{200} = 14$	45°
10	1	$\sqrt{101} = 10$	5.7°

Classroom

Discuss the characteristics of RC circuits.

Discuss series RC circuits.

Audiovisual

Show Transparency 23 (Figure 22).

Classroom

Discuss parallel RC circuits.

Audiovisual

Show Transparency 24 (Figure 23).

ALTERNATING CURRENT 141

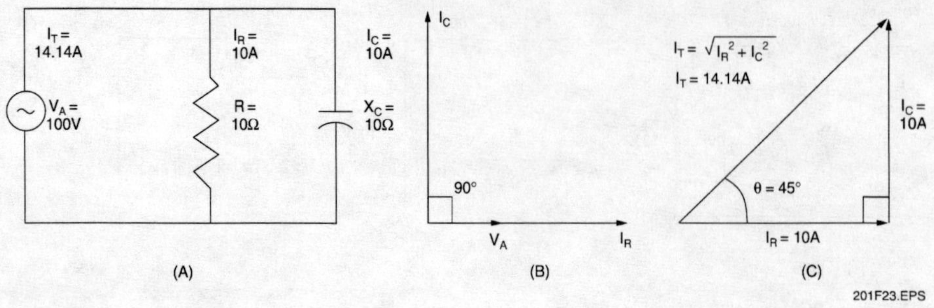

Figure 23 ◈ Parallel RC circuit with vector diagrams.

Classroom

Demonstrate how impedance is calculated in parallel RC circuits.

Discuss the characteristics of LC circuits.

Current within an individual branch of an RC parallel circuit is dependent on the voltage across the branch and on the R or X_C contained in the branch. The current in the resistive branch is in phase with the branch voltage, which is the applied voltage. The current in the capacitive branch leads V_A by 90°. Since the branch voltages are the same, I_C leads I_R by 90°, as shown in *Figure 23(B)*. Since the branch currents are out of phase, they have to be added vectorially to find the line current.

The phase angle, θ, is 45° because R and X_C are equal, resulting in equal branch currents. The phase angle is between the total current I_T and the generator voltage V_A. However, the phase of V_A is the same as the phase of I_R. Therefore, θ is also between I_T and I_R.

The impedance of a parallel RC circuit represents the total opposition to current flow offered by the resistance and capacitive reactance of the circuit. The equation for calculating the impedance of a parallel RC circuit is:

$$Z = \frac{RX_C}{\sqrt{I_R^2 + I_C^2}} \quad \textit{or} \quad Z = \frac{V_A}{I_T}$$

For the example shown in *Figure 23,* Z is:

$$Z = \frac{V_A}{I_T} = \frac{100}{14.14A} = 7.07\Omega$$

This is the opposition in ohms across the generator. This Z of 7.07Ω is equal to the resistance of 10Ω in parallel with the reactance of 10Ω. Notice that the impedance of equal values of R and X_C is not one-half, but equals 70.7% of either one.

When X_C is high relative to R, the parallel circuit is practically resistive because there is little leading capacitive current in the main line. The small value of I_C results from the high reactance of shunt X_C. The total impedance of the parallel circuit is approximately equal to the resistance, since the high value of X_C in a parallel branch has little effect.

As X_C becomes smaller, it provides more leading capacitive current in the main line. When X_C is very small relative to R, practically all the line current is the I_C component. The parallel circuit is practically all capacitive, with a total impedance practically equal to X_C.

The characteristics of different circuit arrangements are shown in *Table 4.*

9.3.0 LC Circuits

An LC circuit consists of an inductance and a capacitance connected in series or in parallel with a voltage source. There is no resistor physically in an LC circuit, but every circuit contains some resistance. Since the circuit resistance of the wiring and voltage source is usually so small, it has little or no effect on circuit operation.

Table 4 Parallel R and X_C Combinations

R (Ω)	X_C (Ω)	I_R (A)	I_C (A)	I_T (A) (Approx.)	Z_T (Ω) (Approx.)	Phase Angle (θ) (°)
1	10	10	1	$\sqrt{101} = 10$	1	5.7°
10	10	1	1	$\sqrt{2} = 1.4$	7.07	45°
10	1	1	10	$\sqrt{101} = 10$	1	84.3°

Instructor's Notes:

142 CHAPTER 4

In a circuit with both X_L and X_C, the opposite phase angles enable one to cancel the effect of the other. For X_L and X_C in series, the net reactance is the difference between the two series reactances, resulting in less reactance than either one. In parallel circuits, the I_L and I_C branch currents cancel. The net line current is then the difference between the two branch currents, resulting in less total line current than either branch current.

9.3.1 Series LC Circuit

As in all series circuits, the current in a series LC circuit is the same at all points. Therefore, the current in the inductor is the same as, and in phase with, the current in the capacitor. Because of this, on the vector diagram for a series LC circuit, the direction of the current vector is the reference or in the 0° direction, as shown in *Figure 24.*

When there is current flow in a series LC circuit, the voltage drops across the inductor and capacitor depend on the circuit current and the values of X_L and X_C. The voltage drop across the inductor leads the circuit current by 90°, and the voltage drop across the capacitor lags the circuit current by 90°. Using Kirchhoff's voltage law, the source voltage equals the sum of the voltage drops across the inductor and capacitor, with respect to the polarity of each.

Since the current through both is the same, the voltage across the inductor leads that across the capacitor by 180°. The method used to add the two voltage vectors is to subtract the smaller vector from the larger, and assign the resultant the direction of the larger. When applied to a series LC circuit, this means the applied voltage is equal to the difference between the voltage drops (E_L and E_C), with the phase angle between the applied voltage (E_T) and the circuit current determined by the larger voltage drop.

In a series LC circuit, one or both of the voltage drops are always greater than the applied voltage. Remember that although one or both of the voltage drops are greater than the applied voltage, they are 180° out of phase. One of them effectively cancels a portion of the other so that the total voltage drop is always equal to the applied voltage.

Recall that X_L is 180° out of phase with X_C. The impedance is then the vector sum of the two reactances. The reactances are 180° apart, so their vector sum is found by subtracting the smaller one from the larger.

Unlike RL and RC circuits, the impedance in an LC circuit is either purely inductive or purely capacitive.

9.3.2 Parallel LC Circuit

In a parallel LC circuit there is an inductance and a capacitance connected in parallel across a voltage source. *Figure 25* shows a parallel LC circuit with its vector diagram.

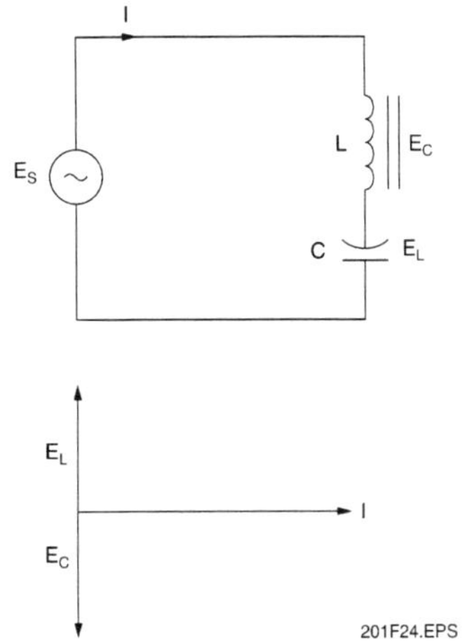

Figure 24 ◈ Series LC circuit with vector diagram.

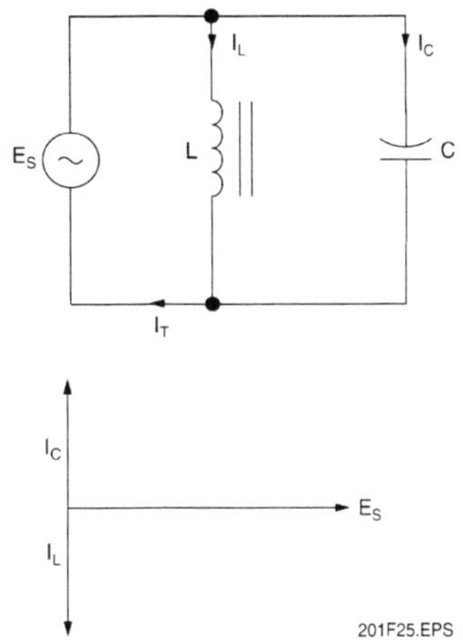

Figure 25 ◈ Parallel LC circuit with vector diagram.

Classroom

Discuss series LC circuits.

Audiovisual

Show Transparency 25 (Figure 24).

Audiovisual

Discuss parallel LC circuits.

Classroom

Show Transparency 26 (Figure 25).

ALTERNATING CURRENT 143

As in any parallel circuit, the voltage across the branches is the same as the applied voltage. Since they are actually the same voltage, the branch voltages and applied voltage are in phase. Because of this, the voltage is used as the 0° phase reference and the phases of the other circuit quantities are expressed in relation to the voltage.

The currents in the branches of a parallel LC circuit are both out of phase with the circuit voltage. The current in the inductive branch (I_L) lags the voltage by 90°, while the current in the capacitive branch (I_C) leads the voltage by 90°. Since the voltage is the same for both branches, currents I_L and I_C are therefore 180° out of phase. The amplitudes of the branch currents depend on the value of the reactance in the respective branches.

With the branch currents being 180° out of phase, the line current is equal to their vector sum. This vector addition is done by subtracting the smaller branch current from the larger.

The line current for a parallel LC circuit, therefore, has the phase characteristics of the larger branch current. Thus, if the inductive branch current is the larger, the line current is inductive and lags the applied voltage by 90°; if the capacitive branch current is the larger, the line current is capacitive, and leads the applied voltage by 90°.

The line current in a parallel LC circuit is always less than one of the branch currents and sometimes less than both. The reason that the line current is less than the branch currents is because the two branch currents are 180° out of phase. As a result of the phase difference, some cancellation takes place between the two currents when they combine to produce the line current. The impedance of a parallel LC circuit can be found using the following equations:

Classroom

Demonstrate how impedance is calculated in parallel LC circuits.

$$Z = \frac{X_L \times X_C}{X_L - X_C} \text{ (for } X_L \text{ larger than } X_C\text{)}$$

or:

Discuss series RLC circuits.

$$Z = \frac{X_L \times X_C}{X_C - X_L} \text{ (for } X_C \text{ larger than } X_L\text{)}$$

When using these equations, the impedance will have the phase characteristics of the smaller reactance.

Audiovisual

Show Transparency 27 (Figure 26).

9.4.0 RLC Circuits

9.4.1 Series RLC Circuit

Circuits in which the inductance, capacitance, and resistance are all connected in series are called *series RLC circuits*. The fundamental properties of series RLC circuits are similar to those for series LC circuits. The differences are caused by the effects of the resistance. Any practical series LC circuit contains some resistance. When the resistance is very small compared to the circuit reactance, it has almost no effect on the circuit and can be considered as zero. When the resistance is appreciable, though, it has a significant effect on the circuit operation and therefore must be considered in any circuit analysis. In a series RLC circuit, the same current flows through each component. The phase relationships between the voltage drops are the same as they were in series RC, RL, and LC circuits. The voltage drops across the inductance and capacitance are 180° out of phase. With current the same throughout the circuit as a reference, the inductive voltage drop (E_L) leads the resistive voltage drop (E_R) by 90°, and the capacitive voltage drop (E_C) lags the resistive voltage drop by 90°.

Figure 26 shows a series RLC circuit and the vector diagram used to determine the applied voltage. The vector sum of the three voltage drops is equal to the applied voltage. However, to calculate this vector sum, a combination of the methods learned for LC, RL, and RC circuits must be used. First, calculate the combined voltage drop of the two reactances. This value is designated E_X and is found as in pure LC circuits by subtracting the smaller reactive voltage drop from the larger. This is shown in *Figure 26* as E_X. The result of this calculation is the net reactive voltage drop and is either inductive or capacitive, depending on which of the individual voltage drops is larger. In *Figure 26*, the net reactive voltage drop is inductive since $E_L > E_C$. Once the net reactive voltage drop is known, it is added vectorially to the voltage drop across the resistance.

The angle between the applied voltage E_A and the voltage across the resistance E_R is the same as the phase angle between E_A and the circuit current. The reason for this is that E_R and I are in phase.

The impedance of a series RLC circuit is the vector sum of the inductive reactance, the capacitive reactance, and the resistance. This is done using the same method as for voltage drop calculations.

When X_L is greater than X_C, the net reactance is inductive, and the circuit acts essentially as an RL circuit. Similarly, when X_C is greater than X_L, the net reactance is capacitive, and the circuit acts as an RC circuit.

The same current flows in every part of a series RLC circuit. The current always leads the voltage across the capacitance by 90° and is in phase with

Instructor's Notes:

144 CHAPTER 4

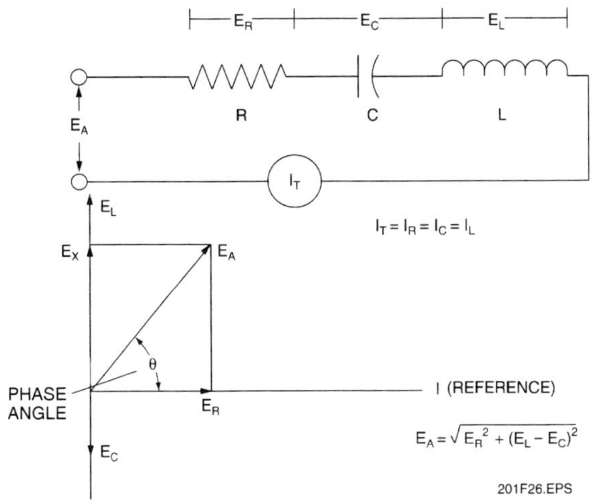

Figure 26 ◈ Series RLC circuit and vector diagram.

the voltage across the resistance. The phase relationship between the current and the applied voltage, however, depends on the circuit impedance. If the impedance is inductive (X_L greater than X_C), the current is inductive and lags the applied voltage by some phase angle less than 90°. If the impedance is capacitive (X_C greater than X_L), the current is capacitive, and leads the applied voltage by some phase angle also less than 90°. The angle of the lead or lag is determined by the relative values of the net reactance and the resistance.

The greater the value of X or the smaller the value of R, the larger the phase angle, and the more reactive (or less resistive) the current. Similarly, the smaller the value of X or the larger the value of R, the more resistive (or less reactive) the current. If either R or X is 10 or more times greater than the other, the circuit will essentially act as though it is purely resistive or reactive, as the case may be.

9.4.2 Series Resonance

Recall that:

$$X_L = 2\pi fL$$

and:

$$X_C = \frac{1}{2\pi fC}$$

With X_L and X_C being frequency sensitive, any change in frequency will affect the operating characteristics of any reactive circuit.

The effects of frequency variations on the input voltage of a series RLC circuit are shown in *Figure 27* and summarized as follows:

- As frequency is increased, X_L will become larger and X_C will become smaller. As a result, the circuit becomes even more inductive, θ increases, and the voltage across L will increase.
- As frequency decreases, X_L will become smaller and X_C larger, and θ will decrease toward zero.
- At a certain frequency, X_L will equal X_C. This is referred to as **resonance.**
- A further decrease in frequency will make X_C larger than X_L. The circuit will become capacitive and will increase in a negative direction.

To determine the resonant frequency at which $X_L = X_C$ requires only a bit of mathematics:

$$X_L = X_C$$

$$X_L = 2\pi fL$$

$$X_C = \frac{1}{2\pi fC}$$

Therefore, at resonance:

$$2\pi fL = \frac{1}{2\pi fC}$$

Discuss series resonance and demonstrate how it is calculated.

Show Transparency 28 (Figure 27).

or:

$$f_r = \text{resonant frequency} = \frac{1}{2\pi\sqrt{LC}}$$

Figure 27 shows a vector diagram of the resistance and reactance at a frequency when $X_L = X_C$, or resonance. Since X_L and X_C are equal and 180° out of phase, the algebraic sum of X_L and X_C is zero. The only opposition to current flow in the circuit is R. At series resonance:

$$Z = R$$

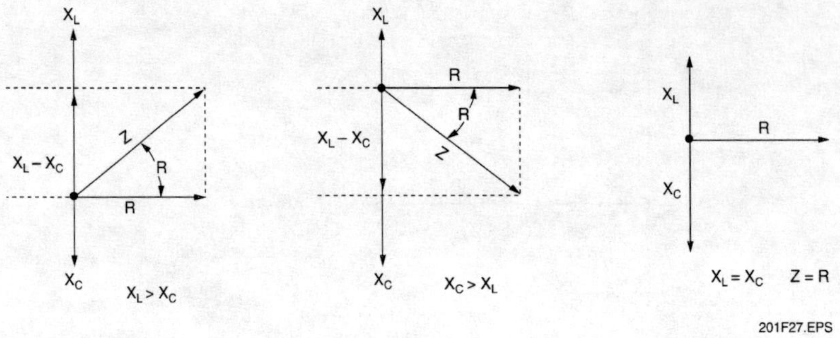

Figure 27 ◆ Effects of frequency variations on an RLC circuit.

Teaching Tip

Discuss the "Think About It." See the Teaching Tip at the end of this module.

THINK ABOUT IT

AC Circuits

The photo below shows a simple series circuit comprised of an ON/OFF switch, small lamp, motor, and capacitor. How would you classify this circuit? When energized, which components insert resistance, inductive reactance, and capacitive reactance into the circuit?

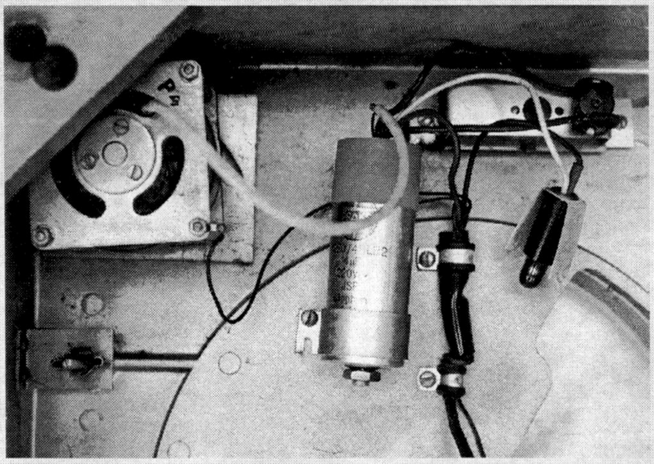

201P0101.EPS

Instructor's Notes:

At frequencies above resonance, X_L is greater than X_C. The circuit becomes inductive and Z increases. At frequencies below resonance, X_C is greater than X_L. The circuit becomes capacitive and Z increases. The point of lowest impedance of the circuit is at resonance. These characteristics are shown in *Figure 28.*

The impedance of a series RLC circuit is minimum at resonance, so the current must therefore be maximum. Both above and below the resonant frequency circuit impedance increases, which means that current decreases. The farther the frequency is from the resonant frequency, the greater the impedance, and the smaller the current becomes. At any frequency, the current can be calculated from Ohm's law for AC circuits using the equation I = E/Z. Since at the resonant frequency the impedance equals the resistance, the equation for current at resonance becomes I = E/R.

The letter Q is used to designate the quality of a tuned circuit. It is an indication of its maximum response as well as its ability to respond within a band of frequencies.

To secure maximum currents and response, the resistance must be kept at a low value. At resonance, R is the only resistance in the circuit. The Q of a circuit is the relationship of the reactance of the circuit to its resistance:

$$Q = \frac{X_L}{R} = \frac{X_C}{R}$$

The bandwidth or bandpass of a tuned circuit is defined as those frequency limits above and below resonant frequency where the response of the circuit will drop to .707 of its peak response. If current or voltage drops to .707 of its peak value, the power drops to 50%.

Bandwidth is frequency above and below resonance where power drops to one-half of its peak value. These are called the *half-power points.*

If the frequency of an RLC circuit is varied and the values of current at the different frequencies are plotted on a graph, the result is a curve known as the *resonance curve* of the circuit, as shown in *Figure 29.*

Actually, *Figure 29* shows two curves: one with a high resistance that results in a lower current flow, low Q, and wide bandwidth; and the second with a low resistance that results in high current flow at resonance, high Q, and low bandwidth.

9.4.3 Parallel RLC Circuit

A parallel RLC circuit is basically a parallel LC circuit with an added parallel branch of resistance. The solution of a parallel circuit involves the solution of a parallel LC circuit, and then the solution of either a parallel RL circuit or a parallel RC circuit. The reason for this is that a parallel combination of L and C appears to the source as a pure L or a pure C. So by solving the LC portion of a parallel RLC circuit first, the circuit is reduced to an equivalent RL or RC circuit.

The distribution of the voltage in a parallel RLC circuit is no different from what it is in a parallel LC circuit, or in any parallel circuit. The branch voltages are all equal and in phase, since they are

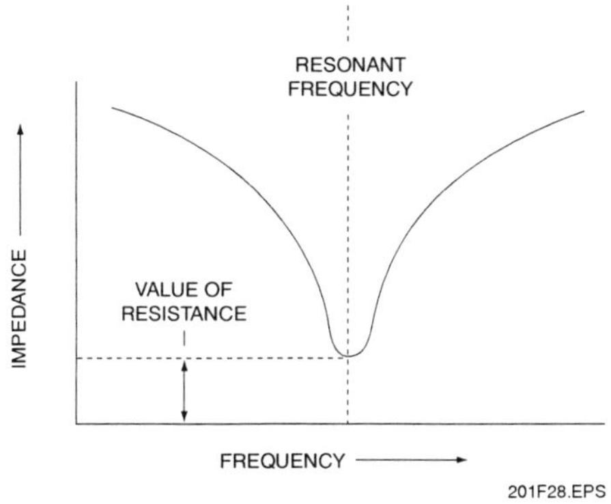

Figure 28 ◈ Frequency-impedance curve.

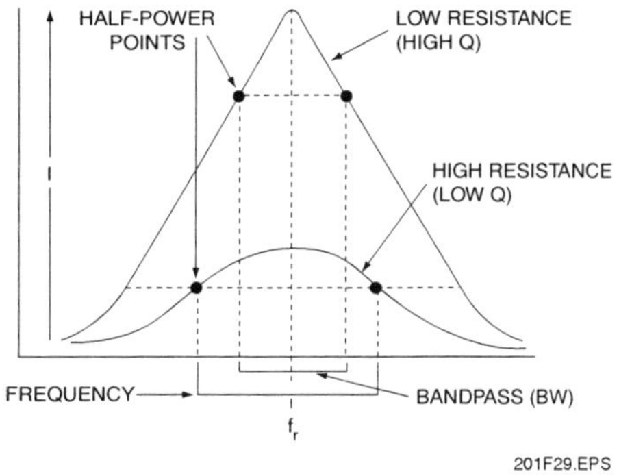

Figure 29 ◈ Typical series resonance curve.

Classroom

Discuss circuit quality and demonstrate how it is calculated in series resonant circuits.

Discuss resonance curves and bandwidths.

Audiovisual

Show Transparencies 29 and 30 (Figures 28 and 29).

Classroom

Discuss parallel RLC circuits.

Audiovisual

Show Transparency 31 (Figure 30).

Classroom

Discuss parallel resonance and demonstrate how it is calculated.

the same as the applied voltage. The resistance is simply another branch across which the applied voltage appears. Because the voltages throughout the circuit are the same, the applied voltage is again used as the θ phase reference.

Figure 30 shows the current relationship in a parallel RLC circuit.

The three branch currents in a parallel RLC circuit are an inductive current I_L, a capacitive current I_C, and a resistive current I_R. Each is independent of the other, and depends only on the applied voltage and the branch resistance or reactance.

The three branch currents all have different phases with respect to the branch voltages. I_L lags the voltage by 90°, I_C leads the voltage by 90°, and I_R is in phase with the voltage. Since the voltages are the same, I_L and I_C are 180° out of phase with each other, and both are 90° out of phase with I_R. Because I_R is in phase with the voltage, it has the same zero-reference direction as the voltage. So I_C leads I_R by 90°, and I_L lags I_R by 90°.

The line current (I_T), or total current, is the vector sum of the three branch currents, and can be calculated by adding I_L, I_C, and I_R vectorially. Whether the line current leads or lags the applied voltage depends on which of the reactive branch currents (I_L or I_C) is the larger. If I_L is larger, I_T lags the applied voltage. If I_C is larger, I_T leads the applied voltage.

To determine the impedance of a parallel RLC circuit, first determine the net reactance X of the inductive and capacitive branches. Then use X to determine the impedance Z, the same as in a parallel RL or RC circuit.

Whenever Z is inductive, the line current will lag the applied voltage. Similarly, when Z is capacitive, the line current will lead the applied voltage.

9.4.4 Parallel Resonance

A parallel resonant circuit is a circuit in which the voltage source is in parallel with L and C. The characteristics of parallel resonance are quite different from those of series resonance. However, the frequency at which parallel resonance takes place is identical to the frequency at which series resonance takes place. Therefore, parallel resonance uses the same formula as series resonance:

$$f_r = \frac{1}{2\pi\sqrt{LC}}$$

The properties of a parallel resonant circuit are based on the action that takes place between the parallel inductance and capacitance, which is often called a *tank circuit* because it has the ability to store electrical energy.

The action of a tank circuit is basically one of interchange of energy between the inductance and capacitance. If a voltage is momentarily applied across the tank circuit, C charges to this voltage. When the applied voltage is removed, C discharges through L, and a magnetic field is built up around L by the discharge current. When C has discharged, the field around L collapses, and in doing so induces a current that is in the same direction as the current that created the field. This current, therefore, charges C in the opposite direc-

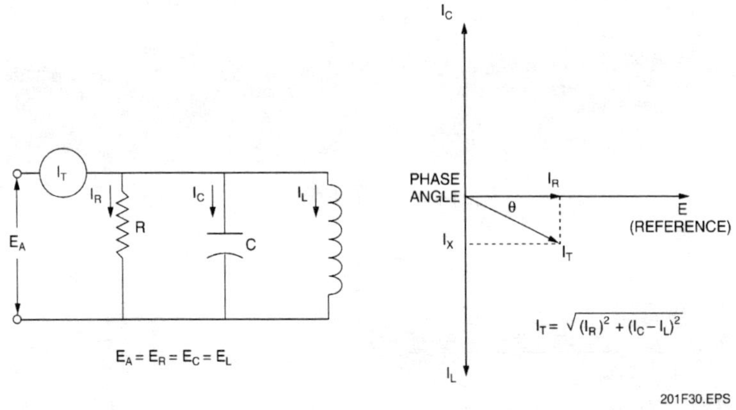

Figure 30 ◈ Parallel RLC circuit and vector diagram.

Instructor's Notes:

tion. When the field around L has collapsed, C again discharges, but this time in the direction opposite to before. The discharge current again causes a magnetic field around L, which, when it collapses, charges C in the same direction in which it was initially charged.

This interchange of energy and the circulating current it produces would continue indefinitely, producing a series of sine waves, if this were an ideal tank circuit with no resistance. However, since some resistance is always present, the circulating current gradually diminishes as the resistance dissipates the energy in the circuit in the form of heat. This causes the sine wave current to be damped out. If a voltage were again momentarily applied across the circuit, the interchange of energy and accompanying circulating current would begin again.

At resonance, X_L equals X_C, so the two currents I_L and I_C are also equal. Because the two currents in a parallel LC circuit are 180° out of phase, the line current, which is their vector sum, must be zero. Thus, the only current is the circulating current in the tank circuit. No line current flows, therefore the circuit has infinite impedance as far as the voltage source is concerned.

These two conditions of zero line current and infinite impedance are characteristic of ideal parallel resonant circuits at resonance. In practical circuits that contain some resistance, the theoretical conditions of zero line current and infinite impedance are not realized. Instead, practical parallel resonant circuits have minimum line current and maximum impedance at resonance. This is the exact opposite of series resonant circuits, which have maximum current and minimum impedance at resonance.

In the ideal parallel resonant circuit at resonance, the branch currents I_L and I_C are equal, so the line current is zero and the circuit impedance is infinite. Above and below the resonant frequency, one of the reactances X_L or X_C is larger than the other. The two branch currents are therefore unequal, and the line current, which equals their vector sum (or arithmetic difference), has some value greater than zero. Since line current flows, the circuit impedance is no longer infinite. The further the frequency is from the resonant frequency, the greater the difference between the values of the reactances. As a result, the line current is larger and the circuit impedance is smaller.

The principal effect of the resistance in a parallel resonant circuit is that it causes the current in the inductive branch to lag the applied voltage by a phase angle of less than 90°, instead of exactly 90° as in the case of the ideal circuit. As a result, the two branch currents are not 180° out of phase. For simplicity, resonance can still be considered as occurring when X_L equals X_C, but now when the two branch currents are added vectorially, their sum is not zero. This means that at resonance, some line current flows. Since there is line current, the impedance cannot be infinite, as it is in the ideal circuit. Thus at resonance, practical parallel resonant circuits have minimum line current and maximum resistance, instead of zero line current and infinite impedance, as do ideal circuits.

For parallel resonance, Q also measures the quality of a circuit. In parallel resonant circuits, Q depends on circuit resistance. The Q of a parallel resonant circuit is defined as:

$$Q = \frac{X_L}{R} \text{ or } \frac{X_C}{R}$$

Recognize this as the same equation used for the Q of a series resonant circuit. As a result, resistance has the same effect on the Q of a parallel resonant circuit as it does on a series resonant circuit. The lower the resistance, the higher the Q of the circuit and the narrower its bandpass. Conversely, the greater the resistance, the lower the Q and the wider the bandpass.

Recall that for every series resonant circuit there is a range of frequencies above and below the resonant frequency at which, for practical purposes, the circuit can be considered as being at resonance. This range of frequencies is called the *bandwidth*, and consists of all the frequencies at which the circuit current was 0.707 or more times its value at resonance. Parallel resonant circuits also have a bandwidth, but it is defined in terms of the frequency-vs.-impedance curve, and consists of all the frequencies that produce a circuit impedance 0.707 or more times the impedance at resonance. *Figure 31* shows the bandpass or bandwidth as all the frequencies between F_1 and F_2.

Circuit resistance affects the width and steepness of the frequency-impedance curve. Therefore, resistance affects the circuit bandpass. A low resistive circuit causes a steep curve and narrow bandpass. A high resistive circuit causes a flatter frequency-impedance curve and therefore, a wide bandpass.

10.0.0 ◆ POWER IN AC CIRCUITS

In DC circuits, the power consumed is the sum of all the I^2R heating in the resistors. It is also equal to the power produced by the source, which is the product of the source voltage and current. In AC

Classroom

Demonstrate how to calculate the quality of a parallel resonant circuit. Note that this is the same calculation used for series resonant circuits.

Discuss bandwidth in parallel resonant circuits.

Audiovisual

Show Transparency 32 (Figure 31).

Classroom

Discuss power in AC circuits.

ALTERNATING CURRENT 149

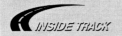

INSIDE TRACK

The Importance of Resonance

A common use of RCL circuits is in resonant filter applications. Because RCL circuits provide a means for frequency selectivity, resonance in electrical circuits is very important to many types of electrical and electronic systems. Series and/or parallel RCL resonant filters are widely used to filter out or reduce the effects of unwanted harmonic frequencies and/or unwanted electromagnetic interference (EMI) signals present in power distribution circuits. Resonant filters are also used to increase the efficiency or power factor of a circuit or system.

In communications, the ability of a radio or television receiver to select a certain frequency transmitted by a certain station, and eliminate frequencies from other stations, is made possible by the use of series and/or parallel RCL resonant circuits.

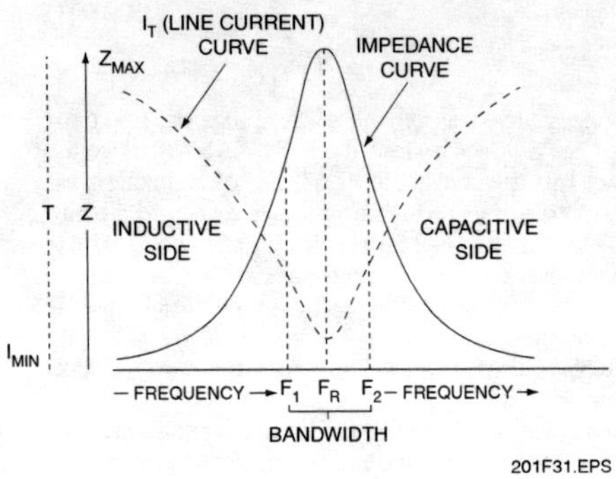

Figure 31 ◆ Tuned parallel circuit curves.

circuits containing only resistors, the above relationship also holds true.

10.1.0 True Power

Classroom

Discuss true power and explain how it is calculated.

Discuss apparent power and explain how it is calculated.

The power consumed by resistance is called *true power* and is measured in units of watts. True power is the product of the resistor current squared and the resistance:

$$P_T = I^2R$$

This formula applies because current and voltage have the same phase across a resistance.

To find the corresponding value of power as a product of voltage and current, this product must be multiplied by the cosine of the phase angle θ:

$$P_T = I^2R \text{ or } P_T = EI\cos\theta$$

Where E and I are in rms values to calculate the true power in watts, multiplying I by the cosine of the phase angle provides the resistive component for true power equal to I^2R.

For example, a series RL circuit has 2A through a 100Ω R in series with the X_L of 173Ω. Therefore:

$$P_T = I^2R$$
$$P_T = 4 \times 100$$
$$P_T = 400W$$

Furthermore, in this circuit the phase angle is 60° with a cosine of 0.5. The applied voltage is 400V. Therefore:

$$P_T = EI\cos\theta$$
$$P_T = 400 \times 2 \times 0.5$$
$$P_T = 400W$$

In both cases, the true power is the same (400W) because this is the amount of power supplied by the generator and dissipated in the resistance. Either formula can be used for calculating the true power.

10.2.0 Apparent Power

In ideal AC circuits containing resistors, capacitors, and inductors, the only mechanism for power consumption is $I^2_{eff}R$ heating in the resistors. Inductors and capacitors consume no power. The only function of inductors and capacitors is to store and release energy. However, because of the phase shifts that are introduced by these elements, the power consumed by the resistors is not equal to the product of the source voltage and current. The product of the source voltage and current is called *apparent power* and has units of volt-amperes (VA).

The apparent power is the product of the source voltage and the total current. Therefore, apparent

Instructor's Notes:

150 CHAPTER 4

power is actual power delivered by the source. The formula for apparent power is:

$$P_A = (E_A)(I)$$

Figure 32 shows a series RL circuit and its associated vector diagram.

This circuit is used to calculate the apparent power and compare it to the circuit's true power:

$P_A = (E_A)(I)$

$P_A = (400V)(2A)$

$P_A = 800VA$

$P_T = EI \cos \theta$

$\theta = \frac{R}{X_L} = \frac{173}{100} = 60°$

$P_T = (400V)(2A)(\cos 60°)$

$P_T = (400V)(2A)(0.5)$

$P_T = 400W$

Note that the apparent power formula is the product of EI alone without considering the cosine of the phase angle.

10.3.0 Reactive Power

Reactive power is that portion of the apparent power that is caused by inductors and capacitors in the circuit. Inductance and capacitance are always present in real AC circuits. No work is performed by reactive power; the power is stored in the inductors and capacitors, then returned to the circuit. Therefore, reactive power is always 90° out of phase with true power. The units for reactive power are volt-amperes-reactive (VARs).

In general, for any phase angle θ between E and I, multiplying EI by sine θ gives the vertical component at 90° for the value of the VARs. In *Figure 32,* the value of sine 60° is 800 × 0.866 = 692.8 VARs.

Note that the factor sine θ for the VARs gives the vertical or reactive component of the apparent power EI. However, multiplying EI by cosine θ as the power factor gives the horizontal or resistive component for the real power.

10.4.0 Power Factor

Because it indicates the resistive component, cosine θ is the power factor (pf) of the circuit, converting the EI product to real power. For series circuits, use the formula:

$$pf = \cos \theta = \frac{R}{Z}$$

For parallel circuits, use the formula:

$$pf = \cos \theta = \frac{I_R}{I_T}$$

In *Figure 32* as an example of a series circuit, R and Z are used for the calculations:

$$pf = \cos \theta = \frac{R}{Z} = \frac{100\Omega}{200\Omega} = 0.5$$

The power factor is not an angular measure but a numerical ratio with a value between 0 and 1, equal to the cosine of the phase angle. With all resistance and zero reactance, R and Z are the same for a series circuit of I_R and I_T and are the same for a parallel circuit. The ratio is 1. Therefore, unity power factor means a resistive circuit. At the opposite extreme, all reactance with zero resistance makes the power factor zero, meaning that the circuit is all reactive.

The power factor gives the relationship between apparent power and true power. The power factor can thus be defined as the ratio of true power to apparent power:

$$pf = \frac{P_T}{P_A}$$

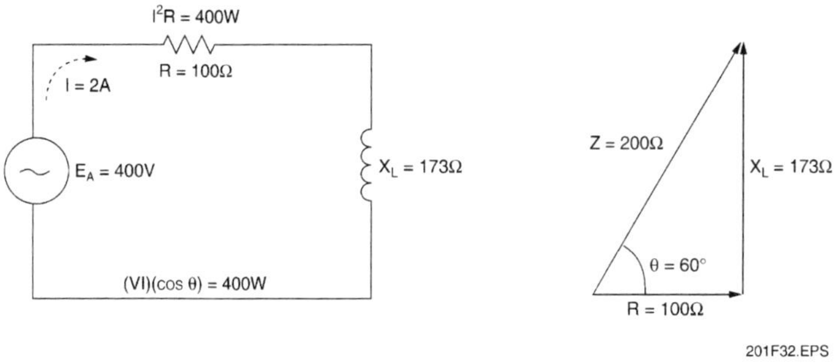

Figure 32 ◈ Power calculations in an AC circuit.

Audiovisual

Show Transparency 33 (Figure 32).

Classroom

Discuss reactive power and explain how it is calculated.

Discuss power factors and explain how they are calculated.

Audiovisual

Show Transparency 34 (Figure 33).

Classroom

Work through the power calculations provided.

Discuss the power triangle and the Pythagorean theorem.

Audiovisual

Show Transparency 35 (Figure 34).

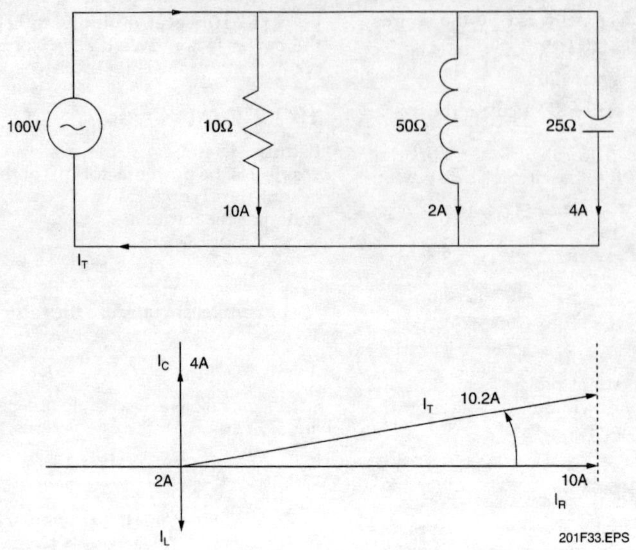

Figure 33 ◆ RLC circuit calculation.

For example, calculate the power factor of the circuit shown in *Figure 33.*

The true power is the product of the resistor current squared and the resistance:

$$P_T = I^2R$$
$$P_T = 10A^2 \times 10\Omega$$
$$P_T = 1{,}000W$$

The apparent power is the product of the source voltage and total current:

$$P_A = (I_T)(E)$$
$$P_A = 10.2A \times 100V$$
$$P_A = 1{,}020VA$$

Calculating total current:

$$I_T = \sqrt{I_R^2 + (I_C - I_L)^2} = \sqrt{10A^2 + (4A - 2A)^2}$$
$$I_T = 10.2A$$

The power factor is the ratio of true power to apparent power:

$$pf = \frac{P_T}{P_A}$$
$$pf = \frac{1{,}000}{1{,}020}$$
$$pf = 0.98$$

As illustrated in the previous example, the power factor is determined by the system load. If the load contained only resistance, the apparent power would equal the true power and the power factor would be at its maximum value of one. Purely resistive circuits have a power factor of unity or one. If the load is more inductive than capacitive, the apparent power will lag the true power and the power factor will be lagging. If the load is more capacitive than inductive, the apparent power will lead the true power and the power factor will be leading. If there is any reactive load on the system, the apparent power will be greater than the true power and the power factor will be less than one.

10.5.0 Power Triangle

The phase relationships among the three types of AC power are easily visualized on the power triangle shown in *Figure 34.* The true power (W) is the horizontal leg, the apparent power (VA) is the hypotenuse, and the cosine of the phase angle between them is the power factor. The vertical leg of the triangle is the reactive power and has units of volt-amperes-reactive (VARs).

As illustrated on the power triangle (*Figure 34*), the apparent power will always be greater

Instructor's Notes:

$$P_A = \sqrt{(P_T^2) + (P_{RX}^2)}$$

$$P_T = \sqrt{(P_A^2) - (P_{RX}^2)}$$

$$P_{RX} = \sqrt{(P_A^2) - (P_T^2)}$$

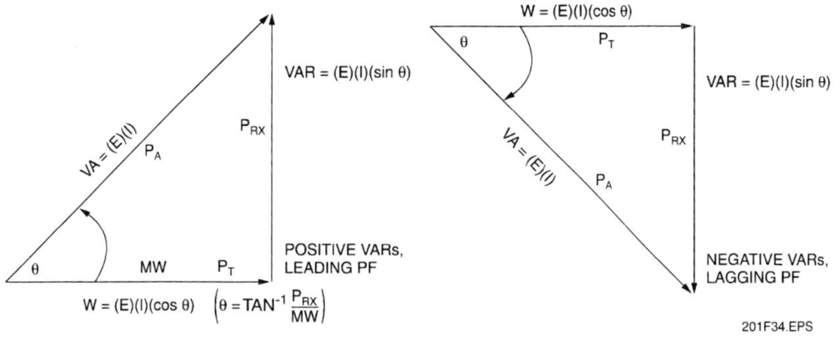

Figure 34 ◈ Power triangle.

than the true power or reactive power. Also, the apparent power is the result of the vector addition of true and reactive power. The power magnitude relationships shown in *Figure 34* can be derived from the Pythagorean theorem for right triangles:

$$c^2 = a^2 + b^2$$

Therefore, c also equals the square root of $a^2 + b^2$, as shown below:

$$c = \sqrt{a^2 + b^2}$$

11.0.0 ◈ TRANSFORMERS

A transformer is a device that transfers electrical energy from one circuit to another by electromagnetic induction (transformer action). The electrical energy is always transferred without a change in frequency, but may involve changes in the magnitudes of voltage and current. Because a transformer works on the principle of electromagnetic induction, it must be used with an input source voltage that varies in amplitude.

11.1.0 Transformer Construction

Figure 35 shows the basic components of a transformer. In its most basic form, a transformer consists of:

- A primary coil or winding
- A secondary coil or winding
- A core that supports the coils or windings

THINK ABOUT IT

Power Factor

In power distribution circuits, it is desirable to achieve a power factor approaching a value of 1 in order to obtain the most efficient transfer of power. In AC circuits where there are large inductive loads such as in motors and transformers, the power factor can be considerably less than 1. For example, in a highly inductive motor circuit, if the voltage is 120V, the current is 12A, and the current lags the voltage by 60°, the power factor is .5 or 50% (cosine of 60° = 0.5). The apparent power is 1,440VA (120V × 12A), but the true power is only 720W [120V × (0.5 × 12A) = 720W]. This is a very inefficient circuit. What would you do to this circuit in order to achieve a circuit having a power factor as close to 1 as possible?

Classroom

Discuss transformers and their components.

Audiovisual

Show Transparency 36 (Figure 35).

Teaching Tip

Discuss the "Think About It." See the Teaching Tip at the end of this module.

ALTERNATING CURRENT 153

Transformers

INSIDE TRACK

Transformers are essential to all electrical systems and all types of electronic equipment. They are especially essential to the operation of AC high-voltage power distribution systems. Transformers are used to both step up voltage and step down voltage throughout the distribution process. For example, a typical power generation plant might generate AC power at 13,800V, step it up to 230,000V for distribution over long transmission lines, step it down to 13,800V again at substations located at different points for local distribution, and finally step it down again to 240V and 120V for lighting and local power use.

Classroom

Explain the operation of a basic transformer.

Audiovisual

Show Transparency 37 (Figure 36).

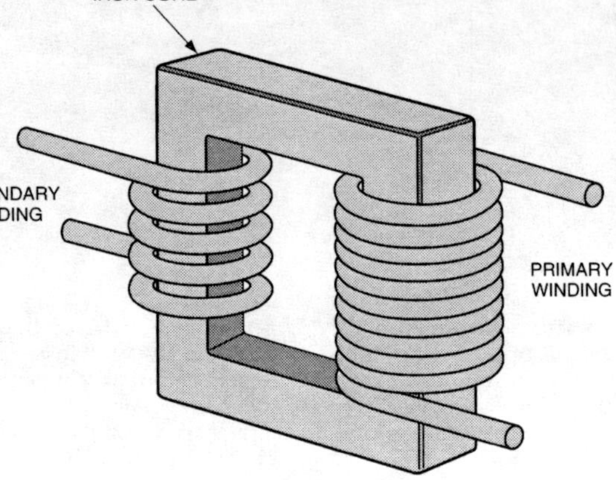

Figure 35 ◆ Basic components of a transformer.

Classroom

Discuss the materials used in core construction.

Audiovisual

Show Transparency 38 (Figure 37).

A simple transformer action is shown in *Figure 36*. The primary winding is connected to a 60Hz AC voltage source. The magnetic field or flux builds up (expands) and collapses (contracts) around the primary winding. The expanding and contracting magnetic field around the primary winding cuts the secondary winding and induces an alternating voltage into the winding. This voltage causes AC to flow through the load. The voltage may be stepped up or down depending on the design of the primary and secondary windings.

11.1.1 Core Characteristics

Commonly used core materials are air, soft iron, and steel. Each of these materials is suitable for particular applications and unsuitable for others. Generally, air-core transformers are used when the voltage source has a high frequency (above 20kHz). Iron-core transformers are usually used when the source frequency is low (below 20kHz). A soft-iron transformer is very useful where the transformer must be physically small yet efficient. The iron-core transformer provides better power transfer than the air-core transformer. Laminated sheets of steel are often used in a transformer to reduce one type of power loss known as *eddy currents*. These are undesirable currents, induced into the core, which circulate around the core. Laminating the core reduces these currents to smaller levels. These steel laminations are insulated with a nonconducting material, such as varnish, and then formed into a core as shown in *Figure 37*. It takes about 50 such laminations to make a core one-inch thick. The most efficient transformer core is one that offers the best path for the most lines of flux, with the least loss in magnetic and electrical energy.

Instructor's Notes:

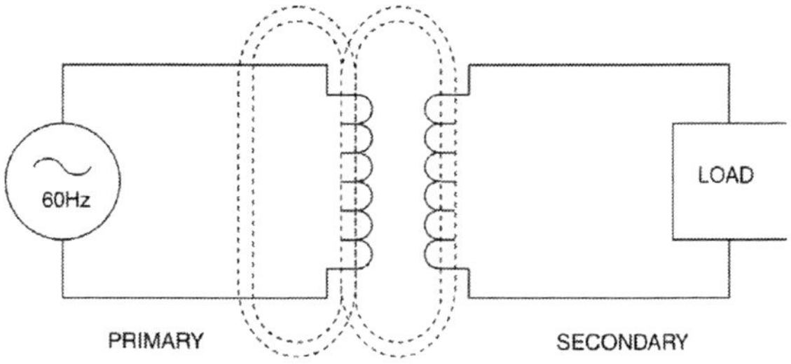

201F36.EPS

Figure 36 ◆ Transformer action.

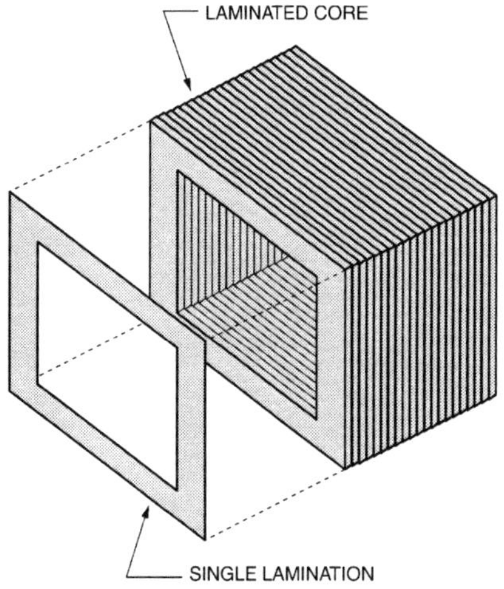

201F37.EPS

Figure 37 ◆ Steel laminated core.

11.1.2 Transformer Windings

A transformer consists of two coils called *windings,* which are wrapped around a core. The transformer operates when a source of AC voltage is connected to one of the windings and a load device is connected to the other. The winding that is connected to the source is called the *primary winding.* The winding that is connected to the load is called the *secondary winding. Figure 38* shows a cutaway view of a typical transformer.

The wire is coated with varnish so that each turn of the winding is insulated from every other turn. In a transformer designed for high-voltage applications, sheets of insulating material such as paper are placed between the layers of windings to provide additional insulation.

When the primary winding is completely wound, it is wrapped in insulating paper or cloth. The secondary winding is then wound on top of the primary winding. After the secondary winding is complete, it too is covered with insulating paper. Next, the iron core is inserted into and around the windings as shown.

Sometimes, terminals may be provided on the enclosure for connections to the windings. The figure shows four leads, two from the primary and two from the secondary. These leads are to be connected to the source and load, respectively.

11.2.0 Operating Characteristics

11.2.1 Energized with No Load

A no-load condition is said to exist when a voltage is applied to the primary, but no load is connected to the secondary. Assume the output of the secondary is connected to a load by a switch that is open. Because of the open switch, there is no current flowing in the secondary winding. With the switch open and an AC voltage applied to the primary, there is, however, a very small amount of current, called *exciting current,* flowing in the primary. Essentially, what this current does is excite the coil of the primary to create a magnetic field. The amount of exciting current is determined by three factors: the amount of voltage applied (E_A); the resistance (R) of the primary coil's wire and core losses; and the X_L, which is dependent on the frequency of the exciting current. These factors are all controlled by transformer design.

Classroom

Discuss the construction of transformer windings.

Audiovisual

Show Transparency 39 (Figure 38).

Classroom

Discuss the operating characteristics of transformers and explain the purpose of the exciting current (E_A).

ALTERNATING CURRENT 155

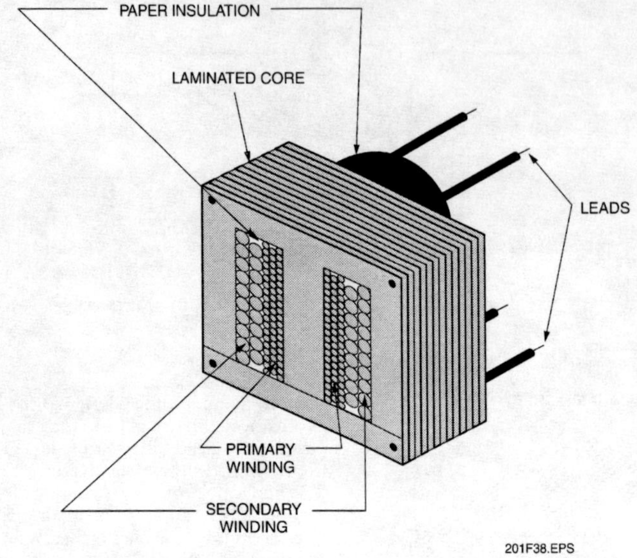

Figure 38 ◆ Cutaway view of a transformer core.

Classroom

Discuss the phase relationships in:
- **Like-wound transformers**
- **Unlike-wound transformers**

Audiovisual

Show Transparency 40 (Figure 39).

This very small amount of exciting current serves two functions:

- Most of the exciting energy is used to support the magnetic field of the primary.
- A small amount of energy is used to overcome the resistance of the wire and core. This is dissipated in the form of heat (power loss).

Exciting current will flow in the primary winding at all times to maintain this magnetic field, but no transfer of energy will take place as long as the secondary circuit is open.

11.2.2 Phase Relationship

The secondary voltage of a simple transformer may be either in phase or out of phase with the primary voltage. This depends on the direction in which the windings are wound and the arrangement of the connection to the external circuit (load). Simply, this means that the two voltages may rise and fall together, or one may rise while the other is falling. Transformers in which the secondary voltage is in phase with the primary are referred to as *like-wound transformers,* while those in which the voltages are 180° out of phase are called *unlike-wound transformers.*

Dots are used to indicate points on a transformer schematic symbol that have the same instantaneous polarity (points that are in phase). The use of phase-indicating dots is illustrated in *Figure 39.* In the first part of the figure, both the primary and secondary windings are wound from top to bottom in a clockwise direction, as viewed from above the windings. When constructed in this manner, the top lead of the primary and the top lead of the secondary have the same polarity. This is indicated by the dots on the transformer symbol.

The second part of the figure illustrates a transformer in which the primary and secondary are wound in opposite directions. As viewed from above the windings, the primary is wound in a clockwise direction from top to bottom, while the secondary is wound in a counterclockwise direction. Notice that the top leads of the primary and secondary have opposite polarities. This is indicated by the dots being placed on opposite ends of the transformer symbol. Thus, the polarity of voltage at the terminals of the transformer secondary depends on the direction in which the secondary is wound with respect to the primary.

Instructor's Notes:

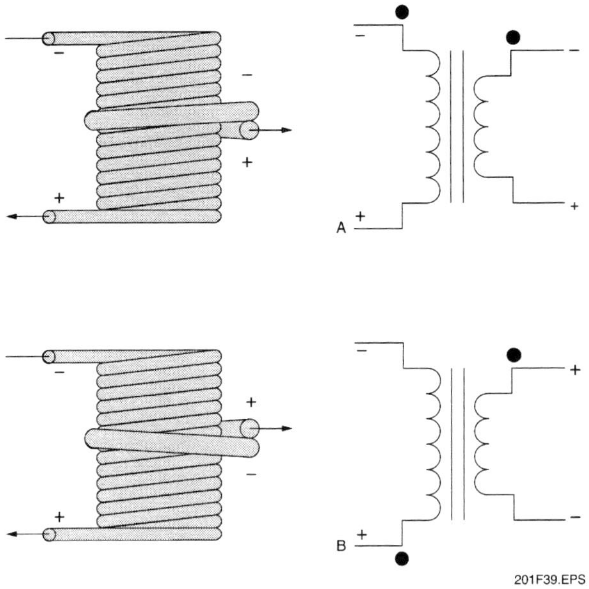

Figure 39 ◆ Transformer winding polarity.

11.3.0 Turns and Voltage Ratios

To understand how a transformer can be used to step up or step down voltage, the term *turns ratio* must be understood. The total voltage induced into the secondary winding of a transformer is determined mainly by the ratio of the number of turns in the primary to the number of turns in the secondary, and by the amount of voltage applied to the primary. Therefore, to set up a formula:

$$\text{Turns ratio} = \frac{\text{Number of turns in the primary}}{\text{Number of turns in the secondary}}$$

The first transformer in *Figure 40* shows a transformer whose primary consists of 10 turns of wire, and whose secondary consists of a single turn of wire. As lines of flux generated by the primary expand and collapse, they cut both the 10 turns of the primary and the single turn of the secondary. Since the length of the wire in the secondary is approximately the same as the length of the wire in each turn of the primary, the EMF induced into the secondary will be the same as the EMF induced into each turn of the primary.

This means that if the voltage applied to the primary winding is 10 volts, the CEMF in the primary is almost 10 volts. Thus, each turn in the primary will have an induced CEMF of approximately $\frac{1}{10}$ of the total applied voltage, or one volt. Since the same flux lines cut the turns in both the secondary and the primary, each turn will have an EMF of one volt induced into it. The first transformer in *Figure 40* has only one turn in the secondary, thus, the EMF across the secondary is one volt.

The second transformer represented in *Figure 40* has a 10-turn primary and a two-turn secondary. Since the flux induces one volt per turn, the total voltage across the secondary is two volts. Notice that the volts per turn are the same for both primary and secondary windings. Since the CEMF in the primary is equal (or almost) to the applied voltage, a proportion may be set up to express the value of the voltage induced in terms of the voltage applied to the primary and the number of

Classroom

Discuss transformer turns ratios and explain how they are calculated.

Audiovisual

Show Transparency 41 (Figure 40).

ALTERNATING CURRENT **157**

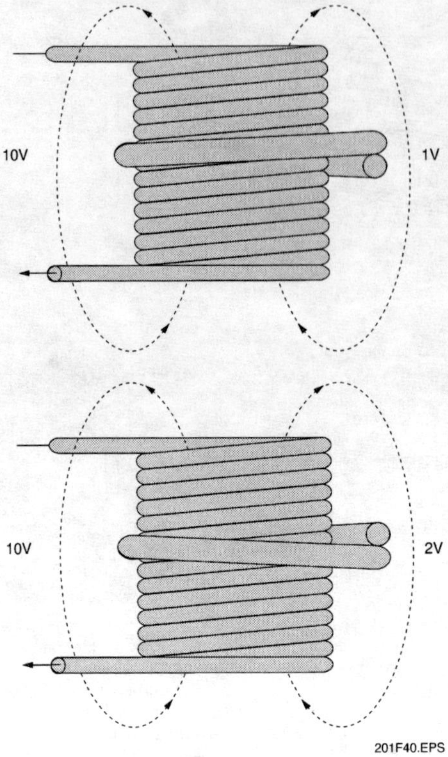

Figure 40 ◆ Transformer turns ratio.

Classroom

Discuss voltage ratios and explain how they are calculated.

turns in each winding. This proportion also shows the relationship between the number of turns in each winding and the voltage across each winding, and is expressed by the equation:

$$\frac{E_S}{E_P} = \frac{N_S}{N_P}$$

Where:

N_P = number of turns in the primary
E_P = voltage applied to the primary
E_S = voltage induced in the secondary
N_S = number of turns in the secondary

The equation shows that the ratio of secondary voltage to primary voltage is equal to the ratio of secondary turns to primary turns. The equation can be written as:

$$E_PN_S = E_SN_P$$

For example, a transformer has 100 turns in the primary, 50 turns in the secondary, and 120VAC applied to the primary (E_P). What is the voltage across the secondary (E_S)?

N_P = 100 turns $\quad E_P$ = 120VAC

N_S = 50 turns

$$\frac{E_SN_S}{E_PN_P} \text{ or } E_S = \frac{E_PN_S}{N_P}$$

$$E_S = \frac{120V \times 50 \text{ turns}}{100 \text{ turns}} = 60VAC$$

The transformers in *Figure 40* have fewer turns in the secondary than in the primary. As a result, there is less voltage across the secondary than across the primary. A transformer in which the

Instructor's Notes:

158 CHAPTER 4

THINK ABOUT IT

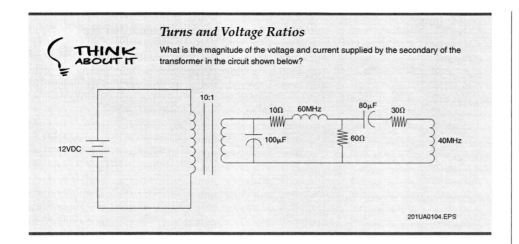

Turns and Voltage Ratios

What is the magnitude of the voltage and current supplied by the secondary of the transformer in the circuit shown below?

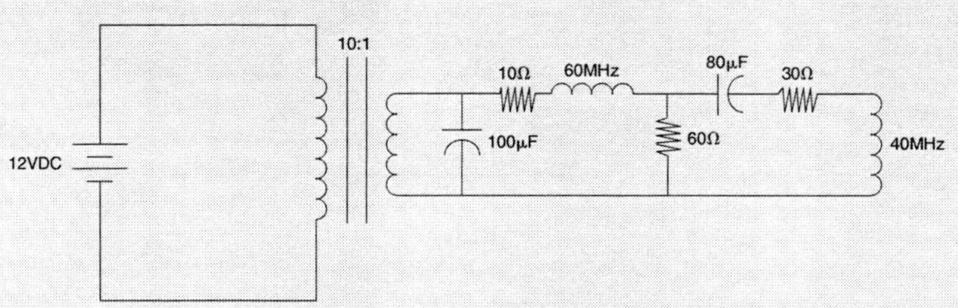

201UA0104.EPS

voltage across the secondary is less than the voltage across the primary is called a *step-down transformer.* The ratio of a 10-to-1 step-down transformer is written as 10:1.

A transformer that has fewer turns in the primary than in the secondary will produce a greater voltage across the secondary than the voltage applied to the primary. A transformer in which the voltage across the secondary is greater than the voltage applied to the primary is called a *step-up transformer.* The ratio of a 1-to-4 step-up transformer should be written 1:4. Notice in the two ratios that the value of the primary winding is always stated first.

11.4.0 Types of Transformers

Transformers are widely used to permit the use of trip coils and instruments of moderate current and voltage capacities and to measure the characteristics of high-voltage and high-current circuits. Since secondary voltage and current are directly related to primary voltage and current, measurements can be made under the low-voltage or low-current conditions of the secondary circuit and still determine primary characteristics. Tripping transformers and instrument transformers are examples of this use of transformers.

The primary or secondary coils of a transformer can be tapped to permit multiple input and output voltages. *Figure 41* shows several tapped transformers. The center-tapped transformer is particularly important because it can be used in conjunction with other components to convert an AC input to a DC output.

11.4.1 Isolation Transformer

Isolation transformers are wound so that their primary and secondary voltages are equal. Their purpose is to electrically isolate a piece of electrical equipment from the power distribution system.

Many pieces of electronic equipment use the metal chassis on which the components are mounted as part of the circuit (*Figure 42*). Personnel working with this equipment may accidentally come in contact with the chassis, completing the circuit to ground, and receive a shock as shown in *Figure 42(A).* If the resistances of their body and the ground path are low, the shock can be fatal. Placing an isolation transformer in the circuit as shown in *Figure 42(B)* breaks the ground current path that includes the worker. Current can no longer flow from the power supply through the chassis and worker to ground; however, the equipment is still supplied with the normal operating voltage and current.

11.4.2 Autotransformer

In a transformer, it is not necessary for the primary and secondary to be separate and distinct windings. *Figure 43* is a schematic diagram of what is known as an *autotransformer.* Note that a single coil of wire is tapped to produce what is electrically both a primary and a secondary winding.

Teaching Tip

Discuss the "Think About It." See the Teaching Tip at the end of this module.

Classroom

Discuss step-down and step-up transformers.

Audiovisual

Show Transparency 42 (Figure 41).

Classroom

Discuss common transformer types:
- **Isolation transformers**
- **Autotransformers**

Audiovisual

Show Transparencies 43 through 46 (Figures 42 through 45).

ALTERNATING CURRENT 159

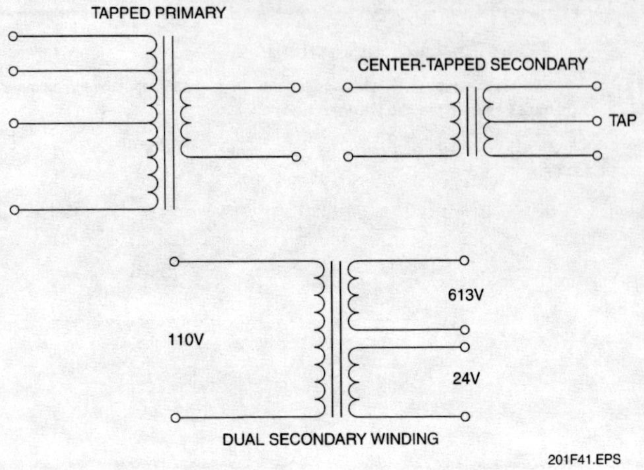

Figure 41 ◈ Tapped transformers.

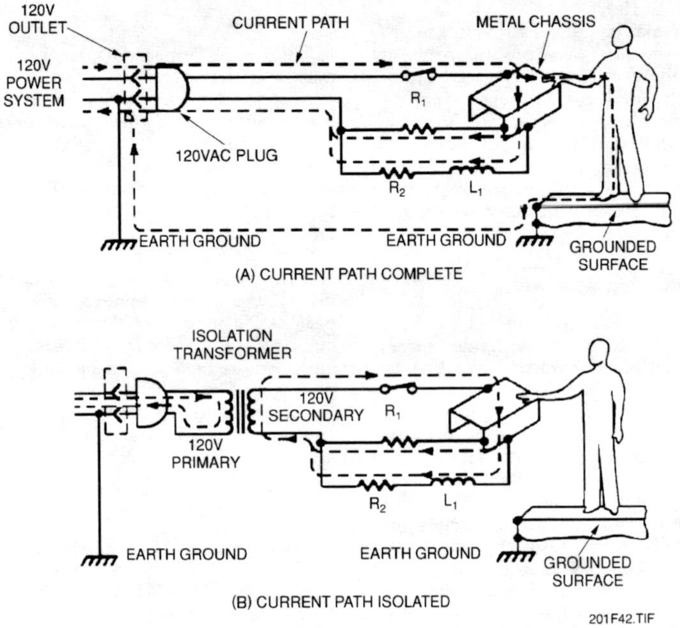

Figure 42 ◈ Importance of an isolation transformer.

Instructor's Notes:

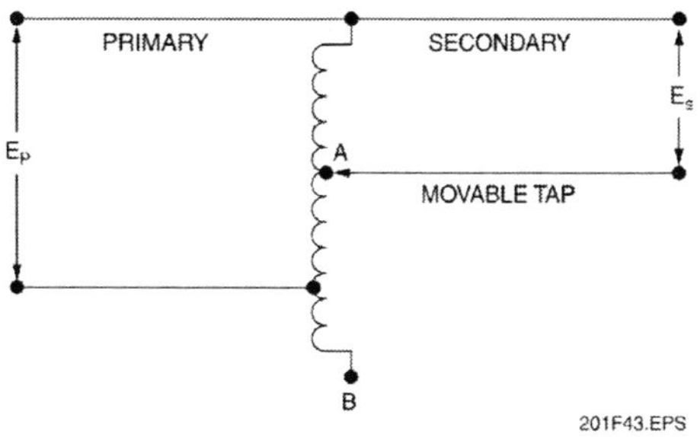

Figure 43 ◈ Autotransformer schematic diagram.

The voltage across the secondary winding has the same relationship to the voltage across the primary that it would have if they were two distinct windings. The movable tap in the secondary is used to select a value of output voltage either higher or lower than E_P, within the range of the transformer. When the tap is at Point A, E_S is less than E_P; when the tap is at Point B, E_S is greater than E_P.

Autotransformers rely on self-induction to induce their secondary voltage. The term *autotransformer* can be broken down into two words: *auto,* meaning self; and *transformer,* meaning to change potential. The autotransformer is made of one winding that acts as both a primary and a secondary winding. It may be used as either a step-up or step-down transformer. Some common uses of autotransformers are as variable AC voltage supplies and fluorescent light ballast transformers, and to reduce the line voltage for various types of low-voltage motor starters.

11.4.3 Current Transformer

A current transformer differs from other transformers in that the primary is a conductor to the load and the secondary is a coil wrapped around the wire to the load. Just as any ammeter is connected in line with a circuit, the current transformer is connected in series with the current to be measured. *Figure 44* is a diagram of a current transformer.

WARNING!
Do not open a current transformer under load.

Since current transformers are series transformers, the usual voltage and current relationships do not apply. Current transformers vary considerably in rated primary current, but are usually designed with ampere-turn ratios such that the secondary delivers five amperes at full primary load.

Current transformers are generally constructed with only a few turns or no turns in the primary. The voltage in the secondary is induced by the changing magnetic field that exists around a single conductor. The secondary is wound on a circular core, and the large conductor that makes up the primary passes through the hole in its center. Because the primary has few or no turns, the secondary must have many turns (providing a high turns ratio) in order to produce a usable voltage. The advantage of this is that you get an output off the secondary proportional to the current flowing through the primary, without an appreciable voltage drop across the primary. This is because the primary voltage equals the current times the impedance. The impedance is kept near zero by using no or very few primary turns. The disadvantage is that you cannot open the secondary circuit with the primary energized. To do so would cause the secondary current to drop rapidly to zero. This would cause the magnetic field generated by the secondary current to collapse rapidly. The rapid collapse of the secondary field through the many turns of the secondary winding would induce a dangerously high voltage in the secondary, creating an equipment and personnel hazard.

INSIDE TRACK

Isolation Transformers

In addition to being used to protect personnel from receiving electrical shocks, shielded isolation transformers are widely used to prevent electrical disturbances on power lines from being transmitted into related load circuits. The shielded isolation transformer has a grounded electrostatic shield between the primary and secondary windings that acts to direct unwanted signals to ground.

Safety

Emphasize that a current transformer should *never* be opened under load.

Classroom

Discuss current transformers.

Current Transformers

INSIDE TRACK

Although the use of a current transformer completely isolates the secondary and the related ammeter from the high-voltage lines, the secondary of a current transformer should never be left open circuited. To do so may result in dangerously high voltage being induced in the secondary.

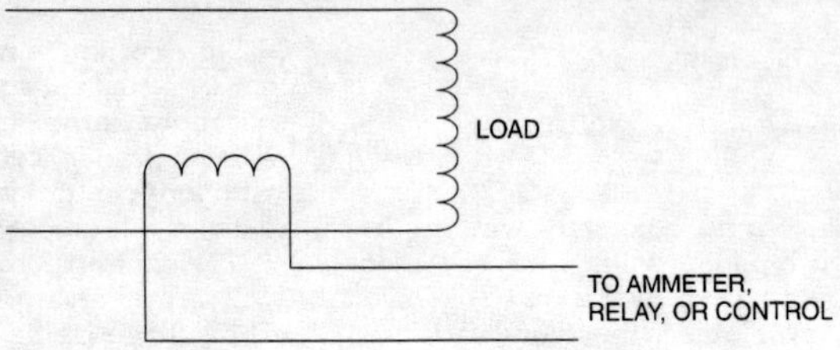

Figure 44 ◆ Current transformer schematic diagram.

Classroom

Discuss potential transformers.

Because the output of current transformers is proportional to the current in the primary, they are most often used to power current-sensing meters and relays. This allows the instruments to respond to primary current without having to handle extreme magnitudes of current.

11.4.4 Potential Transformer

The primary of a potential transformer is connected across or in parallel with the voltage to be measured, just as a voltmeter is connected across a circuit. *Figure 45* shows the schematic diagram for a potential transformer.

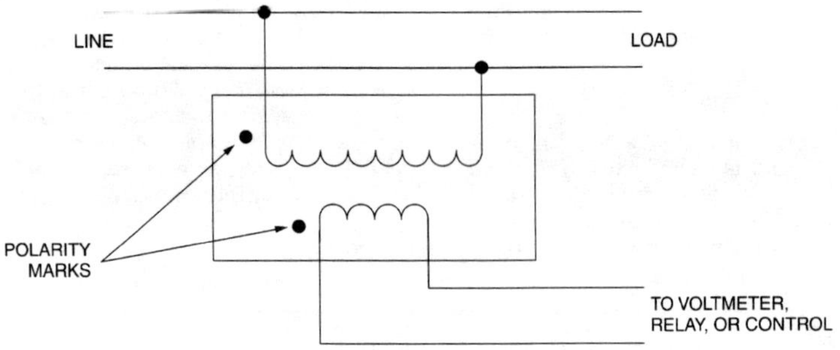

Figure 45 ◆ Potential transformer.

Instructor's Notes:

162 CHAPTER 4

Potential Transformers

In addition to being used to step down high voltages for the purpose of safe metering, potential transformers are widely used in all kinds of control devices where the condition of high voltages must be monitored. One such example involves the use of a potential transformer-operated contactor in emergency lighting standby generator circuits. Under normal conditions with utility power applied, the contactor is energized and its normally closed contacts are open. If the power fails, the contactor deenergizes, causing its contacts to close and activating the standby generator circuit.

AC Power Provides Many Benefits, But Working with It Can Be Dangerous

Working with AC power can be dangerous unless proper safety methods and procedures are followed. The National Institute for Occupational Safety and Health (NIOSH) investigated 224 incidents of electrocutions that resulted in occupational fatalities. One hundred twenty-one of the victims were employed in the construction industry. Two hundred twenty-one of the incidents (99%) involved AC. Of the 221 AC electrocutions, 74 (33%) involved AC voltages less than 600V and 147 (66%) involved 600V or more. Forty of the lower-voltage electrocutions involved 120/240V.

Factors relating to the causes of these electrocutions included the lack of enforcement of existing employer policies including the use of personal protective equipment and the lack of supervisory intervention when existing policies were being violated. Of the 224 victims, 194 (80%) had some type of electrical safety training. Thirty-nine victims had no training at all. It is notable that 100 of the victims had been on the job less than one year.

The Bottom Line: Never assume that you are safe when working at lower voltages. All voltage levels must be considered potentially lethal. Also, safety training does no good if you don't put it into practice every day. Always put safety first.

Putting It All Together

A power company's distribution system has capacitor banks that are automatically switched into the system by a temperature switch during hot weather. Why?

Teaching Tip

Discuss the "Think About It." See the Teaching Tip at the end of this module.

Potential transformers are basically the same as any other single-phase transformer. Although primary voltage ratings vary widely according to the specific application, secondary voltage ratings are usually 120V, a convenient voltage for meters and relays.

Because the output of potential transformers is proportional to the phase-to-phase voltage of the primary, they are often used to power voltage-sensing meters and relays. This allows the instruments to respond to primary voltage while having to handle only 120V. Also, potential transformers are essentially single-phase step-down transformers. Therefore, power to operate low-voltage auxiliary equipment associated with high-voltage switchgear can be supplied off the high-voltage lines that the equipment serves via potential transformers.

ALTERNATING CURRENT 163

Classroom

Have the trainees complete the Review Questions and go over the answers prior to administering the Module Examination.

Summary

The process by which current is produced electromagnetically is called *induction.* As the conductor moves across the magnetic field, it cuts the lines of force, and electrons within the conductor flow, creating an electromotive force (EMF). EMF is also known as *voltage.* There are three conditions that must exist before a current can be produced in this way:

- There must be a magnetic field through which the conductor can pass.
- There must be a conductor in which the voltage will be produced, and the conductor should be perpendicular to the field.
- There must be motion. Either the magnetic field or the conductor must move.

Several factors control the magnitude of the induced current. Voltage will be increased if:

- The speed with which the conductor cuts through the magnetic field is increased (the faster the conductor cuts through the field, the greater the current pulse)
- The strength of the field is increased (the stronger the field, the greater the current pulse)
- The conductor is wound to form a coil (the voltage increases directly with the number of turns of the coil)

A decrease in voltage occurs as the conductor intersects the magnetic field at an angle less than 90°. The greatest current is produced when the conductor intersects the magnetic field at right angles (perpendicular) to the flux lines.

It should be emphasized that a current may be induced by using the magnetic field of a permanent magnet or the magnetic field of another current-carrying conductor (electromagnet).

The magnetic field is among the reasons why phases of current-carrying conductors should not be separated in a raceway; all phase conductors (including the neutral) should be contained in the same raceway. For example, if phase A is separated from phase B and phase C by a metal enclosure, conduit wall, etc., the magnetic field around the conductors will cut across the conduit, causing the conduit to heat up.

Instructor's Notes:

Review Questions

1. An electric current always produces _____.
   a. mutual inductance
   b. a magnetic field
   c. capacitive reactance
   d. high voltage
2. The number of cycles an alternating electric current undergoes per second is known as _____.
   a. amperage
   b. frequency
   c. voltage
   d. resistance
3. What is the peak voltage in a 120VAC circuit?
   a. 117 volts
   b. 120 volts
   c. 150 volts
   d. 170 volts
4. Which of the following conditions exist in a circuit of pure resistance?
   a. The voltage and current are in phase.
   b. The voltage and current are 90° out of phase.
   c. The voltage and current are 120° out of phase.
   d. The voltage and current are 180° out of phase.
5. When the current increases in an AC circuit, what role does inductance play?
   a. It increases the current.
   b. It plays no role at all.
   c. It causes the overcurrent protection to open.
   d. It reduces the current.
6. Which of the following conditions exist in a circuit of pure inductance?
   a. The voltage and current are in phase.
   b. The voltage and current are 90° out of phase.
   c. The voltage and current are 120° out of phase.
   d. The voltage and current are 180° out of phase.
7. The opposition to current flow offered by the capacitance of a circuit is known as _____.
   a. mutual inductance
   b. pure resistance
   c. inductive reactance
   d. capacitive reactance
8. The total opposition to current flow in an AC circuit is known as _____.
   a. resistance
   b. capacitive reactance
   c. inductive reactance
   d. impedance
9. A power factor is not an angular measure, but a numerical ratio with a value between 0 and 1, equal to the _____ of the phase angle.
   a. sine
   b. tangent
   c. cosine
   d. cotangent
10. The two windings of a conventional transformer are known as the _____ windings.
    a. mutual and inductive
    b. high and low voltage
    c. primary and secondary
    d. step-up and step-down

Examination

Administer the Module Examination. Be sure to record the results on Craft Training Report Form 200 and submit the results to the Training Program Sponsor.

ALTERNATING CURRENT 165

GLOSSARY

Trade Terms Introduced in This Module

Capacitance: The storage of electricity in a capacitor; capacitance produces an opposition to voltage change. The unit of measurement for capacitance is the farad (F) or microfarad (μF).

Frequency: The number of cycles an alternating electric current, sound wave, or vibrating object undergoes per second.

Hertz (Hz): A unit of *frequency;* one hertz equals one cycle per second.

Impedance: The opposition to current flow in an AC circuit; impedance includes resistance (R), capacitive *reactance* (X_C), and inductive *reactance* (X_L). Impedance is measured in ohms.

Inductance: The creation of a voltage due to a time-varying current; also, the opposition to current change, causing current changes to lag behind voltage changes. The unit of measure for inductance is the henry (H).

Micro: Prefix designating one-millionth of a unit. For example, one microfarad is one-millionth of a farad.

Peak voltage: The peak value of a sinusoidally varying (cyclical) voltage or current is equal to the root-mean-square (rms) value multiplied by the square root of two (1.414). AC voltages are usually expressed as rms values; that is, 120 volts, 208 volts, 240 volts, 277 volts, 480 volts, etc., are all rms values. The peak voltage, however, differs. For example, the peak value of 120 volts (rms) is actually $120 \times 1.414 = 169.71$ volts.

Radian: An angle at the center of a circle, subtending (opposite to) an arc of the circle that is equal in length to the radius.

Reactance: The imaginary part of *impedance.* Also, the opposition to alternating current (AC) due to capacitance (X_C) and/or inductance (X_L).

Resonance: A condition reached in an electrical circuit when the inductive *reactance* neutralizes the *capacitance reactance,* leaving ohmic resistance as the only opposition to the flow of current.

Root-mean-square (rms): The square root of the average of the square of the function taken throughout the period. The rms value of a sinusoidally varying voltage or current is the effective value of the voltage or current.

Self-inductance: A magnetic field induced in the conductor carrying the current.

Instructor's Notes:

Additional Resources

This module is intended to present thorough resources for task training. The following reference works are suggested for further study. These are optional materials for continued education rather than for task training.

Introduction to Electric Circuits, Latest Edition. New York: Prentice Hall.

Principles of Electric Circuits, Latest Edition. New York: Prentice Hall.

Answers to Review Questions

Answer	Section
1. b	2.0.0
2. b	3.1.0
3. d	3.5.0
4. a	6.0.0
5. d	7.0.0
6. b	7.2.0
7. d	8.5.0
8. d	9.0.0
9. c	10.4.0
10. c	11.1.2

Section 4.2.0 *Think About It—Phase Angles*

The phase angle in *Figure 6* is 90° because waveform B leads waveform A by 90°. In *Figure 7*, waveforms A and B are in phase; therefore, the phase angle is 0°. The vector for *Figure 7* is a straight line because the waveforms are in phase; therefore, the vectors add.

Section 9.4.1 *Think About It—AC Circuits*

The circuit shown is a series RLC circuit. When energized, the lamp inserts resistance into the circuit, the motor inserts inductive reactance, and the capacitor inserts capacitive reactance.

Section 10.5.0 *Think About It—Power Factor*

Add capacitors to the circuit as needed to obtain a resultant capacitive reactance in the circuit that approximately equals the inductive reactance resulting from the various motors in the circuit.

Section 11.3.0 *Think About It—Turns and Voltage Ratios*

The answer is zero. This is because a transformer will not work with a DC source voltage. Because DC is constant, it produces only a momentary pulse when it is initially applied.

Transformers work on the principle of mutual induction. In order to operate, they require the following:

- A conductor (secondary winding of the transformer)
- A magnetic field (generated in the primary winding of the transformer)
- Relative motion (AC current)

AC applied to the primary winding of the transformer produces an expanding/collapsing magnetic field that cuts across and induces AC voltage and current in the secondary winding(s). The magnitude of this induced voltage and current is determined by the turns ratio of the transformer.

Section 11.4.4 *Think About It—Putting It All Together*

In hot weather, everyone turns on air conditioners. Air conditioners have compressors and fans that are driven by motors. These motors are inductive loads. The addition of so many inductive loads in the system can result in a poor system power factor and excessive power losses.

Banks of capacitors are automatically switched into the system so that their capacitive reactance cancels out as near as possible the inductive reactance presented by the motors. Remember that inductors shift the current and voltage away from each other in one direction (ELI), and capacitors shift them away from each other in the opposite direction (ICE). This maintains the system power factor at an acceptable value and keeps system losses to a minimum.

CONTREN™ LEARNING SERIES — USER UPDATES

The NCCER makes every effort to keep these textbooks up-to-date and free of technical errors. We appreciate your help in this process. If you have an idea for improving this textbook, or if you find an error, a typographical mistake, or an inaccuracy in NCCER's Contren™ textbooks, please write us, using this form or a photocopy. Be sure to include the exact module number, page number, a detailed description, and the correction, if applicable. Your input will be brought to the attention of the Technical Review Committee. Thank you for your assistance.

Instructors – If you found that additional materials were necessary in order to teach this module effectively, please let us know so that we may include them in the Equipment/Materials list in the Instructor's Guide.

Write: Curriculum Revision and Development Department
National Center for Construction Education and Research
P.O. Box 141104, Gainesville, FL 32614-1104

Fax: 352-334-0932

E-mail: curriculum@nccer.org

Craft ____________ Module Name ____________

Copyright Date ____________ Module Number ____________ Page Number(s) ____________

Description

____

____

____

____

(Optional) Correction

____

____

____

(Optional) Your Name and Address

____

____

____

Chapter 5

Electrical Test Equipment

26106-02

MODULE OVERVIEW

This course introduces the electrical trainee to the operation and applications of various types of electrical test equipment.

PREREQUISITES

Please refer to the Course Map in the Trainee Module. Prior to training with this module, it is recommended that the trainee shall have successfully completed the following:

Core Curriculum; Residential Electrical I, Chapters 1 through 4.

LEARNING OBJECTIVES

Upon completion of this module, the trainee will be able to:

1. Explain the operation of and describe the following pieces of test equipment:
   - Ammeter
   - Voltmeter
   - Ohmmeter
   - Volt-ohm-milliammeter (VOM)
   - Wattmeter
   - Megohmmeter
   - Frequency meter
   - Power factor meter
   - Continuity tester
   - Voltage tester
   - Recording instruments
   - Cable-length meters
2. Explain how to read and convert from one scale to another using the above test equipment.
3. Explain the importance of proper meter polarity.
4. Define frequency and explain the use of a frequency meter.
5. Explain the difference between digital and analog meters.

PERFORMANCE OBJECTIVES

Under supervision of the instructor, the trainee should be able to:

1. Demonstrate the procedure for safely using a clamp-on ammeter.
2. Connect a multimeter to a circuit and measure the source voltage.
3. Demonstrate the procedure for safely using a voltage tester.

NCCER STANDARDIZED CRAFT TRAINING PROGRAM

The National Center for Construction Education and Research (NCCER) provides a standardized national program of accredited craft training. Key features of the program include instructor certification, competency-based training, and performance testing. The program provides trainees, instructors, and companies with a standard form of recognition through a National Craft Training Registry. The program is described in full in the *Guidelines for Accreditation*, published by the NCCER. For more information on standardized craft training, contact the NCCER at P.O. Box 141104, Gainesville, FL 32614-1104, 352-334-0911, visit our Web site at www.nccer.org, or e-mail info@nccer.org.

HOW TO USE THIS ANNOTATED INSTRUCTOR'S GUIDE

Each page presents two sections of information. The larger section displays each page exactly as it appears in the Trainee Module. The narrow column ties suggested trainee and instructor actions to each page and provides icons to call your attention to material, safety, audiovisual, or testing requirements. The bottom of each page includes space for your notes.

If you see the Teaching Tip icon, that means there is a teaching tip associated with this section. Also refer to the suggested teaching tips at the end of the module.

SAFETY CONSIDERATIONS

Ensure that the trainees are equipped with appropriate personal protective equipment. Stress the importance of following all safety precautions and procedures when working with all types of electrical test equipment.

PREPARATION

Before teaching this module, you should review the Module Outline, Learning and Performance Objectives, and the Materials and Equipment List. Be sure to allow ample time to prepare your own training or lesson plan and gather all required equipment and materials.

MATERIALS AND EQUIPMENT LIST

Materials:

Transparencies

Markers/chalk

Copy of the latest edition of the *National Electrical Code®*

Calculator

Pencils and scratch paper

Module Examinations*

Performance Profile Sheets*

Equipment:

Overhead projector and screen

Whiteboard/chalkboard

Appropriate personal protective equipment

Clamp-on ammeter

Volt-ohm-milliammeters (both analog and digital)

Wattmeter

Megohmmeter

Cable-length meter

Frequency meter

Power factor meter

Continuity tester

Voltage tester

Recording instruments

*Located in the Test Booklet packaged with this Annotated Instructor's Guide.

ADDITIONAL RESOURCES

This module is intended to present thorough resources for task training. The following reference works are suggested for both instructors and motivated trainees interested in further study. These are optional materials for continued education rather than for task training.

Electronics Fundamentals, Latest Edition. Upper Saddle River, NJ: Prentice Hall.

Principles of Electric Circuits, Latest Edition. Upper Saddle River, NJ: Prentice Hall.

NOTES

The designations "National Electrical Code," "NE Code," and "NEC," where used in this document, refer to the *National Electrical Code*®, which is a registered trademark of the National Fire Protection Association, Quincy, MA. All National Electrical Code (NEC) references in this module refer to the 2002 edition of the NEC.

If you feel that additional math instruction would be helpful, Prentice Hall offers a basic math textbook entitled *Fundamentals of Electrical and Mechanical Mathematics.* It covers the basic math requirements for electrical trainees and may be ordered by contacting Prentice Hall Customer Service at 1-800-922-0579.

TEACHING TIME FOR THIS MODULE

An outline for use in developing your lesson plan is presented below. Note that each Roman numeral in the outline equates to one session of instruction. Each session has a suggested time period of 2½ hours. This includes 10 minutes at the beginning of each session for administrative tasks and one 10-minute break during the session. Approximately 7½ hours are suggested to cover *Electrical Test Equipment.* You will need to adjust the time required for hands-on activity and testing based on your class size and resources.

Topic	Planned Time
Session I. Analog Meters; Ammeter; Laboratory; Voltmeter; Ohmmeter; Volt-Ohm-Milliammeter	
A. Analog Meters	______
1. d'Arsonval Meter Movement	______
B. Ammeter	______
C. Laboratory	______
Under instructor supervision, have the trainees demonstrate the procedure for safely using a clamp-on ammeter.	
D. Voltmeter	______
E. Ohmmeter	______
F. Volt-Ohm-Milliammeter	______
1. Specifications	______
2. Overload Protection	______
3. Making Measurements	______
a. Measuring DC Voltage, 0–250 Millivolts	______
b. Measuring DC Voltage, 0–1 Volt	______
c. Measuring DC Voltage, 0–2.5 Through 0–500 Volts	______
d. Measuring DC Voltage, 0–1,000 Volts	______
e. Measuring AC Voltage, 0–2.5 Through 0–500 Volts	______
f. Measuring AC Voltage, 0–1,000 Volts	______
g. Measuring Output Voltage	______
h. Measuring Decibels	______
4. Direct Current Measurements	______
a. Voltage Drop	______
b. Measuring Direct Current, 0–50 Microamperes	______
c. Measuring Direct Current, 0–1 Through 0–500 Milliamperes	______
d. Measuring Direct Current, 0–10 Amperes	______
e. Zero Ohm Adjustment	______
f. Measuring Resistance	______

Session II. Digital Meters; Laboratory; Wattmeter; Megohommeter (Megger); Frequency Meter; Power Factor Meter

A. Digital Meters
   1. Features
   2. Operation
      a. Dual Function ON/Clear Button
      b. Measuring Voltage
      c. Measuring Current
      d. Measuring Resistance
      e. Diode/Continuity Test
      f. Transistor Junction Test
      g. Display Test
   3. Maintenance
      a. Battery Replacement
      b. Fuse Replacement
      c. Calibration

B. Laboratory

   Under instructor supervision, have the trainees connect a multimeter to a circuit and measure the source voltage.

C. Wattmeter

D. Megohmmeter (Megger)
   1. Safety Precautions

E. Frequency Meter
   1. Phase Rotation Tester
   2. Harmonic Test Set

F. Power Factor Meter

Session III. Continuity Tester; Voltage Tester; Laboratory; Recording Instruments; Safety; Module Examination and Performance Testing

A. Continuity Tester
   1. Audio Continuity Tester
   2. Visual Continuity Tester
   3. Cable-Length Meter

B. Voltage Tester
   1. Wiggy®
      a. Principle of Operation

C. Laboratory

   Under instructor supervision, have the trainees demonstrate the procedure for safely using a voltage tester.

D. Recording Instruments
   1. Strip-Chart Recorders
      a. Typical Strip-Chart Recorder

E. Safety
   1. Use of High-Voltage Protection Equipment ____________
   2. General Testing of Electrical Equipment ____________
      a. Example Policy – Work Permits or Test Permits ____________
      b. Example Policy – Authorized Employees for Testing ____________
      c. Example Policy – Circuit Isolation ____________
      d. Example Policy – Clearances ____________
      e. Electrical Circuit Tests ____________
   3. Meter Loading Effects ____________

F. Module Examination ____________
   1. Trainees must score 70% or higher to receive recognition from the NCCER.
   2. Record the testing results on Craft Training Report Form 200 and submit the results to the Training Program Sponsor.

G. Performance Testing ____________
   1. Trainees must perform each task to the satisfaction of the instructor to receive recognition from the NCCER.
   2. Record the testing results on Craft Training Report Form 200 and submit the results to the Training Program Sponsor.

Chapter 5

Module 26106-02

Electrical Test Equipment

Course Map

This course map shows all of the modules in the first level of the Residential Electrical curriculum. The suggested training order begins at the bottom and proceeds up. Skill levels increase as you advance on the course map. The local Training Program Sponsor may adjust the training order.

RESIDENTIAL ELECTRICAL I

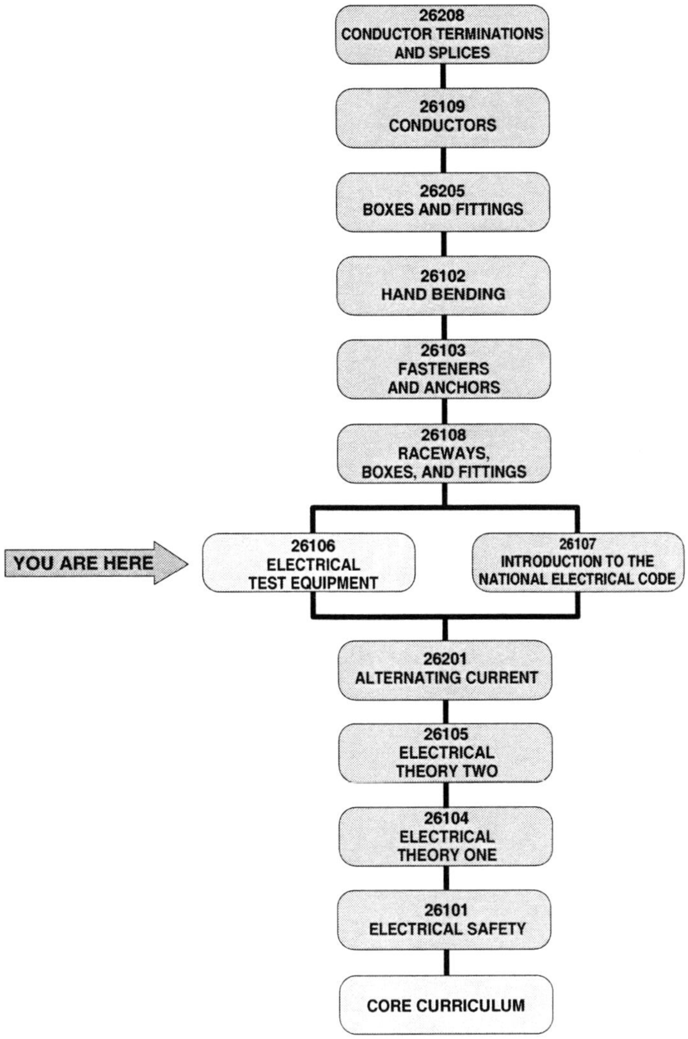

Homework

Assign reading of Module 26106.

MODULE 26106 CONTENTS

Instructor's Notes:

Figures

Table

Chapter 5

MODULE 26106

Electrical Test Equipment

Materials

Ensure you have everything required to teach the course. Check the Materials and Equipment List at the front of this Instructor's Guide.

Objectives

When you have completed this module, you will be able to do the following:

1. Explain the operation of and describe the following pieces of test equipment:
   - Ammeter
   - Voltmeter
   - Ohmmeter
   - Volt-ohm-milliammeter (VOM)
   - Wattmeter
   - Megohmmeter
   - Frequency meter
   - Power factor meter
   - Continuity tester
   - Voltage tester
   - Recording instruments
   - Cable-length meters
2. Explain how to read and convert from one scale to another using the above test equipment.
3. Explain the importance of proper meter polarity.
4. Define frequency and explain the use of a frequency meter.
5. Explain the difference between digital and analog meters.

Prerequisites

Before you begin this module, it is recommended that you successfully complete the following: Core Curriculum; Residential Electrical I, Chapters 1 through 4.

Required Trainee Materials

1. Paper and pencil
2. Copy of the latest edition of the *National Electrical Code*
3. Appropriate personal protective equipment

Audiovisual

Show Transparency 1, Course Objectives.

Show Transparency 2, Performance Profile Tasks.

1.0.0 ◆ INTRODUCTION

Classroom

Discuss the applications of electrical test equipment.

The use of electronic test instruments and meters generally involves these three applications:

- Verifying proper operation of instruments and associated equipment
- Calibrating electronic instruments and associated equipment
- Troubleshooting electrical/electronic circuits and equipment

For these applications, specific test equipment is selected to analyze circuits and to determine specific characteristics of discrete components.

The test equipment an electrician chooses for a specific task depends on the type of measurement and the level of accuracy required. Additional factors that may influence selection include:

- Whether the test equipment is portable
- The amount of information that the test equipment provides

Note: The designations "National Electrical Code," "NE Code," and "NEC," where used in this document, refer to the *National Electrical Code®*, which is a registered trademark of the National Fire Protection Association, Quincy, MA. *All National Electrical Code (NEC) references in this module refer to the 2002 edition of the NEC.*

ELECTRICAL TEST EQUIPMENT 179

Classroom

Explain that terms shown in bold (blue) are defined in the Glossary at the back of this module.

Discuss meter movements.

Audiovisual

Show Transparencies 3 and 4 (Figures 1 and 2).

- The likelihood that the test equipment may damage the circuit or component being tested (some test equipment can generate enough voltage or current to damage an instrument or electronic circuit)

This module will focus on some of the test equipment that you will be required to use in your job as an electrician. The intent is to familiarize you with the use and operation of such equipment and to provide you with practical experience involving that equipment. Upon completion of this module, you should be able to select the appropriate test equipment and effectively use that equipment to perform an assigned task.

2.0.0 ◆ METERS

The functioning of conventional electrical measuring instruments is based upon electromechanical principles. Their mechanical components usually work on direct current (DC). Mechanical **frequency** meters are an exception. A meter that measures alternating current (AC) has a built-in rectifier to change the AC to DC and resistors to correct for the various ranges.

Today, many meters are solid-state digital systems; they are superior because they have no moving parts. These meters will work in any position, unlike mechanical meters, which must remain in one position to be read accurately.

2.1.0 d'Arsonval Meter Movement

In 1882, a Frenchman named Arsene d'Arsonval invented the galvanometer. This meter used a stationary permanent magnet and a moving **coil** *(Figure 1)* to indicate current flow on a calibrated scale. The early galvanometer was very accurate but could only measure very small currents. Over the following years, many improvements were made that extended the range of the meter and increased its ruggedness. The **d'Arsonval meter movement** is the most commonly used meter movement today.

A moving-coil meter movement operates on the electromagnetic principle. In its simplest form, the moving-coil meter uses a coil of very fine wire wound on a light aluminum frame. A permanent magnet surrounds the coil. The aluminum frame is mounted on pivots to allow it and the coil to rotate freely between the poles of the permanent magnet. When current flows through the coil, it becomes magnetized, and the polarity of the coil is such that it is repelled by the field of the permanent magnet. This will cause the coil frame to rotate on its pivots, and the distance it rotates is determined by the amount of current that flows through the coil. By attaching a pointer to the coil frame and adding a calibrated scale, the amount of current flowing through the meter can be measured.

The d'Arsonval meter movement uses this same principle of operation *(Figure 2)*. As the cur-

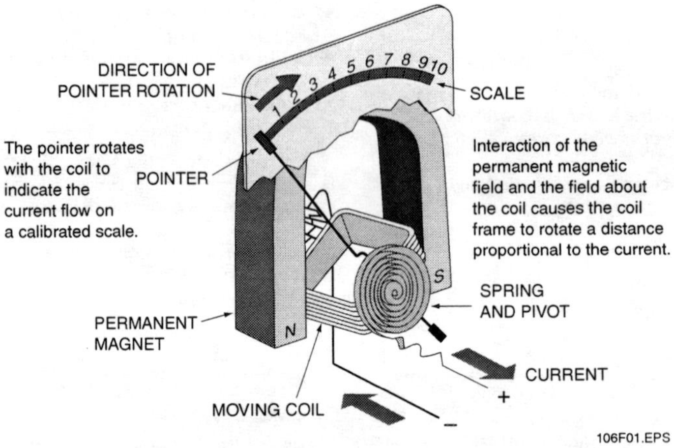

DIRECTION OF POINTER ROTATION
1 2 3 4 5 6 7 8 9 10
SCALE
The pointer rotates with the coil to indicate the current flow on a calibrated scale.
POINTER
Interaction of the permanent magnetic field and the field about the coil causes the coil frame to rotate a distance proportional to the current.
S
SPRING AND PIVOT
PERMANENT MAGNET
N
CURRENT
+
MOVING COIL
–
106F01.EPS

Figure 1 ◆ Moving-coil meter movement.

Instructor's Notes:

180 CHAPTER 5

rent flow increases, the magnetic field around the coil increases, the amount of coil rotation increases, and the pointer swings farther across the meter scale.

3.0.0 ◆ AMMETER

The ammeter is used to measure current. Most models will measure only small amounts of current. The typical range is in microamperes, μA (0.000001A) or milliamperes, mA (0.001A). Very few ammeters can measure more than 10mA. To increase the range to the ampere level, a shunt is used. To measure above 10mA, a shunt with an extremely low resistance is placed in series with the load, and the meter is connected across the shunt to measure the resulting voltage drop proportional to current flow. A shunt has a very large wattage rating in order to carry a large current.

The meter is connected in parallel with the shunt *(Figure 3)*. Shunts located inside the meter case (internal shunts) are generally used to measure values up to 30 amps; shunts located away from the meter case (external shunts) with leads going to the meter are generally used to measure values greater than 30 amps. Above 30 amps of current, the heat generated could damage the meter if an internal shunt were used. The use of a shunt allows the ammeter to derive current in amps by actually measuring the voltage drop across the shunt. Ammeter connections are shown in *Figure 4*.

Never connect an ammeter in parallel with a load. Because of the low resistance in the ammeter, this will cause a short circuit, probable damage to the meter and/or circuit, and personal injury. When connecting an ammeter in a DC circuit, you must observe proper polarity. In other

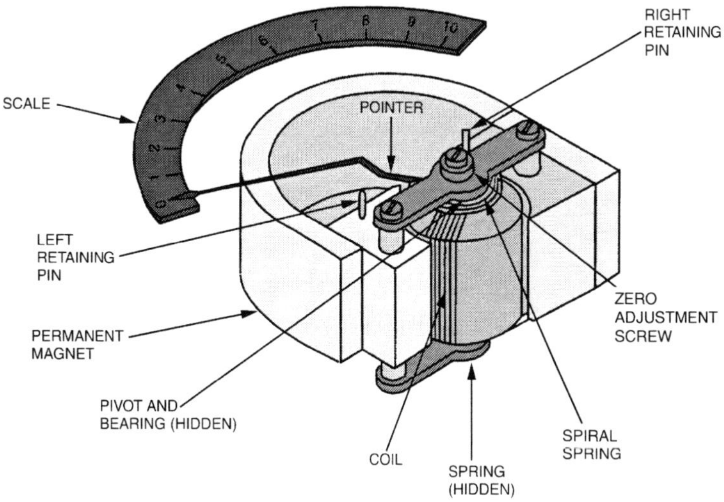

Figure 2 ◆ d'Arsonval meter movement.

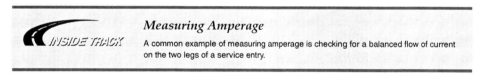

Classroom

Discuss the operation of a basic ammeter.

Audiovisual

Show Transparencies 5 and 6 (Figures 3 and 4).

Safety

Emphasize the importance of *never* connecting an ammeter in parallel with a load.

INSIDE TRACK

Measuring Amperage

A common example of measuring amperage is checking for a balanced flow of current on the two legs of a service entry.

ELECTRICAL TEST EQUIPMENT 181

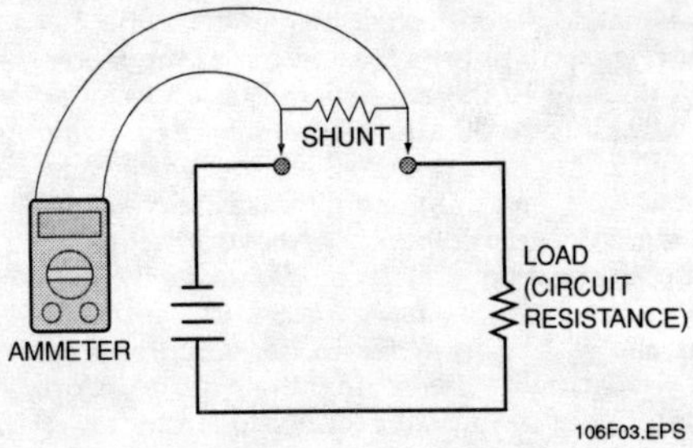

Figure 3 ◆ Ammeter shunt.

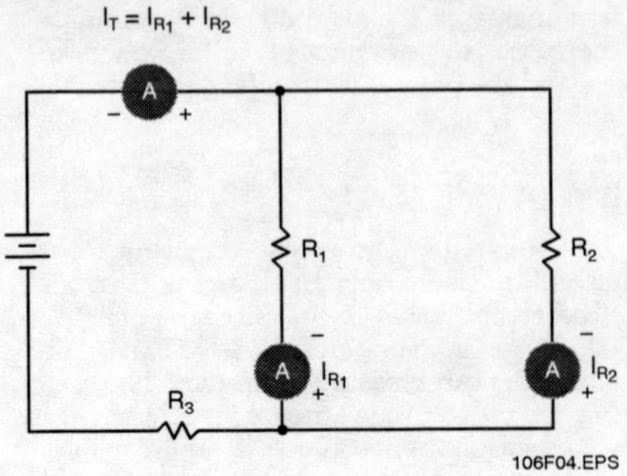

Figure 4 ◆ Ammeter connections.

Teaching Tip

Discuss the "Think About It."

THINK ABOUT IT

Ammeter Connections

Using basic concepts of series and parallel circuits, explain why an ammeter would be damaged when connected in parallel but not when connected in series. What are some practical examples of connecting an ammeter in series with a load?

words, you must connect the negative terminal of the meter to the negative or low potential point in the circuit, and connect the positive terminal of the meter to the positive or high potential point in the circuit (*Figure 5*). Current must flow through the meter from minus (−) to plus (+). If you connect the meter with the polarities reversed, the meter coil will move in the opposite direction, and the pointer might strike the left retaining pin. You will not obtain a current reading, and you might bend the pointer of the meter.

Audiovisual

Show Transparencies 7 through 9 (Figures 5 through 7).

It is not very practical to use an ammeter that has only one range; therefore, a multirange ammeter needs to be discussed. A multirange ammeter is one containing a basic meter movement and several shunts that can be connected across the meter movement. See *Figure 6*.

Classroom

Discuss multirange ammeters.

A range switch is normally used to select the particular shunt for the desired current range. Sometimes, however, separate terminals for each range are mounted on the meter case. Some multirange ammeters have only one set of values on the scale, even though they measure several different current ranges. For example, if the scale is calibrated in values from 0 to 1 milliamp (mA), and the range switch is in the 1mA position, read the current directly. However, if the range switch is in the 10mA position, multiply the scale reading by 10 to find the amount of current flowing through the circuit. See *Figure 7*.

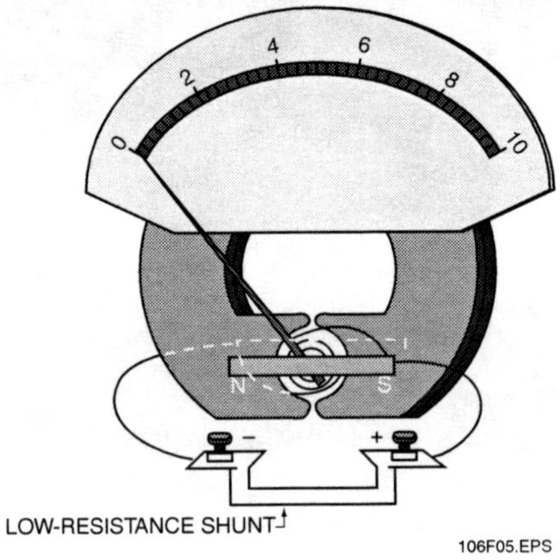

Figure 5 ◆ DC ammeter.

Instructor's Notes:

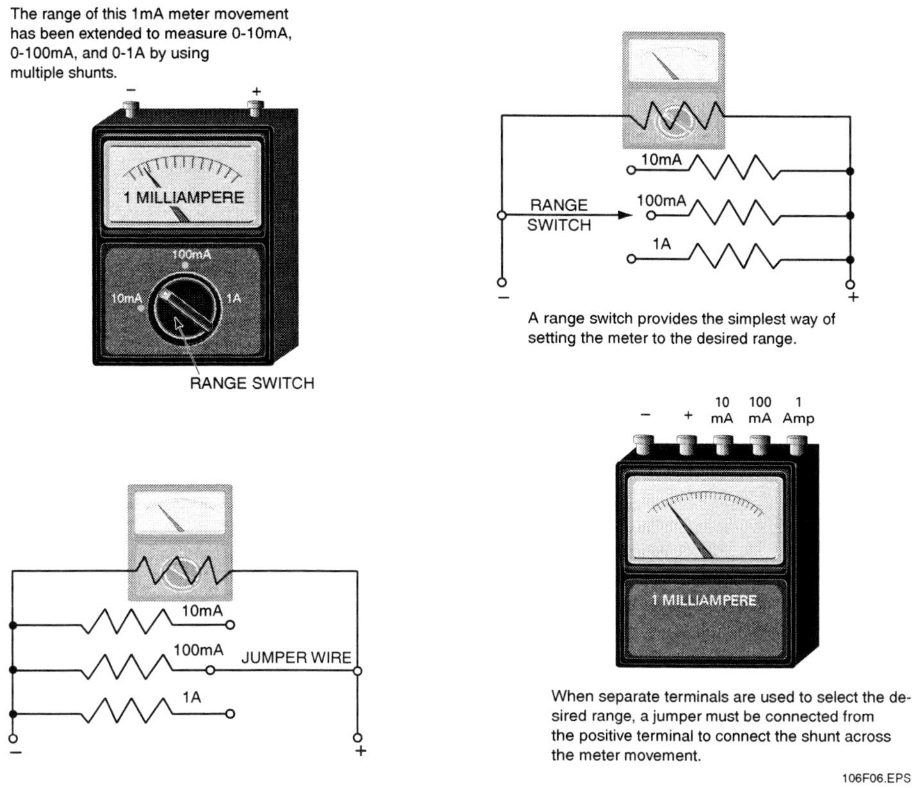
The range of this 1mA meter movement has been extended to measure 0-10mA, 0-100mA, and 0-1A by using multiple shunts.
1 MILLIAMPERE
100mA
10mA
1A
RANGE SWITCH
10mA
100mA
1A
RANGE SWITCH
A range switch provides the simplest way of setting the meter to the desired range.
10mA
100mA
1A
JUMPER WIRE
10 mA
100 mA
1 Amp
1 MILLIAMPERE
When separate terminals are used to select the desired range, a jumper must be connected from the positive terminal to connect the shunt across the meter movement.
106F06.EPS

Figure 6 ◈ Multirange ammeter.

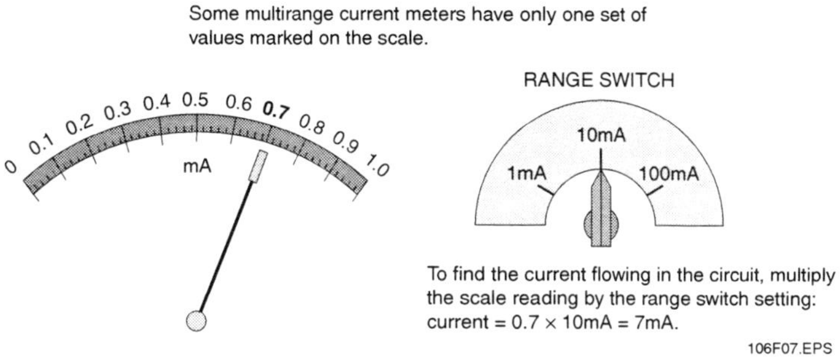
Some multirange current meters have only one set of values marked on the scale.
0 0.1 0.2 0.3 0.4 0.5 0.6 0.7 0.8 0.9 1.0
mA
RANGE SWITCH
10mA
1mA
100mA
To find the current flowing in the circuit, multiply the scale reading by the range switch setting: current = 0.7 × 10mA = 7mA.
106F07.EPS

Figure 7 ◈ Multirange single-scale ammeter.

ELECTRICAL TEST EQUIPMENT 183

Audiovisual

Show Transparency 10 (Figure 8).

Demonstration

Demonstrate the use of a clamp-on ammeter.

Laboratory

Have the trainees practice using a clamp-on ammeter. Note the proficiency of each trainee.

Classroom

Discuss the operation of a basic voltmeter.

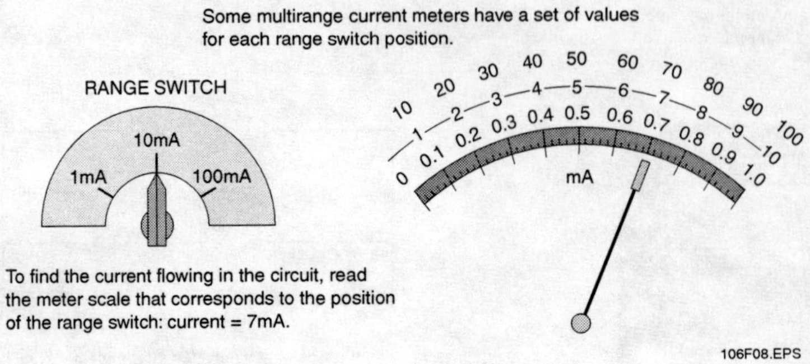
Some multirange current meters have a set of values for each range switch position.
RANGE SWITCH
10mA
1mA
100mA
0 10 20 30 40 50 60 70 80 90 100
1 2 3 4 5 6 7 8 9 10
0.1 0.2 0.3 0.4 0.5 0.6 0.7 0.8 0.9 1.0
mA
To find the current flowing in the circuit, read the meter scale that corresponds to the position of the range switch: current = 7mA.
106F08.EPS

Figure 8 ◆ Multirange multiscale ammeter.

Other current meters have a separate set of values on the calibrated scale that correspond to the different positions of the range switch. In this case, be sure that you read the set of values that correspond to the position of the range switch *(Figure 8)*.

When measuring AC current at levels greater than one ampere, a clamp-on ammeter is often used. This meter also measures DC. The clamp-on ammeter may have a mechanical movement or a digital readout. These meters clamp over the wire and do not break the insulation. This type of ammeter senses current flow by measuring the magnetic field surrounding the conductor. When using this type of ammeter, be sure to measure only one conductor at a time. This type of meter will often measure up to 1,000 amperes at 600 volts. A digital clamp-on multimeter being used to measure current is shown in *Figure 9*.

4.0.0 ◆ VOLTMETER

The basic current meter movement already covered, whether AC or DC, can also be used to measure voltage (electromotive force or emf). The meter coil has a fixed resistance, and, therefore, when current flows through the coil, a voltage drop will be developed across this resistance. According to Ohm's law, the voltage drop will be directly proportional to the amount of current flowing through the coil. Also, the amount of current flowing through the coil is directly proportional to the amount of voltage applied to it. Therefore, by calibrating the meter scale in units of voltage instead of current, the voltage in a circuit can be measured.

Since a basic current meter movement has a low coil resistance and low current-handling capabilities, its use as a voltmeter is very limited. In fact, the maximum voltage that could be measured with a one milliamp meter movement is one volt *(Figure 10)*.

The voltage range of a meter movement can be extended by adding a resistor, called a *multiplier resistor,* in series. The value of this resistor must be such that, when added to the meter coil resistance, the total resistance limits the current to the

106F09.EPS

Figure 9 ◆ Clamp-on multimeter being used to measure current.

Instructor's Notes:

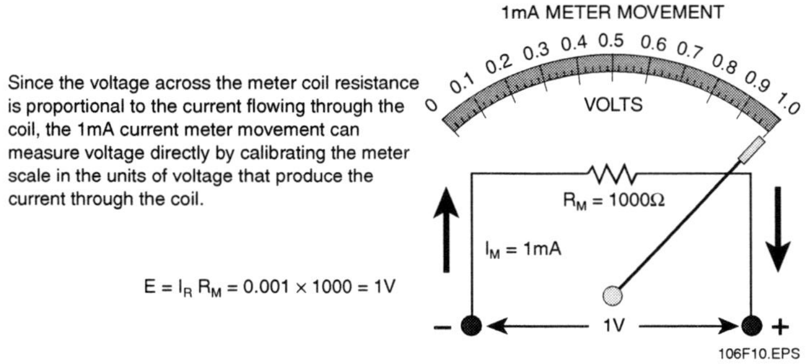
1mA METER MOVEMENT
Since the voltage across the meter coil resistance is proportional to the current flowing through the coil, the 1mA current meter movement can measure voltage directly by calibrating the meter scale in the units of voltage that produce the current through the coil.
0 0.1 0.2 0.3 0.4 0.5 0.6 0.7 0.8 0.9 1.0
VOLTS
$R_M = 1000\Omega$
$I_M = 1mA$
$E = I_R R_M = 0.001 \times 1000 = 1V$
1V
106F10.EPS

Figure 10 ◆ 1mA meter movement.

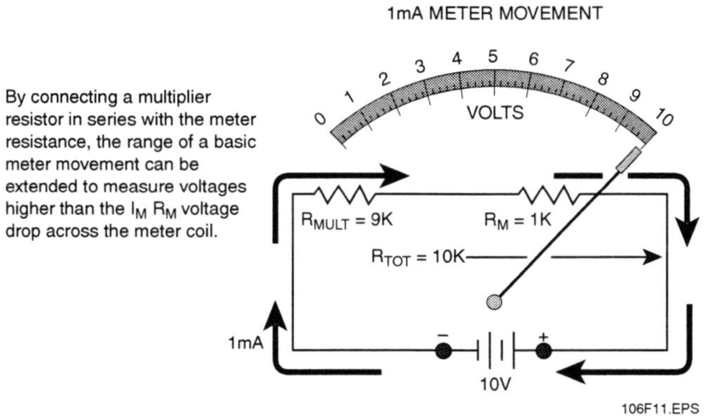
1mA METER MOVEMENT
By connecting a multiplier resistor in series with the meter resistance, the range of a basic meter movement can be extended to measure voltages higher than the $I_M R_M$ voltage drop across the meter coil.
0 1 2 3 4 5 6 7 8 9 10
VOLTS
$R_{MULT} = 9K$
$R_M = 1K$
$R_{TOT} = 10K$
1mA
10V
106F11.EPS

Figure 11 ◆ Adding multiplier resistor.

THINK ABOUT IT

Voltmeters

Again using basic circuit theory, explain why a voltmeter must be connected in parallel with the circuit. What would you be measuring if the voltmeter were connected in series?

full-scale current rating of the meter for any applied voltage *(Figure 11)*.

Voltmeters must be in parallel with the circuit component being measured. On the higher ranges, the amount of current flowing through the meter is much lower due to its very high total resistance. However, an inaccurate reading will result if the voltmeter is placed in series rather than in parallel with a circuit component. When connecting a DC voltmeter, always observe the proper polarity. The negative lead of the meter must be connected to the negative or low potential end of the component, and the positive lead to the positive or high potential end of the component. As was the case

Audiovisual

Show Transparencies 11 and 12 (Figures 10 and 11).

Teaching Tip

Discuss the "Think About It."

ELECTRICAL TEST EQUIPMENT 185

Classroom

Discuss multirange voltmeters.

Audiovisual

Show Transparencies 13 and 14 (Figures 12 and 13).

Classroom

Explain the operation of a basic ohmmeter.

Audiovisual

Show Transparency 15 (Figure 14).

when using an ammeter, if you connect a voltmeter to the component with opposing polarities, the meter coil will move to the left, and the pointer could be bent. In an AC circuit, the voltage constantly reverses polarity, so there is no need to observe polarity when connecting the voltmeter to a component in an AC circuit.

As with ammeters, it is impractical to have a voltmeter that will only measure one range of voltages; therefore, multirange voltmeters also need to be discussed. To make a voltmeter capable of measuring multiple ranges, an electrician needs several multiplier resistors that are switch-selectable for the different ranges desired. Reading the scale of a voltmeter is as simple as reading the scale of an ammeter. Some multirange voltmeters have only one range of values marked on the scale, and the scale reading must be multiplied by the range switch setting to obtain the correct voltage *(Figure 12)*.

Other voltmeters have different ranges on the scale for each setting of the range switch. In this case, be sure that you read the set of values that correspond to the position of the range switch *(Figure 13)*.

Some digital voltmeters are autoranging. These types of voltmeters do not have a range switch. The internal construction of the meter itself will select the proper resistance for the current being detected. However, when using a voltmeter that is *not* autoranging, always start with the highest voltage setting and work down until the indication reads somewhere between half-scale and three-quarter scale. This will give a more accurate reading and will prevent damage to the voltmeter.

5.0.0 ◆ OHMMETER

An ohmmeter is a device that measures the resistance of a circuit or component. It can also be used to locate open circuits or shorted circuits. Basically, an ohmmeter consists of a DC current meter movement, a low-voltage DC power source (usually a battery), and current-limiting resistors, all of which are connected in series *(Figure 14)*.

This combination of devices allows the meter to calculate resistance by deriving it using Ohm's law. Before measuring the resistance of an un-

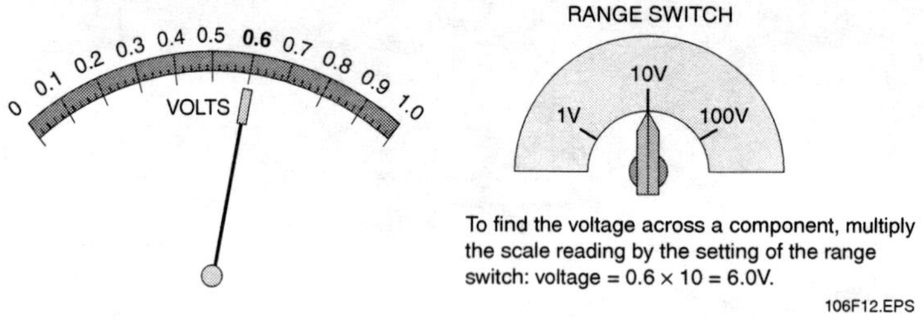

Figure 12 ◆ Multirange single-scale voltmeter.

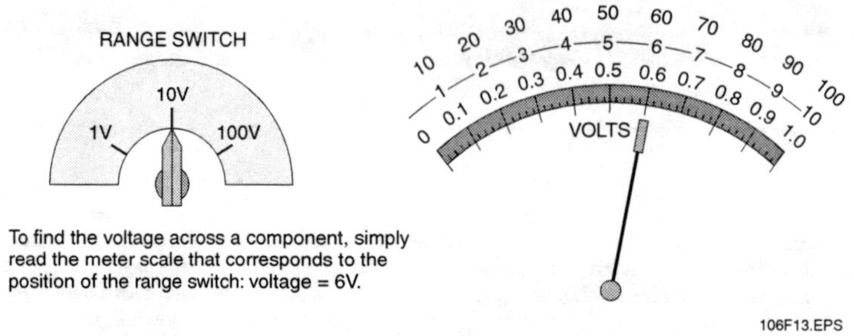

Figure 13 ◆ Multirange multiscale voltmeter.

Instructor's Notes:

186 CHAPTER 5

INSIDE TRACK

Meter Applications

Different electrical systems may have differing levels of available current as well as different voltages. Keep in mind that the meter must be matched to the application. For example, a meter that has adequate protective features to be safely used on interior branch circuits may not be safe to use on the service feeders.

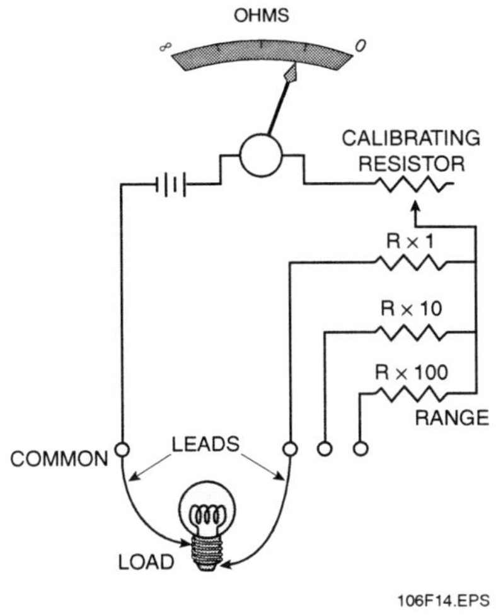

Figure 14 ◆ Ohmmeter schematic.

known resistor or electrical circuit, the test leads are shorted together. This causes current to flow through the meter movement and the pointer deflects to the right. This means that there is zero resistance present across the input terminals of the ohmmeter, and when zero resistance is present, the pointer will deflect full scale. Therefore, full-scale deflection of an ohmmeter indicates zero resistance. Most ohmmeters have a zero adjustment knob. This is used to correct for the fact that as the batteries of the meter age, their output voltage decreases. As the voltage drops, the current through the circuit decreases, and the meter will no longer deflect full scale. By correcting for this change before each use, the internal resistance of the meter, along with the resistance of the leads, is lowered and nulled to deflect the pointer full scale.

After the ohmmeter is adjusted to zero, it is ready to be connected in a circuit to measure resistance. The circuit must be verified as deenergized by using a voltmeter prior to taking a reading with an ohmmeter. If the circuit were energized, its voltage could cause a damaging current to flow through the meter. This can cause damage to the meter and/or circuit as well as personal injury.

When making resistance measurements in circuits, each component in the circuit can be tested individually by removing the component from the circuit and connecting the ohmmeter leads across it. Actually, the component does not have to be totally removed from the circuit. Usually, the part can be effectively isolated by disconnecting one of its leads from the circuit. However, this method can still be somewhat time consuming. Some manufacturers provide charts that list the resistance readings that should be obtained from various test points to a reference point in the equipment. There are usually many parts of the circuit between the test point and the reference point, so if you get an abnormal reading, you must begin checking smaller groups of components, or individual components, to isolate the defective one. If resistance charts are not available, be very careful to ensure that other components are not in parallel with the component being tested.

THINK ABOUT IT

Using an Ohmmeter

Ohmmeters are commonly used to check for the mechanical integrity of circuits. For example, suppose that you wanted to determine if a wire in a house circuit had been broken by a sheetrock screw. How would you connect an ohmmeter to test this house circuit and what reading would you get from a broken circuit?

Teaching Tip

Discuss the "Think About It." Answers will vary from circuit troubleshooting to motor maintenance.

Safety

Emphasize the importance of de-energizing the circuit before using an ohmmeter.

Teaching Tip

Discuss the "Think About It."

ELECTRICAL TEST EQUIPMENT 187

Classroom

Discuss the features of a typical VOM.

Audiovisual

Show Transparency 16 (Figure 15).

6.0.0 ◆ VOLT-OHM-MILLIAMMETER

The volt-ohm-milliammeter (VOM), also known as the *multimeter,* is a multipurpose instrument. It is a combination of the three previous meters discussed: the milliammeter, the voltmeter, and the ohmmeter. One common analog multimeter is shown in *Figure 15.* There are many different models of this basic multimeter. To prevent having to discuss each and every meter, one version will be explained here. Any controls or functions on your meter that are not covered here should be reviewed in the applicable owner's manual.

The typical volt-ohm-milliammeter is a rugged, accurate, compact, easy-to-use instrument. The instrument can be used to make accurate measurements of DC and AC voltages, current, resistance, **decibels,** and output voltage. The output voltage function is used for measuring the AC component of a mixture of AC and DC voltages. This occurs primarily in amplifier circuits.

This meter has the following features: a 0–1 volt DC range, 0–500 volt DC and AC ranges, a TRANSIT position on the range switch, rubber plug bumpers on the bottom of the case to reduce sliding, and an externally accessible battery and fuse compartment.

6.1.0 Specifications

The specifications of the example multimeter are:

- DC voltage:
  Sensitivity: $20K\Omega$ per volt
  Accuracy: $1\frac{3}{4}$% of full scale
- AC voltage:
  Sensitivity: $5K\Omega$ per volt
  Accuracy: 3% of full scale
- DC current:
  250mV to 400mV drop
  Accuracy: $1\frac{3}{4}$% of full scale
- Resistance:
  Accuracy: 1.75° of arc
  Nominal open circuit voltage 1.5V (9V on the $10K\Omega$ ohm range)
- Nominal short circuit current:
  1Ω range: 1.25mA

INSIDE TRACK

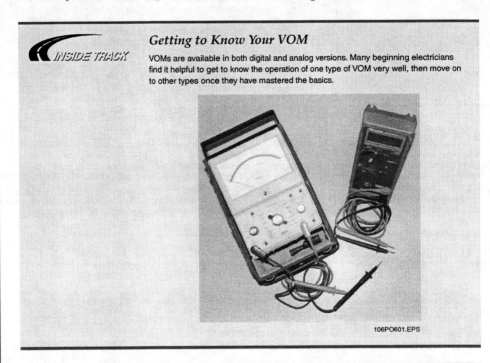

Getting to Know Your VOM

VOMs are available in both digital and analog versions. Many beginning electricians find it helpful to get to know the operation of one type of VOM very well, then move on to other types once they have mastered the basics.

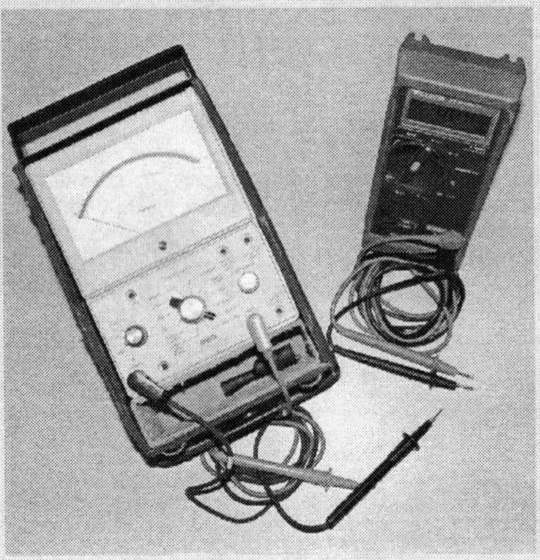

106PO601.EPS

Instructor's Notes:

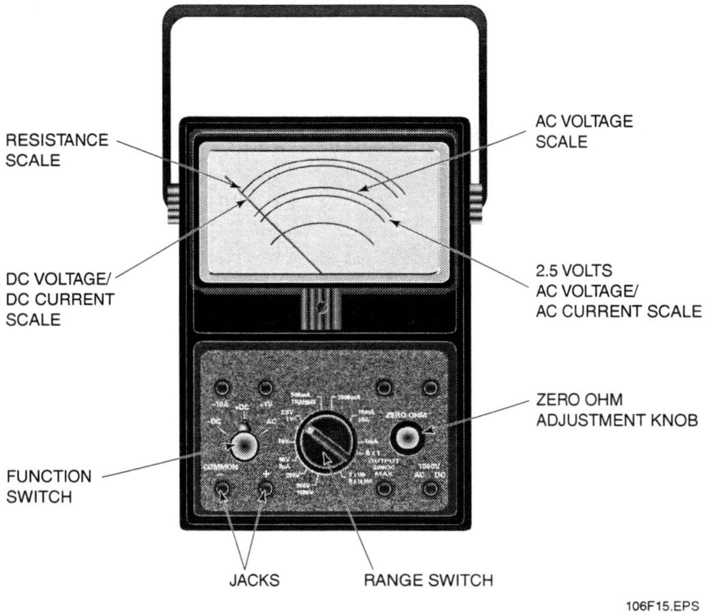

Figure 15 ◆ Multimeter.

INSIDE TRACK

Check Your Connections

When using a VOM, make sure your leads match the function. The meter can be in the voltage function and even display volts, but if the leads are in the amp jack, the meter will short circuit if connected to a voltage source.

100Ω range: 1.25mA
10KΩ range: 75mA

- Meter frequency response:
  Up to 100kHz

6.2.0 Overload Protection

In the example multimeter, a 1A, 250V fuse is provided to protect the circuits on the ohmmeter ranges. It also protects the milliampere ranges from excessive overloads. If the instrument fails to indicate, the fuse may be burned out. The fuse is mounted in a holder in the battery and fuse compartment. A spare fuse is located in a well between the + terminal of the D cell and the side of the case. Access to the compartment is obtained by loosening the single captivating screw on the compartment cover. To replace a burned-out fuse, remove it from the holder and replace it with a fuse of the exact same type. When removing the fuse from its holder, first remove the battery.

In addition to the fuse, a varistor protects the indicating instrument circuit. The varistor limits the current through the moving coil in case of overload.

The fuse and varistor will prevent serious damage to the meter in most cases of accidental overload. However, no overload protection system is completely foolproof, and misapplication on high-voltage circuits can damage the instrument. Care

Classroom

Stress the importance of proper meter connections.

Discuss meter overload protection.

ELECTRICAL TEST EQUIPMENT 189

Classroom

Review the storage and use guidelines for VOMs.

Safety

Review the safety precautions associated with the use of VOMs.

Demonstration

Demonstrate how to make DC voltage tests in various ranges using an analog VOM. (You may need to revise the given procedures based upon the actual meter available for the demonstration.)

and caution should always be exercised to protect both you and the VOM.

Beside the actual steps used in making a measurement, some other points to consider while using a multimeter are:

- Keep the instrument in a horizontal position when storing and away from the edge of a workbench, shelf, or other area where it may be knocked off and damaged.
- Avoid rapid or extreme temperature changes. For example, do not leave the meter in your truck during hot or cold weather. Rapid and extreme temperature changes will advance the aging of the meter components and adversely affect meter life and accuracy.
- Avoid overloading the measuring circuits of the instrument. Develop a habit of checking the range position before connecting the test leads to a circuit. Even slight overloads can damage the meter. Even though it may not be noticeable in blown fuses or a bent needle, damage has been done. Slight overloads will advance the aging of components, again causing changes in meter life and accuracy.
- Place the range switch in the TRANSIT position when the instrument is not in use or when it is being moved. This reduces the swinging of the pointer when the meter is carried. Every meter does not have a TRANSIT position, but if the meter does, it should be used. Random, uncontrolled swings of the meter movement may damage the movement, bend the needle, or reduce its accuracy.
- If the meter has not been used for a long period of time, rotate the function and range switches in both directions to wipe the switch contacts. Most switch contacts are plated with copper or silver. Over a period of time, these materials will oxidize (tarnish). This will create a high resistance through the switch, causing a large inaccuracy. Rotating through the switch positions will clean the tarnish off and provide good electrical contact.

6.3.0 Making Measurements

> **CAUTION**
>
> When using the meter as a millivoltmeter, care must be taken to prevent damage to the indicating instrument from excessive voltage. Before using the 250mV range, first use the 1.0V range to determine that the voltage measured is no greater than 250mV (or 0.25VDC).

6.3.1 Measuring DC Voltage, 0–250 Millivolts

Step 1 Set the function switch to +DC.

Step 2 Plug the black test lead into the − (COMMON) jack and the red test lead into the +50μA/250mV jack.

Step 3 Set the range switch to 50μA (dual position with 50V).

Step 4 Connect the black test lead to the negative side of the circuit being measured and the red test lead to the positive side of the circuit.

Step 5 Read the voltage on the black scale marked DC and use the figures marked 0–250. Read directly in millivolts.

6.3.2 Measuring DC Voltage, 0–1 Volt

Step 1 Set the function switch to −DC. Plug the black test lead into the − (COMMON) jack and the red test lead into the +1V jack.

Step 2 Set the range switch to 1V (dual position with 2.5V).

Step 3 Connect the black test lead to the negative side of the circuit being measured and the red test lead to the positive side of the circuit.

Step 4 Read the voltage on the black scale marked DC. Use the figures marked 0–10 and divide the reading by 10.

6.3.3 Measuring DC Voltage, 0–2.5 Through 0–500 Volts

> **WARNING!**
>
> Be extremely careful when working with higher voltages. Do not touch the instrument test leads while power is on in the circuit being measured.

Step 1 Set the function switch to +DC.

Step 2 Set the range switch to one of the five voltage range positions marked 2.5V, 10V, 50V, 250V, or 500V. When in doubt as to the voltage present, always use the highest voltage range as a protection for the instrument. If the voltage is within a lower range, the switch may be set for the lower range to obtain a more accurate reading.

Instructor's Notes:

Step 3 Plug the black test lead into the −(COMMON) jack and the red test lead into the + jack.

Step 4 Connect the black test lead to the negative side of the circuit being measured and the red test lead to the positive side of the circuit.

Step 5 Read the voltage on the black scale marked DC. For the 2.5V range, use the 0–250 figures and divide by 100. For the 10V, 50V, and 250V ranges, read the figures directly. For the 500V range, use the 0–50 figures and multiply by 10.

6.3.4 Measuring DC Voltage, 0–1,000 Volts

> **WARNING!**
> Be extremely careful when working with higher voltages. Do not touch the instrument test leads while power is on in the circuit being measured.

Step 1 Set the function switch to +DC.

Step 2 Set the range switch to 1,000V (dual position with 500V).

Step 3 Plug the black test lead into the −(COMMON) jack and the red test lead into the 1,000V jack.

Step 4 Be sure power is off in the circuit being measured and all capacitors have been discharged. Connect the black test lead to the negative side of the circuit being measured and the red test lead to the positive side of the circuit.

Step 5 Turn on the power in the circuit being measured.

Step 6 Read the voltage using the 0–10 figures on the black scale marked DC. Multiply the reading by 100.

6.3.5 Measuring AC Voltage, 0–2.5 Through 0–500 Volts

> **WARNING!**
> Be extremely careful when working with higher voltages. Do not touch the instrument or test leads while power is on in the circuit being measured.

> **CAUTION**
> When measuring line voltage such as from a 120V, 240V, or 480V source, be sure that the range switch is set to the proper voltage position.

Step 1 Set the function switch to AC.

Step 2 Set the range switch to one of the five voltage range positions marked 2.5V, 10V, 50V, 250V, or 500V. When in doubt as to the actual voltage present, always use the highest voltage range as a protection to the instrument. If the voltage is within a lower range, the switch may be set for the lower range to obtain a more accurate reading.

Step 3 Plug the black test lead into the − (COMMON) jack and the red test lead into the + jack.

Step 4 Connect the test leads across the voltage source (in parallel with the circuit).

Step 5 Turn on the power in the circuit being measured.

Step 6 For the 2.5V range, read the value directly on the AC scale marked 2.5V. For the 10V, 50V, and 250V ranges, read the red scale marked AC and use the black figures immediately above the scale. For the 500V range, read the red scale marked AC and use the 0–50 figures. Multiply the reading by 10.

6.3.6 Measuring AC Voltage, 0–1,000 Volts

> **WARNING!**
> Be extremely careful when working with higher voltages. Do not touch the instrument or test leads while power is on in the circuit being measured.

Step 1 Set the function switch to AC.

Step 2 Set the range switch to 1,000V (dual position with 500V).

Step 3 Plug the black test lead into the − (COMMON) jack and the red test lead into the 1,000V jack.

Demonstration

Demonstrate how to make AC voltage tests in various ranges using an analog VOM. (Adjust given procedures as necessary to reflect actual meter in use.)

Safety

Point out all applicable safety procedures during the demonstration.

Step 4 Be sure the power is off in the circuit being measured and that all capacitors have been discharged. Connect the test leads to the circuit.

Step 5 Turn on the power in the circuit being measured.

Step 6 Read the voltage on the red scale marked AC. Use the 0–10 figures and multiply by 100.

Demonstration

Demonstrate how to measure output voltage using an analog VOM. (Adjust given procedure as necessary to reflect actual meter in use.)

6.3.7 Measuring Output Voltage

It is often desired to measure the AC component of an output voltage where both AC and DC voltage levels exist. This occurs primarily in amplifier circuits. The meter has a 0.1μF, 400V capacitor in series with the OUTPUT jack. The capacitor blocks the DC component of the current in the test circuit but allows the AC or desired component to pass on to the indicating instrument circuit. The blocking capacitor may alter the AC response at low frequencies but is usually ignored at audio frequencies.

Safety

Warn the trainees not to apply the output jack to a DC circuit with a voltage in excess of 400V.

> **CAUTION**
> When using OUTPUT, do not apply it to a circuit whose DC voltage component exceeds the 400V rating of the blocking capacitor.

Step 1 Set the function switch to AC.

Step 2 Plug the black test lead into the – (COMMON) jack and the red test lead into the OUTPUT jack.

Step 3 Set the range switch to one of the range positions marked 2.5V, 10V, 50V, or 250V.

Step 4 Connect the test leads across the circuit being measured, with the black test lead attached to the ground side.

Step 5 Turn on the power in the test circuit. Read the output voltage on the appropriate AC voltage scale. For the 2.5V range, read the value directly on the AC scale marked 2.5V. For the 10V, 50V, or 250V ranges, use the red scale marked AC and read the black figures immediately above the scale.

Classroom

Describe and/or demonstrate how to make decibel measurements.

6.3.8 Measuring Decibels

For some applications, mockup audio frequency voltages are measured in terms of decibels. The decibel scale (dB) at the bottom of the dial is marked from −20 to +10.

Step 1 To measure decibels, read the dB scale in accordance with instructions for measuring AC. For example, when the range switch is set to the 2.5V position, read the dB scale directly.

Step 2 The dB readings on the scale are referenced to a 0dB power level of .001W across 600Ω, or .775VAC across 600Ω.

Step 3 For the 10V range, read the dB scale and add +12dB to the reading. For the 50V range, read the dB scale and add +26dB to the reading. For the 250V range, read the dB scale and add +40dB to the reading.

Step 4 If the 0dB reference level is .006W across 500Ω, subtract +7dB from the reading.

6.4.0 Direct Current Measurements

6.4.1 Voltage Drop

The voltage drop across the meter on all milliampere current ranges is approximately 250mV measured at the jacks. An exception is the 0–500mA range with a drop of approximately 400mV. This voltage drop will not affect current measurements. In some transistor circuits, however, it may be necessary to compensate for the added voltage drop when making measurements.

6.4.2 Measuring Direct Current, 0–50 Microamperes

> **CAUTION**
> Never connect the test leads directly across voltage when the meter is used as a current-indicating instrument. Always connect the instrument in series with the load across the voltage source.

Step 1 Set the function switch to +DC.

Step 2 Plug the black test lead into the – (COMMON) jack and the red test lead into the +50μA/250mV jack.

Step 3 Set the range switch to 50μA (dual position with 50V).

Instructor's Notes:

INSIDE TRACK

Voltage Testing

Voltage testing covers a very wide range of values. For example, if you test the voltage of a residential oil-fired, hot water furnace, the thermostat circuit will be 24V, the circulating pumps will be 120V, and the ignition transformer output will be 10,000V. Notice that the range of voltages on this one appliance can put you outside the range of many multimeters. The photo shown here depicts a voltage reading being taken of a low-voltage control device.

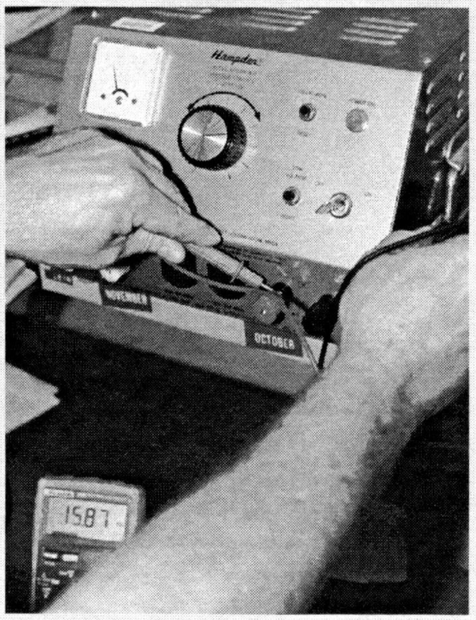

106PO602.EPS

Step 4 Open the circuit in which the current is being measured. Connect the instrument in series with the circuit. Connect the red test lead to the positive side and the black test lead to the negative side.

Step 5 Read the current on the black DC scale. Use the 0–50 figures to read directly in microamperes.

> **Note**
> In all direct current measurements, be certain the power to the circuit being tested has been turned off before disconnecting test leads and restoring circuit **continuity.**

6.4.3 Measuring Direct Current, 0–1 Through 0–500 Milliamperes

Step 1 Set the function switch to +DC.

Step 2 Plug the black test lead into the − (COMMON) jack and the red test lead into the + jack.

Step 3 Set the range switch to one of the four range positions (1mA, 10mA, 100mA, or 500mA).

Step 4 Open the circuit in which the current is being measured. Connect the VOM in series with the circuit. Connect the red test lead to the positive side and the black test lead to the negative side of the part of the circuit you are measuring.

Demonstration

Demonstrate how to make direct current measurements using an analog VOM. (Adjust given procedures as necessary to reflect actual meter in use.)

ELECTRICAL TEST EQUIPMENT 193

Demonstration

Demonstrate how to adjust an analog VOM to zero prior to making a resistance test.

Demonstrate how to measure resistance using an analog VOM.

Safety

Review the applicable safety precautions for making resistance tests.

Step 5 Read the current in milliamperes on the black DC scale. For the 1mA range, use the 0–10 figures and divide by 10. For the 10mA range, use the 0–10 figures and multiply by 10. For the 500mA range, use the 0–50 figures and multiply by 10.

6.4.4 Measuring Direct Current, 0–10 Amperes

Step 1 Plug the black test lead into the −10A jack and the red test lead into the +10A jack.

Step 2 Set the range switch to 10A (dual position with 10mA).

Step 3 Open the circuit in which the current is being measured. Connect the instrument in series with the circuit. Connect the red test lead to the positive side and the black test lead to the negative side.

> **Note**
> The function switch has no effect on polarity for the 10A range.

Step 4 Read the current on the black DC scale. Use the 0–10 figures to read directly in amperes.

> **CAUTION**
> When using the 10A range, never remove a test lead from its panel jack while current is flowing through the circuit. Otherwise, damage may occur to the plug and jack.

6.4.5 Zero Ohm Adjustment

When resistance is measured, the VOM batteries furnish power for the circuit. Since batteries are subject to variation in voltage and internal resistance, the instrument must be adjusted to zero prior to measuring a resistance.

Step 1 Set the range switch to the desired ohms range.

Step 2 Plug the black test lead into the − (COMMON) jack and the red test lead into the + jack.

Step 3 Connect the ends of the test leads to short the VOM resistance circuit.

Step 4 Rotate the ZERO OHM control until the pointer indicates zero ohms. If the pointer cannot be adjusted to zero, one or both of the batteries must be replaced.

Step 5 Disconnect the ends of the test leads and connect them to the component being measured.

6.4.6 Measuring Resistance

> **CAUTION**
> Before measuring resistance, be sure power is off to the circuit being tested. Disconnect the component from the circuit before measuring its resistance.

Step 1 Set the range switch to one of the resistance range positions:
- Use R × 1 for resistance readings from 0 to 200 ohms.
- Use R × 100 for resistance readings from 200 to 20,000 ohms.
- Use R × 10,000 for resistance readings above 20,000 ohms.

Step 2 Set the function switch to either the −DC or +DC position. The operation is the same in either position.

> INSIDE TRACK
>
> ***Current Testing***
>
> Levels of current to be tested vary from milliamperes to thousands of amps. In modern communications and consumer electronic devices, you will be measuring for mA. In household circuits you usually test 15A to 50A circuits or higher when testing the service entrance. A high amperage test would be performed on an industrial service entrance, which might go up to 12,000A.

Instructor's Notes:

194 CHAPTER 5

Step 3 Adjust the ZERO OHM control for each resistance range.
- Observe the reading on the OHMS scale at the top of the dial. Note that the OHMS scale reads from right to left for increasing values of resistance.
- To determine the actual resistance value, multiply the reading by the factor at the switch position (K on the OHMS scale equals one thousand).

Step 4 If there is a forward and backward resistance such as in diodes, the resistance should be relatively low in one direction (for forward polarity) and higher in the opposite direction.

> CAUTION
> Check that the OHMS range being used will not damage any of the semiconductors.

Step 5 If the purpose of the resistance measurement is to check a semiconductor in or out of a circuit (forward and reverse bias resistance measurements), check the following prior to making the measurement:
- The polarity of the voltage at the input jacks is identical to the input jack markings. Therefore, be certain that the polarity of the test leads is correct for the application.
- Ensure that the range selected will not damage the semiconductor (use R × 100 or below).
- Refer to the meter specifications and review the limits of the semiconductor according to the manufacturer's ratings.
  - If the semiconductor is a silicon diode or conventional silicon transistor, no precautions are normally required.
  - If the semiconductor material is germanium, check the ratings of the device and refer to its specifications.

Step 6 Rotate the function switch between the two DC positions to reverse polarity. This will determine if there is a difference between the resistance in the two directions.

Step 7 The resistance of such diodes will measure differently from one resistance range to another on the same VOM with the function switch in a given position. For example, a diode that measures 80Ω on the R × 1 range may measure 300Ω on the R × 100 range. The difference in values is a result of the diode characteristic and does not indicate any fault in the VOM.

7.0.0 ◆ DIGITAL METERS

Digital meters have revolutionized the test equipment world. Improved accuracy is very easily attainable, more functions can be incorporated into one meter, and both autoranging and automatic polarity indication can be used. Technically, digital multimeters are classified as electronic multimeters; however, digital multimeters do not use a meter movement. Instead, a digital meter's input circuit converts a current into a digital signal, which is then processed by electronic circuits and displayed numerically on the meter face.

A major limitation with many meters that use meter movements is that the scale reading must be estimated if the meter pointer falls between scale divisions. Digital multimeters eliminate the need to estimate these readings by displaying the reading as a numerical display.

With digital meters, technicians must revise the way the indications are viewed. For example, if a technician were reading the AC voltage on a normal wall outlet with an analog voltmeter, any indication within the range of 120VAC would be considered acceptable. But, when reading with a digital meter, the technician might think something was wrong if the meter showed a reading of 114.53VAC. Bear in mind that the digital meter is very precise in its reading, sometimes more precise than is called for, or is usable. Also, be aware that the indicated parameter may change with the range used. This is primarily due to the change in accuracy and where the meter is rounding off.

There are many types of digital multimeters. Some are bench-type multimeters, while others are designed to be handheld. Most types of digital multimeters have an input impedance of 10 megohms and above. They are very sensitive to small changes in current and are therefore very accurate.

An example of a digital meter is shown in *Figure 16.* The internal operation of this meter is basically the same as other digital meters. The following paragraphs discuss the operation and use of this particular meter. For specific instructions, always refer to the owner's manual supplied with your meter.

Classroom

Provide an overview of digital meters and their applications.

Audiovisual

Show Transparency 17 (Figure 16).

ELECTRICAL TEST EQUIPMENT 195

Digital Multimeter Classifications

INSIDE TRACK

Newer digital multimeters are rated for safety according to voltage and current limitations and fault interrupting capacity. Never use a multimeter outside of the limits specified by the manufacturer.

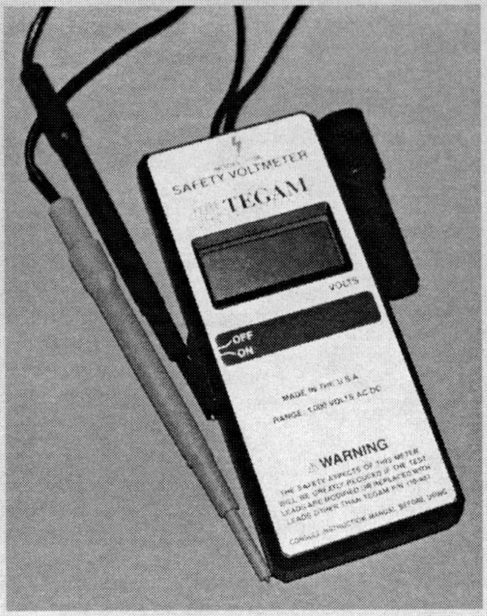

106PO603.EPS

Check Your Meter Before Using It

INSIDE TRACK

Never assume that a blank meter represents a reading of zero volts. Some digital meters automatically cut off the power to preserve the battery.

Similarly, if you press the HOLD button on some meters when you are reading 0V, it will lock on that reading and will continue to read 0V, regardless of the actual voltage present. Always check the meter for proper operation before using it.

Digital Meters

THINK ABOUT IT

Most applications do not require the precision of a digital meter. Identify some applications in which precision is important. When wouldn't it be important?

Teaching Tip

Discuss the "Think About It."

Instructor's Notes:

106F16.EPS

Figure 16 ◈ Digital meter.

7.1.0 Features

The example meter offers the following features:

- *Autorange/manual range modes*—The meter features autoranging for all measurement ranges. Press the RANGE button to enter manual range mode. A flashing symbol may be used to show that you are in the manual range mode. Press the RANGE button as required to select the desired range. To switch back to auto range, press the ON/CLEAR button once (clear mode) or select another function.
- *Automatic off*—The example meter turns itself off after one hour of non-use. The current draw while the meter is turned off does not affect battery life. If the meter turns itself off while a parameter is being monitored, press the ON/CLEAR button to turn it on again. To protect against electrical damage, the meter also turns itself off if a test lead is inserted into the 10A jack while the meter is in any mode other than A ⎓ or A ~.
- *Dangerous voltage indication*—The meter shows the symbol for any range over 20V. In the autoranging mode, the meter also beeps when it changes to any range over 20V.
- *Out of Limits (OL)*—The meter displays OL and a rapidly flashing decimal point (position determined by range) when the measured value is greater than the limit of the instrument or selected range.
- *Audible acknowledgment*—The meter acknowledges each press of a button or actuation of the selector switch with a beep.

7.2.0 Operation

This section will discuss the use of various controls and explain how measurements should be taken.

7.2.1 Dual Function ON/CLEAR Button

Press the ON/CLEAR button to turn the meter on. Operation begins in the autorange mode, and the range for maximum resolution is selected automatically. Press the ON/CLEAR button again to turn the meter off.

7.2.2 Measuring Voltage

Step 1 Select V ⎓ or V ~.

Step 2 Connect the test leads as shown in *Figure 17*.

Step 3 Observe the voltage reading on the display. Depending on the range, the meter displays units in mV or V.

To avoid shock hazard or meter damage, do not apply more than 1,500VDC or 1,000VAC to the meter input or between any input jack and earth ground.

7.2.3 Measuring Current

Step 1 Select A ⎓ or A ~.

Step 2 Insert the meter in series with the circuit with the red lead connected to either:
- The mA jack for input up to 200 milliamps
- The 10A jack for input up to 20 amps

Step 3 Make hookups as shown in *Figures 18* and *19*.

Step 4 Observe the current reading.

> **Note**
> The meter shuts itself off if a test lead is inserted into the 10A jack when the meter is in any function other than A ⎓ or A~.

Classroom

Review the features of a typical digital VOM.

Demonstration

Demonstrate how to make voltage and current measurements using a digital VOM. (Revise procedures as necessary to reflect actual meter in use.)

Audiovisual

Show Transparencies 18 through 20 (Figures 17 through 19).

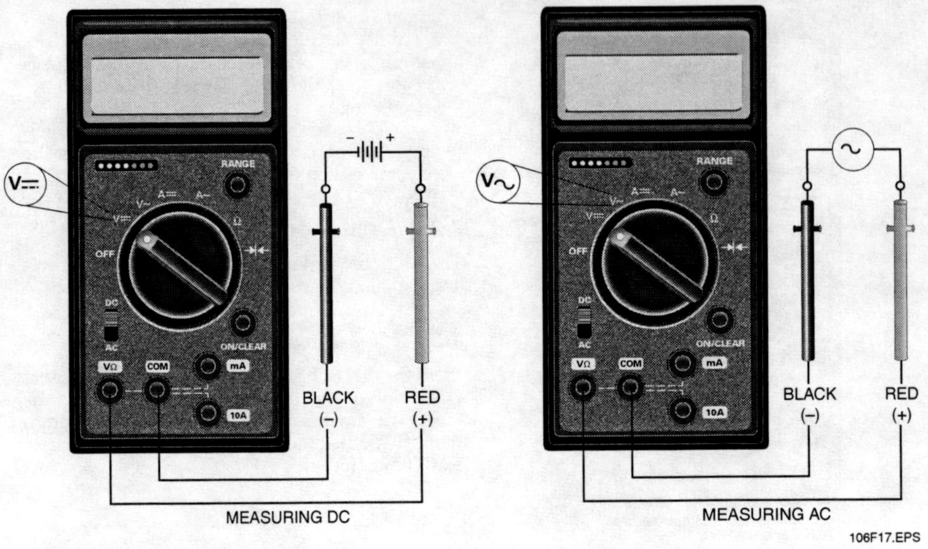

Figure 17 ◆ Measuring voltage.

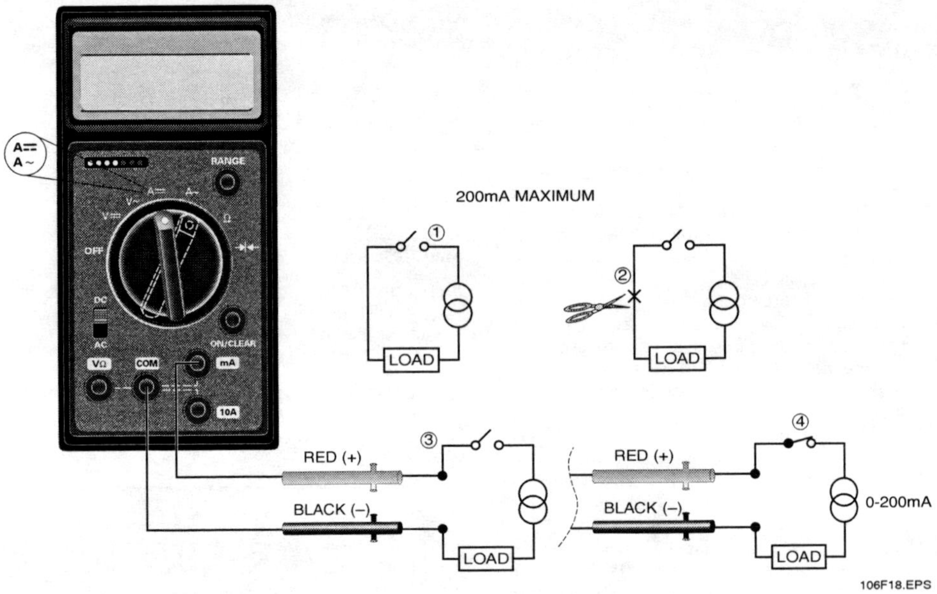

Figure 18 ◆ Measuring current (mA).

Instructor's Notes:

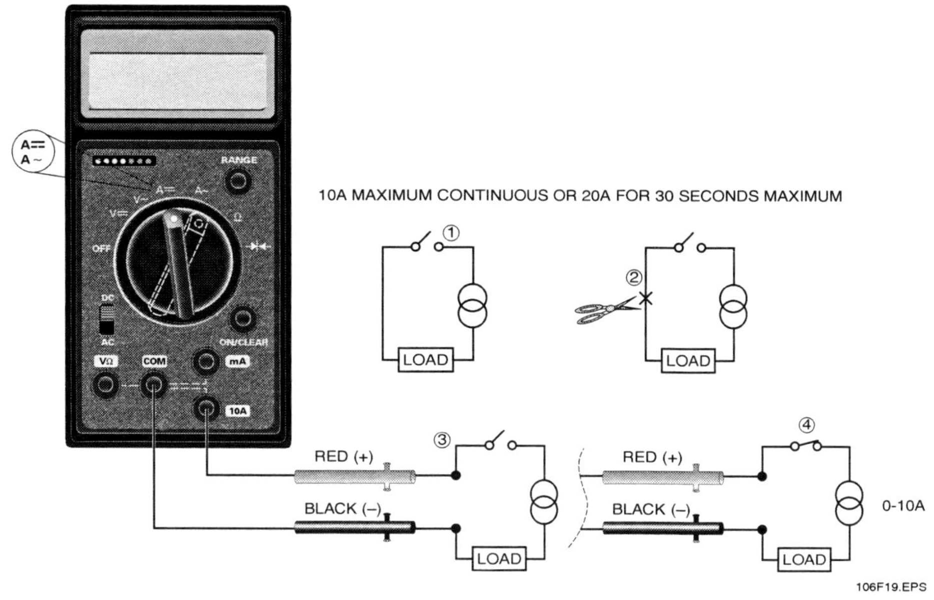

Figure 19 ◆ Measuring current (amps).

7.2.4 Measuring Resistance

When measuring resistance, any voltage present will cause an incorrect reading. For this reason, the capacitors in a circuit in which resistance measurements are about to be taken should first be discharged.

Step 1 Select Ω (ohms).

Step 2 Connect the test leads as shown in *Figure 20.*

Step 3 Observe the resistance reading.

7.2.5 Diode/Continuity Test

Step 1 Select ⇥)) (diode/continuity).

Step 2 Choose one of the following:

- *Forward bias*—Connect to the diode as shown in *Figure 21(A).* The meter will display one of the following: the forward voltage drop (V_F) of a good diode (<0.7V), a very low reading for a shorted diode (<0.3V), or OL for an open diode.
- *Reverse bias or open circuit*—Reverse the leads to the diode. The meter displays OL, as shown in *Figure 21(B).* It does not beep.
- *Continuity*—The meter beeps once if the circuit resistance is less than 150 ohms, as shown in *Figure 21(C).*

7.2.6 Transistor Junction Test

Test transistors in the same manner as diodes by checking the two diode junctions formed between the base and emitter, and the base and collector of the transistor. *Figure 22* shows the orientation of these effective diode junctions for PNP and NPN transistors. Also check between the collector and emitter to determine if a short is present.

7.2.7 Display Test

To test the LCD display, hold the ON/CLEAR button down when turning on the meter. Verify that the display shows all segments (see *Figure 23*).

Demonstrate how to measure resistance using a digital VOM. (Revise given procedure as necessary to reflect actual meter in use).

Show Transparency 21 (Figure 20).

Under careful supervision, have the trainees practice using a digital VOM to measure voltage, current, and resistance. Note the proficiency of each trainee.

Describe and/or demonstrate the other tests that may be made using a digital VOM.

Show Transparencies 22 through 24 (Figures 21 through 23).

ELECTRICAL TEST EQUIPMENT 199

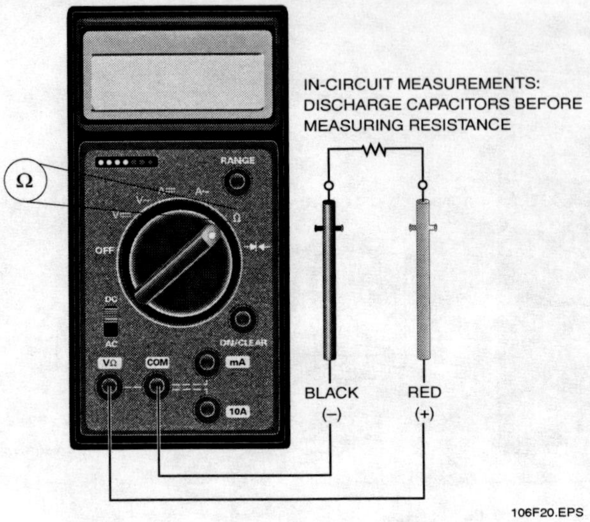

Figure 20 ◆ Measuring resistance.

> **INSIDE TRACK**
>
> ***Understanding Diodes***
>
> A diode is a two-terminal semiconductor device that behaves like a check valve. *Forward bias* means that you are testing that the diode conducts properly when it is closed. *Reverse bias* means that current does not flow through the diode.

Demonstration

Demonstrate the maintenance procedures required for the proper care of a VOM.

Audiovisual

Show Transparency 25 (Figure 24).

7.3.0 Maintenance

The following sections discuss the necessary maintenance for a multimeter.

7.3.1 Battery Replacement

Replace the battery as soon as the meter's decimal point starts blinking during normal use; this indicates that <100 hours of battery life remain. Remove the case back and replace the battery with the same or equivalent 9V alkaline battery.

7.3.2 Fuse Replacement

The meter uses two input protection fuses for the mA and 10A inputs. Remove the case back to gain access to the fuses. Replace with the same type only. The large fuse should be readily available. A spare for the smaller fuse is included in the case. If necessary, this fuse must be reordered from the factory.

7.3.3 Calibration

Have a qualified technician calibrate the meter once a year.

Step 1 Remove the case back (*Figure 24*). Turn the meter ON and select V ⎓.

Step 2 Apply +1.900VDC +/−0.0001V to the V-n input (negative to COM).

Step 3 Adjust the DC control through the hole in the circuit board for a display of +1.900V.

Step 4 Select V.

Instructor's Notes:

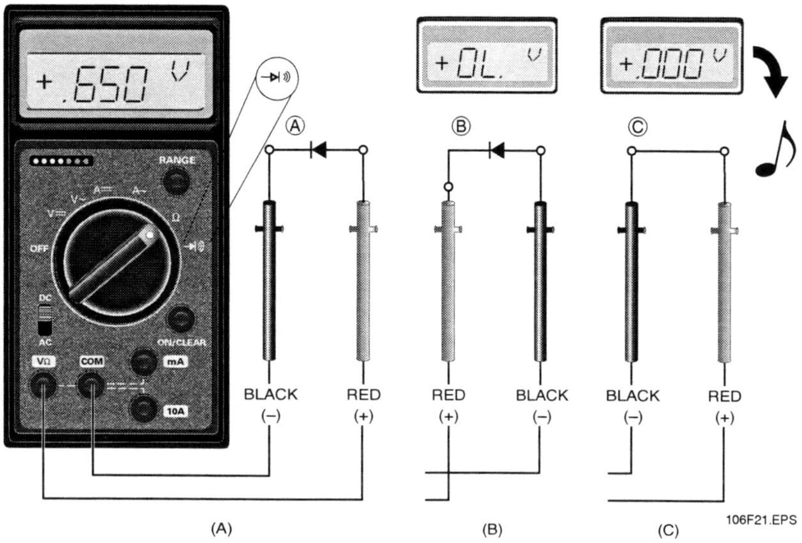

Figure 21 ◆ Diode/continuity test.

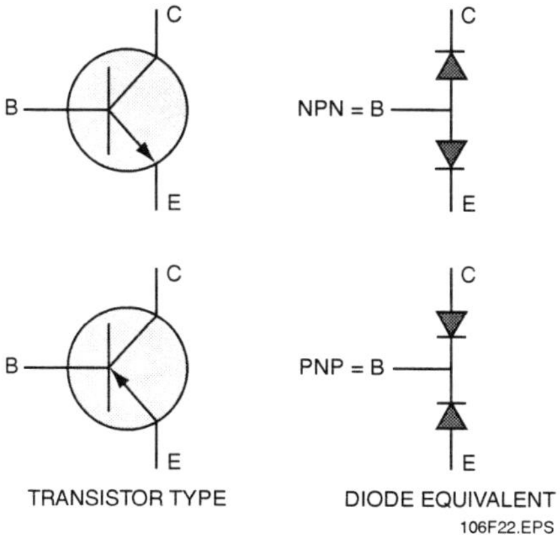

Figure 22 ◆ Transistor junction test.

Step 5 Apply +1.900VAC +/−0.002VAC @ 60Hz to the V-n input.

Step 6 Adjust the AC control through the hole in the circuit board for a display of +1.900V.

Step 7 Reassemble the meter.

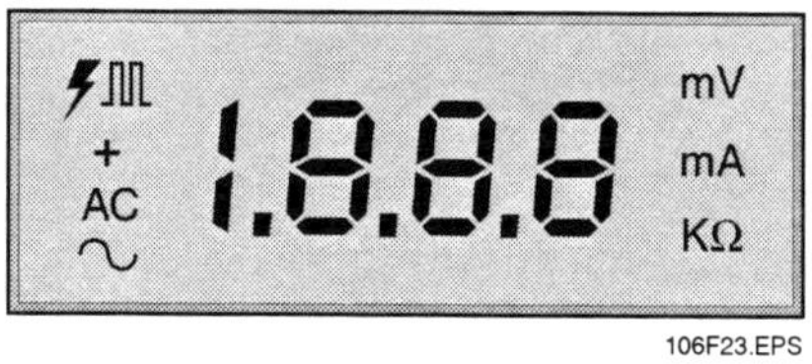

Figure 23 ◆ Display test.

8.0.0 ◆ WATTMETER

Rather than performing two measurements and then calculating power, a power-measuring meter called a *wattmeter* can be connected into a circuit to measure power. The power can be read directly from the scale of this meter. Not only does a wattmeter simplify power measurements, but it has two other advantages:

First, voltage and current in an AC circuit are not always in phase; current sometimes either leads or lags the voltage (this is known as the *power factor*). When this happens, multiplying the voltage times the current results in apparent power, not true power. Therefore, in an AC circuit, measuring the voltage and current and then multiplying them can

Classroom

Discuss the purpose and operation of a basic wattmeter.

ELECTRICAL TEST EQUIPMENT 201

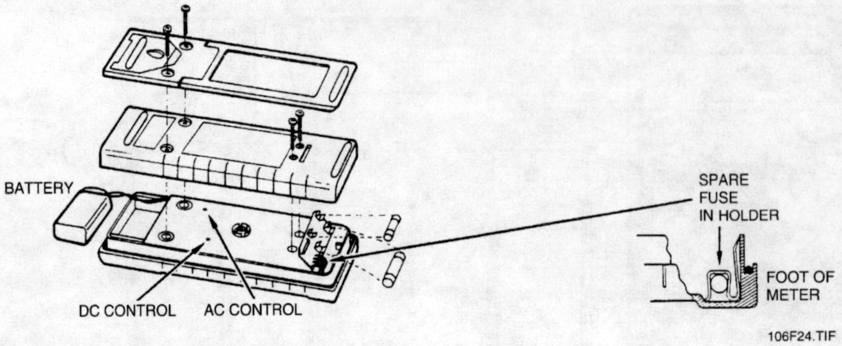

Figure 24 ◆ Meter maintenance.

often result in an incorrect value of power dissipation by the circuit. However, the wattmeter takes the power factor into account and always indicates true power.

Second, voltmeters and ammeters consume power. The amount consumed depends on the levels of the voltage and the current in the circuit, and it cannot be accurately predicted. Therefore, very accurate power measurements cannot be made by measuring voltage and current and then calculating power. However, some wattmeters compensate for their own power losses so that only the power dissipated in the circuit is measured. If the wattmeter is not compensated, the power that is dissipated is sometimes marked on the meter, or it can easily be determined so that a very accurate measurement can be made. Typically, the accuracy of a wattmeter is within 1 percent.

The basic wattmeter consists of two stationary coils connected in series and one movable coil (*Figure 25*). The moving coil, wound with many turns of fine wire, has a high resistance.

The stationary coils, wound with a few turns of a larger wire, have a low resistance. The interaction of the magnetic fields around the different coils will cause the movable coil and its pointer to

Audiovisual

Show Transparency 26 (Figure 25).

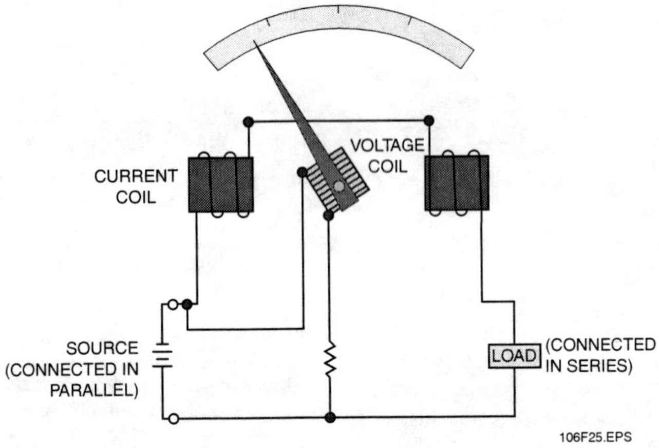

Figure 25 ◆ Wattmeter schematic.

Instructor's Notes:

Wattmeter

INSIDE TRACK

Wattmeters are less commonly used than multimeters, since the electrician is usually interested in the existence of electrical potential or the flow of current. The meter used to measure power consumption is the watt-hour meter attached to the service entrance.

106PO604.EPS

rotate in proportion to the voltage across the load and the current through the load. Thus, the meter indicates E times I, or power.

The two circuits in the wattmeter will be damaged if too much current passes through them. This fact is of special importance because the reading on the meter does not tell the user that the coils are being overheated. If an ammeter or voltmeter is overloaded, the pointer will indicate beyond full-scale deflection. In a wattmeter, both the current and potential (voltage) circuits may be carrying such an overload that their insulation is burning, and yet the pointer may only be partway up the scale. A low-power factor circuit will give a low reading on the wattmeter even when the current and potential circuits are loaded to their maximum safe limits.

9.0.0 ◆ MEGOHMMETER (MEGGER)

An ordinary ohmmeter cannot be used for measuring resistances of several million ohms, such as those found in conductor insulation or between motor or transformer windings, and so on. To adequately test these types of very high resistances, it is necessary to use a much higher potential than is furnished by the battery of an ohmmeter. For this purpose, a megger is used. There are three types of meggers: hand, battery, and electric.

The megger is similar to a moving-coil meter except that it has two windings (coils). See *Figure 26*. Coil A is in series with resistor R_2 across the output of the generator.

This coil is wound so that it causes the pointer to move toward the high-resistance end of the scale when the generator is in operation. Winding B is in series with R_1 and R_X (the unknown resistance to be measured). This winding is wound so that it causes the pointer to move toward the low or zero-resistance end of the scale when the generator is in operation.

When an extremely high resistance appears across the input terminals of the megger, the current through coil A causes the pointer to read infinity. Conversely, when a relatively low resistance appears across the input terminals, the current through coil B causes the pointer to deflect toward zero. The pointer stops at a point on the

Classroom

Describe the basic operation and applications of a megohmmeter.

Audiovisual

Show Transparencies 27 and 28 (Figures 26 and 27).

ELECTRICAL TEST EQUIPMENT 203

INSIDE TRACK

Caring for Your Meter

Like the other meters discussed here, a megger is a sensitive instrument. Treat it with care and keep it in its case when not in use.

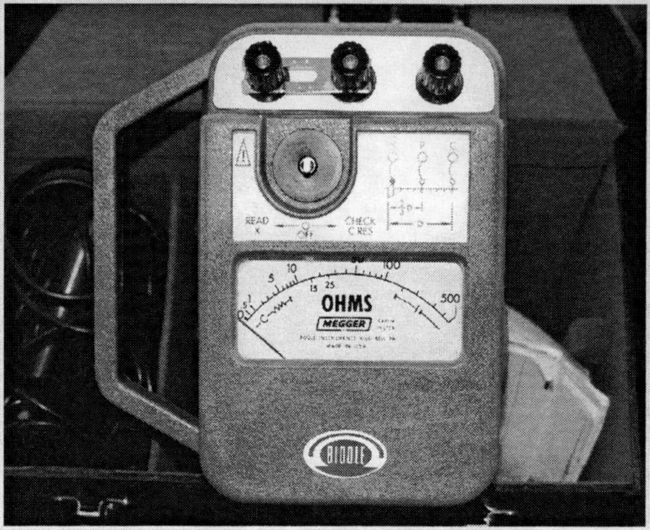

106PO605.EPS

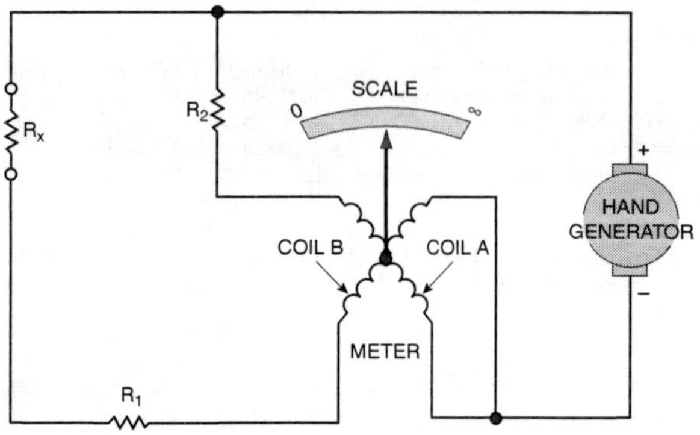

106F26.EPS

Figure 26 ◆ Megger schematic.

Instructor's Notes:

204 CHAPTER 5

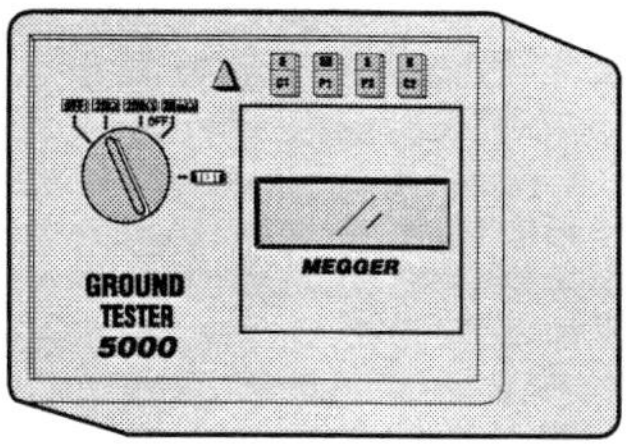

106F27.EPS

Figure 27 ◆ Digital readout megger.

scale determined by the current through coil B, which is controlled by R_X.

Digital meggers *(Figure 27)* use the same operational principles. Instead of having a scaled meter movement, these meters give the value of resistance in a digital readout display. The digital readout makes reading the measurement much easier and helps to eliminate errors.

To avoid excessive test voltages, most hand meggers are equipped with friction clutches. When the generator is cranked faster than its rated speed, the clutch slips, and the generator speed and output voltage are maintained at their rated values.

9.1.0 Safety Precautions

> **WARNING!**
> When a megger is used, the generator voltage is present on the test leads. This voltage could be hazardous to you or the equipment you are testing. *NEVER TOUCH THE TEST LEADS WHILE THE TESTER IS BEING USED.* Isolate the item you are testing from the circuit before using the megger. Protect all parts of the test subject from contact by others.

When using a megger, you could be injured or cause damage to the equipment being worked on if you do not observe the following minimum safety precautions:

- Use meggers on high-resistance measurements *only* (such as insulation measurements or to check two separate conductors in a cable).
- *Never* touch the test leads while the handle is being cranked.
- Deenergize and verify the deenergization of the circuit completely before connecting the meter.
- Disconnect the item being checked from other circuitry, if possible, *before* using the meter.
- After the test, ground the tested circuit to discharge any energy that may be left in the circuit.

10.0.0 ◆ FREQUENCY METER

Frequency is the number of cycles completed each second by a given AC voltage, and it is usually expressed in hertz (one hertz = one cycle per second). The frequency meter is used in AC power-producing devices such as generators to ensure that the correct frequency is being produced. Failure to produce the correct frequency will result in excess heat and component damage.

There are two common types of frequency meters. One operates with a set of reeds having natural vibration frequencies that respond in the range being tested. The reed with a natural frequency closest to that of the current being tested will vibrate most strongly when the meter operates. The frequency is read from a calibrated scale.

A moving-disk frequency meter works with two coils, one of which is a magnetizing coil whose current varies inversely with the frequency. A disk with a pointer mounted between the coils turns in the direction determined by the stronger coil. Solid-state frequency meters are also available.

10.1.0 Phase Rotation Tester

A phase rotation tester is an indicator that allows you to see the phase rotation of incoming current.

10.2.0 Harmonic Test Set

A harmonic is a sine wave whose frequency is a whole number multiple of the original base frequency. For example, a standard 60Hz sine wave may have second and third harmonics at 120Hz and 180Hz, respectively. Harmonics may be caused by various circuit loads such as fluorescent lights and by certain three-phase transformer connections.

The results of harmonics are heating in wiring and a voltage or current that cannot be detected by most digital meters.

11.0.0 ◆ POWER FACTOR METER

The power factor is the ratio of true (actual) power to apparent power. The power factor of a circuit or piece of equipment may be found by using an

Safety

Stress the safety precautions associated with the use of megohmmeters.

Classroom

Discuss the operation and applications of frequency meters.

Classroom

Discuss the operation and applications of power factor meters. Demonstrate how to calculate the power factor if a power factor meter is not available.

ELECTRICAL TEST EQUIPMENT 205

Audiovisual

Show Transparency 29 (Figure 28).

Classroom

Discuss the operation and applications of continuity testers.

ammeter, wattmeter, and voltmeter. To calculate the power factor, divide the wattmeter reading (true power) by the product of the ammeter and voltmeter readings (apparent power or EI). The ideal power factor is 1.

$$\text{Power Factor} = \frac{\text{True Power (Wattmeter Reading)}}{\text{Apparent Power (EI)}}$$

It is not necessary to calculate these readings if a power factor meter is available to read the power factor directly (*Figure 28*). This meter indicates the equivalent of pure resistance or unit power factor, which is a one-to-one ratio.

12.0.0 ◆ CONTINUITY TESTER

An ohmmeter can be used to test continuity, but this means carrying an expensive, often bulky test instrument with you. Pocket-type continuity testers are just as reliable and much more compact and portable.

There are two types of continuity testers: audio and visual. These are specialized devices for identifying conductors in a conduit by checking continuity.

12.1.0 Audio Continuity Tester

This type of tester is used to ring out wires in conduit runs. At one end of the conduit run, select one wire, strip off a little insulation, and connect that wire to the conduit. At the other end of the conduit run, clip one lead of the tester to the conduit. Touch the other lead to one wire at a time until the audible alarm sounds, which indicates continuity (a closed circuit). Then, put matching tags on the wire. Continue this procedure, one wire at a time, to identify the other wires.

12.2.0 Visual Continuity Tester

This type of tester is useful if you are working in an area where background noise might make it hard to hear the audio tester. The procedure is the same. When the proper wire is tested, the light will come on.

12.3.0 Cable-Length Meter

A cable-length meter measures the length and condition of a cable by sending a signal down the cable and then reading the signal that is reflected back.

13.0.0 ◆ VOLTAGE TESTER

A voltage tester is a simple aid that determines whether there is a potential difference between two points. It *does not* calculate the value of that difference. If the actual value of the potential difference is needed, use a voltmeter.

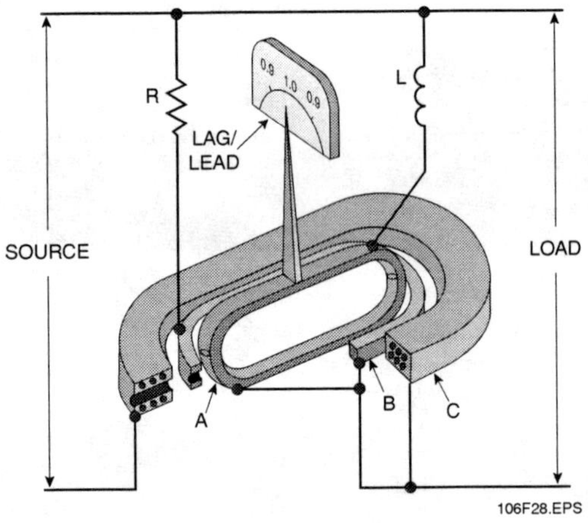

Figure 28 ◆ Power factor meter.

Instructor's Notes:

INSIDE TRACK

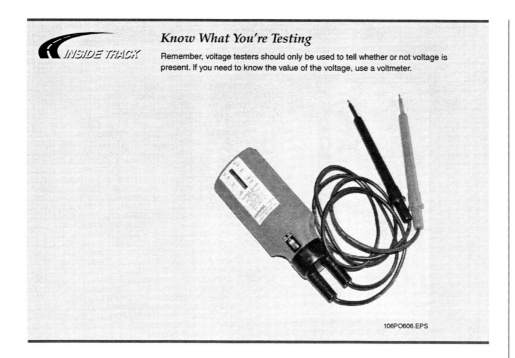

Know What You're Testing

Remember, voltage testers should only be used to tell whether or not voltage is present. If you need to know the value of the voltage, use a voltmeter.

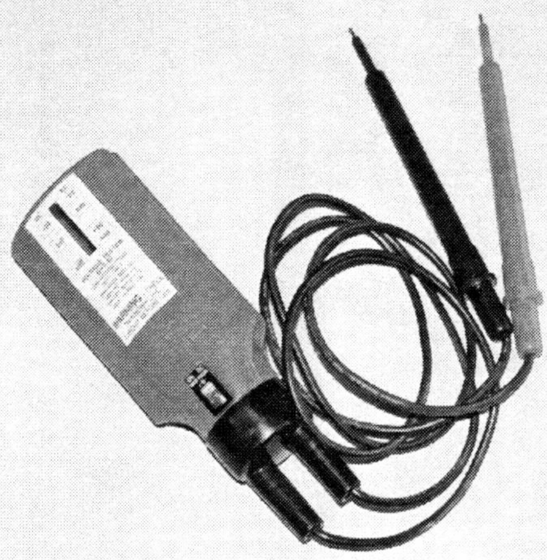

106PO606.EPS

13.1.0 Wiggy®

Pocket-type voltage testers are inexpensive and portable. They can easily be carried in your tool pouch, eliminating the need to carry a delicate voltmeter on the job. Simple neon testers are becoming quite popular. Another type is known as a *Wiggy®* (see *Figure 29*).

13.1.1 Principle of Operation

The operation of a Wiggy® is fairly simple. The basic component is a solenoid. When a current flows through the coil, it will produce a magnetic field, which will pull the plunger down against a spring. The spring will limit how far the plunger can be drawn into the cylinder. As the current increases, the plunger will move farther. The amount of current depends upon the potential difference applied across the coil. A pointer on the plunger indicates the potential difference on the scale.

The scale on the tester has voltage indications for AC on one side of the pointer and DC on the other side. The AC scale indicates 120, 240, 480, and 600 volts. The DC scale indicates 125, 240, and 600 volts.

To use a voltage tester, place the probes across a possible source of voltage. If there is voltage present, current flows through the coil inside the tester, creating a magnetic field. The magnetized coil pulls the indicator along the scale until it reaches a point corresponding to the approximate

CAUTION

This is an approximation only. If you need to know the actual value of the voltage, use a voltmeter.

CAUTION

To avoid damage to the tester, do not use a Wiggy® above its stated range.

Classroom

Discuss the operation and applications of voltage testers.

Audiovisual

Show Transparency 30 (Figure 29).

Demonstration

Demonstrate how to use a voltage tester to check for the presence of voltage.

Safety

Emphasize the safety precautions associated with the use of voltage testers.

Laboratory

Have the trainees practice using a voltage tester. Note the proficiency of each trainee.

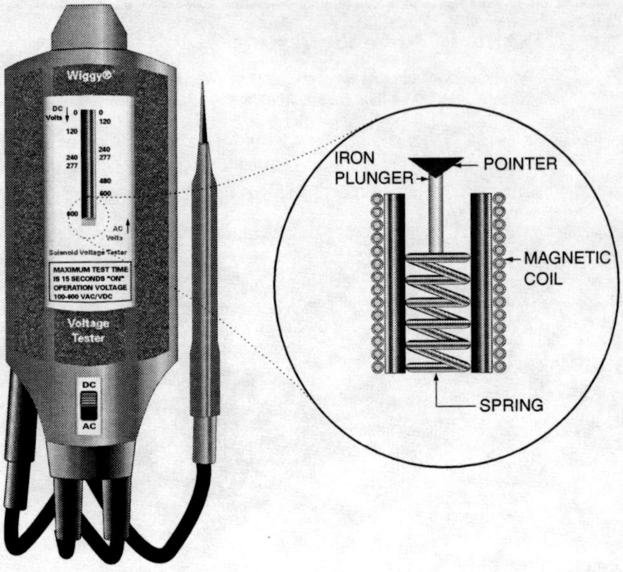

Figure 29 ◆ Wiggy® voltage tester.

voltage. If there is no voltage, there is no current flow, so the indicator on the scale does not move, and no voltage reading is displayed.

The range of voltage and the type of current (AC and/or DC) that a voltage tester is capable of measuring are usually indicated on the scales that display the reading. The scales for each type of current are marked accordingly.

The methods used to show voltage readings vary from one type of voltage tester to another. For example, there are voltage testers that have lights to indicate the approximate amount of voltage registered. Both lights and scales have relatively broad readout ranges because these voltage testers can only indicate approximate values, not precise voltage measurements.

A voltage tester should always be checked before each use to make sure that it is in good condition and is operating correctly. The external check of the tester should include a careful inspection of the insulation on the leads for cracks or frayed areas. Faulty leads constitute a safety hazard, so they must be replaced. As a check to make sure that the voltage tester is operating correctly, the probes of the tester are connected to a power source that is known to be energized. The voltage indicated on the tester should match the voltage of the power source. If there is no indication, the voltage tester is not operating correctly, and it must be repaired or replaced. It must also be repaired or replaced if it indicates a voltage different from the known voltage of the source.

It is essential to check a voltage tester before use. A faulty voltage tester can be dangerous to the electrician using it and to other personnel. For example, damaged insulation or a cracked casing could expose the electrician to electrical shock. Also, a faulty voltage tester might indicate that power is off when it is really on. This would create a serious safety hazard for personnel involved in equipment repair. The face plate or the scale on the front of the tester should be checked before a voltage tester is used to be sure that the tester can handle the amount of voltage that the power source may contain. Care should be taken when placing the probes of the tester across the power source. A voltage tester is designed to take a quick reading. If the probes are left in the circuit too long, the tester will burn out. A voltage tester should never be connected for more than a few seconds at a time.

Safety

Emphasize the importance of carefully checking a voltage tester prior to each use.

Instructor's Notes:

CAUTION

Do not use a voltage tester in hazardous locations.

Voltage testers are used to make sure that power is available when it is needed and to make sure that power has been cut off when it should have been. In a troubleshooting situation, it might be necessary to verify that power is available in order to be sure that lack of power is not the problem. For example, if there were a problem with a power tool, such as a drill, a voltage tester might be used to make sure that power is available to run the drill. A voltage tester might also be used to verify that there is power available to a three-phase motor that will not start.

For safety reasons, it is always necessary to make sure that the power is turned off before working on any electrical equipment. A voltage tester can be used for such a test.

Keep the following in mind when using a voltage tester:

- Check the tester before each use.
- Handle the tester as if it were a calibrated instrument.
- Use good safety practices when operating the tester.
- Do not use circuits expected to be above the scale on the tester.
- Do not use if damage is indicated to the tester.
- Do not use in classified, hazardous areas or in high-frequency circuits.

14.0.0 ◈ RECORDING INSTRUMENTS

The term *recording instrument* describes many instruments that make a permanent record of measured quantities over a period of time. Recording instruments can be divided into three general groups:

- Instruments that record electrical quantities, including potential difference, current, power, resistance, and frequency.
- Instruments that record nonelectrical quantities by electrical means (such as a temperature recorder that uses a potentiometer system to record thermocouple output).
- Instruments that record nonelectrical quantities by mechanical means (such as a temperature recorder that uses a bimetallic element to move a pen across an advancing strip of paper).

It is often necessary to know the conditions that exist in an electrical circuit over a period of time to determine such things as peak loads, voltage fluctuations, etc. It may be neither practical nor economical to assign a worker to watch an indicating instrument and record its readings. An automatic recording instrument can be connected to take continuous readings, and the record can be collected for review and analysis.

Recording instruments are basically the same as the indicating meters already covered, but they have recording mechanisms attached to them. They are generally made of the same parts, use the same electrical mechanisms, and are connected in the same way. The only basic difference is the permanent record.

14.1.0 Strip-Chart Recorders

Strip-chart recorders are the most widely used recording instruments for electrical measurement. Their name comes from the fact that the record is made on a strip of paper, usually four to six inches wide and perhaps up to 60 feet long. These can be used to record either voltage or current. A recording ammeter is shown in *Figure 30*.

Strip-chart recorders offer several advantages in electrical measurement. The long charts allow the recording to cover a considerable length of time with little attention, and strip-chart recorders can be operated at a relatively high speed to provide very detailed records.

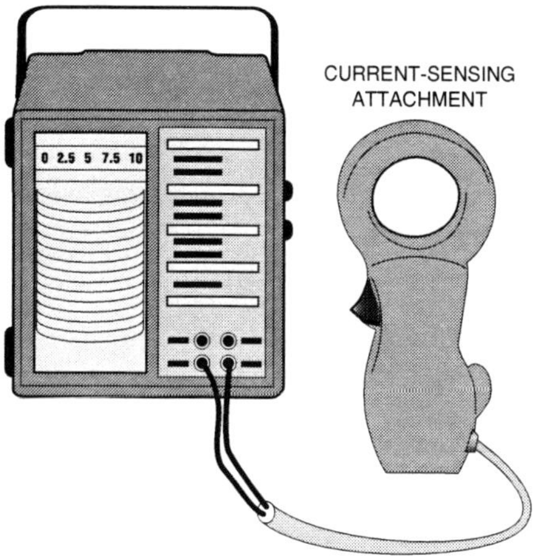

Figure 30 ◈ Recording ammeter. 106F30.EPS

Classroom

Discuss the operation and applications of various recording instruments.

Audiovisual

Show Transparency 31 (Figure 30).

ELECTRICAL TEST EQUIPMENT 209

Safety

Review the basic safety precautions that apply to the use of test equipment.

14.1.1 Typical Strip-Chart Recorder

A typical strip-chart recorder includes the following parts:

- The frame, which supports the other parts
- The moving system, which consists of the parts that move as a direct result of variations in the quantity being measured
- The graphic record, which is the line traced on the chart
- The chart carriage, which consists of the clock mechanism, timing gears, timing drum, chart spool, rewinding mechanism, and writing table
- A scale on the paper strip, which makes it possible to read the values of the quantity being measured
- The inking system, which consists of a special pen with an ink reservoir
- A case, which provides convenience in installation and removal, as well as in replacement of the chart paper

The moving parts used in recording instruments are basically the same as those used in the indicating instrument used to measure the same electrical quantity. However, the parts of a recording instrument are larger and require more power because of the added friction of the pen on the chart paper.

The paper chart is graduated in two directions. In one direction, the graduations correspond to the indicating scale of the instrument. In the other, they indicate time (seconds, minutes, hours, or days depending on the clock mechanism). The time graduations are uniformly spaced because the chart moves at a constant speed.

The chart carriage includes the entire mechanism for handling the chart paper. At the top, it holds the roll of paper. Just above the roll, and directly beneath the pen, is the writing table. The table is a metal plate that provides a flat surface under the chart paper at all points on the arc of the pen's motion. This prevents the pen from puncturing or tearing the paper.

The heart of the chart carriage is the clock mechanism, which drives the timing drum. The mechanism may be driven by an electric clock. Some of these mechanisms also contain a motor-wound spring clock or a battery-powered clock for a backup in the event of a loss of power.

The timing drum is connected to the clock through a gear unit, which determines the speed of the drum. Drive pins around the timing drum fit into holes at the edges of the chart paper.

The inking system usually includes an inkwell containing enough ink for a considerable time.

15.0.0 ◆ SAFETY

In the interest of safety, all test equipment must be inspected and tested before use. A thorough visual inspection, checking for broken meters or knobs, damaged plugs, or frayed cords is important.

Perform an operational check. For example:

- On an ohmmeter, short the probes and ensure that you can zero the meter.
- With an oscilloscope, make sure that you can obtain a trace, and if a test signal is available, connect a test probe and check it.
- A voltmeter can be checked against an AC wall receptacle or a battery.

If a meter has a calibration sticker, check to see if it has been calibrated recently. For precise measurements, a recently calibrated meter is a more reliable instrument.

Every person who works with electronic equipment should be constantly alert to the hazards to which personnel may be exposed, and should also be capable of rendering first aid. The hazards considered in this section are: electric shock, burns, and related hazards.

Safety must be the primary responsibility of all personnel. The installation, maintenance, and operation of electrical equipment enforces a stern safety code. Carelessness on the part of the technician or operator can result in serious injury or death due to electrical shock, falls, burns, flying objects, etc. After an accident has occurred, investigation almost invariably shows that it could have been prevented by the exercise of simple safety precautions and procedures. Each person concerned with electrical equipment is responsible for reading and becoming thoroughly familiar with the safety practices and procedures contained in all safety codes and equipment technical manuals *before* performing work on electrical equipment. It is your personal responsibility to identify and eliminate unsafe conditions and unsafe acts that cause accidents.

You must bear in mind that deenergizing main supply circuits by opening supply switches will not necessarily deenergize all circuits in a given piece of equipment. A source of danger that has often been neglected or ignored, sometimes with tragic results, is the input to electrical equipment

Instructor's Notes:

210 CHAPTER 5

from other sources, such as backfeeds. Moreover, the rescue of a victim shocked by the power input from a backfeed is often hampered because of the time required to determine the source of power and isolate it. Therefore, turn off *all* power inputs before working on equipment and tag and lock out, then check with an operating tester (e.g., wattmeter) to be sure that the equipment is safe to work on. Take the time to be safe when working on electrical circuits and equipment. Carefully study the schematics and wiring diagrams of the entire system, noting what circuits must be deenergized in addition to the main power supply. Remember, electrical equipment commonly has more than one source of power. Be certain that all power sources are deenergized before servicing the equipment. Do not service any equipment with the power on unless absolutely necessary. Remember that the 115V power supply voltage is not a low, relatively harmless voltage but is the voltage that has caused more deaths than any other medium.

Safety can never be stressed enough. There are times when your life literally depends on it. The following is a listing of common-sense safety precautions that must be observed at all times:

- Use only one hand when turning power switches on or off. Keep the doors to switch and fuse boxes closed except when working inside or replacing fuses. Use a fuse puller to remove cartridge fuses, after first making certain that the circuit is dead.
- Your company will make the determination as to whether or not you are qualified to work on an electrical circuit.
- Do not work with energized equipment by yourself; have another person (safety observer), qualified in first aid for electrical shock, present at all times. The person stationed nearby should also know which circuits and switches control the equipment, and that person should be given instructions to pull the switch immediately if anything unforeseen happens.
- Always be aware of the nearness of high-voltage lines or circuits. Use rubber gloves where applicable and stand on *approved* rubber matting. Not all rubber mats are good insulators.
- Inform those in charge of operations as to the circuit on which work is being performed.
- Keep clothing, hands, and feet dry. When it is necessary to work in wet or damp locations, use a dry platform and place a rubber mat or other nonconductive material on top of the wood. Use insulated tools and insulated flashlights of the molded type when required to work on exposed parts.
- Do *not* work on energized circuits unless absolutely necessary.
- All power supply switches or cutout switches from which power could possibly be fed must be secured in the OPEN (safety) position and locked and tagged.
- Never short out, tamper with, or block open an interlock switch.
- Keep clear of exposed equipment; when it is absolutely necessary to work on it, use only one hand as much as possible.
- Avoid reaching into enclosures except when absolutely necessary. When reaching into an enclosure, use rubber blankets to prevent accidental contact with the enclosure.
- Do not use bare hands to remove hot vacuum tubes from their sockets. Wear protective gloves or use a tube puller.
- Use a shorting stick to discharge all high-voltage capacitors.
- Make certain that the equipment is properly grounded. Ground all test equipment to the equipment under test.
- Turn off the power before connecting alligator clips to any circuit.
- When measuring circuits over 300V, do not hold the insulated test prods with bare hands.

15.1.0 Use of High-Voltage Protection Equipment

Anyone working on or near energized circuitry must use special equipment to provide protection from electrical shock. Protective equipment includes gloves, leather sleeves, rubber blankets, and rubber mats. It should be noted that this electrical protective equipment is in addition to the regular protective equipment normally required for maintenance work. Regular protective equipment typically includes hard hats that are rated for electrical resistance, eye protection, safety shoes, and long sleeves.

Gloves that are approved for protection from electrical shock are made of rubber. A separate leather cover protects the rubber from punctures or other damage. They protect the worker by insulating the hands from electrical shock. Gloves are rated as providing protection from certain amounts of voltage. Whenever an individual is going to be working around exposed conductors, the gloves chosen should be rated for at least as much voltage as the conductors are carrying.

Safety

Emphasize the importance of appropriate protective equipment when conducting tests on energized circuits. Stress that this type of work is to be done only by qualified personnel.

Rubber sleeves are used along with gloves to provide additional protection. The combination of sleeves and gloves protects the hands and arms from electrical shock.

Rubber blankets and floor mats have many uses. Blankets are used to cover energized conductors while work is going on around them. They might be used to cover the energized main buses in a breaker panel before working on a deenergized breaker. Rubber floor mats are used to insulate workers from the ground. If a worker is standing on a rubber mat and then contacts an energized conductor, the current cannot flow through the body to the ground, so the worker will not get shocked.

Classroom

Review example testing policies.

15.2.0 General Testing of Electrical Equipment

As stated earlier, personnel must be familiar with the proper use of available test equipment. Also, personnel must be familiar with and use the local instructions governing the testing of electrical circuits to protect both the person and the equipment.

The next section includes an example of instructions from a local utility and covers work permits, test permits, authorized employees for testing, circuit isolation, and relay, instrument, and meter testing. Each topic is discussed individually.

15.2.1 Example Policy—Work Permits or Test Permits

A written work permit must be obtained by an authorized technician or supervisor from the operator in charge before testing equipment under the operator's control. A test permit is required to make a dielectric proof test.

The following must be on the work or test permit:

- Designation of the equipment to be tested
- Scope and limits of tests to be made
- Method of isolation and protection provided

If it is necessary to alter the nature or to extend the scope of tests for which a work permit or test permit has been granted, return the permit to the operator and request a new permit before proceeding with the revised tests. When the test is completed, return the work permit or test permit to the operator, and report whether the results of the test were satisfactory or unsatisfactory.

Whenever a work permit or test permit remains in effect beyond the working hours of the person to whom it was issued and a different person will be in charge of the tests, consult the operator for the proper turnover procedure. The person assuming the responsibility for continuance of the work must verify that the items on the work or test permit are correct.

If a work permit or test permit is issued for testing that is to continue over a period of several days, the permit must be returned to the operator at the end of each working day and picked up at the beginning of the next working day.

15.2.2 Example Policy—Authorized Employees for Testing

Only employees authorized to perform operations at a given station shall:

- Open, remove, close, or replace doors or covers on electrical compartments.
- Remove or replace potential transformer fuses.
- Open or close switches.
- Open or close fuse cutouts or disconnecting potheads.
- Make tests for the presence of potential.
- Place or remove blocks and protective tags.
- Apply or remove grounds or short circuits.
- Attach or remove leads to high- or intermediate-voltage conductors.
- Make operating tests.
- Test links, fuses, switches associated with generators, buses, feeders, transformers, etc. normally shall be removed, replaced, or operated only under the direct supervision of an authorized employee when the associated equipment is live or is available for normal operation.

In some cases, when it is necessary for the equipment to remain in service, or in any case when the associated equipment has been removed from service, the operator-in-charge may authorize the person to whom the work permit was issued to remove and replace test links and fuses and to operate the heel and toe switches and potential switches in low-voltage circuits.

15.2.3 Example Policy—Circuit Isolation

All equipment to be tested that is rated at 350V phase-to-ground or higher must be isolated from all sources of supply by either two breaks, or a single break and a ground, unless test equipment is used that can be operated safely with one break. Where possible, two breaks in series are preferred over one break.

If no potential is to be applied to the equipment, the ground may be applied on the equipment side of the single break.

Instructor's Notes:

212 CHAPTER 5

If a test potential is to be applied to the equipment, the ground has to be on the far side of the single break. If a test potential of less than 400V is to be applied to the equipment, the ground may be replaced with an approved protective discharge gap on the equipment to be tested. The protective gaps and test leads are to be connected or disconnected while the equipment is grounded. If disconnect switches are used to provide the breaks, the switches shall be locked open, and the operating motor fuses removed.

On equipment operating at 350V or less, a test potential may be applied with only a single break. An open truck-type or elevating-type circuit breaker withdrawn or lowered to the disconnect position or a swinging-type disconnect switch shall be considered as a single break.

If any of the following are the only type of disconnecting device available, short circuits and grounds must be applied at the nearest available point to the device.

- A disconnect switch with the blade prevented from falling closed by an insulating block placed over the jaw
- A disconnect switch in which the separation of the blade and jaw cannot be verified visually
- An oil circuit breaker or oil switch having no associated disconnecting device
- An oil-filled fusible cutout having no provision for disconnecting the leads

Potential transformers that may become energized from the low-voltage side shall be isolated on that side by opening switches, removing links, removing fuses, or by removing connections temporarily. The low-voltage side of the potential transformer shall also be short circuited and grounded.

Equipment that has been subjected to DC high-voltage tests shall be grounded to discharge the residual voltage before any person is permitted to touch the test cables or any current-carrying part of the equipment tested. If the equipment tested has considerable capacitance, then it should be grounded for at least 30 minutes after the test potential is removed. Equipment of low capacitance should be grounded for a length of time equivalent to the time that the potential was applied.

15.2.4 Example Policy—Clearances

Adequate clearances are to be maintained between energized and exposed conductors and personnel (refer to ***NEC Section 110.26***). Where DC voltages are involved, clearances specified shall be used with specified voltages considered as DC line-to-ground values.

If adequate clearances cannot be maintained from exposed live parts of apparatus in the normal course of free movement within the area during test, then access to that area shall be restricted by fences and barricades. Signs clearly indicating the hazard shall be posted in conspicuous locations. This requirement applies to equipment in service as well as to equipment to which test voltages are applied.

Whenever there is any question of the adequacy of clearance between the specific area in which work is to be done and exposed live parts of adjacent equipment, a field inspection shall be made by management representatives of the group involved before starting the job. The result of this inspection should be to outline the protection necessary to complete the work safely, including watchers where needed.

15.2.5 Electrical Circuit Tests

Testing of electrical circuits can be required at new installations or existing installations where wiring has been modified or new wiring has been interfaced with existing circuits. It is broken down into two sections:

- Electrical circuit tests performed on AC relay protection and metering circuits include:
  - Point-to-point wire checks to verify that the wiring has been installed according to the diagram of connections
  - Current transformer polarity, impedance, continuity, and where applicable, tap progression and intercore coupling tests
  - Voltage transformer polarity, ratio, and self-induced high-potential (hi-pot) tests
  - Insulation resistance tests
- Electrical circuit tests performed on control, power, alarm, and annunciator circuits include:
  - Insulation resistance tests
  - Operation tests

An important piece of information is the minimum distance allowed when working near energized electrical circuits because large voltages can arc across an air gap. Personnel must maintain a distance that is greater than that arc distance. This is especially true when using a hot stick to open a disconnect. These distances are listed in *Table 1.*

15.3.0 Meter Loading Effects

It is important to understand how a piece of test equipment will affect the circuit to which it is

ELECTRICAL TEST EQUIPMENT 213

Classroom

Discuss the effects of meter loading.

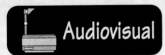

Audiovisual

Show Transparencies 32 and 33 (Figures 31 and 32).

Table 1 OSHA Table

Voltage Range (Phase-to-Phase) Kilovolts	Minimum Working and Clear Hot Stick Distance
2.1 to 15	2 ft. 0 in.
15.2 to 35	2 ft. 4 in.
35.1 to 46	2 ft. 6 in.
46.1 to 72.5	3 ft. 0 in.
72.6 to 121	3 ft. 4 in.
138 to 145	3 ft. 6 in.
161 to 169	3 ft. 8 in.
230 to 242	5 ft. 0 in.
345 to 362	*7 ft. 0 in.
500 to 552	*11 ft. 0 in.
700 to 765	*15 ft. 0 in.

*For voltages above 345kV, the minimum working and clear hot stick distances may be reduced provided that such distances are not less than the shortest distance between the energized part and a grounded surface.

> **Note**
> Clearances listed in *Table 1* apply to qualified personnel.

attached. Even when properly used, a piece of test equipment can put a load on the circuit under testing.

An ammeter becomes a part of the circuit when it is used. Because of this, the meter can sometimes significantly alter the voltages and currents in the circuit. The following example illustrates how this can happen.

In Circuit A on the left in *Figure 31,* the current is calculated to be 10µA. To measure this current using a meter that has an internal resistance of 2,000Ω, the meter would be placed as shown in Circuit B. The 2,000Ω resistance of the meter adds to the 10,000Ω resistance, reducing the current in the circuit to 8.33µA. The presence of the meter, therefore, alters the current flowing in the circuit, resulting in an incorrect reading.

A voltmeter can also alter circuit parameters. The following example illustrates how this can happen.

In Circuit A in *Figure 32,* the voltage across the 40,000Ω resistor is to be measured using a voltmeter with a 1,000Ω internal resistance. We can determine the effect that the meter will have on the circuit when it is connected as shown in Circuit B.

Connecting the voltmeter in parallel with the 40,000Ω resistor, as shown in Circuit B, significantly changes the total resistance in the circuit. The parallel combination of 40,000Ω and 200,000Ω is:

$$R = \frac{40{,}000 \times 200{,}000}{40{,}000 + 200{,}000} = 33{,}333\Omega$$

Add the 10KΩ resistance in series for a total circuit resistance of 43,333Ω.

Therefore, with the voltmeter in the circuit, the current is:

$$E \div R = I \text{ or } 10 \div 43{,}333 = 2.31 \times 10^{-4} \text{ A}$$

This current causes a voltage drop across the combination of the meter and the 40,000Ω resistor of:

$$E = IR \text{ or } (2.31 \times 10^{-4}) \times 33{,}333 = 7.7\text{V}$$

The voltmeter will read 7.7V, whereas the voltage drop across the 40,000Ω resistor without the voltmeter connected is 8V.

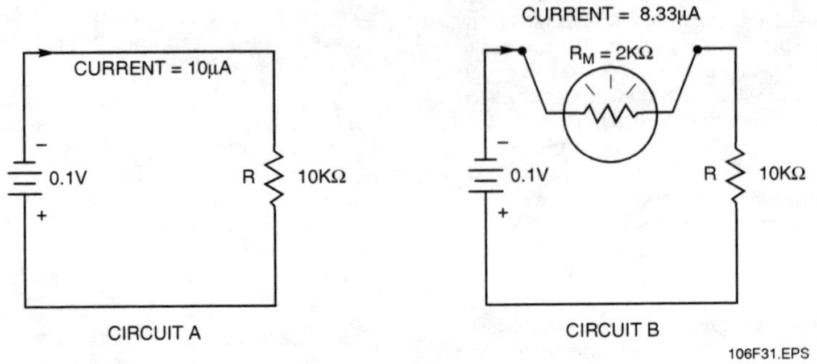

Figure 31 ◆ Meter loading—Example 1.

Instructor's Notes:

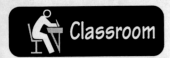

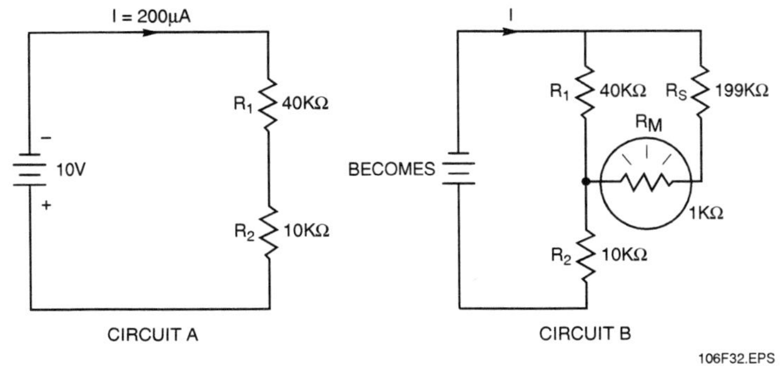

Figure 32 ◈ Meter loading—Example 2.

THINK ABOUT IT

Putting It All Together

What kind of measuring device would you select or need for the following tasks, and how would you apply the device?

- A short circuit in house wiring
- The output from an AC transformer
- A fuse in house wiring
- A relay

Teaching Tip

Discuss the "Think About It."

In most cases, the current flowing through the voltmeter movement is negligible compared to the current flowing through the element whose voltage is being measured. When this is the case, the voltmeter has a negligible effect on the circuit.

In most commercial voltmeters, the internal resistance is expressed as the ohms-per-volt rating of the meter. This quantity tells what the internal resistance of the meter is on any particular full-scale setting. In general, the meter's internal resistance is the ohms-per-volt rating times the full-scale voltage. The higher the ohms-per-volt rating, the higher the internal resistance of the meter, and the smaller the effect of the meter on the circuit.

Care should be taken when selecting either an ammeter or a voltmeter in which the meter does not load down the circuit under test. Also, realize that the same circuit under test with a different meter will indicate differently due to the separate loading effects of each meter.

Summary

The use of test equipment is an important part of your job as an electrician. Selecting the proper instrument to be used in a specific application will help you to fully perform your task. You will use the information provided by these instruments to help evaluate the work being accomplished.

Classroom

Have the trainees complete the Review Questions and go over the answers prior to administering the Module Examination.

Examination

Administer the Module Examination. Record the results on Craft Training Report Form 200 and submit the results to the Training Program Sponsor.

Performance Testing

Administer the Performance Test and fill out Performance Profile Sheets for each trainee. If desired, trainee proficiency noted during laboratory sessions may be used to complete the Performance Test. Be sure to record the results on Craft Training Report Form 200 and submit the results to the Training Program Sponsor.

Review Questions

1. An ammeter is used to measure _____.
   a. current
   b. voltage
   c. resistance
   d. insulation value
2. Ammeters use a ________ to measure values higher than 10mA.
   a. multiplier resistor
   b. rectifier bridge
   c. transformer coil
   d. shunt
3. Voltmeters use a _____ to measure values higher than one volt.
   a. multiplier resistor
   b. rectifier bridge
   c. transformer coil
   d. shunt
4. Electromotive force (emf) is measured using a(n) _____.
   a. ammeter
   b. wattmeter
   c. voltmeter
   d. ohmmeter
5. A voltmeter is connected _____ with the circuit being tested.
   a. in parallel
   b. in series
   c. in a sequential configuration
   d. in a looped configuration
6. Which of the following is *not* true regarding an ohmmeter?
   a. It is powered by a DC source, usually a battery.
   b. It contains current-limiting resistors.
   c. It must be used in an energized circuit.
   d. It should be adjusted to zero before each use.
7. All of the following values can be directly measured using a VOM, *except* _____.
   a. decibels
   b. voltage
   c. resistance
   d. wattage
8. When in doubt as to the voltage present, always use the _____ range to protect the VOM.
   a. lowest
   b. highest
   c. middle
   d. It does not matter with today's precision instruments.
9. Wattmeters are used to measure _____.
   a. power
   b. voltage
   c. resistance
   d. impedance
10. To check the resistance between motor or transformer windings, use a _____.
    a. standard ohmmeter
    b. wattmeter
    c. megger
    d. power factor meter
11. A Wiggy® is used to _____.
    a. measure wattage
    b. measure impedance
    c. test for potential difference
    d. provide an exact voltage reading
12. To provide long-term voltage monitoring, it would be best to use _____.
    a. a VOM
    b. constant supervision by experienced maintenance personnel
    c. a power factor meter
    d. a strip chart recorder
13. The minimum working distance for a 15kV circuit is _____.
    a. 2 ft. 0 in.
    b. 5 ft. 0 in.
    c. 7 ft. 0 in.
    d. 15 ft. 0 in.
14. Connecting a voltmeter may affect the total _____ in a circuit.
    a. voltage
    b. current
    c. resistance
    d. electromotive force
15. Which of the following is *not* true regarding meter loading?
    a. It may cause different meters to produce different readings.
    b. The higher the ohms-per-volt rating, the smaller the effect of the meter on the circuit.
    c. It may impact both voltage and resistance.
    d. It can be prevented by removing power to the circuit being tested.

Instructor's Notes:

216 CHAPTER 5

PROFILE IN SUCCESS

Joe Sullivan, Bay Harbour Electric

Nothing is more important to the electrician than safety. Joe Sullivan recognized this fact and chose to point his career in that direction. Today, he is the safety compliance officer for his company and an OSHA Outreach Program trainer.

What do you think it takes to be a success in the electrical trade?
Training and education. In the electrical trade, there are many industry standards and codes that must be adhered to. Knowledge of these codes and standards is essential. Math skills, organizational skills, and the ability to plan ahead are all required for the electrician. The way you develop these skills is to make a commitment to learning.

Why is good equipment care important in electrical work?
Accidents can be prevented. Keeping your tools and equipment in good condition and following safe work practices in the use of tools and equipment will dramatically reduce the chances of work-related injuries. A person who takes pride in his tools will also take pride in his work.

What do you like most about your job?
It is gratifying to work for a company like Bay Harbour Electric. The company is committed to being a leader of distinction in our industry and our community, and they care about the success of their employees. Knowing that I have the full support of the chairman and president in developing and implementing safety policies and procedures assures me that my company is concerned about the safety, health, and welfare of its employees. When I travel from job site to job site to conduct safety audits and safety training seminars, it is very satisfying to see the safety policies and procedures I developed being put into practice.

Why did you make a career change from field electrician to safety officer?
When I started with Bay Harbour Electric, I had to take the 10-hour safety course, learn CPR and first aid, and attend frequent safety courses and workshops. I also worked on a project that involved international travel and required me to become familiar with the safety codes and practices of other countries. All of this sparked my interest in safety. When the safety compliance officer job opened at Bay Harbour, I applied for it. My employer thought my interest in safety, combined with my electrical experience, was the right mix for the job.

What would you say to someone entering the trade today?
If you are still in high school, and your school has a school-to-work program, take full advantage of it. Doing hands-on work in the trade will help you learn what the trade is really about and what will be expected of you. Also, choose a company that has an accredited apprenticeship program. This will ensure that you have an opportunity to work in the field while you are still in school.

ELECTRICAL TEST EQUIPMENT 217

GLOSSARY

Trade Terms Introduced in This Module

Coil: A number of turns of wire, especially in spiral form, used for electromagnetic effects or for providing electrical resistance.

Continuity: An uninterrupted electrical path for current flow.

d'Arsonval meter movement: A meter movement that uses a permanent magnet and moving coil arrangement to move a pointer across a scale.

Decibel: A unit used to express a relative difference in power between electric signals.

Frequency: The number of cycles completed each second by a given AC voltage; usually expressed in hertz; one hertz = one cycle per second.

Instructor's Notes:

Additional Resources

This module is intended to present thorough resources for task training. The following reference works are suggested for further study. These are optional materials for continued education rather than for task training.

Electronics Fundamentals, Latest Edition. New York: Prentice Hall.
Principles of Electric Circuits, Latest Edition. New York: Prentice Hall.

ELECTRICAL TEST EQUIPMENT 219

ANSWER KEY

Answers to Review Questions

Answer	Section
1. a	3.0.0
2. d	3.0.0
3. a	4.0.0
4. c	4.0.0
5. a	4.0.0
6. c	5.0.0
7. d	6.0.0
8. b	6.3.3
9. a	8.0.0
10. c	9.0.0
11. c	13.1.1
12. d	14.1.0
13. a	15.2.5/Table 1
14. c	15.3.0
15. d	15.3.0

Section 3.0.0 *Think About It—Ammeter Connections*

Connecting an ammeter in parallel with a load (resistance) in an energized circuit applies the voltage developed across that load directly across the parallel-connected meter movement and shunt resistor internal to the ammeter. This voltage, which is higher than that for which the ammeter movement is designed, can cause excessive current to flow through the meter movement, damaging the movement.

One example of an ammeter connected in series with a load is the *AMPS* meter installed in the instrument panel of some automobiles. Another example is the ammeter installed in a battery charger used to monitor the rate of current flow as a battery is being charged.

Section 4.0.0 *Think About It—Voltmeters*

A voltmeter must be connected in parallel with the circuit under test so that the series-connected multiplier resistor and meter coil resistance internal to the voltmeter are connected across the voltage being measured. In this manner, the current flow through the multiplier and meter coil resistance produces a meter indication that is directly proportional to the amplitude of the measured voltage. If the voltmeter were connected in series in the circuit like an ammeter, the meter would indicate a voltage equal to the voltage drop developed across the meter resistance by the circuit current flowing through it. Also, when connected in series, it would limit the amount of current flow through the circuit as a result of the large resistance of the multiplier resistor built into the voltmeter.

Section 5.0.0 *Think About It—Using an Ohmmeter*

The instructor should demonstrate this circuit measurement. Connect one end of the de-energized wire to ground using a jumper wire. Measure between the other end of the wire and ground using an ohmmeter. If the wire is broken, the ohmmeter will indicate infinite resistance. If the wire is good, the ohmmeter will indicate a short (0Ω).

Section 7.0.0 *Think About It—Digital Meters*

Precision measurements are required when adjusting and calibrating equipment such as power supplies, or when determining the percentage of voltage or current imbalance that exists in a three-phase system. Precision is not important if making measurements just to determine the absence or presence of voltage or current, as is commonly done when troubleshooting.

Section 15.3.0 *Think About It—Putting It All Together*

- Ohmmeter
- Voltmeter
- Ohmmeter or voltmeter
- Ohmmeter, voltmeter, or ammeter

The instructor should discuss and/or demonstrate ways to use different meters to test the same device under different circumstances.

CONTREN™ LEARNING SERIES — USER UPDATES

The NCCER makes every effort to keep these textbooks up-to-date and free of technical errors. We appreciate your help in this process. If you have an idea for improving this textbook, or if you find an error, a typographical mistake, or an inaccuracy in NCCER's Contren™ textbooks, please write us, using this form or a photocopy. Be sure to include the exact module number, page number, a detailed description, and the correction, if applicable. Your input will be brought to the attention of the Technical Review Committee. Thank you for your assistance.

Instructors – If you found that additional materials were necessary in order to teach this module effectively, please let us know so that we may include them in the Equipment/Materials list in the Instructor's Guide.

Write: Curriculum Revision and Development Department
National Center for Construction Education and Research
P.O. Box 141104, Gainesville, FL 32614-1104

Fax: 352-334-0932

E-mail: curriculum@nccer.org

Craft ______________________ Module Name ______________________

Copyright Date ______________ Module Number ______________ Page Number(s) ______________

Description

____

____

____

____

(Optional) Correction

____

____

____

(Optional) Your Name and Address

____

____

____

Chapter 6

Introduction to the *National Electrical Code®*

26107-02

MODULE OVERVIEW

This course introduces the electrical trainee to the requirements of the *National Electrical Code*®.

PREREQUISITES

Please refer to the Course Map in the Trainee Module. Prior to training with this module, it is recommended that the trainee shall have successfully completed the following:

Core Curriculum; Residential Electrical I, Chapters 1 through 5

LEARNING OBJECTIVES

Upon completion of this module, the trainee will be able to:

1. Explain the purpose and history of the *National Electrical Code* (NEC).
2. Describe the layout of the NEC.
3. Explain how to navigate the NEC.
4. Describe the purpose of the National Electrical Manufacturers' Association (NEMA) and the National Fire Protection Association (NFPA).
5. Explain the role of testing laboratories.

PERFORMANCE OBJECTIVES

Under supervision of the instructor, the trainee should be able to:

1. Use ***NEC Article 90*** to determine the scope of the NEC. State what is covered by the NEC and what is not.
2. Find the definition of the term *feeder* in the NEC.
3. Look up the NEC specifications that you would need to follow if you were installing an outlet near a swimming pool.
4. Find the minimum wire bending space required if two No. 1/0 AWG conductors were to be installed in a junction box or cabinet.

NCCER STANDARDIZED CRAFT TRAINING PROGRAM

The National Center for Construction Education and Research (NCCER) provides a standardized national program of accredited craft training. Key features of the program include instructor certification, competency-based training, and performance testing. The program provides trainees, instructors, and companies with a standard form of recognition through a National Craft Training Registry. The program is described in full in the *Guidelines for Accreditation,* published by the NCCER. For more information on standardized craft training, contact the NCCER at P.O. Box 141104, Gainesville, FL 32614-1104, 352-334-0911, visit our Web site at www.nccer.org, or e-mail info@nccer.org.

HOW TO USE THIS ANNOTATED INSTRUCTOR'S GUIDE

Each page presents two sections of information. The larger section displays each page exactly as it appears in the Trainee Module. The narrow column ties suggested trainee and instructor actions to each page and provides icons to call your attention to material, safety, audiovisual, or testing requirements. The bottom of each page includes space for your notes.

Teaching Tip

If you see the Teaching Tip icon, that means there is a teaching tip associated with this section. Also refer to the suggested teaching tips at the end of the module.

SAFETY CONSIDERATIONS

Ensure that the trainees are equipped with appropriate personal protective equipment. Emphasize that the primary purpose of the *National Electrical Code®* is to safeguard persons and property from the hazards arising from the use of electricity.

PREPARATION

Before teaching this module, you should review the Module Outline, Learning and Performance Objectives, and the Materials and Equipment List. Be sure to allow ample time to prepare your own training or lesson plan and gather all required equipment and materials.

MATERIALS AND EQUIPMENT LIST

Materials:

Transparencies

Markers/chalk

Copy of the latest edition of the *National Electrical Code®*

Module Examinations*

Performance Profile Sheets*

Equipment:

Overhead projector and screen

Whiteboard/chalkboard

Appropriate personal protective equipment

*Located in the Test Booklet packaged with this Annotated Instructor's Guide.

ADDITIONAL RESOURCES

This module is intended to present thorough resources for task training. The following reference is suggested for both instructors and motivated trainees interested in further study. This is optional material for continued education rather than for task training.

National Electrical Code Handbook, Latest Edition. Quincy, MA: National Fire Protection Association.

NOTES

The designations "National Electrical Code," "NE Code," and "NEC," where used in this document, refer to the *National Electrical Code®*, which is a registered trademark of the National Fire Protection Association, Quincy, MA. All National Electrical Code (NEC) references in this module refer to the 2002 edition of the NEC.

If you feel that additional math instruction would be helpful, Prentice Hall offers a basic math textbook entitled *Fundamentals of Electrical and Mechanical Mathematics.* It covers the basic math requirements for electrical trainees and may be ordered by contacting Prentice Hall Customer Service at 1-800-922-0579.

TEACHING TIME FOR THIS MODULE

An outline for use in developing your lesson plan is presented below. Note that each Roman numeral in the outline equates to one session of instruction. Each session has a suggested time period of 2½ hours. This includes 10 minutes at the beginning of each session for administrative tasks and one 10-minute break during the session. Approximately 2½ hours are suggested to cover *Introduction to the National Electrical Code.* You will need to adjust the time required for hands-on activity and testing based on your class size and resources.

Topic	Planned Time
Session I. Purpose and History of the NEC; Layout of the NEC; Navigating the NEC; Other Organizations; Module Examination and Performance Testing	
A. Purpose and History of the NEC	______
1. History	______
a. Who is Involved?	______
B. Layout of the NEC	______
1. Types of Rules	______
2. NEC Introduction	______
3. Body of the NEC	______
a. NEC Definitions	______
4. Reference Portion of the NEC	______
a. Organization of the Chapters	______
5. Text in the NEC	______
C. Navigating the NEC	______
1. Equipment for General Use	______
2. Special Occupancies	______
3. Special Equipment	______
4. Special Conditions	______
5. Examples of Navigating the NEC	______
a. Installing Type SE Cable	______
b. Installing Track Lighting	______
D. Other Organizations	______
1. Role of Testing Laboratories	______
2. Nationally Recognized Testing Laboratories	______
3. National Electrical Manufacturers' Association	______
E. Module Examination	______
1. Trainees must score 70% or higher to receive recognition from the NCCER.	
2. Record the testing results on Craft Training Report Form 200 and submit the results to the Training Program Sponsor.	
F. Performance Testing	______
1. Trainees must perform each task to the satisfaction of the instructor to receive recognition from the NCCER.	
2. Record the testing results on Craft Training Report Form 200 and submit the results to the Training Program Sponsor.	

Chapter 6

Module 26107-02

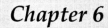

Introduction to the National Electrical Code®

Course Map

This course map shows all of the modules in the first level of the Residential Electrical curriculum. The suggested training order begins at the bottom and proceeds up. Skill levels increase as you advance on the course map. The local Training Program Sponsor may adjust the training order.

RESIDENTIAL ELECTRICAL I

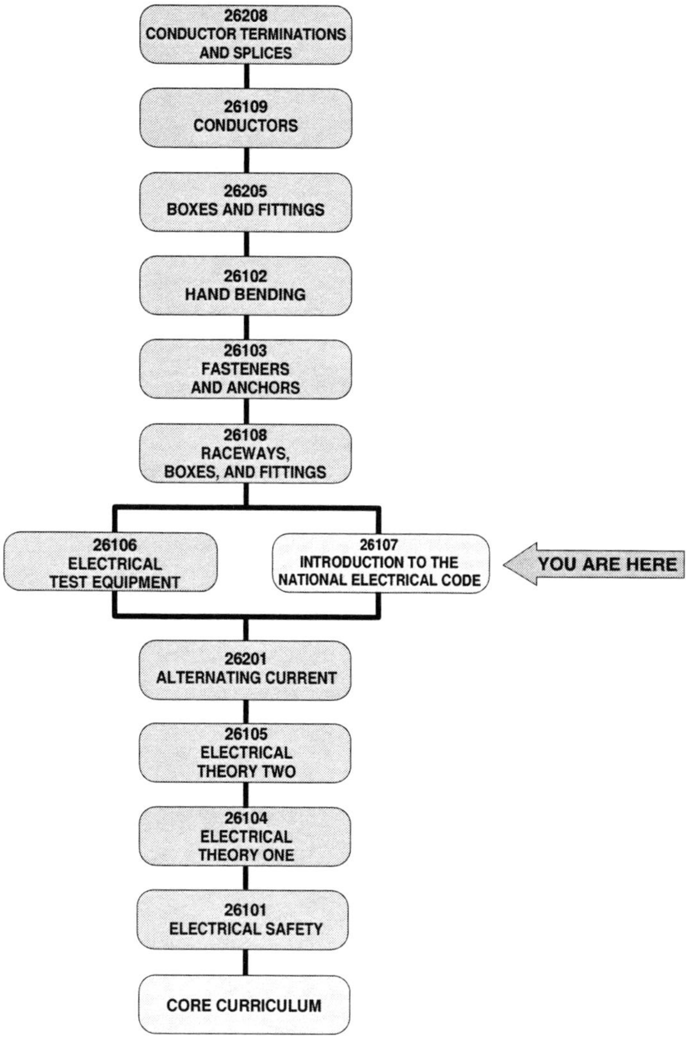

Homework

Assign reading of Module 26107.

INTRODUCTION TO THE NATIONAL ELECTRICAL CODE 225

MODULE 26107 CONTENTS

Figures

Instructor's Notes:

Chapter 6

MODULE 26107

Introduction to the National Electrical Code®

Objectives

When you have completed this module, you will be able to do the following:

1. Explain the purpose and history of the *National Electrical Code* (NEC).
2. Describe the layout of the NEC.
3. Explain how to navigate the NEC.
4. Describe the purpose of the National Electrical Manufacturers' Association (NEMA) and the National Fire Protection Association (NFPA).
5. Explain the role of testing laboratories.

Prerequisites

Before you begin this module, it is recommended that you successfully complete the following: Core Curriculum; Residential Electrical I, Chapters 1 through 5.

Required Trainee Materials

1. Paper and pencil
2. Copy of the latest edition of the *National Electrical Code*

1.0.0 ◆ INTRODUCTION

The *National Electrical Code* (NEC) is published by the **National Fire Protection Association (NFPA).** The NEC is one of the most important tools for the electrician. When used together with the electrical code for your local area, the NEC provides the minimum requirements for the installation of electrical systems. Unless otherwise specified, always use the latest edition of the NEC as your on-the-job reference. It specifies the minimum provisions necessary for protecting people and property from electrical hazards. In some areas, however, local laws may specify different editions of the NEC, so be sure to use the edition specified by your employer. Also, bear in mind that the NEC only specifies *minimum* requirements, so local or job requirements may be more stringent.

2.0.0 ◆ PURPOSE AND HISTORY OF THE NEC

The primary purpose of the NEC is the practical safeguarding of persons and property from hazards arising from the use of electricity [***NEC Section 90.1(A)***]. A thorough knowledge of the NEC is one of the first requirements for becoming a trained electrician. The NEC is probably the most widely used and generally accepted code in the world. It has been translated into several languages. It is used as an electrical installation, safety, and reference guide in the United States. Many other parts of the world use it as well. Compliance with NEC standards increases the safety of electrical installations—the reason the NEC is so widely used.

Note: The designations "National Electrical Code," "NE Code," and "NEC," where used in this document, refer to the *National Electrical Code®*, which is a registered trademark of the National Fire Protection Association, Quincy, MA. *All National Electrical Code (NEC) references in this module refer to the 2002 edition of the NEC.*

Materials

Ensure you have everything required to teach the course. Check the Materials and Equipment List at the front of this Instructor's Guide.

Audiovisual

Show Transparency 1, Course Objectives.

Show Transparency 2, Performance Profile Tasks.

Classroom

Describe the purpose of the NEC.

Explain that items shown in bold (blue) are defined in the Glossary at the back of this module.

Teaching Tip

Discuss "What's Wrong with These Pictures?"

What's Wrong with These Pictures?

107PO701.EPS

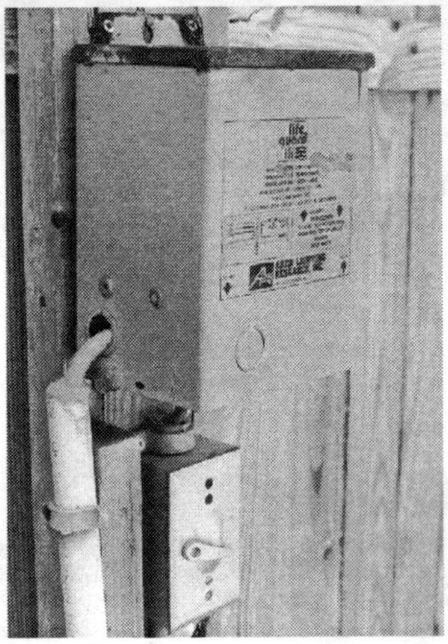

107PO702.EPS

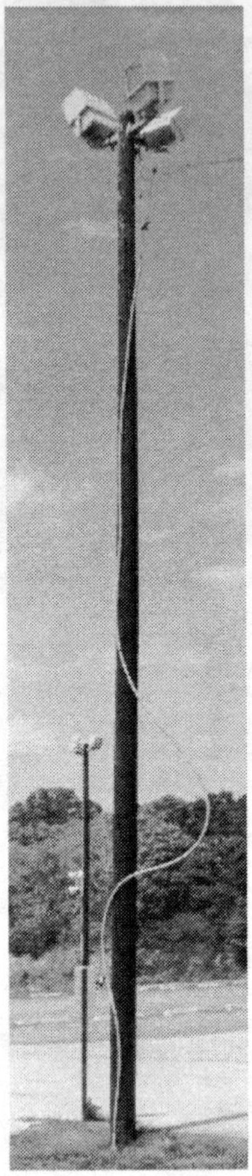

107PO703.EPS

Classroom

Stress the importance of becoming familiar with NEC requirements.

Although the NEC itself states "This Code is not intended as a design specification nor an instruction manual for untrained persons," it does provide a sound basis for the study of electrical installation procedures—under the proper guidance. The NEC has become the standard reference of the electrical construction industry. Anyone involved in electrical work should obtain an up-to-date copy and refer to it frequently.

Whether you are installing a new electrical system or altering an existing one, all electrical work must comply with the current NEC and all local ordinances. Like most laws, the NEC is easier to work with once you understand the language

Instructor's Notes:

THINK ABOUT IT

The NEC

Why do you think it's necessary to have a standard set of procedures for electrical installations? Find out who does the electrical inspection in your area. Who determines what will be inspected, when it will be inspected, and who will do the inspection?

Teaching Tip

Discuss the "Think About It."

Note

This module is *not* a substitute for the NEC. You need to acquire a copy of the most recent edition and keep it handy at all times. The more you know about the NEC, the better an electrician you will become.

and know where to look for the information you need.

2.1.0 History

In 1881, the National Association of Fire Engineers met in Richmond, Virginia. From this meeting came the idea to draft the first *National Electrical Code.* The first nationally recommended electrical code was published by the National Board of Fire Underwriters (now the American Insurance Association) in 1895.

In 1896, the National Electric Light Association (NELA) was working to make the requirements of the fire insurance organizations and electrical utilities fit together. NELA succeeded in promoting a conference that would result in producing a standard national code. The NELA code would serve the interests of the insurance industry, operating concerns, manufacturing, and industry.

The conference produced a set of requirements that was unanimously accepted. In 1897, the first edition of the NEC was published, and the NEC became the first cooperatively produced national code. The organization that produced the NEC was known as the *National Conference on Standard Electrical Rules.* This group became a permanent organization, and its job was to develop the NEC.

In 1911, the NFPA took over administration and control of the NEC. However, the National Bureau of Fire Underwriters (NBFU) continued to publish the NEC until 1962. During the period from 1911 until now, the NEC has experienced several major changes, as well as regular three-year updates. In 1923, the NEC was rearranged and rewritten, and in 1937, it was editorially revised. In 1959, the NEC was revised to include a numbering system under which each **section** of each **article** was identified with an article/section number.

2.1.1 Who Is Involved?

The creation of a universally accepted set of rules is an involved and complicated process. Rules made by a committee have the advantage that they usually do not leave out the interests of any of the groups represented on the committee. However, since the rules must represent the interests and requirements of an assortment of groups, they are often quite complicated and wordy.

In 1949, the NFPA reorganized the NEC committee into its present structure. The present structure consists of a Correlating Committee and twenty Code Making Panels (CMPs).

The Correlating Committee consists of ten principal voting members and six alternates. The principal function of the Correlating Committee is to ensure that:

- No conflict of requirements exists
- Correlation has been achieved
- NFPA regulations governing committee projects have been followed
- A practical schedule of revision and publication is established and maintained

Each of the twenty CMPs have members who are experts on particular subjects and have been assigned certain articles to supervise and revise as required. Members of the CMPs represent such special interest groups as trade associations, electrical contractors, electrical designers and engineers, electrical inspectors, electrical manufacturers and suppliers, electrical testing laboratories, and insurance organizations.

Each panel is structured so that not more than one-third of its members are from a single interest group. The members of the NEC Committee create or revise requirements for the NEC through probing, debating, analyzing, weighing, and reviewing new input. Anyone, including you, can submit proposals to amend the NEC. Sample

Classroom

Discuss the history and structure of the NEC committee.

Teaching Tip

Discuss the "Think About It."

THINK ABOUT IT

Code Changes

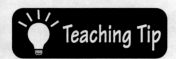

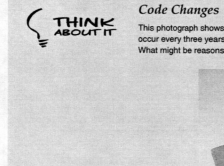

This photograph shows the first code book and the current code book. Code changes occur every three years. Who can suggest changes to the *National Electrical Code?* What might be reasons for submitting changes?

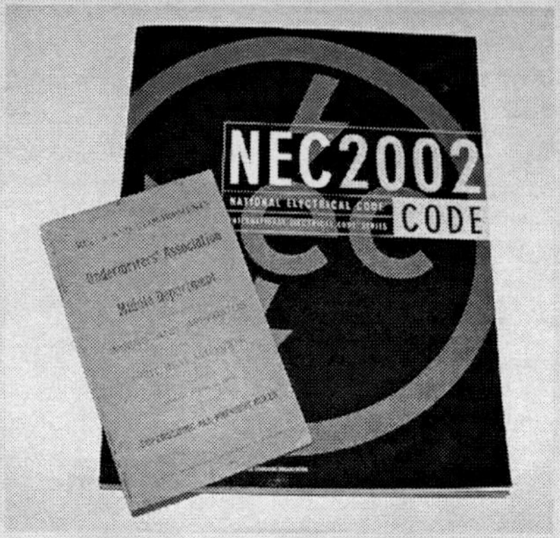

107PO704.EPS

Classroom

Discuss the layout of the NEC.

Audiovisual

Show Transparency 3 (Figure 1).

forms for this purpose may be obtained from the Secretary of the Standards Council at NFPA Headquarters. In addition to written proposals, the NFPA also holds meetings to discuss code changes and proposals.

The NFPA membership is drawn from the fields listed above. In addition to publishing the NEC, the duties of the NFPA include the following:

- Developing, publishing, and distributing standards that are intended to minimize the possibility and effects of fire and explosion
- Conducting fire safety education programs for the general public
- Providing information on fire protection, prevention, and suppression
- Compiling annual statistics on causes and occupancies of fires, large-loss fires (over one million dollars), fire deaths, and firefighter casualties
- Providing field service by specialists on electricity, flammable liquids and gases, and marine fire problems
- Conducting research projects that apply statistical methods and operations research to develop computer models and data management systems

3.0.0 ◆ THE LAYOUT OF THE NEC

The NEC begins with a brief history. *Figure 1* shows how the NEC is organized.

3.1.0 Types of Rules

There are two basic types of rules in the NEC: mandatory rules and permissive rules. It is important to understand these rules as they are defined in *NEC Section 90.5*. Mandatory rules contain the words *shall* or *shall not* and must be adhered to. Permissive rules identify actions that are allowed but not required and typically cover options or alternative methods. Be aware that local ordinances may amend requirements of the NEC. This means that a city or county may have additional requirements or prohibitions that must be followed in that jurisdiction.

Instructor's Notes:

232 CHAPTER 6

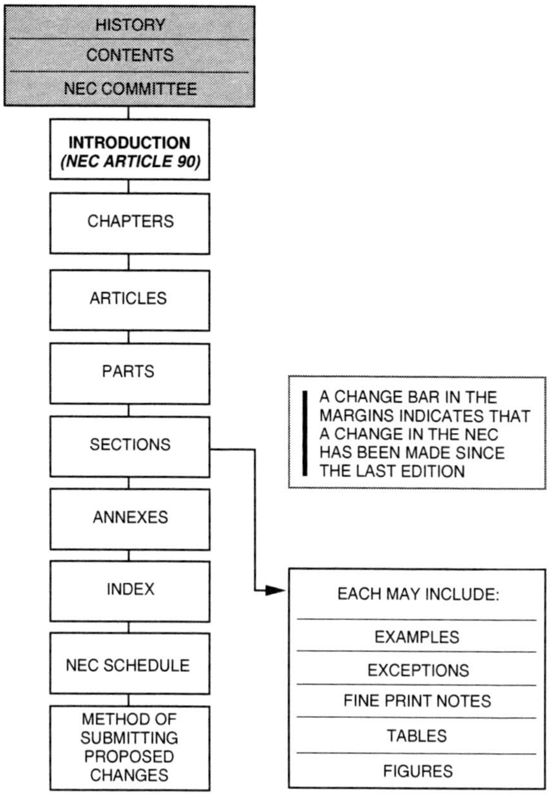

107F01.EPS

Figure 1 ◈ The layout of the NEC.

INSIDE TRACK

NEC Layout

Remember, chapters contain a group of articles relating to a broad category. An article is a specific subject within that category, such as ***NEC Article 250,*** *Grounding,* which is in Chapter 2 relating to wiring and protection. When an article applies to different installations in the same category, it will be divided into parts using roman numerals (for example, ***NEC Article 250*** is divided into ten parts). Any specific requirements in any of the articles may also have exception(s) to the main rules.

INTRODUCTION TO THE NATIONAL ELECTRICAL CODE **233**

Classroom

Discuss *NEC Article 90.*

Discuss *NEC Article 100.*

Discuss Underwriters Laboratories, Inc.

Audiovisual

Show Transparency 4 (Figure 2).

3.2.0 NEC Introduction

The main body of the text begins with an *Introduction,* also entitled ***NEC Article 90.*** This introduction gives you an overview of the NEC. Items included in this section are:

- Purpose of the NEC
- Scope of the code book
- Code arrangement
- Code enforcement
- Mandatory rules and explanatory material
- Formal interpretation
- Examination of equipment for safety
- Wiring planning
- Metric units of measurement

3.3.0 The Body of the NEC

The remainder of the book is organized into nine **chapters.** See *Figure 1* for an overview of the layout. ***NEC Chapter 1*** contains a list of definitions used in the NEC. These definitions are referred to as ***NEC Article 100. NEC Article 110*** gives the general requirements for electrical installations. It is important for you to be familiar with this general information and the definitions.

3.3.1 NEC Definitions

There are many definitions included in ***NEC Article 100.*** You should become familiar with the definitions. Here are two that you should become especially familiar with:

- *Labeled*—"Equipment or materials to which has been attached a label or other identifying mark of an organization that is acceptable to the authority having jurisdiction and concerned with product evaluation, that maintains periodic inspection of production of labeled equipment or materials, and by whose labeling the manufacturer indicates compliance with appropriate standards or performance in a specified manner."
- *Listed*—"Equipment or materials included in a list published by an organization acceptable to the authority having jurisdiction and concerned with product evaluation, that maintains periodic inspection of production of listed equipment or materials, and whose listing states either that the equipment or material meets appropriate designated standards or has been tested and found suitable for use in a specified manner."

Besides installation rules, you will also have to be concerned with the type and quality of materials

INSIDE TRACK

The NEC as a Reference Tool

The NEC, although carefully organized, is a highly technical document; as such, it uses many technical words and phrases. It's crucial for you to understand their meaning—if you don't, you'll have trouble understanding the rules themselves. When using the NEC as a reference tool, you may have to refer to a number of articles to find your answer(s). NEC definitions are covered in ***NEC Article 100.*** Many issues concerning the NEC may be resolved by simply reviewing the definitions.

INSIDE TRACK

Underwriters Laboratories, Inc.

Underwriters Laboratories, Inc. is an internationally recognized authority on product safety testing and safety certification and standards development. It was established in 1893. The Chicago World's Fair was opened and thanks to Edison's introduction of the electric light bulb, the World's Fair lit up the world. But all was not perfect—wires soon sputtered and crackled, and, ironically, the Palace of Electricity caught fire. The fair's insurance company called in a troubleshooting engineer, who, after careful inspection, found faulty and brittle insulation, worn out and deteriorated wiring, bare wires, and overloaded circuits.

He called for standards in the electrical industry, and then set up a testing laboratory above a Chicago firehouse to do just that. Hence, Underwriters Laboratories, Inc. (UL), an independent testing organization, was born.

Instructor's Notes:

234 CHAPTER 6

that are used in electrical wiring systems. Nationally recognized testing laboratories are product safety certification laboratories. Underwriters Laboratories, also called *UL,* is one such laboratory. These laboratories establish and operate product safety certification programs to make sure that items produced under the service are safeguarded against reasonable foreseeable risks. Some of these organizations maintain a worldwide network of field representatives who make unannounced visits to manufacturing facilities to counter-check products bearing their seal of approval. The UL label is shown in *Figure 2.*

3.4.0 The Reference Portion of the NEC

The annexes included with the NEC provide reference sources that can be used to determine the proper application of the NEC requirements.

3.4.1 Organization of the Chapters

NEC Chapters 1 through 8 each contain numerous articles. Each chapter focuses on a general category of electrical application, such as ***NEC Chapter 2****, Wiring and Protection*. Each article emphasizes a more specific **part** of that category, such as ***NEC Article 210****, Branch Circuits, Part I General Provisions*. Each section gives examples of a specific application of the NEC, such as ***NEC Section 210.4****, Multiwire Branch Circuits*. ***NEC Chapter 9*** contains tables that are used when referenced by any of the articles in ***NEC Chapters 1 through 7. Annexes A through D*** provide informational material and examples that are helpful when applying NEC requirements.

The chapters of the NEC are organized into four major categories:

- ***NEC Chapters 1, 2, 3, and 4***—The first four chapters present the rules for the design and installation of electrical systems. They generally apply to all electrical installations.
- ***NEC Chapters 5, 6, and 7***—These chapters are concerned with special occupancies, equipment, and conditions. Rules in these chapters may modify or amend those in the first four chapters.
- ***NEC Chapter 8***—This chapter covers communications systems, such as the telephone and telegraph, as well as radio and television receiving equipment. It may also reference other articles, such as the installation of grounding electrode conductor connections as covered in ***NEC Section 250.52.***
- ***NEC Chapter 9***—This chapter contains tables which are applicable when referenced by other chapters in the NEC.
- ***Annexes A through F***—Annexes A through F contain helpful information that is not mandatory.
  - –***Annex A*** contains a list of product safety standards. These standards provide further references for requirements that are in addition to the NEC requirements for the electrical components mentioned.
  - –***Annex B*** contains information for determining ampacities of conductors under engineering supervision.
  - –***Annex C*** contains the conduit fill tables for multiple conductors of the same size and type within the accepted raceways.
  - –***Annex D*** contains examples of calculations for branch circuits, feeders, and services as well as other loads such as motor circuits.
  - –***Annex E*** contains information on types of building construction.
  - –***Annex F*** contains a cross reference index for the articles that were renumbered in the 2002 code.

107F02.EPS

Figure 2 ◈ Underwriters Laboratories label.

THINK ABOUT IT

Other Testing Laboratories

In addition to UL, there are several other recognized testing laboratories. How many can you name?

Classroom

Discuss the chapter organization of the NEC.

Teaching Tip

Discuss the "Think About It."

INTRODUCTION TO THE NATIONAL ELECTRICAL CODE **235**

Classroom

Review the different types of text in the NEC.

Demonstrate how to locate information in the NEC.

Have the trainees practice finding information in the NEC.

Discuss the NEC chapters dealing with servicing electrical systems.

Teaching Tip

Discuss the "Think About It" features.

3.5.0 Text in the NEC

When you open the NEC, you will notice several different types of text or printing used. Here is an explanation of each type of text:

- *Bold black letters*—Headings for each NEC application.
- *Exceptions*—**Exceptions** explain the circumstances under which a specific part of the NEC does not apply. Exceptions are written in *italics* under that part of the NEC to which they pertain.
- *Fine print notes*—**Fine print notes (FPNs)** explain something in an application, suggest other sections to read about the application, or provide tips about the application. These are defined in the text by the term *(FPN)* shown in parentheses before a paragraph in smaller print.
- *Figures*—These may be included with explanations to give you a picture of what your application may look like.
- *Tables*—Tables are often included when there is more than one possible application of the NEC. You would use a table to look up the specifications of your application.

4.0.0 ◆ NAVIGATING THE NEC

To locate information for a particular procedure being performed, use the following steps:

Step 1 Familiarize yourself with *NEC Articles 90, 100, and 110* to gain an understanding of the material covered in the NEC and the definitions used in it.

Step 2 Turn to the *Table of Contents* at the beginning of the NEC.

Step 3 Locate the chapter that focuses on the desired category.

Step 4 Find the article pertaining to your specific application.

Step 5 Turn to the page indicated. Each application will begin with a bold heading.

Note

An index is provided at the end of the NEC. The index lists specific topics and provides a reference to the location of the material within the NEC. The index is helpful when you are looking for a specific topic rather than a general category.

Once you are familiar with *NEC Articles 90, 100, and 110,* you can move on to the rest of the NEC. There are several key sections used often in servicing electrical systems.

- *Wiring design and protection—NEC Chapter 2* discusses wiring design and protection, the information electrical technicians need most often. It covers the use and identification of grounded conductors, branch circuits, feeders, calculations, services, overcurrent protection, and grounding. This is essential information for all types of electrical systems. If you run into a problem related to the design or installation of a conventional electrical system, you can probably find a solution for it in this chapter.
- *Wiring methods and materials—NEC Chapter 3* lists the rules on wiring methods and materials.

THINK ABOUT IT

Junction Boxes

Find the rule in the NEC that explains whether a junction box without devices can be supported solely by two or more lengths of RMC. Explain the technical terminology in everyday language.

THINK ABOUT IT

NEC Article 90

After you've familiarized yourself with ***NEC Article 90,*** explain its intent. What part of electrical installation does it not cover?

Instructor's Notes:

The materials and procedures to use on a particular system depend on the type of building construction, the type of occupancy, the location of the wiring in the building, the type of atmosphere in the building or in the area surrounding the building, mechanical factors, and the relative costs of different wiring methods. The general requirements for conductors and wiring methods that form an integral part of manufactured equipment are not included in the requirements of *NEC Chapter 3.*

NEC Article 300 provides the general requirements for all wiring methods, including information such as minimum burial depths and permitted wiring methods for areas above suspended ceilings.

NEC Article 310 contains a complete description of the acceptable conductors for the wiring methods contained in *NEC Chapter 3.*

NEC Articles 312 and 314 give rules for raceways, boxes, cabinets, and raceway fittings. Outlet boxes vary in size and shape, depending on their use, the size of the raceway, the number of conductors entering the box, the type of building construction, and the atmospheric conditions of the area. These articles should answer most questions on the selection and use of these items.

The NEC does not describe in detail all types and sizes of outlet boxes. However, the manufacturers of outlet boxes provide excellent catalogs showing their products. Collect these catalogs, since these items are essential to your work.

NEC Articles 320 through 340 cover sheathed cables of two or more conductors, such as nonmetallic-sheathed and metal-clad cable.

NEC Articles 342 through 356 cover raceway wiring systems, such as rigid and flexible metal and nonmetallic conduit.

NEC Articles 358 through 362 cover tubing wiring methods, such as electrical metallic and nonmetallic tubing.

NEC Articles 366 through 390 cover other wiring methods, such as busways and wireways.

Cable trays are a system of support for the wiring methods found not only in *NEC Chapter 3,* but also in *NEC Chapters 4, 7, and 8.*

4.1.0 Equipment for General Use

NEC Chapter 4 begins with the use and installation of flexible cords and cables, including the trade name, type letter, wire size, number of conductors, conductor insulation, outer covering, and use of each. This chapter also covers fixture wires, again giving the trade name, type letter, and other important details.

NEC Article 404 covers the requirements for the uses and installation of switches, switching devices, and circuit breakers where used as switches.

NEC Article 406 gives the rules for the ratings, types, and installation of receptacles, cord connectors, and attachment plugs (cord caps).

NEC Article 410 on lighting fixtures is especially important. It gives installation procedures for fixtures in specific locations. For example, it covers fixtures near combustible material and fixtures in closets. However, the NEC does not describe how many fixtures will be needed in a given area to provide a certain amount of illumination.

NEC Article 430 covers electric motors, including mounting the motor and making electrical connections to it. Motor controls and overload protection are also covered.

NEC Articles 440 through 460 cover air conditioning and heating equipment, transformers, and capacitors.

THINK ABOUT IT

Disconnects

How would you proceed to find the NEC rule for the maximum number of disconnects permitted for a service?

THINK ABOUT IT

Sealing Requirements

Study ***NEC Section 501.5*** and summarize the key points. Discuss your interpretation with the rest of the class.

Classroom

Discuss *NEC Chapter 4.*

Teaching Tip

Discuss the "Think About It" features.

Classroom

Discuss the three classes of special occupancy location and provide examples of each type. Refer the trainees to *NEC Chapter 5.*

Discuss *NEC Chapter 6.*

NEC Article 480 provides requirements related to battery-operated electrical systems. Storage batteries are seldom thought of as part of a conventional electrical system, but they often provide standby emergency lighting service. They may also supply power to security systems that are separate from the main AC electrical system.

4.2.0 Special Occupancies

NEC Chapter 5 covers special occupancy areas. These are areas where the sparks generated by electrical equipment may cause an explosion or fire. The hazard may be due to the atmosphere of the area or the presence of a volatile material in the area. Commercial garages, aircraft hangers, and service stations are typical special occupancy locations.

NEC Article 500 covers the different types of special occupancy atmospheres where an explosion is possible. The atmospheric groups were established to make it easy to test and approve equipment for various types of uses.

NEC Articles 501.4, 502.4, and 503.3 cover the installation of explosion-proof wiring. An explosion-proof system is designed to prevent the ignition of a surrounding explosive atmosphere when arcing occurs within the electrical system.

There are three main classes of special occupancy location:

- *Class I*—Areas containing flammable gases or vapors in the air. Class I areas include paint spray booths, dyeing plants where hazardous liquids are used, and gas generator rooms (*NEC Article 501*).
- *Class II*—Areas where combustible dust is present, such as grain-handling and storage plants, dust and stock collector areas, and sugar-pulverizing plants (*NEC Article 502*). These are areas where, under normal operating conditions, there may be enough combustible dust in the air to produce explosive or ignitable mixtures.
- *Class III*—Areas that are hazardous because of the presence of easily ignitable fibers or other particles in the air, although not in large enough quantities to produce ignitable mixtures (*NEC Article 503*). Class III locations include cotton mills, rayon mills, and clothing manufacturing plants.

NEC Articles 511 and 514 regulate garages and similar locations where volatile or flammable liquids are used. While these areas are not always considered critically hazardous locations, there may be enough danger to require special precautions in the electrical installation. In these areas, the NEC requires that volatile gases be confined to an area not more than four feet above the floor. So in most cases, conventional raceway systems are permitted above this level. If the area is judged to be critically hazardous, explosion-proof wiring (including seal-offs) may be required.

NEC Article 520 regulates theaters and similar occupancies where fire and panic can cause hazards to life and property. Drive-in theaters do not present the same hazards as enclosed auditoriums, but the projection rooms and adjacent areas must be properly ventilated and wired for the protection of operating personnel and others using the area.

NEC Chapter 5 also covers service stations, bulk storage plants, health care facilities, mobile homes and parks, and temporary installations.

4.3.0 Special Equipment

Residential electrical workers will seldom need to refer to the articles in *NEC Chapter 6*, but this chapter is of great concern to commercial and industrial electrical workers.

NEC Article 600 covers electric signs and outline lighting. *NEC Article 610* applies to cranes and hoists. *NEC Article 620* covers the majority of the electrical work involved in the installation and operation of elevators, dumbwaiters, escalators, and moving walks. The manufacturer is responsible for most of this work. The electrician usually just furnishes a feeder terminating in a disconnect means in the bottom of the elevator shaft. The electrician may also be responsible for a lighting circuit to a junction box midway in the elevator shaft for connecting the elevator cage lighting cable and exhaust fans. The articles in this chapter list most of the requirements for these installations.

NEC Article 630 regulates electric welding equipment. It is normally treated as a piece of industrial power equipment requiring a special power outlet, but there are special conditions that apply to the circuits supplying welding equipment. These are outlined in detail in this chapter.

NEC Article 640 covers wiring for sound recording and similar equipment. This type of equipment normally requires low-voltage wiring. Special outlet boxes or cabinets are usually provided with the equipment, but some items may be mounted in or on standard outlet boxes. Some sound recording systems require direct current. It is supplied from rectifying equipment, batteries, or motor generators. Low-voltage alternating current comes from relatively small transformers connected on the primary side to a 120V circuit within the building.

Instructor's Notes:

238 CHAPTER 6

Other items covered in *NEC Chapter 6* include X-ray equipment (*NEC Article 660*), induction and dielectric heat-generating equipment (*NEC Article 665*), and machine tools (*NEC Article 670*).

If you ever have work that involves *NEC Chapter 6,* study the chapter before work begins. That can save a lot of installation time. Another way to cut down on labor hours and prevent installation error is to acquire a set of rough-in drawings of the equipment being installed. It is easy to install the wrong outlet box or to install the right box in the wrong place. Having a set of rough-in drawings can prevent these simple but costly errors.

4.4.0 Special Conditions

In most commercial buildings, the NEC and local ordinances require a means of lighting public rooms, halls, stairways, and entrances. There must be enough light to allow the occupants to exit from the building if the general building lighting is interrupted. Exit doors must be clearly indicated by illuminated exit signs.

NEC Chapter 7 covers the installation of emergency systems. These circuits should be arranged so that they can automatically transfer to an alternate source of current, usually storage batteries or gasoline-driven generators. As an alternative in some types of occupancies, you can connect them to the supply side of the main service, so disconnecting the main service switch would not disconnect the emergency circuits. This chapter also covers fire alarms and a variety of other equipment, systems, and conditions that are not easily categorized elsewhere in the NEC.

NEC Chapter 8 is a special category for wiring associated with electronic communications systems including telephone and telegraph, radio and TV, and community antenna systems.

4.5.0 Examples of Navigating the NEC

4.5.1 Installing Type SE Cable

Suppose you are installing Type SE (service-entrance) cable on the side of a home. You know that this cable must be secured, but you are not sure of the spacing between cable clamps. To find out this information, use the following procedure:

Step 1 Look in the NEC *Table of Contents* and follow down the list until you find an appropriate category. (Or you can use the index at the end of the book.)

Step 2 ***NEC Article 230*** will probably catch your eye first, so turn to the page where it begins.

Step 3 Scan down through the section numbers until you come to ***NEC Section 230.51,*** *Mounting Supports.* Upon reading this section, you will find in paragraph *(a) Service-Entrance Cables* that "Service-entrance cable shall be supported by straps or other approved means within 12 inches (305mm) of every service head, gooseneck, or connection to a raceway or enclosure, and at intervals not exceeding 30 inches (762mm)."

After reading this section, you will know that a cable strap is required within 12 inches of the service head and within 12 inches of the meter base. Furthermore, the cable must be secured in between these two termination points at intervals not exceeding 30 inches.

4.5.2 Installing Track Lighting

Assume that you are installing track lighting in a residential occupancy. The owners want the track located behind the curtain of their sliding glass patio doors. To determine if this is an NEC violation, follow these steps:

Step 1 Look in the NEC *Table of Contents* and find the chapter that contains information about the general application you are working on. ***NEC Chapter 4,*** *Equipment for General Use,* covers track lighting.

Step 2 Now look for the article that fits the specific category you are working on. In this case, ***NEC Article 410*** covers lighting fixtures, lampholders, lamps, and receptacles.

Step 3 Next locate the section within ***NEC Article 410*** that deals with the specific application. For this example, refer to *Part XV, Lighting Track.*

Step 4 Turn to the page listed.

Step 5 Read ***NEC Section 410.100,*** *Definitions,* to become familiar with track lighting. Continue down the page with ***NEC Section 410.101*** and read the information contained therein. Note that paragraph *(c) Locations Not Permitted* under ***NEC Section 410.101*** states the following: "Lighting track shall not be installed in the following locations: (1) where likely to be subjected to physical damage; (2) in wet or damp locations;

Classroom

Review *NEC Chapters 7 and 8.*

Demonstrate how to find the NEC requirements for example installations.

INTRODUCTION TO THE NATIONAL ELECTRICAL CODE 239

(3) where subject to corrosive vapors; (4) in storage battery rooms; (5) in hazardous (classified) locations; (6) where concealed; (7) where extended through walls or partitions; (8) less than 1.5m (5 feet) above the finished floor except where protected from physical damage or track operating at less than 30 volts RMS open-circuit voltage; (9) within the zone measured 900mm (3 feet) horizontally and 2.5m (8 feet) vertically from the top of the bathtub rim."

Step 6 Read ***NEC Section 410.101(C)*** carefully. Do you see any conditions that would violate any NEC requirements if the track lighting is installed in the area specified? In checking these items, you will probably note condition (6), "where concealed." Since the track lighting is to be installed behind a curtain, this sounds like an NEC violation. We need to check further.

Step 7 You need the NEC definition of concealed. Therefore, turn to ***NEC Article 100,*** *Definitions* and find the main term *concealed.* It reads: "Concealed: Rendered inaccessible by the structure or finish of the building . . ."

Step 8 Although the track lighting may be out of sight if the curtain is drawn, it will still be readily accessible for maintenance. Consequently, the track lighting is really not concealed according to the NEC definition.

When using the NEC to determine electrical installation requirements, keep in mind that you will nearly always have to refer to more than one section. Sometimes the NEC itself refers the reader to other articles and sections. In some cases, the user will have to be familiar enough with the NEC to know what other sections pertain to the installation at hand. It can be a confusing situation, but time and experience in using the NEC will make it much easier. A pictorial road map of some NEC topics is shown in *Figures 3* and *4.*

Audiovisual

Show Transparencies 5 and 6 (Figures 3 and 4).

5.0.0 ◆ OTHER ORGANIZATIONS

5.1.0 The Role of Testing Laboratories

Classroom

Discuss the roles of various testing laboratories.

As mentioned earlier, testing laboratories are an integral part of the development of the NEC. The NFPA and other organizations provide testing laboratories to conduct research into electrical equipment and its safety. These laboratories perform extensive testing of new products to make sure they are built to NEC standards for electrical and fire safety. These organizations receive statistics and reports from agencies all over the United States concerning electrical shocks and fires and their causes. Upon seeing trends developing concerning the association of certain equipment and dangerous situations or circumstances, this equipment will be specifically targeted for research. All the reports from these laboratories are used in the generation of changes or revisions to the NEC.

INSIDE TRACK

Different Interpretations

For as many trainees in class with you today, there will be as many different interpretations of the NEC. However, a difference of opinion is not always a problem—discussing the NEC with your peers will allow you to expand your own understanding of it. In your previous discussion about ***NEC Section 501.5,*** what differences of interpretation did you have to resolve?

Teaching Tip

Discuss the "Think About It."

THINK ABOUT IT

Conformance and Electrical Equipment

What other resources are available for finding information about the use of electrical equipment and materials?

Instructor's Notes:

240 CHAPTER 6

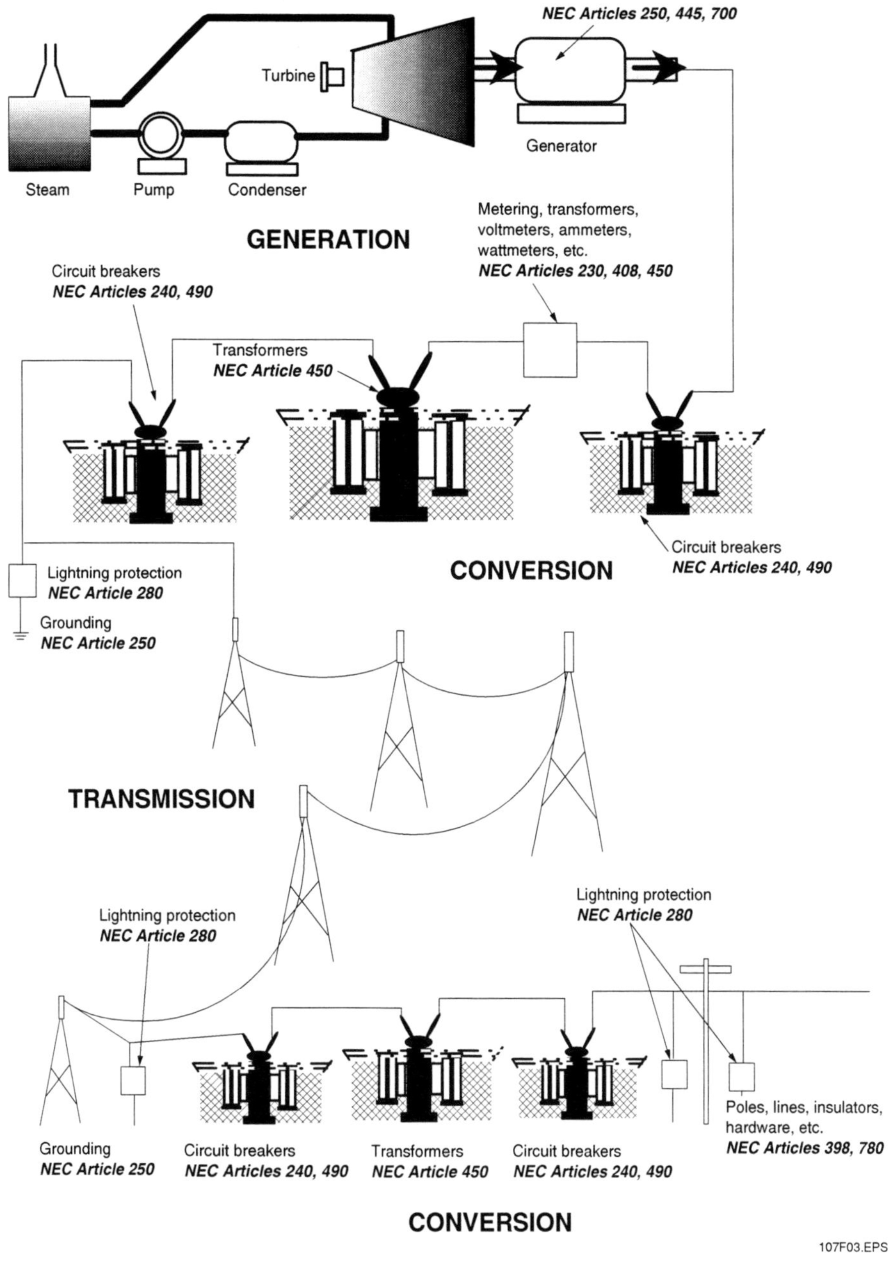

Figure 3 ◈ NEC references for power generation and transmission.

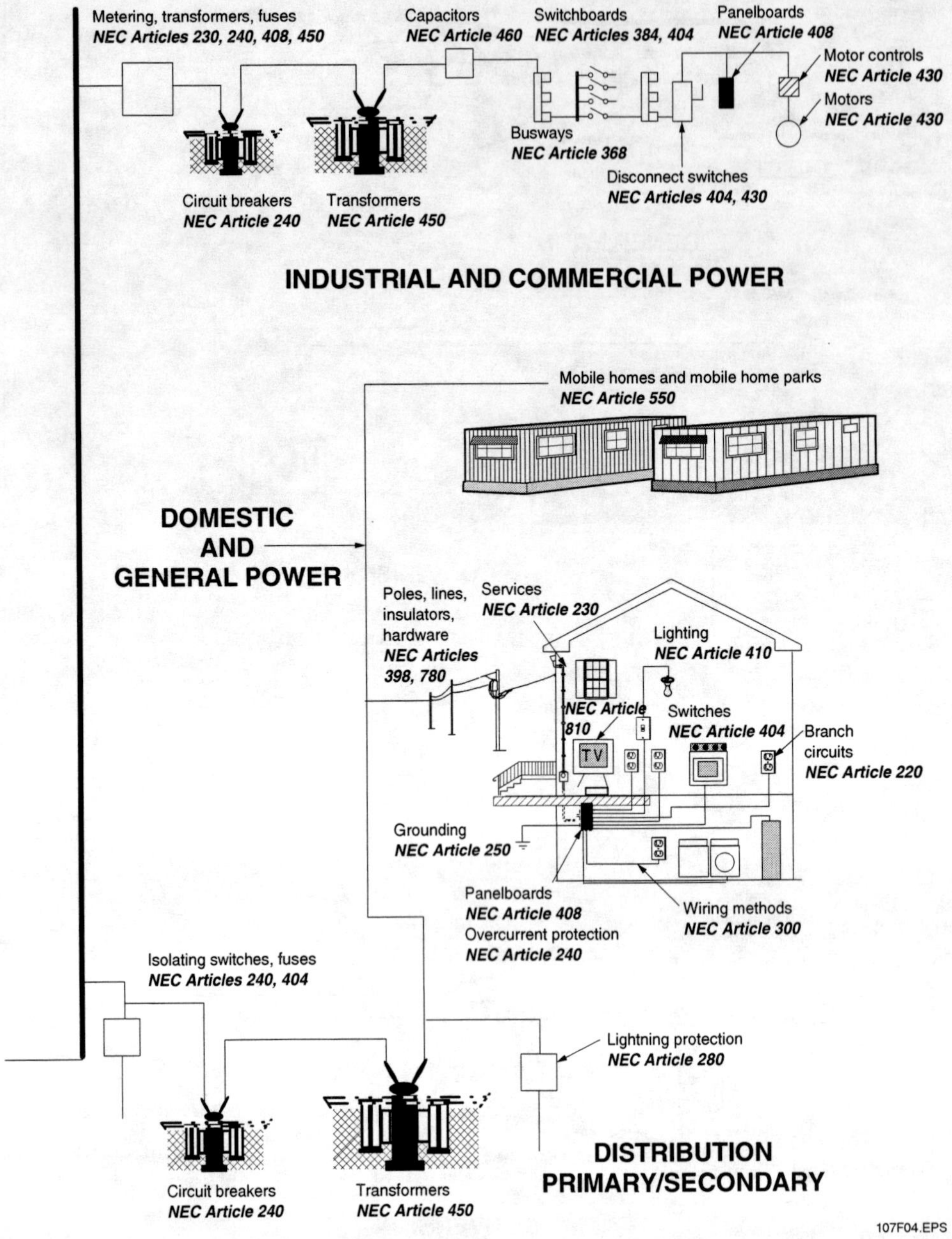

Figure 4 ◆ NEC references for industrial, commercial, and residential power.

Instructor's Notes:

THINK ABOUT IT

Putting It All Together

Look around you at the electrical components and products used and the quality of the work. Do you see any components or products that have not been listed or labeled? If so, how might these devices put you in harm's way? Do you see any code violations?

5.2.0 Nationally Recognized Testing Laboratories

Nationally Recognized Testing Laboratories (NRTL) are product safety certification laboratories. They establish and operate product safety certification programs to make sure that items produced under the service are safeguarded against reasonably foreseeable risks. NRTLs maintain a worldwide network of field representatives who make unannounced visits to factories to check products bearing their safety marks.

5.3.0 National Electrical Manufacturers' Association

The **National Electrical Manufacturers' Association (NEMA)** was founded in 1926. It is made up of companies that manufacture equipment used for generation, transmission, distribution, control, and utilization of electric power. The objectives of NEMA are to maintain and improve the quality and reliability of products; to ensure safety standards in the manufacture and use of products; and to develop product standards covering such matters as naming, ratings, performance, testing, and dimensions. NEMA participates in developing the NEC and the *National Electrical Safety Code* and advocates their acceptance by state and local authorities.

Summary

The NEC specifies the minimum provisions necessary for protecting people and property from hazards arising from the use of electricity and electrical equipment. As an electrician, you must be aware of how to use and apply the NEC on the job site. Using the NEC will help you to safely install and maintain the electrical equipment and systems you come into contact with.

Review Questions

1. The NEC provides the _____ requirements for the installation of electrical systems.
   a. minimum
   b. most stringent
   c. design specification
   d. complete
2. All of the following groups are usually represented on the Code Making Panels, *except* _____ .
   a. trade associations
   b. electrical inspectors
   c. insurance organizations
   d. government lobbyists
3. Mandatory and permissive rules are defined in _____ .
   a. *NEC Article 90*
   b. *NEC Article 100*
   c. *NEC Article 110*
   d. *NEC Article 200*
4. *NEC Article 110* covers _____ .
   a. branch circuits
   b. definitions
   c. general requirements for electrical installations
   d. wiring design and protection
5. The design and installation of electrical systems is covered in _____ .
   a. *NEC Chapters 1, 2, and 7*
   b. *NEC Chapters 1, 2, 3, 4, and 9*
   c. *NEC Chapters 6, 7, and 9*
   d. *NEC Chapters 5, 6, 7, and 9*
6. Devices such as radios, televisions, and telephones are covered in _____ .
   a. *NEC Chapter 8*
   b. *NEC Chapter 7*
   c. *NEC Chapter 6*
   d. *NEC Chapter 5*

Teaching Tip

Discuss the "Think About It."

Classroom

Discuss NEMA.

Have the trainees complete the Review Questions and go over the answers prior to administering the Module Examination.

Examination

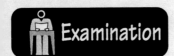

Administer the Module Examination. Record the results on Craft Training Report Form 200 and submit the results to the Training Program Sponsor.

Performance Testing

Administer the Performance Test and fill out Performance Profile Sheets for each trainee. If desired, trainee proficiency noted during laboratory sessions may be used to complete the Performance Test. Be sure to record the results on Craft Training Report Form 200 and submit the results to the Training Program Sponsor.

7. Busways are covered in _____ .
   a. *NEC Articles 330 through 339*
   b. *NEC Articles 342 through 358*
   c. *NEC Article 368*
   d. *NEC Article 318*

8. Installation procedures for lighting fixtures in specific locations are provided in _____ .
   a. *NEC Article 410*
   b. *NEC Article 501*
   c. *NEC Article 364*
   d. *NEC Article 460*

9. Theaters are covered in _____ .
   a. *NEC Article 339*
   b. *NEC Article 110*
   c. *NEC Article 430*
   d. *NEC Article 520*

10. *NEC Article 600* covers _____ .
    a. track lighting
    b. electric signs and outline lighting
    c. X-ray equipment
    d. emergency lighting systems

Instructor's Notes:

PROFILE IN SUCCESS

James G. Stallcup, Greyboy & Associates

Today, the name James Stallcup is recognized throughout the electrical industry as an author, publisher, teacher, lecturer, and expert on electrical codes.

James Stallcup didn't start out his career as a writer and teacher. Like most others who have achieved success in the industry, he began in a four-year electrical apprenticeship program, which he entered because his father-in-law was an electrical contractor. In addition to his four-year apprenticeship program, James completed 756 hours of industrial electrical training.

Later on, he applied for and obtained a job as an electrical inspector for the city of Fort Worth, Texas. Because of his ability and his extensive training, James was elevated to the position of chief electrical inspector within one year.

James began teaching at the local union apprentice school to supplement his city income, and later began his own continuing education school, where he trained many electricians, corporate maintenance people, and municipal inspectors.

The handouts for his course became the embryo of his first book. Since then, James has written and published more than a dozen books. He also serves on numerous industry committees and has received many industry awards.

This remarkable success story could not have been told were it not for three key factors: proper training, the desire to continue learning as a lifetime endeavor, and James's determination to stretch his personal horizons.

INTRODUCTION TO THE NATIONAL ELECTRICAL CODE 245

GLOSSARY

Trade Terms Introduced in This Module

Articles: The articles are the main topics of the NEC, beginning with ***NEC Article 80,*** *Administration and Enforcement,* and ending with ***NEC Article 830,*** *Network-Powered Broadband Communications Systems.*

Chapters: Nine chapters form the broad structure of the NEC.

Exceptions: Exceptions follow the applicable sections of the NEC and allow alternative methods to be used under specific conditions.

Fine Print Note (FPN): Explanatory material that follows specific NEC sections.

National Electrical Manufacturers' Association (NEMA): The association that maintains and improves the quality and reliability of electrical products.

National Fire Protection Association (NFPA): The publishers of the NEC. The NFPA develops standards to minimize the possibility and effects of fire.

Nationally Recognized Testing Laboratories (NRTL): Product safety certification laboratories that are responsible for testing and certifying electrical equipment.

Parts: Certain articles in the NEC are subdivided into parts. Parts have letter designations (e.g., Part A).

Sections: Parts and articles are subdivided into sections. Sections have numeric designations that follow the article number and are preceded by a period (e.g., 501.4).

Instructor's Notes:

Additional Resources

This module is intended to present thorough resources for task training. The following reference work is suggested for further study. This is optional materials for continued education rather than for task training.

National Electrical Code Handbook, Latest Edition. Quincy, MA: National Fire Protection Association.

INTRODUCTION TO THE NATIONAL ELECTRICAL CODE 247

ANSWER KEY

Answers to Review Questions

Answer	Section
1. a	1.0.0
2. d	2.1.1
3. a	3.1.0
4. c	3.3.0
5. b	3.4.1
6. c	4.0.0
7. a	4.1.0
8. d	4.2.0
9. b	4.3.0
10. a	3.4.1/4.4.0

Section 2.0.0 *What's Wrong With These Pictures?*

Top left—Fluorescent fixture is not supported (wire is simply wrapped around the overhead beam).

Bottom Left—Improper box entry (no fittings).

Right—This unsupported PVC conduit melted in the summer heat.

Section 2.0.0 *Think About It—The NEC*

Describe the reasons for standardizing installation procedures, including convenience and the obvious interest to public safety. Have the trainees talk to practicing electricians about their use of the NEC, including what is to be inspected, when it is to be inspected, and who does the inspecting. (Inspection procedures will vary by location.)

Section 2.1.1 *Think About It—Code Changes*

Anyone can suggest a change to the *National Electrical Code*® by submitting a proposal. Suggestions for change are typically a result of safety considerations, but may also be influenced by convenience, cost, or the introduction of new materials or devices.

Section 3.3.0 *Think About It—Other Testing Laboratories*

There are many testing laboratories other than UL. For example, the Canadian Standards Association (CSA) performs product evaluations and tests similar to those performed by UL. Through international agreements, both UL and CSA are authorized to test and certify products for use in both the U.S. and Canadian markets. Other testing laboratories include ITS (Intertek Testing Services), which encompasses various labels including ETL, Warnock Hersey, and Semco; Metlabs; TÜV; Factory Mutual; and others. Encourage the trainees to examine various electrical devices for these and other marks of compliance.

Section 4.0.0 *Think About It—Junction Boxes*

NEC Section 314.23(E) covers raceway-supported enclosures. Have the trainees interpret the main points of this section, including the exception.

Section 4.0.0 *Think About It—NEC Article 90*

NEC Article 90 covers the purpose and scope of the NEC and provides a brief overview of its arrangement, enforcement, types of rules, formal interpretation procedures, and other points.

The NEC does not cover certain types of installations, as listed in ***NEC Section 90.2(B).*** For example, the NEC does not cover installations in ships, automobiles, or aircraft. It also does not cover installations under the control of electrical utilities, including metering equipment.

Section 4.0.0 *Think About It—Disconnects*

The trainees should follow the steps outlined in Section 4.0.0 of the Trainee Module for navigating the NEC. ***NEC Section 230.71*** covers the maximum number of disconnects.

Section 4.2.0 *Think About It—Sealing Requirements*

NEC Section 501.5 covers sealing conduit and cable systems and drainage of control equipment, motors, generators, etc. in Class I hazardous locations. Class I locations are those in which flammable gases or vapors are, or may be, present in the air in quantities sufficient to produce explosive or ignitable mixtures. The trainees should read this section carefully and restate it in their own words.

Section 5.3.0 *Think About It—Conformance and Electrical Equipment*

Other resources for finding information about the use of electrical equipment and materials can include master electricians, manufacturers' catalogs, Web sites, electrical distributors, application engineers, and manufacturers' sales representatives.

Section 5.3.0 *Think About It—Putting It All Together*

Answers will depend upon the building(s) in question. Unlisted devices and/or improper installation can result in short circuits or ground faults, as well as other problems, including faulty operation, environmental dangers, and financial losses, such as rework and possible liability.

The NCCER makes every effort to keep these textbooks up-to-date and free of technical errors. We appreciate your help in this process. If you have an idea for improving this textbook, or if you find an error, a typographical mistake, or an inaccuracy in NCCER's Contren™ textbooks, please write us, using this form or a photocopy. Be sure to include the exact module number, page number, a detailed description, and the correction, if applicable. Your input will be brought to the attention of the Technical Review Committee. Thank you for your assistance.

Instructors – If you found that additional materials were necessary in order to teach this module effectively, please let us know so that we may include them in the Equipment/Materials list in the Instructor's Guide.

Write: Curriculum Revision and Development Department
National Center for Construction Education and Research
P.O. Box 141104, Gainesville, FL 32614-1104

Fax: 352-334-0932

E-mail: curriculum@nccer.org

Craft ____________________ Module Name ____________________

Copyright Date __________ Module Number __________ Page Number(s) __________

Description

____

____

____

____

(Optional) Correction

____

____

____

(Optional) Your Name and Address

____

____

____

Chapter 7

Raceways, Boxes, and Fittings

26108-02

Raceways, Boxes, and Fittings
Annotated Instructor's Guide

Module 26108

MODULE OVERVIEW

This course introduces the electrical trainee to the various types of raceways, boxes, and fittings, including their installation procedures and NEC requirements.

PREREQUISITES

Please refer to the Course Map in the Trainee Module. Prior to training with this module, it is recommended that the trainee shall have successfully completed the following:

Core Curriculum; Residential Electrical I, Chapters 1 through 6

LEARNING OBJECTIVES

Upon completion of this module, the trainee will be able to:

1. Describe various types of cable trays and raceways.
2. Identify and select various types and sizes of raceways.
3. Identify and select various types and sizes of cable trays.
4. Identify and select various types of raceway fittings.
5. Identify various methods used to install raceways.
6. Demonstrate knowledge of NEC raceway requirements.
7. Describe procedures for installing raceways and boxes on masonry surfaces.
8. Describe procedures for installing raceways and boxes on concrete surfaces.
9. Describe procedures for installing raceways and boxes in a metal stud environment.
10. Describe procedures for installing raceways and boxes in a wood frame environment.
11. Describe procedures for installing raceways and boxes on drywall surfaces.
12. Recognize safety precautions that must be followed when working with boxes and raceways.

PERFORMANCE OBJECTIVES

Under supervision of the instructor, the trainee should be able to:

1. Identify fittings, boxes, and types of raceways. Reference the appropriate NEC section(s) for these items.
2. Make a conduit-to-box connection.

NCCER STANDARDIZED CRAFT TRAINING PROGRAM

The National Center for Construction Education and Research (NCCER) provides a standardized national program of accredited craft training. Key features of the program include instructor certification, competency-based training, and performance testing. The program provides trainees, instructors, and companies with a standard form of recognition through a National Craft Training Registry. The program is described in full in the *Guidelines for Accreditation,* published by the NCCER. For more information on standardized craft training, contact the NCCER at P.O. Box 141104, Gainesville, FL 32614-1104, 352-334-0911, visit our Web site at www.nccer.org, or e-mail info@nccer.org.

HOW TO USE THIS ANNOTATED INSTRUCTOR'S GUIDE

Each page presents two sections of information. The larger section displays each page exactly as it appears in the Trainee Module. The narrow column ties suggested trainee and instructor actions to each page and provides icons to call your attention to material, safety, audiovisual, or testing requirements. The bottom of each page includes space for your notes.

Teaching Tip

If you see the Teaching Tip icon, that means there is a teaching tip associated with this section. Also refer to the suggested teaching tips at the end of the module.

SAFETY CONSIDERATIONS

Ensure that the trainees are equipped with appropriate personal protective equipment. Emphasize the use of the proper safety procedures when working with the tools and materials used in the installation of raceways, boxes, and fittings.

PREPARATION

Before teaching this module, you should review the Module Outline, Learning and Performance Objectives, and the Materials and Equipment List. Be sure to allow ample time to prepare your own training or lesson plan and gather all required equipment and materials.

MATERIALS AND EQUIPMENT LIST

Materials:

Transparencies

Markers/chalk

Copy of the latest edition of the *National Electrical Code®*

Module Examinations*

Performance Profile Sheets*

Section(s) of EMT

EMT compression fittings

EMT setscrew fittings

Section(s) of rigid metal conduit

Section(s) of plastic-coated RMC

Section(s) of aluminum conduit

Section(s) of rigid black conduit

Section(s) of IMC

Sections of EB and DB RNC conduit

Section(s) of LFNC and connectors

Section(s) of flexible metal conduit

Various conduit couplings

Type C, Type L, Type T, and Type X conduit bodies

Threaded weatherproof hubs

Insulating bushings

Offset nipples

Metal boxes

Nonmetallic boxes

Bushings and locknuts

Sealing fittings and packing material

Liquid sealing compound

Various straps

Standoff support

Hammer

Knockout punch

Screwdriver

Equipment:

Overhead projector and screen

Whiteboard/chalkboard

Appropriate personal protective equipment

Access to a job site where trainees can observe a variety of wireway components, including:
- Connectors
- End plates
- Closing plates
- Tee fittings
- Crosses
- Elbows
- Nipples
- Slip fittings

Access to a job site where trainees can observe a variety of cable tray support systems, including:
- Direct rod
- Trapeze mounting
- Center hung support
- Wall mounting
- Pipe rack mounting

*Located in the Test Booklet packaged with this Annotated Instructor's Guide.

ADDITIONAL RESOURCES

This module is intended to present thorough resources for task training. The following reference works are suggested for both instructors and motivated trainees interested in further study. These are optional materials for continued education rather than for task training.

Benfield Conduit Bending Manual, Latest Edition. New York, NY: McGraw-Hill Publishing Company.

National Electrical Code Handbook, Latest Edition. Quincy, MA: National Fire Protection Association.

NOTES

The designations "National Electrical Code," "NE Code," and "NEC," where used in this document, refer to the *National Electrical Code®*, which is a registered trademark of the National Fire Protection Association, Quincy, MA. All National Electrical Code (NEC) references in this module refer to the 2002 edition of the NEC.

If you feel that additional math instruction would be helpful, Prentice Hall offers a basic math textbook entitled *Fundamentals of Electrical and Mechanical Mathematics.* It covers the basic math requirements for electrical trainees and may be ordered by contacting Prentice Hall Customer Service at 1-800-922-0579.

TEACHING TIME FOR THIS MODULE

An outline for use in developing your lesson plan is presented below. Note that each Roman numeral in the outline equates to one session of instruction. Each session has a suggested time period of 2½ hours. This includes 10 minutes at the beginning of each session for administrative tasks and one 10-minute break during the session. Approximately 12½ hours are suggested to cover *Raceways, Boxes, and Fittings.* You will need to adjust the time required for hands-on activity and testing based on your class size and resources.

Topic	Planned Time
Session I. Raceways; Conduit	
A. Raceways	____________
B. Conduit	____________
1. Conduit as a Ground Path	____________
2. Types of Conduit	____________
a. Electrical Metallic Tubing	____________
b. Rigid Metal Conduit	____________
c. Plastic-Coated RMC	____________
d. Aluminum Conduit	____________
e. Black Enamel Steel Conduit	____________
f. Intermediate Metal Conduit	____________
g. Rigid Nonmetallic Conduit	____________
h. Liquidtight Flexible Nonmetallic Conduit	____________
i. Flexible Metal Conduit	____________
Session II. Metal Conduit Fittings; Boxes; Bushings and Locknuts; Sealing Fittings	
A. Metal Conduit Fittings	____________
1. Couplings	____________
2. Conduit Bodies	____________
a. Type C Conduit Bodies	____________
b. Type L Conduit Bodies	____________
c. Type T Conduit Bodies	____________
d. Type X Conduit Bodies	____________
e. Threaded Weatherproof Hubs	____________

3. Insulating Bushings ____________
   a. Nongrounding Insulating Bushings ____________
   b. Grounding Insulating Bushings ____________
4. Offset Nipples ____________

B. Boxes ____________
1. Metal Boxes ____________
   a. Pryouts ____________
   b. Knockouts ____________
2. Nonmetallic Boxes ____________

C. Bushings and Locknuts ____________

D. Sealing Fittings ____________

Session III. Raceway Supports; Wireways; Field Trip

A. Raceway Supports ____________
1. Straps ____________
2. Standoff Supports ____________
3. Electrical Framing Channels ____________
4. Beam Clamps ____________

B. Wireways ____________
1. Auxiliary Gutters ____________
2. Types of Wireways ____________
   a. Wireway Fittings ____________
   b. Connectors ____________
   c. End Plates ____________
   d. Tees ____________
   e. Crosses ____________
   f. Elbows ____________
   g. Telescopic Fittings ____________
3. Wireway Supports ____________
   a. Suspended Hangers ____________
   b. Gusset Brackets ____________
   c. Standard Hangers ____________
   d. Wireway Hangers ____________
4. Other Types of Raceways ____________
   a. Surface Metal and Nonmetallic Raceways ____________
   b. Plugmold Multi-Outlet Systems ____________
   c. Pole Systems ____________
   d. Underfloor Systems ____________
   e. Cellular Metal Floor Raceways ____________
   f. Cellular Concrete Floor Raceways ____________

C. Field Trip ____________

If possible, take the trainees to a commercial/industrial job site where they may observe various types of wireway components and raceway systems.

Session IV. Cable Trays; Field Trip; Storing Raceways; Handling Raceways; Ducting

A. Cable Trays ____________

1. Cable Tray Fittings ____________
2. Cable Tray Supports ____________
   a. Direct Rod Suspension ____________
   b. Trapeze Mounting and Center Hung Support ____________
   c. Wall Mounting ____________
   d. Pipe Rack Mounting ____________

B. Field Trip ____________

If possible, take the trainees to a commercial/industrial job site where they may observe various types of cable tray systems.

C. Storing Raceways ____________

D. Handling Raceways ____________

E. Ducting ____________

Session V. Underground Systems; Making a Conduit-to-Box Connection; Laboratory; Various Construction Procedures; Module Examination and Performance Testing

A. Underground Systems ____________

1. Duct Line ____________
2. Duct Materials ____________
3. Rigid Nonmetallic Conduit ____________
4. Monolithic Concrete Duct ____________
5. Cable-in-Duct ____________

B. Making a Conduit-to-Box Connection ____________

C. Laboratory ____________

Under instructor supervision, have the trainees practice making a conduit-to-box connection.

D. Various Construction Procedures ____________

1. Masonry and Concrete Flush-Mount Construction ____________
2. Metal Stud Environment ____________
3. Wood Frame Environment ____________
4. Steel Environment ____________

E. Module Examination ____________

1. Trainees must score 70% or higher to receive recognition from the NCCER.
2. Record the testing results on Craft Training Report Form 200 and submit the results to the Training Program Sponsor.

F. Performance Testing ____________

1. Trainees must perform each task to the satisfaction of the instructor to receive recognition from the NCCER.
2. Record the testing results on Craft Training Report Form 200 and submit the results to the Training Program Sponsor.

Chapter 7

Module 26108-02

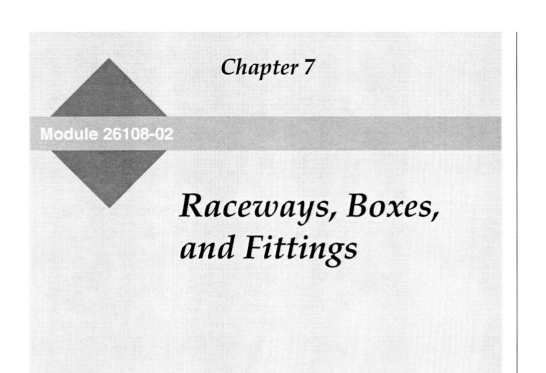

Raceways, Boxes, and Fittings

Course Map

This course map shows all of the modules in the first level of the Residential Electrical curriculum. The suggested training order begins at the bottom and proceeds up. Skill levels increase as a trainee advances on the course map. The training order may be adjusted by the local Training Program Sponsor.

RESIDENTIAL ELECTRICAL I

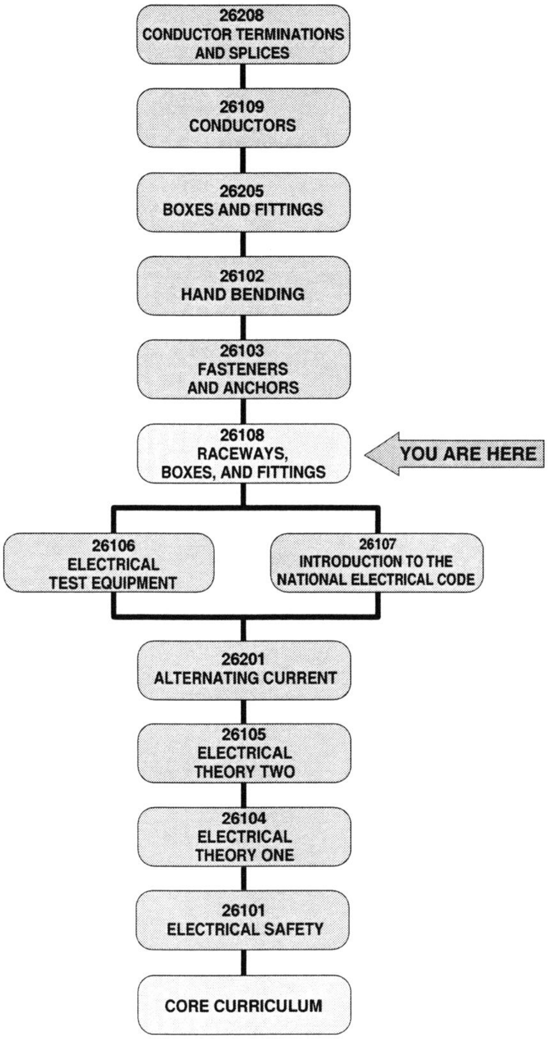

Homework

Assign reading of Module 26108.

RACEWAYS, BOXES, AND FITTINGS 253

MODULE 26108 CONTENTS

RACEWAYS, BOXES, AND FITTINGS 255

Instructor's Notes:

256 CHAPTER 7

Figures

Table

Instructor's Notes:

MODULE 26108

Raceways, Boxes, and Fittings

Objectives

When you have completed this module, you will be able to do the following:

1. Describe various types of cable trays and raceways.
2. Identify and select various types and sizes of raceways.
3. Identify and select various types and sizes of cable trays.
4. Identify and select various types of raceway fittings.
5. Identify various methods used to install raceways.
6. Demonstrate knowledge of NEC raceway requirements.
7. Describe procedures for installing raceways and boxes on masonry surfaces.
8. Describe procedures for installing raceways and boxes on concrete surfaces.
9. Describe procedures for installing raceways and boxes in a metal stud environment.
10. Describe procedures for installing raceways and boxes in a wood frame environment.
11. Describe procedures for installing raceways and boxes on drywall surfaces.
12. Recognize safety precautions that must be followed when working with boxes and raceways.

Prerequisites

Before you begin this module, it is recommended that you successfully complete the following: Core Curriculum; Residential Electrical I, Chapters 1 through 6.

Required Trainee Materials

1. Pencil and paper
2. Copy of the latest edition of the *National Electrical Code*
3. Appropriate personal protective equipment

1.0.0 ◆ INTRODUCTION

Electrical **raceways** present challenges and new requirements involving proper installation techniques, general understanding of raceway systems, and applications of the NEC to raceway systems. Acquiring quality installation skills for raceway systems requires practice, knowledge, and training.

A presentation of the various types of raceway systems and fittings, basic raceway installation skills, and NEC requirements applicable to raceway systems are included in this module. This module also covers raceway supports and environmental considerations for raceway systems, as well as general raceway information.

Note: The designations "National Electrical Code," "NE Code," and "NEC," where used in this document, refer to the *National Electrical Code*®, which is a registered trademark of the National Fire Protection Association, Quincy, MA. *All National Electrical Code (NEC) references in this module refer to the 2002 edition of the NEC.*

Materials

Ensure you have everything required to teach the course. Check the Materials and Equipment List at the front of this Instructor's Guide.

Audiovisual

Show Transparency 1, Course Objectives.

Show Transparency 2, Performance Profile Tasks.

Classroom

Introduce raceway systems.

Explain that items shown in bold (blue) are defined in the Glossary at the back of this module.

Classroom

Review the NEC requirements for raceway systems.

Define *raceway.*

Define *conduit.*

Explain how equipment is grounded.

Along with the study of this module, the following NEC Articles should be referenced:

- *NEC Article 250—Grounding*
- *NEC Article 342—Intermediate Metal Conduit*
- *NEC Article 344—Rigid Metal Conduit*
- *NEC Article 348—Flexible Metal Conduit*
- *NEC Article 350—Liquidtight Flexible Metal Conduit*
- *NEC Article 352—Rigid Nonmetallic Conduit*
- *NEC Article 356—Liquidtight Flexible Nonmetallic Conduit*
- *NEC Article 358—Electrical Metallic Tubing*
- *NEC Article 376—Metal Wireways*
- *NEC Article 378—Nonmetallic Wireways*
- *NEC Article 392—Cable Trays*

> **Note**
> Mandatory rules in the NEC are characterized by the use of the word *shall.* Explanatory material is in the form of Fine Print Notes (FPNs). When referencing specific sections of the NEC, always check to see if any exceptions apply.

2.0.0 ◆ RACEWAYS

Raceway is a general term referring to a wide range of circular and rectangular enclosed channels used to house electrical wiring. Raceways can be metallic or nonmetallic and come in different shapes. Depending on the particular purpose for which they are intended, raceways include enclosures such as underfloor raceways, flexible metal **conduit, wireways,** surface metal raceways, and surface nonmetallic raceways and support systems such as cable trays.

3.0.0 ◆ CONDUIT

Conduit is a raceway with a circular cross section, similar to pipe, that contains wires or cables. Conduit is used to provide protection for **conductors** and route them from one place to another. In addition, conduit makes it easier to replace or add wires to existing structures. Metal conduit also provides a permanent electrical path to ground. This equipment should be listed per the NEC.

3.1.0 Conduit as a Ground Path

For safety reasons, most equipment that receives electrical power and has a metallic frame is grounded. In order to ground the equipment, an electrical connection must be made to connect the metal frame of the electrically powered equipment to the grounding point at the service-entrance equipment. This is usually done in one or both of the following ways:

- The frame of the equipment is connected to a wire (equipment grounding conductor), which is directly connected to the ground point at the grounding terminal.
- The frame of the equipment is connected (bonded) to a metal conduit or other type of raceway system, which provides an uninterrupted and low-impedance circuit to the ground point at the service-entrance equipment. The metal raceway or conduit acts as the equipment grounding conductor.

> **Note**
> According to ***NEC Section 250.96,*** metal raceways, cable trays, cable armor, cable sheath, enclosures, frames, fittings, and other metal noncurrent-carrying parts that are to serve as grounding conductors with or without the use of supplementary equipment grounding conductors shall be effectively bonded where necessary to ensure electrical continuity and the capacity to safely conduct any fault current likely to be imposed on them. The purpose of the equipment grounding conductor is to provide a low resistance path to ground for all equipment that receives power. This is done so that if an ungrounded conductor comes in contact with the frame of a piece of equipment, the circuit overcurrent device immediately acts to open the circuit. It also reduces the voltage to ground that would be present on the faulted equipment if a person came in contact with the equipment frame.

3.2.0 Types of Conduit and Tubing

There are many types of conduit used in the construction industry. The size of conduit to be used is determined by engineering specifications, local codes, and the NEC. Refer to *NEC Chapter 9, Tables 1 through 8 and Annex C* for conduit fill with various conductors. There are several common types of conduit to examine.

3.2.1 Electrical Metallic Tubing

Electrical metallic tubing (EMT) is the lightest duty tubing available for enclosing and protecting electrical wiring. EMT is widely used for residential,

Instructor's Notes:

commercial, and industrial wiring systems. It is lightweight, easily bent and/or cut to shape, and is the least costly type of metallic conduit. Because the wall thickness of EMT is less than that of rigid conduit, it is often referred to as *thinwall conduit.* A comparison of inside and outside diameters of EMT to rigid metal conduit and intermediate metal conduit (IMC) is shown in *Figure 1.*

NEC Section 358.10(A) permits the installation of EMT for either exposed or concealed work where it will not be subject to severe physical damage during installation or after construction. Installation of EMT is permitted in wet locations such as outdoors or indoors in dairies, laundries, and canneries using waterproof fittings.

Note

Refer to ***NEC Section 358.12*** for restrictions that apply to the use of EMT.

EMT shall not be used (1) where, during installation or afterward, it will be subject to severe physical damage; (2) where protected from corrosion solely by enamel; (3) in cinder concrete or cinder fill where subject to permanent moisture unless protected on all sides by a layer of non-cinder concrete at least two inches thick or unless the tubing is at least 18 inches under the fill; (4) in any hazardous (classified) locations except as permitted by *NEC Sections 502.4, 503.3, and 504.20;* or (5) for the support of fixtures or other equipment.

In a wet area, EMT and other conduit must be installed to prevent water from entering the conduit system. In locations where walls are subject to regular wash-down (see *NEC Section 300.6*), the entire conduit system must be installed to provide a ¼-inch air space between it and the wall or supporting surface. The entire conduit system is considered to include conduit, boxes, and fittings. To ensure resistance to corrosion caused by wet environments, EMT is galvanized. The term *galvanized* is used to describe the procedure in which the interior and exterior of the conduit are coated with a corrosion-resistant zinc compound.

EMT, being a good conductor of electricity, may be used as an equipment grounding conductor. In order to qualify as an equipment grounding conductor [see *NEC Section 250.118(4)*], the conduit system must be tightly connected at each joint and provide a continuous grounding path from each electrical load to the service equipment. The connectors used in an EMT system ensure electrical and mechanical continuity throughout the system (see *NEC Sections 250.96, 300.10, and 358.42*).

EMT fittings are manufactured in two basic types. One type of fitting is the compression coupling. (See *Figure 2.*)

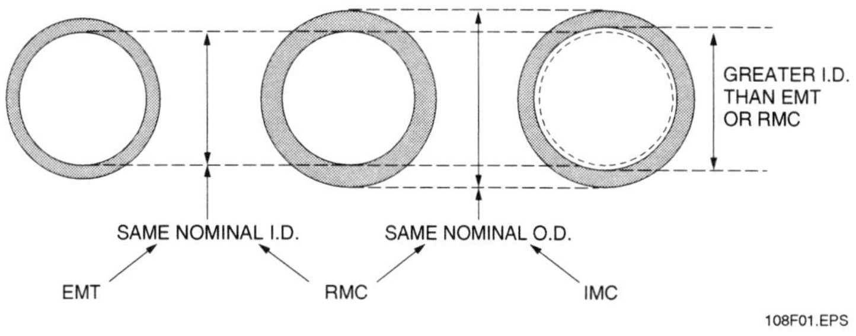

Figure 1 ◈ Conduit comparison.

EMT Use

THINK ABOUT IT

Where would you use EMT? Are there any circumstances where EMT cannot be run through a suspended ceiling? What are some differences between EMT and rigid conduit?

Classroom

Discuss EMT and its applications. Pass around examples for the trainees to examine.

Audiovisual

Show Transparency 3 (Figure 1).

Classroom

Stress NEC requirements that apply to EMT.

Teaching Tip

Discuss the "Think About It."

RACEWAYS, BOXES, AND FITTINGS 261

Audiovisual

Show Transparency 4 (Figure 2).

Classroom

Discuss EMT fittings. Pass around examples for the trainees to examine.

Discuss rigid metal conduit and its applications. Pass around an example for the trainees to examine.

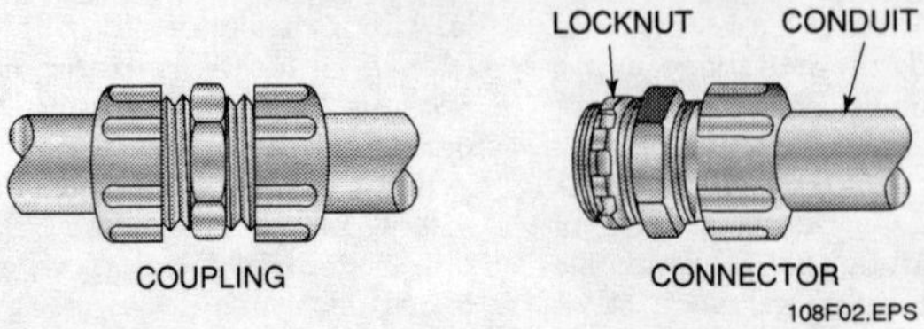

Figure 2 ◆ Compression fittings.

> **Note**
>
> Support requirements for EMT are also covered in ***NEC Section 358.30.*** The types of supports will be discussed later in this training module.

Because EMT is too thin for threads, special fittings must be used. For wet or damp locations, compression fittings such as those shown in *Figure 2* are used. These fittings contain a compression ring made of metal that forms a watertight seal.

When EMT compression couplings are used, they must be securely tightened, and when installed in masonry concrete, they must be of the concrete-tight type. If installed in a wet location, they must be the raintight type. Refer to ***NEC Section 358.42.***

EMT fittings for dry locations can be either the setscrew type or the indenting type. To use the setscrew type, the ends of the EMT are inserted into the sleeve and the setscrews are tightened to make the connection. Various types of setscrew fittings are shown in *Figure 3.*

EMT sizes of 2½ inches and larger have the same outside diameter as corresponding sizes of galvanized rigid metal conduit (RMC). RMC threadless connectors may be used to connect EMT.

> **Note**
>
> EMT connectors, although they are the same size as RMC threadless connectors, may not be used to connect RMC.

Both setscrew and compression couplings are available in die-cast or steel construction. Steel couplings are stronger than die-cast couplings and have superior quality.

Support requirements for EMT are presented in ***NEC Section 358.30.*** As with most other metal conduit, EMT must be supported at least every 10 feet and within 3 feet of each outlet box, junction box, cabinet, fitting, or terminating end of the conduit. An exception to ***NEC Section 358.30(A), Exception 1*** allows the fastening of unbroken lengths of EMT to be increased to a distance of 5 feet (1.52m) where structural members do not readily permit fastening within 3 feet (914mm).

Figure 3 ◆ Setscrew fittings.

Electrical nonmetallic tubing (ENT) is also available. It provides an economical alternative to EMT, but it can only be used in certain applications. See ***NEC Article 362.***

3.2.2 Rigid Metal Conduit

Rigid metal conduit is conduit that is constructed of metal having sufficient thickness to permit the cutting of pipe threads at each end. Rigid metal conduit provides the best physical protection for conductors of any of the various types of conduit. Rigid metal conduit is supplied in 10-foot lengths including a threaded coupling on one end.

Rigid metal conduit may be fabricated from steel or aluminum. Rigid metal steel conduit may be galvanized, or enamel-coated inside and out.

Instructor's Notes:

262 CHAPTER 7

EMT Installation

INSIDE TRACK

When installing EMT, hook your index finger up through the box to check that the conduit is seated in the connector. If you feel a lip between the conduit and the connector, the conduit is not properly seated.

108PO801.EPS

> **Note**
> Specific information on rigid metal conduit may be found in ***NEC Article 344.***

108F04.EPS

Figure 4 ◆ Rigid metal conduit.

Because of its threaded fittings, rigid metal conduit provides an excellent equipment grounding conductor as defined in *NEC Section 250.118(2)*. A piece of rigid metal conduit is shown in *Figure 4*. The support requirements for rigid metal conduit are presented in *NEC Table 344.30(B)(2)*.

RMC is mostly used in industrial applications. RMC is heavier than EMT and IMC. It is more difficult to cut and bend, usually requires threading of cut ends, and has a higher purchase price

RACEWAYS, BOXES, AND FITTINGS 263

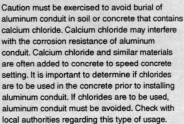

Rigid Metal Conduit Installations

INSIDE TRACK

Use rigid metal conduit in hazardous environments such as feed mills or in areas such as outdoor environments where there is a chance of physical abuse or extreme moisture. The NEC does allow EMT to be buried in the ground or in concrete, but galvanized rigid metal conduit is more commonly used.

Classroom

Discuss plastic-coated RMC and its applications. Pass around an example for the trainees to examine.

Discuss aluminum conduit and its applications. Pass around an example for the trainees to examine.

Discuss black enamel steel conduit and its applications. Pass around an example for the trainees to examine.

than EMT and IMC. As a result, the cost of installing RMC is generally higher than the cost of installing EMT and IMC.

3.2.3 Plastic-Coated RMC

Plastic-coated RMC has a thin coating of PVC over the RMC. This combination is useful when an environment calls for the ruggedness of RMC along with the corrosion resistance of RNC. Typical installations where plastic-coated RMC may be required are:

- Chemical plants
- Food plants
- Refineries
- Fertilizer plants
- Paper mills
- Wastewater treatment plants

Plastic-coated RMC requires special threading and bending techniques.

3.2.4 Aluminum Conduit

Aluminum conduit has several characteristics that distinguish it from steel conduit. Because it has better resistance to wet environments and some chemical environments, aluminum conduit generally requires less maintenance in installations such as sewage treatment plants.

Direct burial of aluminum conduit results in a self-stopping chemical reaction on the conduit surface, which forms a coating on the conduit. This coating acts to prevent further corrosion, increasing the life of the installation.

Note

Caution must be exercised to avoid burial of aluminum conduit in soil or concrete that contains calcium chloride. Calcium chloride may interfere with the corrosion resistance of aluminum conduit. Calcium chloride and similar materials are often added to concrete to speed concrete setting. It is important to determine if chlorides are to be used in the concrete prior to installing aluminum conduit. If chlorides are to be used, aluminum conduit must be avoided. Check with local authorities regarding this type of usage.

Since aluminum conduit is lighter than steel conduit, there are some installation advantages to using aluminum. For example, a 10-foot section of 3-inch aluminum conduit weighs about 23 pounds, compared to the 68-pound weight of its steel counterpart.

3.2.5 Black Enamel Steel Conduit

Rigid black enamel steel conduit (often called *black conduit*) is steel conduit that is coated with a black enamel. In the past, this type of conduit was used exclusively for indoor wiring. Black enamel steel conduit is no longer manufactured for sale in the United States. It is mentioned only because it may still be found in existing installations.

Use of Aluminum Conduit

INSIDE TRACK

Aluminum conduit is used for special purposes such as high-cycle lines (400 cycles or above); around cooling towers, food service areas, and other applications in which corrosion is a factor; or where magnetic induction is a concern, such as near magnetic resonance imaging (MRI) equipment in hospitals.

Instructor's Notes:

264 CHAPTER 7

3.2.6 Intermediate Metal Conduit

Intermediate metal conduit (IMC) has a wall thickness that is less than rigid metal conduit but greater than that of EMT. The weight of IMC is approximately ⅔ that of rigid metal conduit. Because of its lower purchase price, lighter weight, and thinner walls, IMC installations are generally less expensive than comparable rigid metal conduit installations. However, IMC installations still have high strength ratings.

> **Note**
> Additional information on intermediate metal conduit may be found in ***NEC Article 342.***

The outside diameter of a given size of IMC is the same as that of the comparable size of rigid metal conduit. Therefore, rigid metal conduit fittings may be used with IMC. Since the threads on IMC and rigid metal conduit are the same size, no special threading tools are needed to thread IMC. Some electricians feel that threading IMC is more difficult than threading rigid metal conduit because IMC is somewhat harder.

The internal diameter of a given size of IMC is somewhat larger than the internal diameter of the same size of rigid metal conduit because of the difference in wall thickness. Bending IMC is considered easier than bending rigid metal conduit because of the reduced wall thickness. However, bending is sometimes complicated by kinking, which may be caused by the increased hardness of IMC.

The NEC requires that intermediate metal conduit be identified along its length at 5-foot intervals with the letters *IMC*. ***NEC Sections 110.21 and 342.120*** describe this marking requirement.

Like RMC, IMC is permitted to act as an equipment grounding conductor, as defined in ***NEC Section 250.118(3).*** The use of IMC may be restricted in some jurisdictions. It is important to investigate the requirements of each jurisdiction before selecting any materials.

3.2.7 Rigid Nonmetallic Conduit

The most common type of rigid nonmetallic conduit (RNC) is manufactured from polyvinyl chloride (PVC). Because RNC is noncorrosive, chemically inert, and non-aging, it is often used for installation in wet or corrosive environments. Corrosion problems found with steel and aluminum rigid metal conduit do not occur with RNC. However, RNC may deteriorate under some conditions, such as extreme sunlight, unless marked sunlight resistant.

All RNC is marked according to standards established by the National Electrical Manufacturers' Association (NEMA) or **Underwriters Laboratories (UL).** A section of RNC is shown in *Figure 5.*

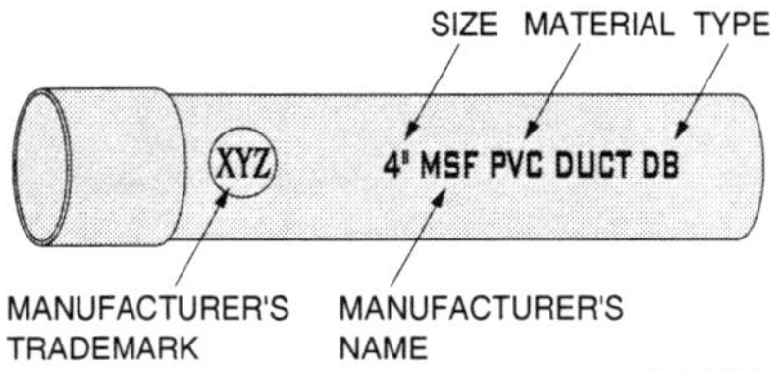

Figure 5 ◈ Rigid nonmetallic conduit.

Since RNC is lighter than steel or aluminum rigid conduit, IMC, or EMT, it is considered easier to handle. RNC can usually be installed much faster than other types of conduit because the joints are made up with cement and require no threading.

RNC contains no metal. This characteristic reduces the voltage drop of conductors carrying alternating current in RNC compared to identical conductors in steel conduit.

Because RNC is nonconducting, it cannot be used as an equipment grounding conductor. An equipment grounding conductor sized in accordance with ***NEC Table 250.122*** must be pulled in each RNC conductor run (except for underground service-entrance conductors).

RNC is available in lengths up to 20 feet. However, some jurisdictions require it to be cut to 10-foot lengths prior to installation. RNC is subject to expansion and contraction directly related to the difference in temperature, plus any radiating effects on the conduit. In moderate climates, even a 10-foot installation of RNC would require an expansion joint per the NEC. Each straight section of conduit run must be treated independently from other sections when connected by elbows. To avoid damage to RNC caused by temperature changes, expansion couplings are used. (See *Figure 6.*) The inside of the coupling is sealed with one or more O-rings. This type of coupling may allow up to six inches of movement. Check the requirements of the local jurisdiction prior to installing RNC.

Classroom

Discuss IMC and its applications. Pass around an example for the trainees to examine.

Audiovisual

Show Transparency 5 (Figure 5). Discuss RNC and its applications. Pass around examples of Types EB and DB for the trainees to examine.

Show Transparency 6 (Figure 6). Discuss RNC expansion couplings.

RACEWAYS, BOXES, AND FITTINGS 265

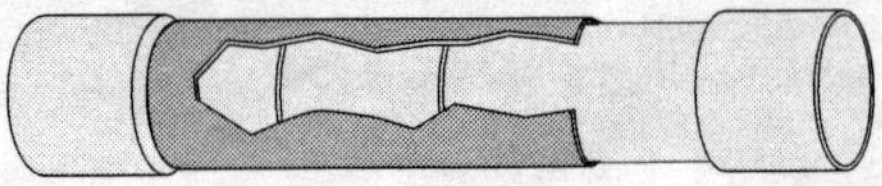

108F06.EPS

Figure 6 ◆ RNC expansion coupling.

RNC is manufactured in two types:

- *Type EB*—Thin wall for underground use only when encased in concrete. Also referred to as *Type I.*
- *Type DB*—Thick wall for underground use without encasement in concrete. Also referred to as *Type II.*

Type DB is available in two wall thicknesses, Schedule 40 and Schedule 80.

- Schedule 40 is heavy wall for direct burial in the earth and above-ground installations.
- Schedule 80 is extra heavy wall for direct burial in the earth, above-ground installations for general applications, and installations where the conduit is subject to physical damage.

RNC is affected by higher-than-usual ambient temperatures. Support requirements for RNC are found in ***NEC Table 352.30(B).*** As with other conduit, it must be supported within three feet of each termination, but the maximum spacing between supports depends upon the size of the conduit. Some of the regulations for the maximum spacing of supports are:

- ½- to 1-inch conduit: every 3 feet
- 1¼- to 2-inch conduit: every 5 feet
- 2½- to 3-inch conduit: every 6 feet
- 3½- to 5-inch conduit: every 7 feet
- 6-inch conduit: every 8 feet

3.2.8 Liquidtight Flexible Nonmetallic Conduit

Liquidtight flexible nonmetallic conduit (LFNC) was developed as a raceway for industrial equipment where flexibility was required and protection of conductors from liquids was also necessary. This is covered by ***NEC Article 356.*** Usage of LFNC has been expanded from industrial applications to outside and direct burial usage where listed.

Several varieties of LFNC have been introduced. The first product (LFNC-A) is commonly referred to as *hose.* It consists of an inner and outer layer of neoprene with a nylon reinforcing web between the layers. A second-generation product (LFNC-B), and most widely used, consists of a smooth wall, flexible PVC with a rigid PVC integral reinforcement rod. The third product (LFNC-C) is a nylon corrugated shape without any integral reinforcements. These three permitted LFNC raceway designs must be flame resistant with fittings **approved** for installation of electrical conductors. Nonmetallic connectors are listed for use and some liquidtight metallic flexible conduit connectors are dual-listed for both metallic and nonmetallic liquidtight flexible conduit.

LFNC is sunlight-resistant and suitable for use at conduit temperatures of 80°C dry and 60°C wet. It is available in ⅜-inch through 4-inch sizes. ***NEC Section 350.12*** states that LFNC cannot be used where subject to physical damage or LFNC-A in lengths longer than six feet, except where properly secured, where flexibility is required, or as permitted by ***NEC Section 350.10.*** Also, it cannot be used to contain conductors in excess of 600 volts nominal except as permitted by ***NEC Section 600.7(A).***

Liquidtight flexible metal conduit is a raceway of circular cross section having an outer liquidtight, nonmetallic, sunlight-resistant jacket over an inner flexible metal core with associated couplings and connectors covered by ***NEC Article 350.***

Flex connectors are used to connect flexible conduit to boxes or equipment. They are available in straight, 45°, and 90° configurations (*Figure 7*).

3.2.9 Flexible Metal Conduit

Flexible metal conduit, also called *flex,* may be used for many kinds of wiring systems. Flexible metal conduit is made from a single strip of steel or aluminum, wound and interlocked. It is typically available in sizes from ⅜ inch to 4 inches in

THINK ABOUT IT

Applications of RNC

What installations would be suitable for the use of RNC? For what situations would RNC be a poor choice?

Classroom

Discuss LFNC and its applications. Pass around an example for the trainees to examine.

Audiovisual

Show Transparency 7 (Figure 7).

Teaching Tip

Discuss the "Think About It."

Instructor's Notes:

266 CHAPTER 7

INSIDE TRACK

Liquidtight Conduit

Liquidtight conduit protects conductors from vapors, liquids, and solids. Liquidtight conduit that includes an inner metal core is widely used in commercial construction.

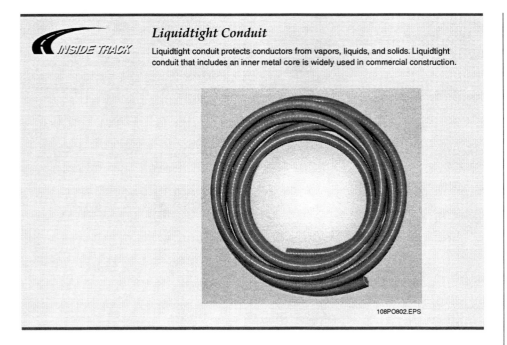

108PO802.EPS

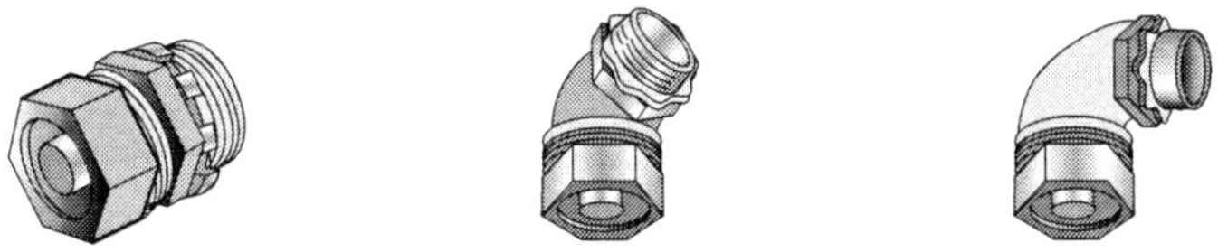

108F07.EPS

Figure 7 ◆ Flex connectors.

diameter. An illustration of flexible metal conduit is shown in *Figure 8.*

Flexible metal conduit is often used to connect equipment or machines that vibrate or move slightly during operation. Also, final connection to equipment having an electrical connection point that is marginally **accessible** is often accomplished with flexible metal conduit.

Flexible metal conduit is easily bent, but the minimum bending radius is the same as for other types of conduit. It should not be bent more than the equivalent of four quarter bends (360° total) between pull points (e.g., conduit bodies and boxes). It can be connected to boxes with a flexible conduit connector and to rigid conduit or EMT by using a combination coupling.

Two types of combination couplings are shown in *Figure 9.*

Flexible metal conduit is generally available in two types: nonliquidtight and liquidtight. *NEC Articles 348 and 350* cover the uses of flexible metal conduit.

Liquidtight flexible metal conduit has an outer covering of liquidtight, sunlight-resistant flexible

Classroom

Discuss flexible metal conduit and its applications. Pass around an example for the trainees to examine.

Audiovisual

Show Transparency 8 (Figure 9).

RACEWAYS, BOXES, AND FITTINGS 267

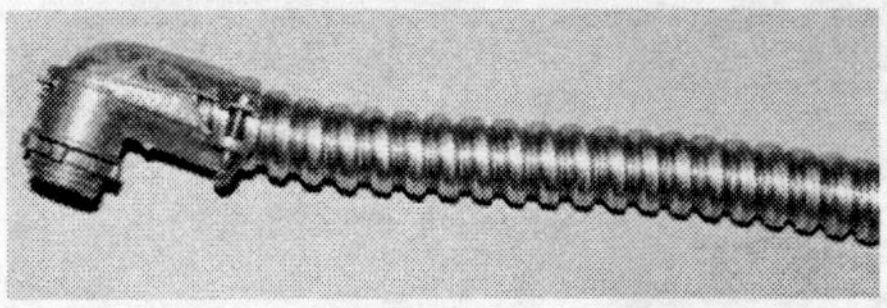

108F08.EPS

Figure 8 ◆ Flexible metal conduit.

> **Note**
> When using a combination coupling, be sure the flexible conduit is pushed as far as possible into the coupling. This covers the end and protects the conductors from damage.

Classroom

Discuss couplings and their applications. Pass around examples for the trainees to examine.

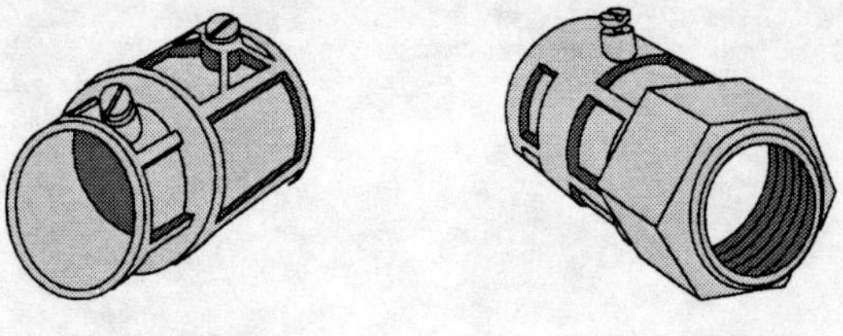

108F09.EPS

Figure 9 ◆ Combination couplings.

material that acts as a moisture seal. It is intended for use in wet locations. It is used primarily for equipment and motor connections when movement of the equipment is likely to occur. The number of bends, size, and support requirements for liquidtight conduit are the same as for all flexible conduit. Fittings used with liquidtight conduit must also be of the liquidtight type.

Support requirements for flexible metal conduit are found in ***NEC Sections 348.30 and 350.30.*** Straps or other means of securing the flexible metal conduit must be spaced every 4½ feet and within 12 inches of each end. (This spacing is closer together than for rigid conduit.) However, at terminals where flexibility is necessary, lengths of up to 36 inches without support are permitted. Failure to provide proper support for flexible conduit can make pulling conductors difficult.

4.0.0 ◆ METAL CONDUIT FITTINGS

A large variety of conduit fittings are available to do electrical work. Manufacturers design and construct fittings to permit a multitude of applications. The type of conduit fitting used in a particular application depends upon the size and type of conduit, the type of fitting needed for the application, the location of the fitting, and the installation method. The requirements and proper applications of boxes and fittings (conduit bodies)

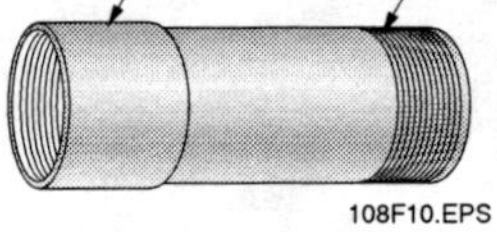

108F10.EPS

Figure 10 ◆ Conduit and coupling.

are found in ***NEC Section 300.15.*** Some of the more common types of fittings are examined in the following sections.

4.1.0 Couplings

Couplings are sleeve-like fittings that are typically threaded inside to join two male threaded pieces of rigid conduit or IMC. A piece of conduit with a coupling is shown in *Figure 10.*

Other types of couplings may be used depending upon the location and type of conduit. Several types are shown in *Figure 11.*

4.2.0 Conduit Bodies

Conduit bodies are a separate portion of a conduit or tubing system that provide access through a removable cover(s) to the interior of the system at a

Audiovisual

Show Transparencies 9 and 10 (Figures 10 and 11).

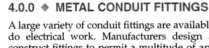

Instructor's Notes:

268 CHAPTER 7

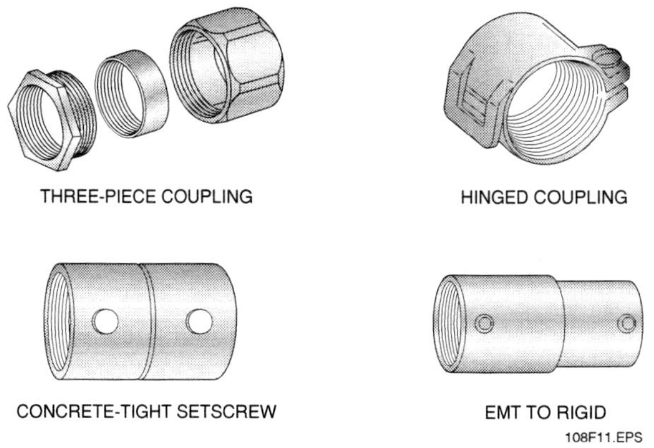

Figure 11 ◆ Metal conduit couplings.

junction of two or more sections of the system, a pull point, or at a terminal point of the system. They are usually cast and are significantly higher in cost than the stamped steel boxes permitted with EMT. However, there are situations in which conduit bodies are preferable, such as in outdoor locations, for appearance's sake in an **exposed location,** or to change types or sizes of raceways. Also, conduit bodies do not have to be supported, as do stamped steel boxes. They are also used when elbows or bends would not be appropriate.

NEC Section 314.16(C)(2) states that conduit bodies cannot contain **splices, taps,** or devices unless they are durably and legibly marked by the manufacturer with their cubic inch capacity. The maximum number of conductors permitted in a conduit body is found using *NEC Table 314.16(B).* (See *Table 1.*)

Table 1 Volume Required per Conductor

Size of Conductor	Free Space Within Box for Each Conductor
No. 18	1.5 cubic inches
No. 16	1.75 cubic inches
No. 14	2.0 cubic inches
No. 12	2.25 cubic inches
No. 10	2.5 cubic inches
No. 8	3.0 cubic inches
No. 6	5.0 cubic inches

4.2.1 Type C Conduit Bodies

Type C conduit bodies may be used to provide a pull point in a long conduit run or a conduit run that has bends totaling more than 360°. A Type C conduit body is shown in *Figure 12.*

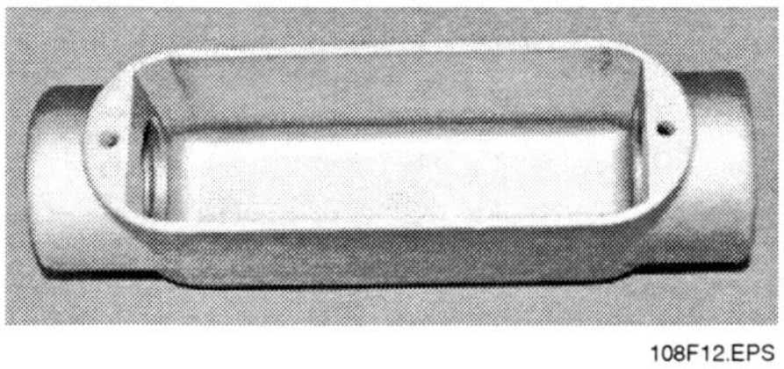

Figure 12 ◆ Type C conduit body.

4.2.2 Type L Conduit Bodies

When referring to conduit bodies, the letter *L* represents an *elbow.* A Type L conduit body is used as a pulling point for conduit that requires a 90° change in direction. The cover is removed, then the wire is pulled out, coiled on the ground or floor, reinserted into the other conduit body's opening, and pulled. The cover and its associated gasket are then replaced. Type L conduit bodies are available with the cover on the back (Type LB), on the sides (Type LL or LR), or on both sides (Type LRL). Several Type L conduit bodies are shown in *Figure 13.*

Classroom

Explain how to find the maximum number of conductors permitted in a conduit body. Pass around various conduit bodies for the trainees to examine.

Discuss Type C conduit bodies and their applications.

Discuss Type L conduit bodies and their applications.

RACEWAYS, BOXES, AND FITTINGS 269

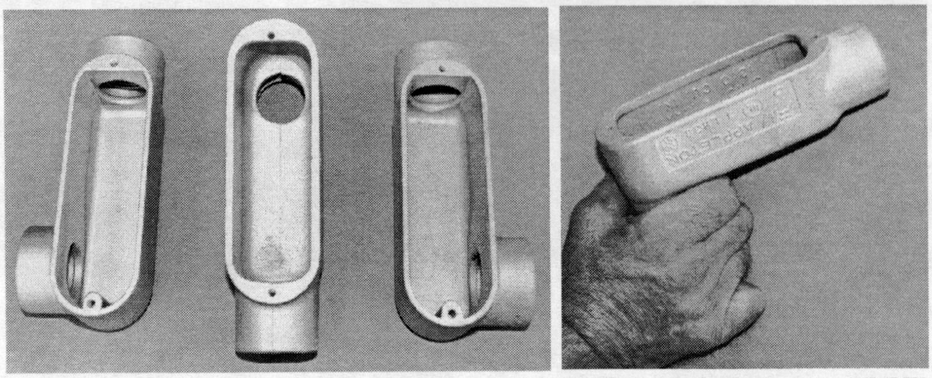

108F13A.EPS 108F13B.EPS

Figure 13 ◆ Type L conduit bodies and how to identify them.

Classroom

Explain how to identify Type L conduit bodies.

Discuss Type T conduit bodies and their applications.

Discuss Type X conduit bodies and their applications.

Discuss weatherproof hubs and their applications. Pass around examples for the trainees to examine.

> **Note**
> The cover and gasket must be ordered separately. Do not assume that these parts come with conduit bodies when they are ordered.

To identify Type L conduit bodies, use the following method:

Step 1 Hold the body like a pistol.

Step 2 Locate the opening on the body:
- If the opening is to the left, it is a Type LL.
- If the opening is to the right, it is a Type LR.
- If the opening is on top (back), it is a Type LB.
- If there are openings on both the left and the right, it is a Type LRL.

4.2.3 Type T Conduit Bodies

Type T conduit bodies are used to provide a junction point for three intersecting conduits and are used extensively in conduit systems. A Type T conduit body is shown in *Figure 14*.

4.2.4 Type X Conduit Bodies

Type X conduit bodies are used to provide a junction point for four intersecting conduits. The removable cover provides access to the interior of the X so that wire pulling and splicing may be performed. A Type X conduit body is shown in *Figure 15*.

108F14.EPS

Figure 14 ◆ Type T conduit body.

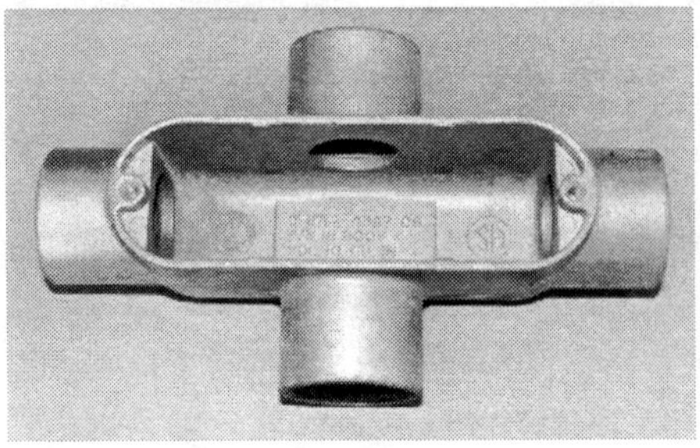

108F15.EPS

Figure 15 ◆ Type X conduit body.

4.2.5 Threaded Weatherproof Hub

Threaded weatherproof hubs are used for conduit entering a box in a wet location. *Figure 16* shows a typical threaded weatherproof hub.

Instructor's Notes:

270 CHAPTER 7

INSIDE TRACK

Installation of Conduit Bodies

It will be much easier to identify conduit bodies once you begin to see them in use. Here we show liquidtight nonmetallic conduit entering a Type T conduit body (A) and a Type LB conduit body in an outdoor commercial application (B).

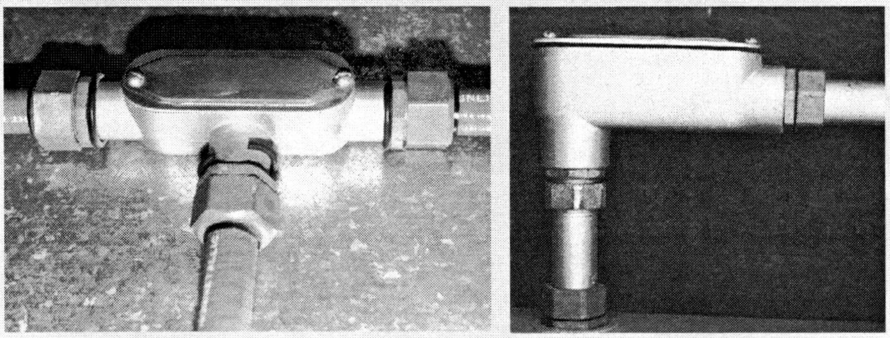

(A) 108PO803.EPS (B) 108PO804.EPS

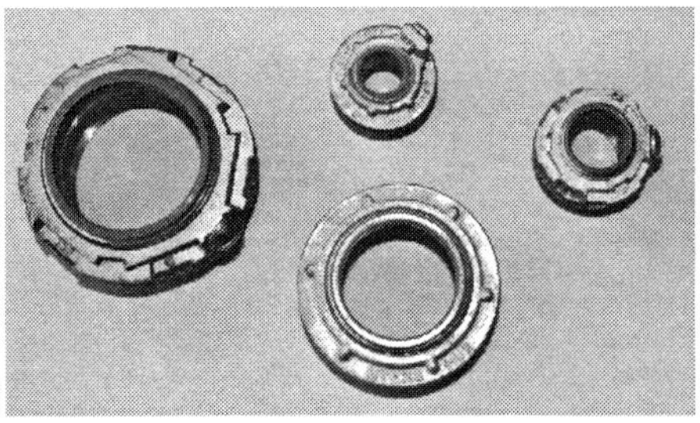

108F16.EPS

Figure 16 ◆ Threaded weatherproof hubs.

4.3.0 Insulating Bushings

An insulating bushing is either made of nonmetal or has an insulated throat. Insulating bushings are installed on the threaded end of conduit that enters a sheet metal enclosure.

4.3.1 Nongrounding Insulating Bushings

The purpose of a nongrounding insulating bushing is to protect the conductors from being damaged by the sharp edges of the threaded conduit end. *NEC Chapter 3* states that where a conduit enters a box, fitting, or other enclosure, a bushing must be provided to protect the wire from abrasion unless the design of the box, fitting, or enclosure is such as to afford equivalent protection. *NEC Section 312.6(C)* references *Section 300.4(F),* which states that where ungrounded conductors of No. 4 or larger enter a raceway in a cabinet or box enclosure, the conductors shall be protected by a substantial fitting providing a smoothly rounded insulating surface, unless the conductors are separated from the raceway fitting by substantial insulating material securely fastened in place. An exception is where threaded hubs or bosses that are an integral part of a cabinet, box enclosure, or raceway provide a smoothly rounded or flared entry for conductors. Insulating bushings are shown in *Figure 17.*

4.3.2 Grounding Insulating Bushings

Grounded insulating bushings, usually called *grounding bushings,* are used to protect conductors and also have provisions for connection of an equipment grounding conductor. The ground wire, once connected to the grounding bushing, may be connected to the enclosure to which the conduit is connected. Grounding insulating bushings are shown in *Figure 18.*

4.4.0 Offset Nipples

Offset nipples are used to connect two pieces of electrical equipment in close proximity where a slight offset is required. They come in sizes ranging from ½" to 2" in diameter. See *Figure 19.*

Classroom

Discuss nongrounding and grounding insulating bushings and their applications. Pass around examples for the trainees to examine.

Discuss offset nipples and their applications. Pass around examples for the trainees to examine.

RACEWAYS, BOXES, AND FITTINGS 271

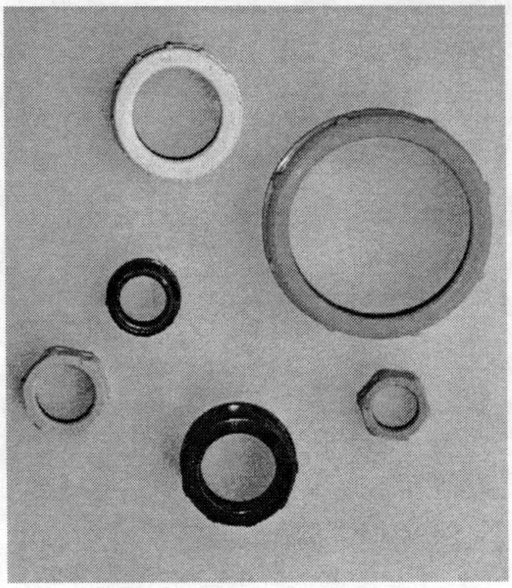

108F17.EPS

Figure 17 ◆ Insulating bushings.

108F18.EPS

Figure 18 ◆ Grounding insulating bushings.

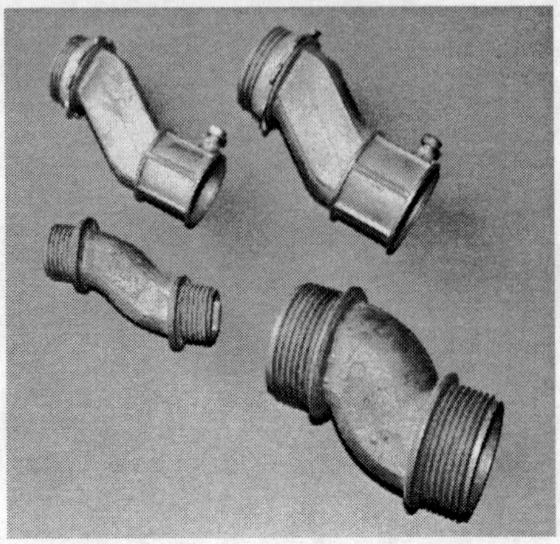

108F19.EPS

Figure 19 ◆ Offset nipples.

5.0.0 ◆ BOXES

A box is installed at each outlet, switch, or junction point for all wiring installations and branch circuits. Boxes are made from either metallic or nonmetallic material. *Figure 20* shows various boxes used in raceway systems.

5.1.0 Metal Boxes

Metal boxes are made from sheet steel. The surface is galvanized to resist corrosion and provide a continuous ground. Refer to ***NEC Section 314.40*** for information on thickness and grounding pro-

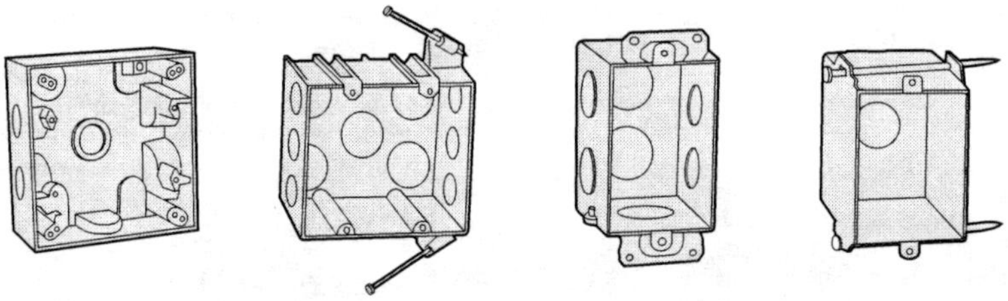

108F20.EPS

Figure 20 ◆ Various boxes used in raceway systems.

Audiovisual

Show Transparency 11 (Figure 20).

Classroom

Discuss metal boxes and their applications. Pass around examples for the trainees to examine.

Instructor's Notes:

272 CHAPTER 7

visions. Metal boxes are made with removable circular sections called *pryouts* or *knockouts*. These circular sections are removed to make openings for conduit or cable connections.

5.1.1 Pryouts

In a pryout, a section is cut completely through the metal but only part of the way around, leaving solid metal tabs at two points. A slot is cut in the center of the pryout. To remove the pryout, a screwdriver is inserted into the slot and twisted to break the solid tabs (*Figure 21*).

5.1.2 Knockouts

Knockouts are pre-punched circular sections that do not include a pryout slot. The knockout is easily removed when sharply hit by a hammer and punch, as shown in *Figure 22.*

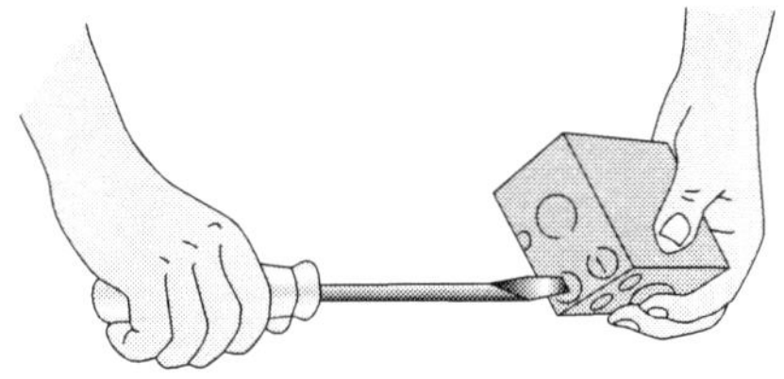
108F21.EPS

Figure 21 ◈ Pryout removal.

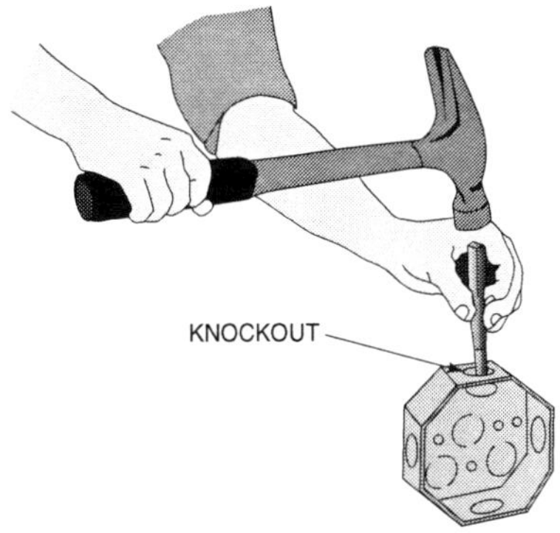

108F22.EPS

Figure 22 ◈ Knockout removal.

Often conduit must enter boxes, cabinets, or panels that do not have pre-cut knockouts. In these cases, a knockout punch can be used to make a hole for the conduit connection. A knockout punch kit is shown in *Figure 23.*

5.2.0 Nonmetallic Boxes

Nonmetallic boxes are made of PVC or Bakelite (a fiber-reinforced plastic). Nonmetallic boxes are often used in corrosive environments. ***NEC Section 314.3*** covers the use of nonmetallic boxes and the types of conduit, fittings, and grounding requirements for specific applications.

6.0.0 ◈ BUSHINGS AND LOCKNUTS

Conduit is joined to boxes by connectors, adapters, threaded hubs, or locknuts.

Bushings protect the wires from the sharp edges of the conduit. Bushings are usually made of plastic or metal. Some metal bushings have a grounding screw to permit a **bonding wire** to be installed. Some different types of plastic and metal bushings are shown in *Figure 24.*

Locknuts are used on the inside and outside walls of the box to which the conduit is connected.

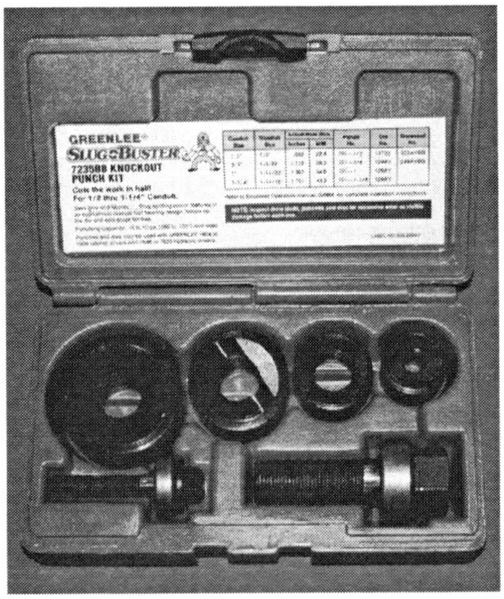

108F23.EPS

Figure 23 ◈ Knockout punch kit.

Audiovisual

Show Transparencies 12 and 13 (Figures 21 and 22).

Demonstration

Demonstrate how to remove knockouts and pryouts.

Audiovisual

Show Transparencies 14 and 15 (Figures 24 and 25).

Classroom

Discuss nonmetallic boxes and their applications. Pass around examples for the trainees to examine.

Discuss bushings and locknuts and their applications. Pass around examples for the trainees to examine.

Demonstration

Demonstrate how to use a knockout punch.

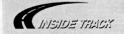

INSIDE TRACK

Using a Punch

To cut a hole with a knockout punch, first measure for the center and drill a pilot hole large enough to insert the drive screw. Turn the drive nut with a wrench until the punch cuts through the box wall.

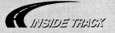

INSIDE TRACK

Removing Knockouts

For concentric or eccentric knockouts, first drive down one section of the smallest ring and cut it in half, then drive down the next section of the ring and cut it in half. Finally, twist off the attached portion of the knockout.

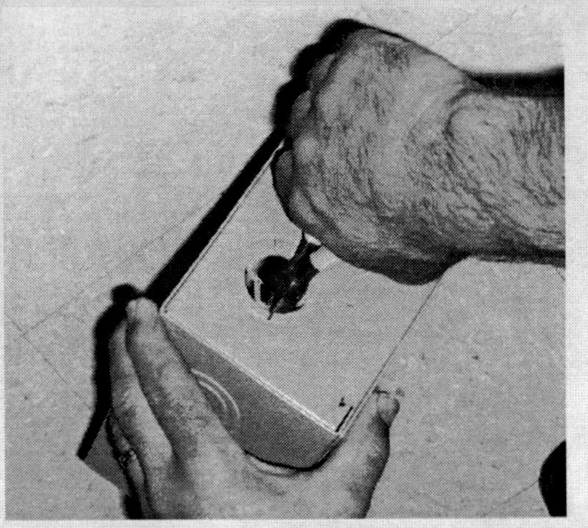

108PO805.EPS

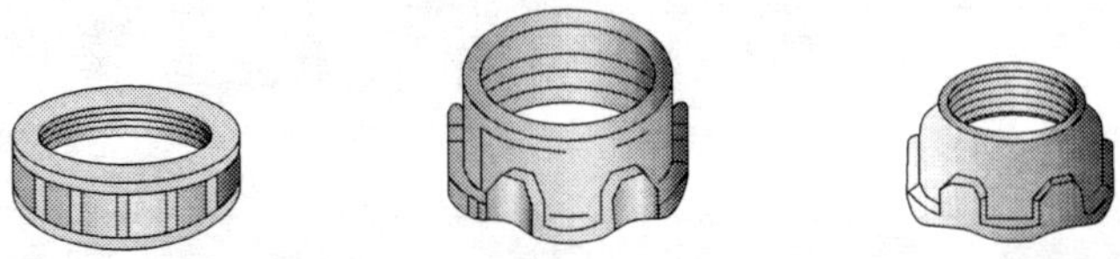

Figure 24 ◆ Bushings.

Instructor's Notes:

274 CHAPTER 7

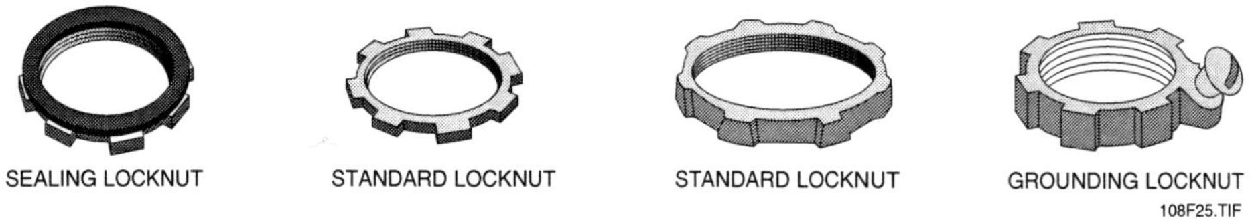

Figure 25 ◆ Locknuts.

A grounding locknut may be needed if a bonding wire is to be installed. Special sealing locknuts are also used in wet locations. Several types of locknuts are shown in *Figure 25.*

7.0.0 ◆ SEALING FITTINGS

Hazardous locations in manufacturing plants and other industrial facilities involve a wide variety of flammable gases and vapors and ignitable dusts. These hazardous substances have widely different flash points, ignition temperatures, and flammable limits requiring fittings that can be sealed. Sealing fittings are installed in conduit runs to minimize the passage of gases, vapors, or flames through the conduit and reduce the accumulation of moisture. They are required by ***NEC Article 500*** in hazardous locations where explosions may occur. They are also required where conduit passes from a hazardous location of one classification to another or to an unclassified location. Several types of sealing fittings are shown in *Figure 26.*

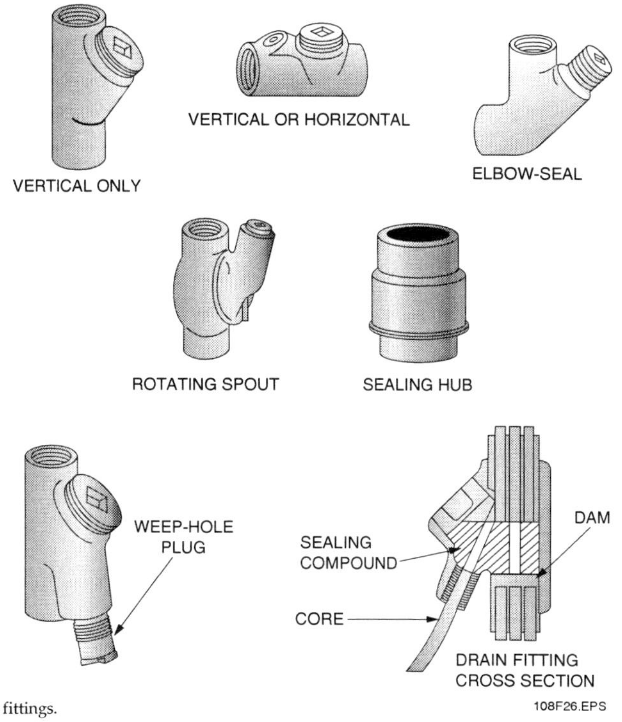

Figure 26 ◆ Sealing fittings.

Classroom

Explain the purpose of sealing fittings. Pass around examples for the trainees to examine.

Audiovisual

Show Transparency 16 (Figure 26).

Demonstration

Demonstrate how to install sealing fittings.

Classroom

Discuss various types of raceway supports, including straps and standoff supports. Pass around examples for the trainees to examine.

Audiovisual

Show Transparency 17 (Figure 27).

INSIDE TRACK

Installing Sealing Fittings

These fittings must be sealed after the wires are pulled. A fiber dam is first packed into the base of the fitting between and around the conductors, then the liquid sealing compound is poured into the fitting.

108PO806.EPS

8.0.0 ◆ RACEWAY SUPPORTS

Raceway supports are available in many types and configurations. This section discusses the most common conduit supports found in electrical installations. *NEC Section 300.11* discusses the requirements for branch circuit wiring that is supported from above suspended ceilings. Electrical equipment and raceways must have their own supporting methods and may not be supported by the supporting hardware of a fire-rated roof/ceiling assembly.

8.1.0 Straps

Straps are used to support conduit to a surface (see *Figure 27*). The spacing of these supports must conform to the minimum support spacing requirements for each type of conduit. One- and two-hole straps are used for all types of conduit: EMT, RMC, IMC, RNC, and flex. The straps can be flexible or rigid. Two-part straps are used to secure conduit to electrical framing channels (struts). Parallel and right angle beam clamps are also used to support conduit from structural members.

Clamp back straps can also be used with a backplate to maintain the ¼-inch spacing from the surface required for installations in wet locations.

8.2.0 Standoff Supports

The standoff support, often referred to as a *Minerallac* (the name of a manufacturer of this type of support), is used to support conduit away from the supporting structure. In the case of the one-hole and two-hole straps, the conduit must be kicked up wherever a fitting occurs. If standoff supports are used, the conduit is held away from the supporting surface, and no offsets (**kicks**) are required in the conduit at the fittings. Standoff supports may be used to support all types of conduit including RMC, IMC, EMT, RNC, and flex, as well as tubing installations. A standoff support is shown in *Figure 28.*

Instructor's Notes:

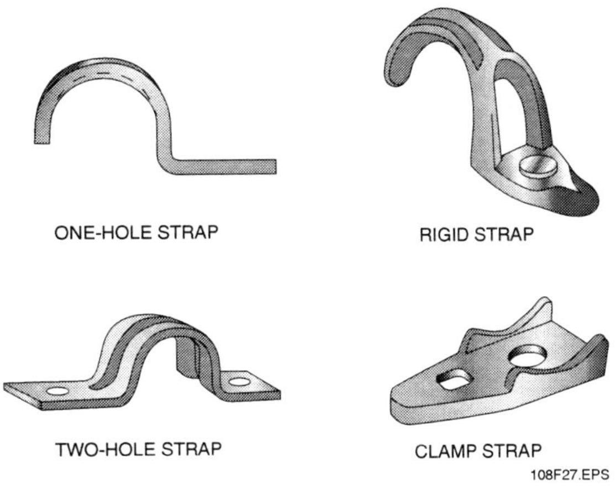

Figure 27 ◈ Straps.

8.3.0 Electrical Framing Channels

Electrical framing channels or other similar framing materials are used together with Unistrut-type conduit clamps to support conduit (see *Figure 29*). They may be attached to a ceiling, wall, or other surface or be supported from a trapeze hanger.

8.4.0 Beam Clamps

Beam clamps are used with suspended hangers. The raceway is attached to or laid in the hanger. The hanger is suspended by a threaded rod. One end of the threaded rod is attached to the hanger and the other end is attached to a beam clamp. The beam clamp is then attached to a beam. A beam clamp with wireway support assembly is shown in *Figure 30*.

Figure 28 ◈ Standoff support.

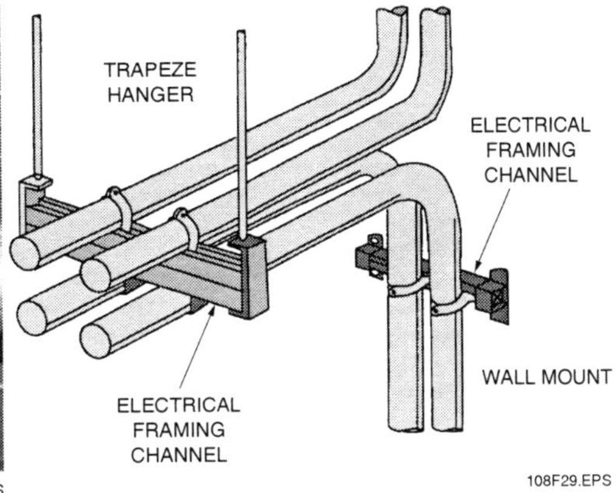

Figure 29 ◈ Electrical framing channels.

Audiovisual

Show Transparencies 18 and 19 (Figures 29 and 30).

Classroom

Discuss framing channels and beam clamps and their applications.

Bundling Conductors

INSIDE TRACK

When conductors are bundled together in a wireway their magnetic fields tend to cancel, thus minimizing inductive heating in the conductors.

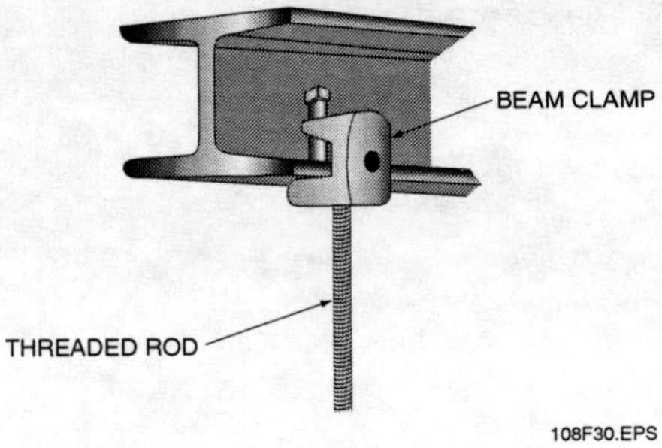

Figure 30 ◆ Beam clamp.

Classroom

Discuss various types of wireways and their applications.

Discuss the purpose of auxiliary gutters.

Stress NEC requirements.

9.0.0 ◆ WIREWAYS

Wireways are sheet metal **troughs** provided with hinged or screw-on removable covers. Like other types of raceways, wireways are used for housing electric wires and cables. Wireways are available in various lengths, including 1, 2, 3, 4, 5, and 10 feet. The availability of various lengths allows runs of any exact number of feet to be made without cutting the wireway ducts. Wireways are dealt with specifically in *NEC Article 376.*

As listed in *NEC Section 376.22,* the sum of the cross-sectional areas of all contained conductors at any cross section of a wireway shall not exceed 20% of the interior cross-sectional area of the wireway. The derating factors in *NEC Table 310.15(B)(2)* shall be applied only where the number of current-carrying conductors exceeds 30, including neutral conductors classified as current-carrying under the provisions of *NEC Section 310.15(B)(4).* Conductors for signaling or controller conductors between a motor and its starter used only for starting duty shall not be considered current-carrying conductors.

It is also noted in *NEC Section 376.56* that conductors, together with splices and taps, must not fill the wireway to more than 75% of its cross-sectional area. No conductor larger than that for which the wireway is designed shall be installed in any wireway. Be sure to check *NEC Article 378* for the requirements of nonmetallic wireways.

9.1.0 Auxiliary Gutters

Strictly speaking, an auxiliary gutter is a wireway that is intended to add to wiring space at switchboards, meters, and other distribution locations. Auxiliary gutters are dealt with specifically in *NEC Article 366.* Even though the component parts of wireways and auxiliary gutters are identical, you should be familiar with the differences in their use. Auxiliary gutters are used as parts of complete assemblies of apparatus such as switchboards, distribution centers, and control equipment. However, an auxiliary gutter may only contain conductors or busbars, even though it looks like a surface metal raceway that may contain devices and equipment. Unlike auxiliary gutters, wireways represent a type of wiring because they are used to carry conductors between points located considerable distances apart.

The allowable ampacities for insulated conductors in wireways and gutters are given in *NEC Tables 310.16 and 310.18.* It should be noted that these tables are used for raceways in general. These NEC tables and the notes are often used to determine if the correct materials are on hand for an installation. They are also used to determine if it is possible to add conductors in an existing wireway or gutter.

In many situations, it is necessary to make extensions from the wireways to wall receptacles and control devices. In these cases, *NEC Section 376.70* specifies that these extensions be made using any wiring method presented in *NEC Article*

Instructor's Notes:

300 that includes a means for equipment grounding. Finally, as required in *NEC Section 376.120,* wireways must be marked in such a way that their manufacturer's name or trademark will be visible.

As you can see in *Figure 31,* a wide range of fittings is required for connecting wireways to one another and to fixtures such as switchboards, power panels, and conduit.

9.2.0 Types of Wireways

Rectangular duct-type wireways come as either hinged-cover or screw-cover troughs. Typical lengths are 1, 2, 3, 4, 5, and 10 feet. Shorter lengths are also available. Raintight troughs are permitted to be used in environments where moisture is not permitted within the raceway. However, the raintight trough should not be confused with the raintight lay-in wireway, which has a hinged cover. *Figure 32* shows a raintight trough with a removable side cover.

Wireway troughs are exposed when first installed. Whenever possible, they are mounted on the ceilings or walls, although they may sometimes be suspended from the ceiling. Note that in *Figure 33,* the trough has knockouts similar to those found on junction boxes. After the wireway system has been installed, branch circuits are brought from the distribution panels using conduit. The conduit is joined to the wireway at the most convenient knockout possible.

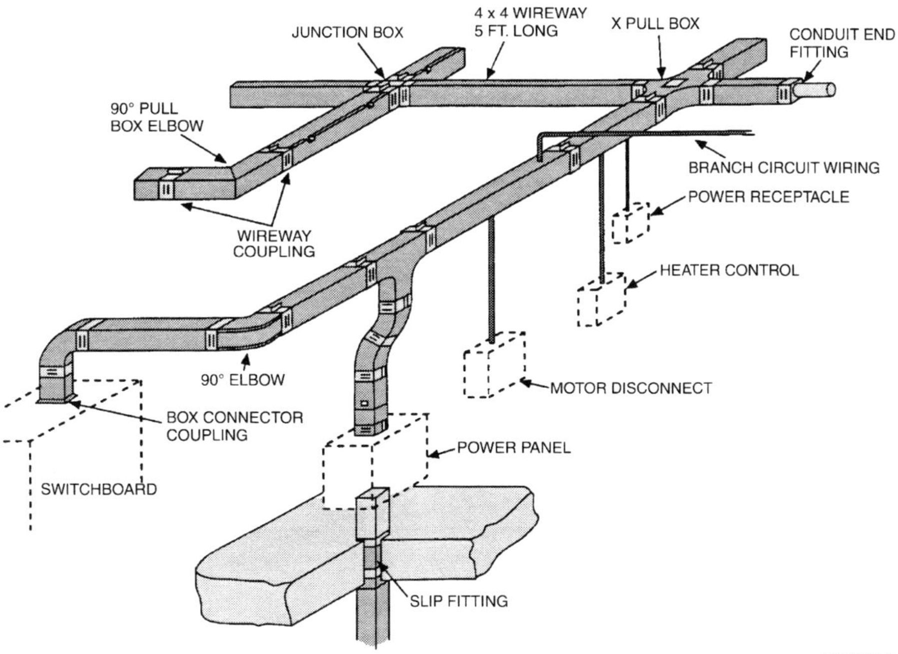

Figure 31 ◈ Wireway system layout.

Classroom

Discuss the various components in a typical wireway system.

Audiovisual

Show Transparency 20 (Figure 31).

Classroom

Discuss duct-type wireways.

Audiovisual

Show Transparencies 21 and 22 (Figures 32 and 33).

Classroom

Discuss wireway fittings.

Audiovisual

Show Transparency 23 (Figure 34).

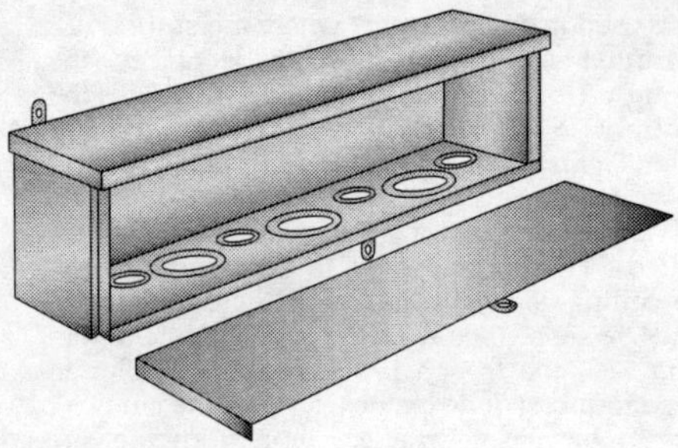

108F32.EPS

Figure 32 ◆ Raintight trough.

Wireway components such as trough crosses, 90° internal elbows, and tee connectors serve the same function as fittings on other types of raceways. The fittings are attached to the duct using slip-on connectors. All attachments are made with nuts and bolts or screws. When assembling wireways, always place the head of the bolt on the inside and the nut on the outside so that the conductors will not be resting against a sharp edge. It is usually best to assemble sections of the wireway system on the floor, and then raise the sections into position. An exploded view of a section of wireway is shown in *Figure 34*. Both the wireway fittings and the duct come with screw-on, hinged, or snap-on covers to permit conductors to be laid in or pulled through.

The NEC specifies that wireways may be used only for exposed work. Therefore, they cannot be used in underfloor installations. If they are used for outdoor work, they must be of an approved raintight construction. It is important to note that wireways must not be installed where they are subject to severe physical damage, corrosive vapors, or hazardous locations.

Wireway troughs must be installed so that they are supported at distances not exceeding 5 feet. When specially approved supports are used, the distance between supports must not exceed 10 feet.

9.2.1 Wireway Fittings

Many different types of fittings are available for wireways, especially for use in exposed, dry locations. The following sections explain fittings commonly used in the electrical craft.

9.2.2 Connectors

Connectors are used to join wireway sections and fittings. Connectors are slipped inside the end of a wireway section and are held in place by small

INSIDE TRACK

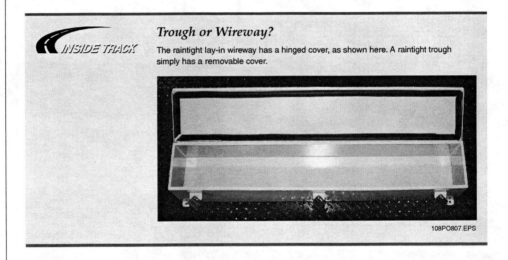

Trough or Wireway?

The raintight lay-in wireway has a hinged cover, as shown here. A raintight trough simply has a removable cover.

108PO807.EPS

Instructor's Notes:

280 CHAPTER 7

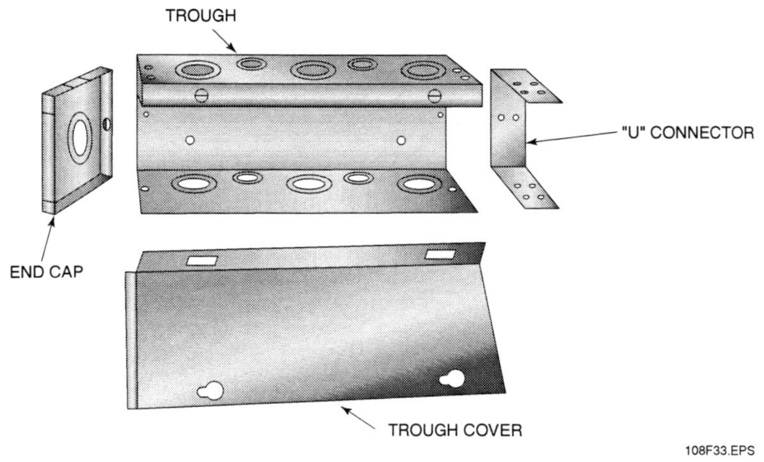

Figure 33 ◆ Trough.

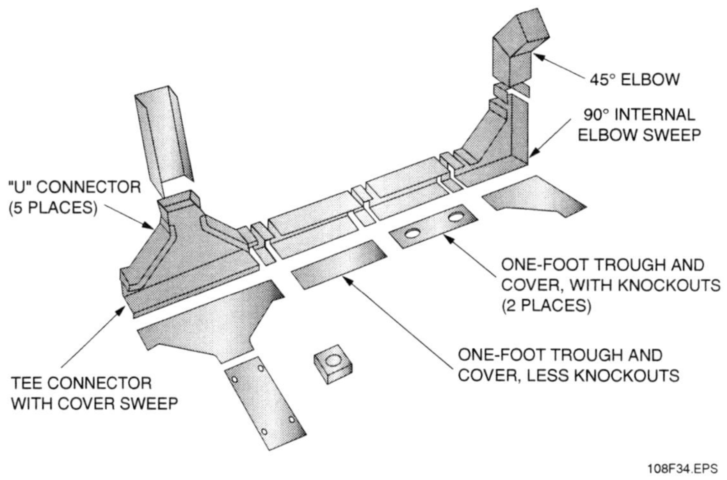

Figure 34 ◆ Wireway sections.

bolts and nuts. Alignment slots allow the connector to be moved until it is flush with the inside surface of the wireway. After the connector is in position, it can be bolted to the wireway. This helps to ensure a strong rigid connection. Connectors have a friction hinge that helps hold the wireway cover open when needed. A connector is shown in *Figure 35*.

Classroom

Discuss wireway connectors.

Audiovisual

Show Transparency 24 (Figure 35).

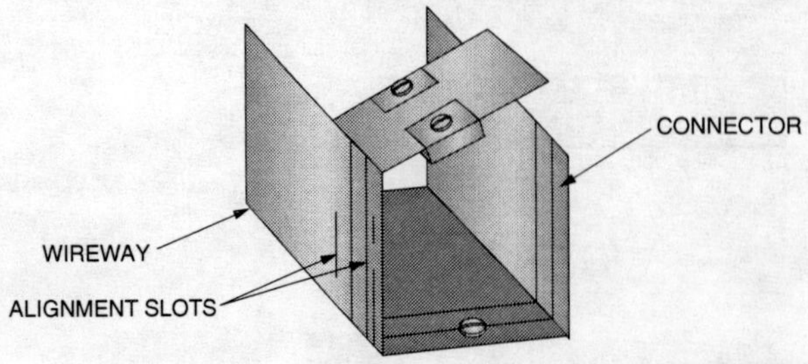

Figure 35 ◆ Connector.

9.2.3 End Plates

End plates, or closing plates, are used to seal the ends of wireways. They are inserted into the end of the wireway and fastened by screws and bolts. End plates contain knockouts so that conduit or cable may be extended from the wireway. An end plate is shown in *Figure 36*.

9.2.4 Tees

Tee fittings are used when a tee connection is needed in a wireway system. A tee connection is used where circuit conductors may branch in different directions. The tee fitting's covers and sides can be removed for access to splices and taps. Tee fittings are attached to other wireway sections using standard connectors. A tee is shown in *Figure 37*.

9.2.5 Crosses

Crosses have four openings and are attached to other wireway sections with standard connectors. The cover is held in place by screws and can be easily removed for laying in wires or for making connections. A cross is shown in *Figure 38*.

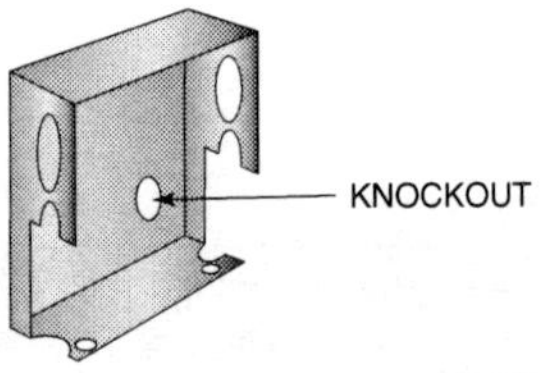

Figure 36 ◆ End plate.

9.2.6 Elbows

Elbows are used to make a bend in the wireway. They are available in angles of 22½°, 45°, or 90°, and are either internal or external. They are attached to wireway sections with standard connectors. Covers and sides can be removed for wire installation. The inside corners of elbows are rounded to prevent damage to conductor insulation. An inside elbow is shown in *Figure 39*.

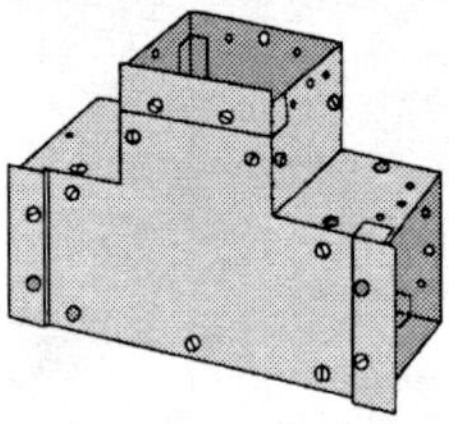

Figure 37 ◆ Tee.

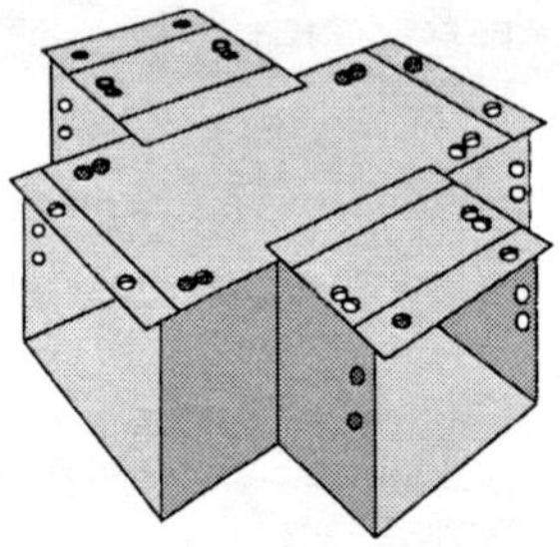

Figure 38 ◆ Cross.

Audiovisual

Show Transparencies 25 through 28 (Figures 36 through 39). Discuss end plates, tees, crosses, and elbows.

Instructor's Notes:

282 CHAPTER 7

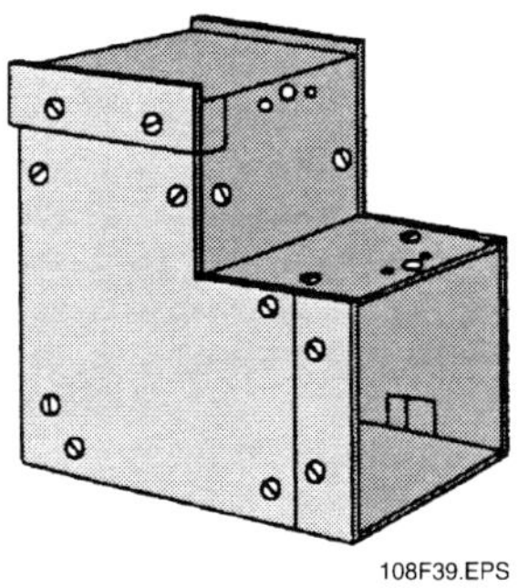

Figure 39 ◈ 90° inside elbow.

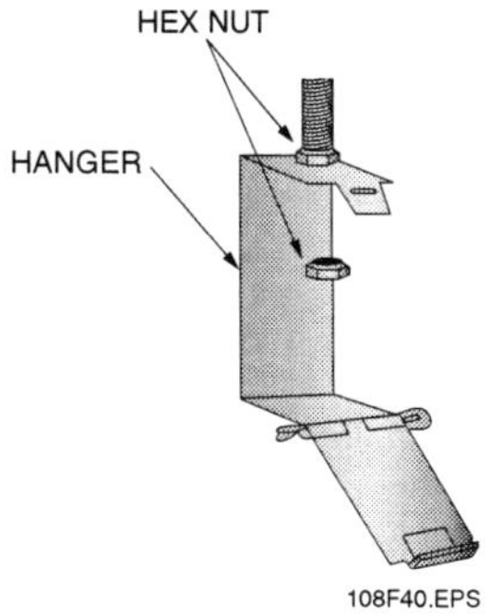

Figure 40 ◈ Suspended hanger.

9.2.7 Telescopic Fittings

Telescopic or slip fittings may be used between lengths of wireway. Slip fittings are attached to standard lengths by setscrews and usually adjust from ½ inch to 11½ inches. Slip fittings have a removable cover for installing wires and are similar in appearance to a nipple.

9.3.0 Wireway Supports

Wireways should be securely supported where run horizontally at each end and at intervals of no more than 5 feet or for individual lengths greater than 5 feet at each end or joint, unless listed for other support intervals. In no case shall the support distance be greater than 10 feet, in accordance with ***NEC Section 376.30.*** If possible, wireways can be mounted directly to a surface. Otherwise, wireways are supported by hangers or brackets.

9.3.1 Suspended Hangers

In many cases, the wireway is supported from a ceiling, beam, or other structural member. In such installations, a suspended hanger (*Figure 40*) may be used to support the wireway.

The wireway is attached to or laid in the hanger. The hanger is suspended by a threaded rod. One end of the rod is attached to the hanger with hex nuts. The other end of the rod is attached to a beam clamp or anchor.

9.3.2 Gusset Brackets

Another type of support used to mount wireways is a gusset bracket. This is an L-type bracket that is mounted to a wall. The wireway rests on the bracket and is attached by screws or bolts. A gusset bracket is shown in *Figure 41.*

9.3.3 Standard Hangers

Standard hangers are made in two pieces. The two pieces are combined in different ways for different installation requirements. The wireway is attached to the hanger by bolts and nuts. A standard hanger is shown in *Figure 42.*

9.3.4 Wireway Hangers

When a larger wireway must be suspended, a wireway hanger may be used. A wireway hanger is made by suspending a piece of strut from a ceiling, beam, or other structural member. The strut is suspended by threaded rods attached to beam clamps or other ceiling anchors, as shown in *Figure 43.*

9.4.0 Other Types of Raceways

In this section, other types of raceways will be discussed. Depending on the particular purpose for which they are intended, raceways include enclosures such as surface metal and nonmetallic raceways, and underfloor raceways.

9.4.1 Surface Metal and Nonmetallic Raceways

Surface metal raceways consist of a wide variety of special raceways designed primarily to carry power and communications wiring to locations on the surface of ceilings or walls of building interiors.

Classroom

Discuss slip fittings.

Discuss wireway supports, including hangers and brackets.

Audiovisual

Show Transparencies 29 through 32 (Figures 40 through 43).

Classroom

Discuss surface metal and nonmetallic raceways and their applications.

Stress NEC requirements.

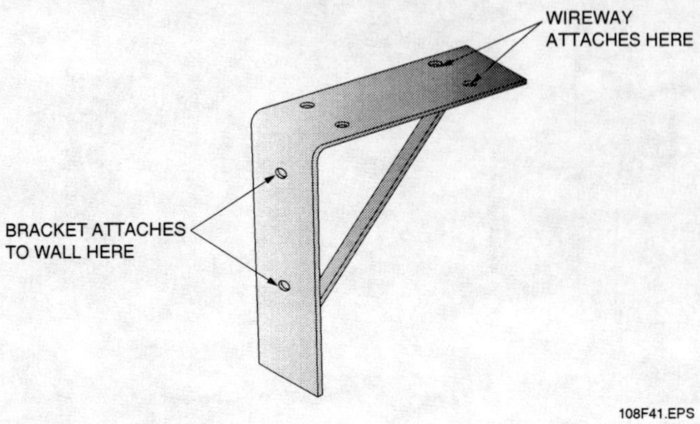

Figure 41 ◆ Gusset bracket.

Audiovisual

Show Transparencies 33 and 34 (Figures 44 and 45).

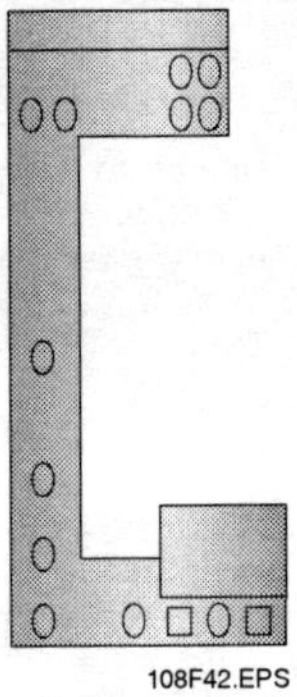

Figure 42 ◆ Standard hanger.

Installation specifications of both surface metal raceways and surface nonmetallic raceways are listed in detail in *NEC Articles 386 and 388,* respectively. All these raceways must be installed in dry, interior locations. The number of conductors, their amperage, and the allowable cross-sectional area of the conductors, as well as regulations for combination raceways, are specified in ***NEC Tables 310.16 and 310.18*** and ***NEC Articles 386 and 388.***

One use of surface metal raceways is to protect conductors that run to non-accessible outlets.

Surface metal and nonmetallic raceways have been divided into subgroups based on the specific purpose for which they are intended. There are three small surface raceways that are primarily used for extending power circuits from one point to another. In addition, there are six larger surface raceways that have a much wider range of applications. Typical cross sections of the first three smaller raceways are shown in *Figure 44.*

Additional surface metal raceway designs are referred to as *pancake raceways,* because their flat cross sections resemble pancakes. Their primary use is to extend power, lighting, telephone, or signal wire to locations away from the walls of a room without embedding them under the floor. A pancake raceway is shown in *Figure 45.*

There are also surface metal raceways available that house two or three different conductor raceways. These are referred to as *twinduct* or *tripleduct.* These raceways permit different circuits, such as power and signal, to be placed within the same raceway.

The number and types of conductors permitted to be installed and the capacity of a particular surface raceway must be calculated and matched with NEC requirements, as discussed previously. ***NEC Tables 310.16 through 310.19*** are used for surface raceways in the same manner in which they are used for wireways. For surface raceway installations with more than three conductors in each raceway, particular reference must be made to ***NEC Table 310.15(B)(2)(a).***

Classroom

Discuss plugmold systems and their applications.

9.4.2 Plugmold Multi-Outlet Systems

Plugmold multi-outlet systems are covered in ***NEC Article 380.*** Manufacturers offer a wide variety of plugmold multi-outlet surface raceways.

Instructor's Notes:

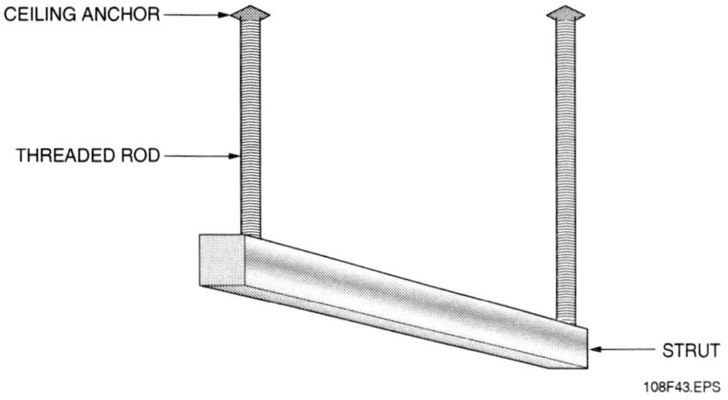

Figure 43 ◈ Wireway hanger.

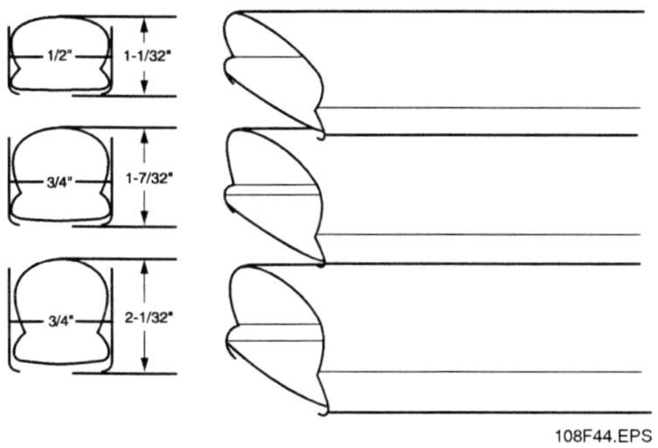

Figure 44 ◈ Smaller surface raceways.

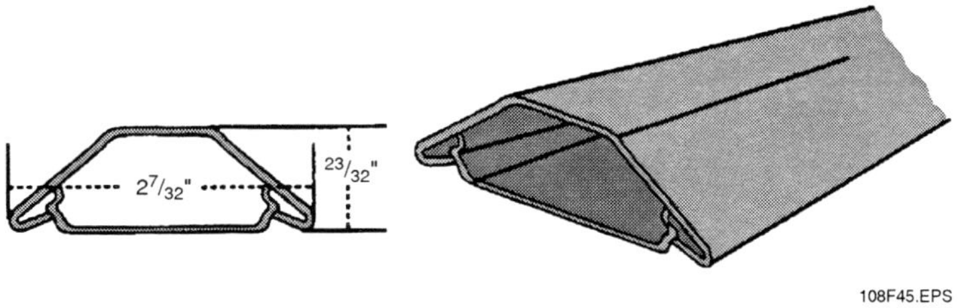

Figure 45 ◈ Pancake raceway.

RACEWAYS, BOXES, AND FITTINGS 285

Surface Raceways

INSIDE TRACK

Surface raceways with multiple channels are commonly used in computer networking applications to provide conductors for AC power to the computers, as well as telephone and other specialized wiring.

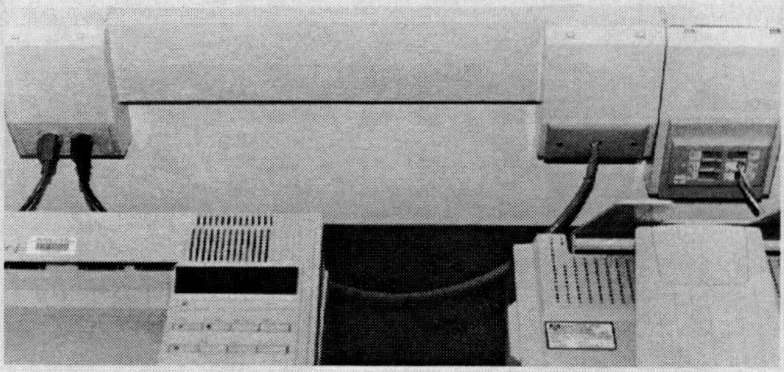

108PO808.EPS

Audiovisual

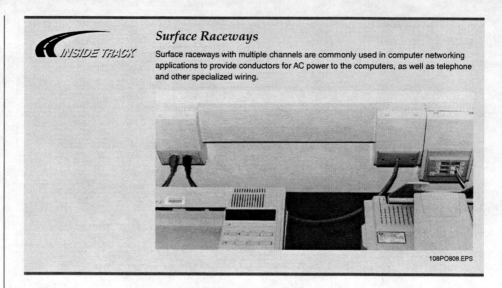

Show Transparency 35 (Figure 46).

Classroom

Discuss pole, underfloor, and cellular metal raceways and their applications.

Audiovisual

Show Transparency 36 (Figure 48).

Their function is to hold receptacles and other devices within the raceway. When the surface raceways are used in this manner, the assembly is referred to as a *multi-outlet assembly.*

Plugmold systems are either wired in the field or come pre-wired from the factory. *Figure 46* shows typical plugmold cross sections, wiring configurations, and some of the available fittings for the system.

9.4.3 Pole Systems

There are many situations in which power and other electric circuits have to be carried from overhead wiring systems to devices that are not located near existing wall outlets or control circuits. This type of wiring is typically used in open office spaces where cubicles are provided by temporary dividers. Poles are used to accomplish this. Some common manufacturers' names for these poles include *Tele-Power poles, Quick-E poles,* and *Walkerpoles.* The poles usually come in lengths suitable for 10-, 12-, or 15-foot ceilings. *Figure 47* shows a typical pole base.

9.4.4 Underfloor Systems

Underfloor raceway systems were developed to provide a practical means of bringing conductors for lighting, power, and signaling to cabinets and consoles. Underfloor raceways are available in 10-foot lengths and widths of 4 and 8 inches. The sections are made with inserts spaced every 24 inches. The inserts can be removed for outlet installation. These are explained in ***NEC Article 390.***

> **Note**
> Inserts must be installed so that they are flush with the finished grade of the floor.

Junction boxes are used to join sections of underfloor raceways. Conduit is also used with underfloor raceways by using a raceway-to-conduit connector (conduit adapter). A typical underfloor raceway duct with fittings is shown in *Figure 48.*

This wiring method makes it possible to place a desk or table in any location where it will always be over, or very near to, a duct line. The wiring method for lighting and power between cabinets and the raceway junction boxes may be conduit, underfloor raceway, wall elbows, and cabinet connectors. ***NEC Article 390*** covers the installation of underfloor raceways.

9.4.5 Cellular Metal Floor Raceways

A cellular metal floor raceway is a type of floor construction designed for use in steel-frame

Instructor's Notes:

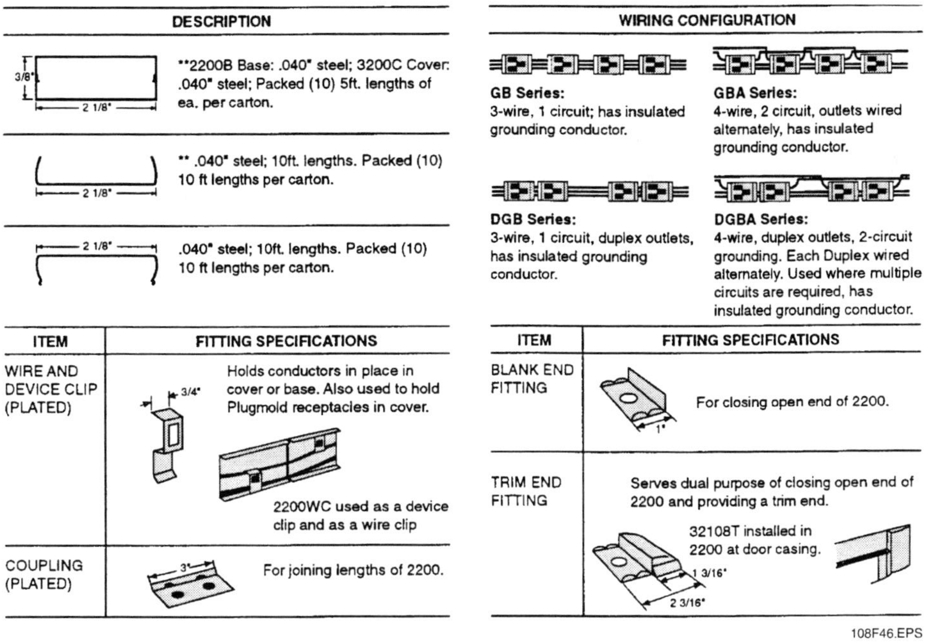
DESCRIPTION
3/8"
2 1/8"
**2200B Base: .040" steel; 3200C Cover: .040" steel; Packed (10) 5ft. lengths of ea. per carton.
2 1/8"
** .040" steel; 10ft. lengths. Packed (10) 10 ft lengths per carton.
2 1/8"
.040" steel; 10ft. lengths. Packed (10) 10 ft lengths per carton.
WIRING CONFIGURATION
GB Series:
3-wire, 1 circuit; has insulated grounding conductor.
GBA Series:
4-wire, 2 circuit, outlets wired alternately, has insulated grounding conductor.
DGB Series:
3-wire, 1 circuit, duplex outlets, has insulated grounding conductor.
DGBA Series:
4-wire, duplex outlets, 2-circuit grounding. Each Duplex wired alternately. Used where multiple circuits are required, has insulated grounding conductor.
ITEM
FITTING SPECIFICATIONS
WIRE AND DEVICE CLIP (PLATED)
3/4"
Holds conductors in place in cover or base. Also used to hold Plugmold receptacles in cover.
2200WC used as a device clip and as a wire clip
COUPLING (PLATED)
3"
For joining lengths of 2200.
ITEM
FITTING SPECIFICATIONS
BLANK END FITTING
1"
For closing open end of 2200.
TRIM END FITTING
Serves dual purpose of closing open end of 2200 and providing a trim end.
32108T installed in 2200 at door casing.
1 3/16"
2 3/16"
108F46.EPS

Figure 46 ◆ Plugmold system.

buildings. In these buildings, the members supporting the floor between the beams consist of sheet steel rolled into shapes. These shapes are combined to form cells, or closed passageways, which extend across the building. The cells are of various shapes and sizes, depending upon the structural strength required. The cells of this type of floor construction form the raceways, as shown in *Figure 49*.

Connections to the cells are made using headers that extend across the cells. A header connects only to those cells to be used as raceways for conductors. A junction box or access fitting is necessary at each joint where a header connects to a cell. Two or three separate headers, connecting to different sets of cells, may be used for different systems. For example, light and power, signaling systems, and public telephones would each have a separate header. A special elbow fitting is used to extend the headers up to the distribution equipment on a wall or column. *NEC Article 374* covers the installation of cellular metal floor raceways.

9.4.6 Cellular Concrete Floor Raceways

The term *precast cellular concrete floor* refers to a type of floor used in steel-frame, concrete-frame, and wall-bearing construction. In this type of system, the floor members are precast with hollow voids, which form smooth, round cells. The cells form raceways, which can be adapted, using fittings, for use as underfloor raceways. A precast cellular concrete floor is fire-resistant and requires no further fireproofing. The precast reinforced concrete floor members form the structural floor and are supported by beams or bearing walls. Connections to the cells are made with headers that are secured to the precast concrete floor. *NEC Article 372* covers the installation of cellular concrete floor raceways.

10.0.0 ◆ CABLE TRAYS

Cable trays function as a support for conductors and tubing (see *NEC Article 392*). A cable tray has the advantage of easy access to conductors, and thus lends itself to installations where the addition

Audiovisual

Show Transparency 37 (Figure 49).

Classroom

Discuss cellular concrete floor raceways and their applications.

Teaching Tip

If possible, take the trainees to a commercial/industrial job site where they can observe various types of wireway components and raceway systems.

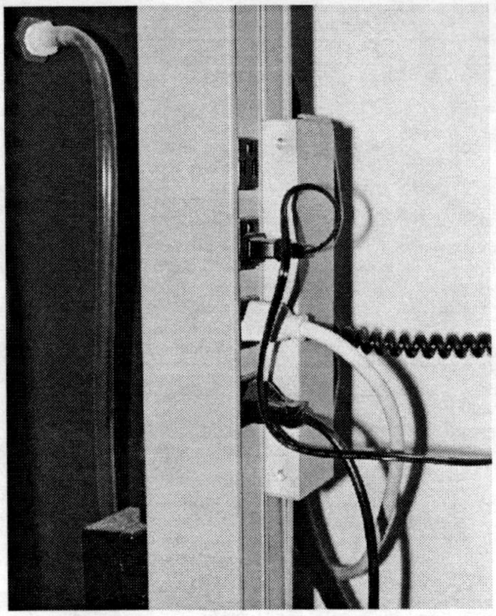

Figure 47 ◆ Power pole.

or removal of conductors is a common practice. Cable trays are fabricated from aluminum, steel, and fiberglass. Cable trays are available in two basic forms: ladder and trough. Ladder tray, as the name implies, consists of two parallel channels connected by rungs. Trough consists of two parallel channels (side rails) having a corrugated, ventilated bottom, or a corrugated, solid bottom. (There is also a special center rail cable tray available for use in light-duty applications such as telephone and sound wiring. We will discuss this type of cable tray in more detail in Level 2.)

Cable trays are commonly available in 12- and 24-foot lengths. They are usually available in widths of 6, 9, 12, 18, 24, 30, and 36 inches, and load depths of 4, 6, and 8 inches.

Cable trays may be used in most electrical installations. Cable trays may be used in air handling ceiling space, but only to support the wiring methods permitted in such spaces by ***NEC Section 300.22(C)(1).*** Also, cable trays may be used in Class 1, Division 2 locations according to ***NEC Section 501.4(B).*** Cable trays may also be used above a suspended ceiling that is not used as an air handling space. Some manufacturers offer an aluminum cable tray that is coated with PVC for installation in caustic environments. A typical cable tray system with fittings is shown in *Figure 50.*

Wire and cable installation in cable trays is defined by the NEC. Read ***NEC Article 392*** to become familiar with the requirements and restrictions made by the NEC for safe installation of wire and cable in a cable tray.

Metallic cable trays that support electrical conductors must be grounded as required by ***NEC Article 250.*** Where steel and aluminum cable tray systems are used as an equipment grounding conductor, all of the provisions of ***NEC Section 392.7*** must be complied with.

> **WARNING!**
> Do not stand on, climb in, or walk on a cable tray.

Audiovisual

Show Transparency 38 (Figure 50). Discuss cable trays and their applications.

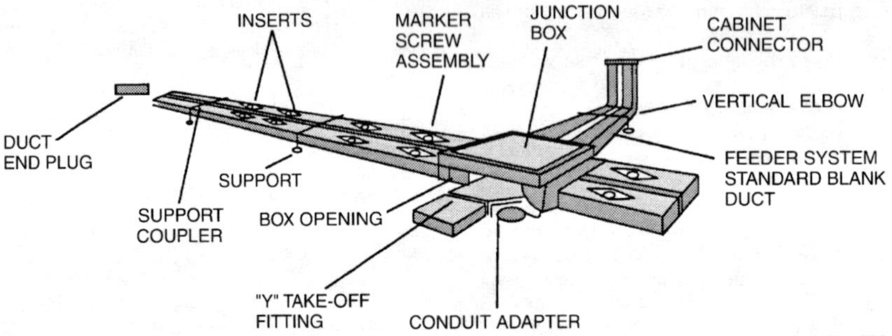

Figure 48 ◆ Underfloor raceway duct.

Instructor's Notes:

288 CHAPTER 7

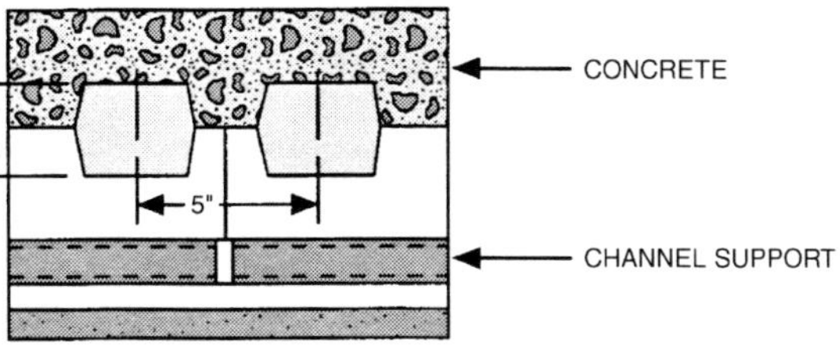

108F49.EPS

Figure 49 ◈ Cross section of a cellular floor.

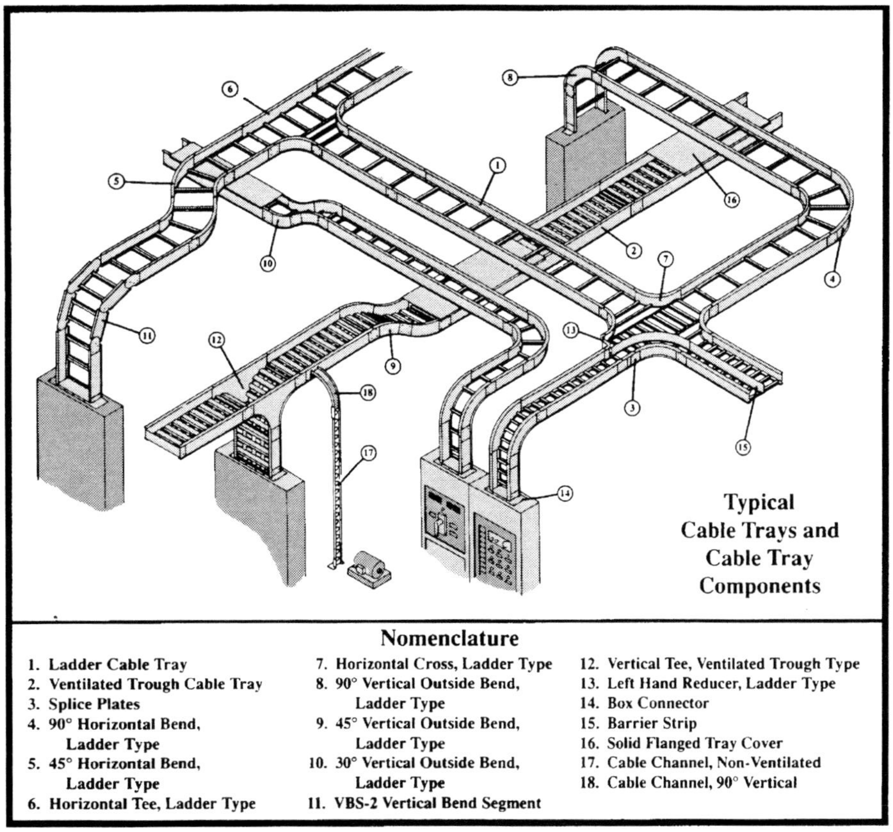

108F50.EPS

Figure 50 ◈ Cable tray system.

RACEWAYS, BOXES, AND FITTINGS 289

Teaching Tip

Discuss the "Think About It."

Classroom

Discuss cable tray fittings and their applications.

Discuss various types of cable tray supports and their applications.

Audiovisual

Show Transparencies 39 through 41 (Figures 51 through 53).

THINK ABOUT IT

Cable Trays and Wireways

What is the difference between a wireway and a cable tray? What kinds of conductors would you expect to find in a cable tray as compared to a wireway?

INSIDE TRACK

Cable Tray Systems

Cable tray systems must be continuous and grounded. One of the advantages of using a cable tray system is that it makes it easy to expand or modify the wiring system following installation. Unlike conduit systems, wires can be added or changed by simply laying them into (or lifting them out of) the tray.

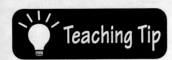

10.1.0 Cable Tray Fittings

Cable tray fittings are part of the cable tray system and provide a means of changing the direction or dimension of the different trays. Some of the uses of horizontal and vertical tees, horizontal and vertical bends, horizontal crosses, reducers, barrier strips, covers, and box connectors are shown in *Figure 50.*

10.2.0 Cable Tray Supports

Cable trays are usually supported in one of five ways: direct rod suspension, trapeze mounting, center hung, wall mounting, and pipe rack mounting.

10.2.1 Direct Rod Suspension

The direct rod suspension method of supporting cable tray uses threaded rods and hanger clamps. One end of the threaded rod is connected to the ceiling or other overhead structure. The other end is connected to hanger clamps that are attached to the cable tray side rails. A direct rod suspension assembly is shown in *Figure 51.*

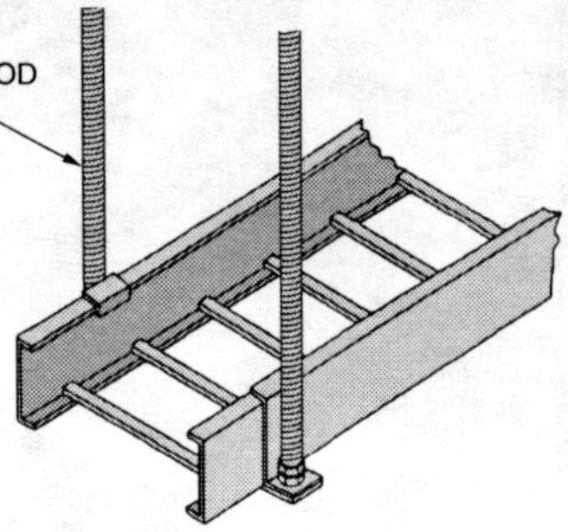

Figure 51 ◆ Direct rod suspension.

10.2.2 Trapeze Mounting and Center Hung Support

Trapeze mounting of cable tray is similar to direct rod suspension mounting. The difference is in the method of attaching the cable tray to the threaded rods. A structural member, usually a steel channel or strut, is connected to the vertical supports to provide an appearance similar to a swing or trapeze. The cable tray is mounted to the structural member. Often, the underside of the channel or strut is used to support conduit. A trapeze mounting assembly is shown in *Figure 52.*

A method that is similar to trapeze mounting is a center hung tray support (*Figure 52*). In this case, only one rod is used and it is centered between the cable tray side rails.

10.2.3 Wall Mounting

Wall mounting is accomplished by supporting the cable tray with structural members attached to the wall. This method of support is often used in tunnels and other underground or sheltered installations where large numbers of conductors interconnect equipment that is separated by long distances. A wall mounting assembly is shown in *Figure 53.*

Instructor's Notes:

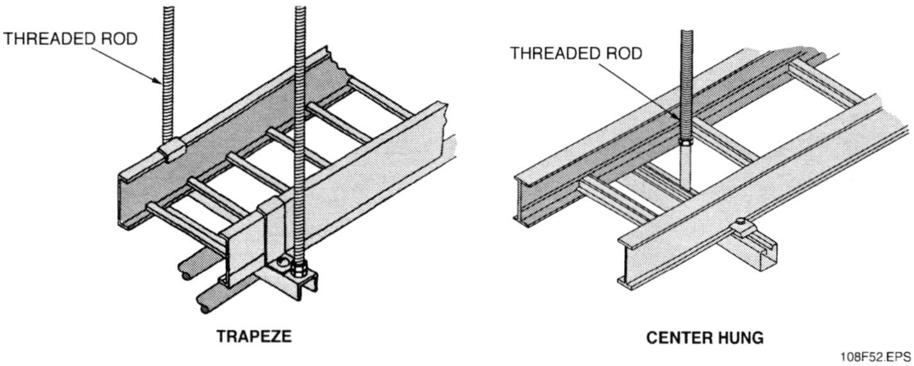

Figure 52 ◆ Trapeze mounting and center hung support.

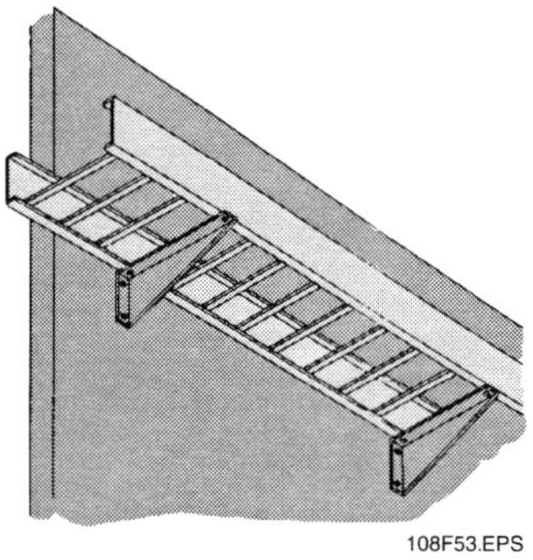

Figure 53 ◆ Wall mounting.

10.2.4 Pipe Rack Mounting

Pipe racks are structural frames used to support piping that interconnects equipment in outdoor industrial facilities. Usually, some space on the rack is reserved for conduit and cable tray. Pipe rack mounting of cable tray is often used when power distribution and electrical wiring is routed over a large area.

11.0.0 ◆ STORING RACEWAYS

Proper and safe methods of storing conduit, wireways, raceways, and cable trays may sound like a simple task, but improper storage techniques can result in wasted time and damage to the raceways, as well as personal injury. There are correct ways to store raceways that will help avoid costly damage, save time in identifying stored raceways, and reduce the chance of personal injury.

Pipe racks are commonly used for storing conduit. The racks provide support to prevent bending, sagging, distorting, scratching, or marring of conduit surfaces. Most racks have compartments where different types and sizes of conduit can be separated for ease of identification and selection. The storage compartments in racks are usually elevated to help avoid damage that might occur at floor level. Conduit that is stored at floor level is easily damaged by people and other materials or equipment in the area.

The ends of stored conduit should be sealed to help prevent contamination and damage. Conduit ends can be capped, taped, or plugged.

Always inspect raceway before storing it to make sure that it is clean and not damaged. It is discouraging to get raceway for a job and find that it is dirty or damaged. Also, make sure that the raceway is stored securely so that when someone comes to get it for a job, it will not fall in any way that could cause injury.

To prevent contamination and corrosion of stored raceway, it should be covered with a tarpaulin or other suitable covering. It should also be separated from non-compatible materials such as hazardous chemicals.

Wireways, surface metal raceways, and cable trays should always be stored off the ground on boards in an area where people will not step on it and equipment will not run over it. Stepping or running over raceway bends the metal and makes it unusable.

Teaching Tip

If possible, take the trainees to a commercial/industrial job site where they can observe various types of cable tray support systems.

Classroom

Discuss the proper method of raceway storage.

RACEWAYS, BOXES, AND FITTINGS 291

![Classroom]

Discuss the proper method of handling raceways.

Discuss ducting and list the various types of underground duct systems and their applications.

Discuss duct lines.

12.0.0 ◆ HANDLING RACEWAYS

Raceway is made to strict specifications. It can be easily damaged by careless handling. From the time raceway is delivered to a job site until the installation is complete, use proper and safe handling techniques. These are a few basic guidelines for handling raceway that will help avoid damaging or contaminating it:

- Never drag raceway off a delivery truck or off other lengths of raceway.
- Never drag raceway on the ground or floor. Dragging raceway can cause damage to the ends.
- Keep the thread protection caps on when handling or transporting conduit raceway.
- Keep raceway away from any material that might contaminate it during handling.
- Flag the ends of long lengths of raceway when transporting it to the job site.
- Never drop or throw raceway when handling it.
- Never hit raceway against other objects when transporting it.
- Always use two people when carrying long pieces of raceway. Make sure that you both stay on the same side and that the load is balanced. Each person should be about ¼ of the length of the raceway from the end. Lift and put down the raceway at the same time.

13.0.0 ◆ DUCTING

In the common vocabulary of the electrical trade, a duct is a single enclosed raceway, or runway, through which conductors or cables can be led. Basically, ducting is a system of ducts. However, underground duct systems include manholes, transformer vaults, and risers.

There are several reasons for running power lines underground rather than overhead. In some situations, an overhead high-voltage line would be dangerous, or the space may not be adequate. For aesthetic reasons, architectural plans may require buried lines throughout a subdivision or a planned community. Tunnels may already exist, or be planned, for carrying steam or water lines. In any of these situations, underground installations are appropriate. Underground cables may be buried directly in the ground or run through conduit.

In underground construction, a duct system provides a safe passageway for power lines, communication cables, or both. In buildings, underfloor raceways and cellular floor raceways are built to provide ducting so that electricity will be available throughout a large area. As an electrician, you need to know the approved methods of constructing underground ducting. You also need to know how to avoid potential electrical hazards in both original construction and maintenance. It is essential to understand the requirements and limitations imposed on running wires through underfloor and cellular floor raceways and ducts.

14.0.0 ◆ UNDERGROUND SYSTEMS

There are five different ways to install cable underground:

- Duct line
- Conductors located in tunnels
- Conductors buried directly in the earth
- RNC
- RMC

The method used will depend on the materials that are available, the number of conductors to be pulled, and the type of wiring to be done (service drop, branch circuit, commercial, etc.).

14.1.0 Duct Line

A duct line consists of at least one subway placed in a trench and covered with earth. Conduit, in some cases, can be classified as a *subway.* Subways may come in a single duct line or multiple duct lines (2, 3, 4, and 6 subways per section). The depth at which the duct will be placed is determined using ***NEC Table 300.5.*** The conduit subways are encased in concrete or other materials. This provides good mechanical strength and allows for power losses from the cable to be dissipated into the earth.

In underground cable installations, a duct is a buried conduit through which a cable passes. Manholes are set at intervals in an underground duct run. Manholes provide access through throats (sometimes called *chimneys*). At ground level, or street surface level, a manhole cover closes off the manhole area tightly. An individual cable length running underground normally terminates at a manhole, where it is spliced to another length of cable. A duct line may consist of a single conduit or several, each carrying a cable length from one manhole to the next.

Manholes provide room for installing lengths of cable in conduit lines. They are also used for maintenance work and for performing tests. Workers enter a manhole from above. In a two-way manhole, cables enter and leave in only two directions. There are also three-way and four-way manholes. Often manholes are located at the in-

Instructor's Notes:

tersection of two streets so that they can be used for cables leaving in four directions. Manholes are usually constructed of brick or concrete. Their design must provide room for drainage and for workers to move around inside them. A similar opening known as a *handhole* is sometimes provided for splicing on lateral two-way duct lines.

Transformer vaults house power transformers, voltage regulators, network protectors, meters, and circuit breakers. A cable may end at a transformer vault. Other cables end at a customer's substation or terminate as risers that connect with overhead lines.

14.2.0 Duct Materials

Underground duct lines can be made of fiber, vitrified tile, iron conduit, plastic, or poured concrete. The inside diameter of the ducting for a specific job is determined by the size of the cable that will be drawn into the duct. Sizes from two to six inches (inside diameter) are available for most types of ducting.

> **WARNING!**
> Be careful when working with unfamiliar duct materials. In older installations, asbestos/cement duct may have been used. You must be certified to remove or disturb asbestos.

14.3.0 Rigid Nonmetallic Conduit

Rigid nonmetallic conduit (RNC) may be made of PVC (polyvinyl chloride), PE (polyethylene), or styrene. Since this type of conduit is available in lengths up to 20 feet, fewer couplings are needed than with other types of ducting. RNC is popular because it is easy to install, requires less labor than other types of conduit, and is low in cost.

14.4.0 Monolithic Concrete Duct

Monolithic concrete duct is poured at the job site. Multiple duct lines can be formed using rubber tubing cores on spacers. The cores may be removed after the concrete has set. A die containing steel tubes, known as a *boat,* can also be used to form ducts. It is pulled slowly through the trench on a track as concrete is poured from the top. Poured concrete ducting made by either method is relatively expensive, but offers the advantage of creating a very clean duct interior with no residue that can decay. The rubber core method is especially useful for curving or turning part of a duct system.

14.5.0 Cable-in-Duct

One of the most popular duct types is the cable-in-duct. This type of duct comes from the manufacturer with cables already installed. The duct comes in a reel and can be laid in the trench with ease. The installed cables can be withdrawn in the future, if necessary. This type of duct, because of the form in which it comes, reduces the need for fittings and couplings. It is most frequently used for street lighting systems.

15.0.0 ◆ MAKING A CONDUIT-TO-BOX CONNECTION

A proper conduit-to-box connection is shown in *Figure 54.*

In order to make a good connection, use the following procedure:

Step 1 Thread the external locknut onto the conduit. Run the locknut to the bottom of the threads.

Step 2 Insert the conduit into the box opening.

Step 3 If an inside locknut or grounding locknut is required, screw it onto the conduit inside the box opening.

Step 4 Screw the bushing onto the threads projecting into the box opening. Make sure the bushing is tightened as much as possible.

Step 5 Tighten the external locknut to secure the conduit to the box.

It is important that the bushings and locknuts fit tightly against the box. For this reason, the conduit must enter straight into the box (*Figure 55*). This may require that a box offset or kick be made in the conduit.

16.0.0 ◆ VARIOUS CONSTRUCTION PROCEDURES

16.1.0 Masonry and Concrete Flush-Mount Construction

In a reinforced concrete construction environment, the conduit and boxes must be embedded in the concrete to achieve a flush surface. Ordinary boxes may be used, but special concrete boxes are preferred and are available in depths up

Classroom

Discuss the various types of material used in underground duct lines.

Discuss RNC, monolithic concrete, and cable-in-duct systems and list the advantages and disadvantages of each type.

Demonstration

Demonstrate how to make a conduit-to-box connection.

Audiovisual

Show Transparencies 42 and 43 (Figures 54 and 55).

Laboratory

Have the trainees practice making conduit-to-box connections. Note the proficiency of each trainee.

Classroom

Discuss the various procedures for installing boxes in concrete/masonry construction.

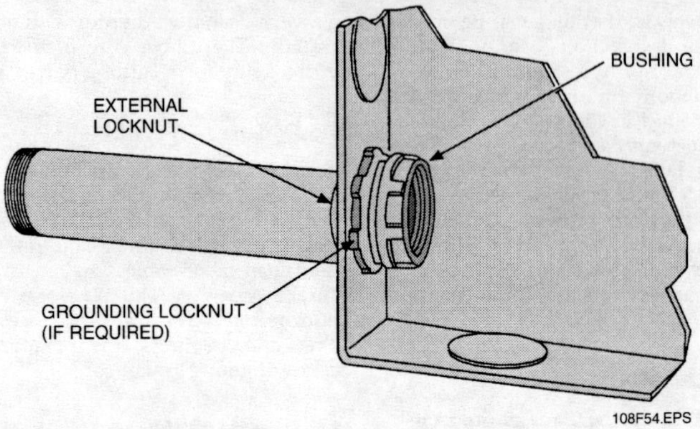

Figure 54 ◆ Conduit-to-box connection.

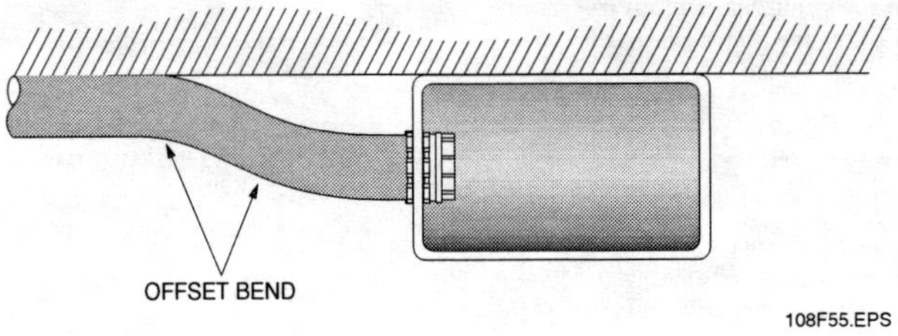

Figure 55 ◆ Correct entrance angle.

to six inches. These boxes have special ears by which they are nailed to the wooden forms for the concrete. When installing them, stuff the boxes tightly with paper to prevent concrete from seeping in. *Figure 56* shows an installed box.

Flush construction can also be done on existing concrete walls, but this requires chiseling a channel and box opening, anchoring the box and conduit, and then resealing the wall.

To achieve flush construction with masonry walls, the most acceptable method is for the electrician to work closely with the mason laying the blocks. When the construction blocks reach the convenience outlet elevation, boxes are made up as shown in *Figure 57*. The figure shows a raised tile ring or box device cover.

Figure 58 shows the use of a 4-S extension ring installed to bring the box to the masonry surface. *Figure 59* shows a masonry box that needs no extension or deep plaster ring to bring it to the surface.

Note

The electrician must work with the mason to ensure the box is properly grouted and sealed.

Note

Conduit should be installed in the rear knockout of this masonry switch box.

Sections of conduit are then coupled in short (four- or five-foot) lengths. This is done because it is impractical for the mason to maneuver blocks over ten-foot sections of conduit.

Audiovisual

Show Transparencies 44 through 47 (Figures 56 through 59).

Instructor's Notes:

294 CHAPTER 7

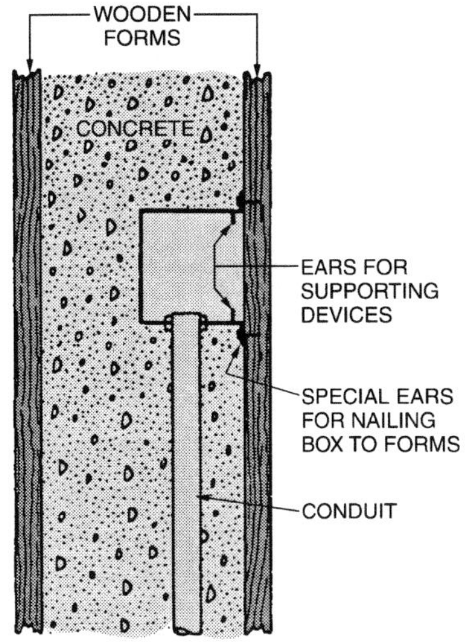

108F56.EPS

Figure 56 ◈ Concrete flush-mount installation.

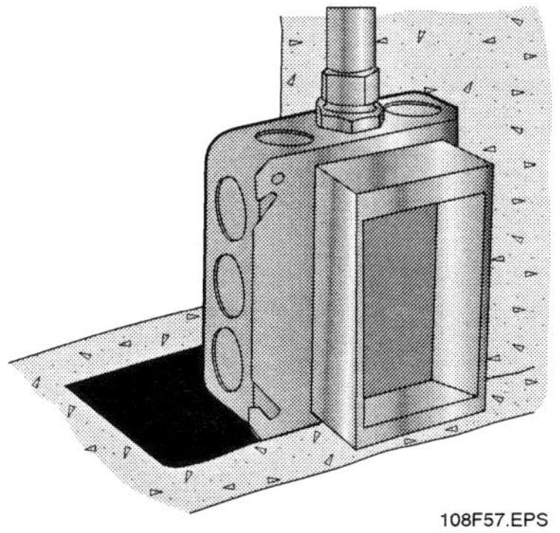

108F57.EPS

Figure 57 ◈ Concrete outlet box.

16.2.0 Metal Stud Environment

Metal stud walls are a popular method of construction for the interior walls of commercial buildings. Metal stud framing consists of relatively thin metal channel studs, usually constructed of galvanized steel and with an overall dimension the same as standard 2 × 4 wooden

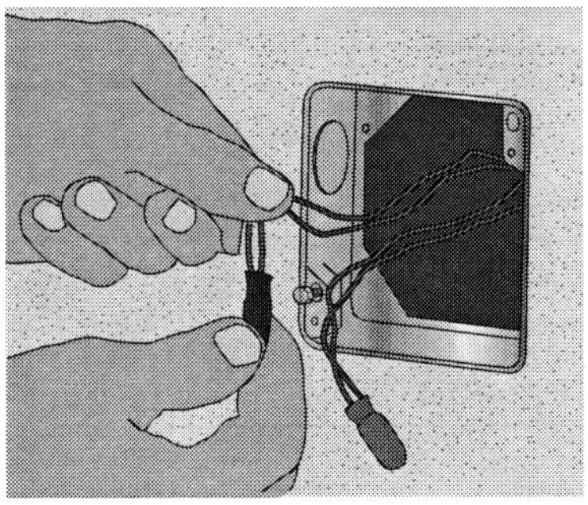

108F58.EPS

Figure 58 ◈ 4-S extension ring used to bring the box to the masonry surface.

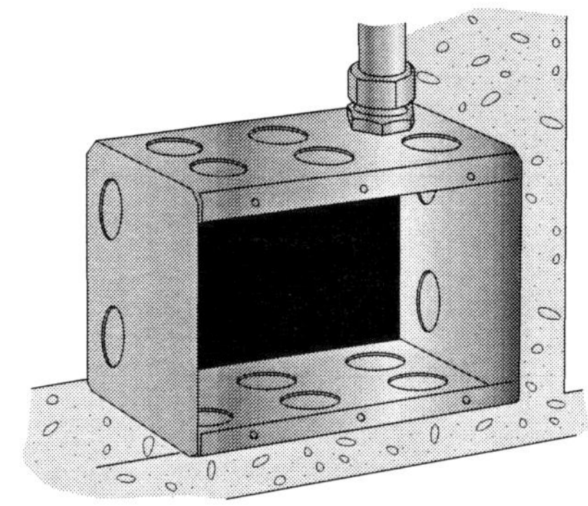

108F59.EPS

Figure 59 ◈ Three-gang switch box.

studs. Wiring in this type of construction is relatively easy when compared to masonry.

EMT conduit is the most common type of raceway specified for metal stud wiring. Metal studs usually have some number of pre-punched holes that can be used to route the conduit. If a pre-punched hole is not located where it needs to be, holes can be easily punched in the metal stud with a hole cutter or knockout punch.

> **CAUTION**
> Cutting or punching metal studs can create sharp edges. Avoid contact that can result in cuts.

Classroom

Explain the procedure for routing conduit and installing boxes in metal stud walls.

Safety

Stress the importance of wearing gloves when working with metal.

Conductors Through Metal Framing

INSIDE TRACK

Single cables should not be run through holes drilled in metal studs unless they are fully grommeted. This protects the cables from the friction of pulling during installation and from the weight of the cable and vibrations following the installation.

108F60.EPS

Figure 60 ◆ Caddy-fastening devices.

Boxes can be secured to the metal stud using self-tapping screws or one of the many types of box supports available. EMT conduit is supported by the metal studs using conduit straps or other approved methods. It is important that the conduit be properly supported to facilitate pulling the conductors through the tubing. Boxes are mounted on the metal studs so that the box will be flush with the finished walls. You must know what the finished wall thickness is going to be to properly secure the boxes to the metal studs. For example, if the finished wall will be ⅝-inch drywall, then the box must be fastened so that it protrudes ⅝ of an inch from the metal stud.

> **WARNING!**
> When using a screw gun or cordless drill to mount boxes to studs, keep the hand holding the box away from the gun/drill to avoid injury.

Figure 60 shows several examples of clips known as *caddy-fastening devices* that are used in metal stud environments.

16.3.0 Wood Frame Environment

At one time, the use of rigid conduit in partitions and ceilings was a laborious and time-consuming operation. Thinwall conduit makes an easier and far quicker job, largely because of the types of fittings that are specially adapted to it.

Figure 61 shows two methods of running thinwall conduit in these locations: boring timbers and notching them. When boring, holes must be drilled large enough for the tubing to be inserted between the studs. The tubing is cut rather short, calling for multiple couplings. EMT can be bowed quite a bit while threading through holes in studs. Boring is the preferred method.

> **WARNING!**
> Always wear safety goggles when boring wood.

NEC Section 300.4 addresses the requirements to prevent physical damage to conductors and cabling in wood members. By keeping the edge of the drilled hole 1¼ inches from the closest edge of the stud, nails are not likely to penetrate the stud far enough to damage the cables. The building codes provide maximum requirements for bored or notched holes in studs.

The exception in the NEC permits IMC, RMC, RNC, and EMT to be installed through bored holes or laid in notches less than 1¼ inches from the nearest edge without a steel plate or bushing.

Because of its weakening effect upon the structure, notching should be resorted to only where absolutely necessary. Notches should be as narrow as possible and in no case deeper than ⅕ the stock of a bearing timber. A bearing timber sup-

Safety

Emphasize the importance of proper safety procedures and protective equipment during conduit and box installation.

Audiovisual

Show Transparency 48 (Figure 60).

Classroom

Discuss the procedures for running conduit and installing boxes in wood frame construction.

Audiovisual

Show Transparency 49 (Figure 61).

Instructor's Notes:

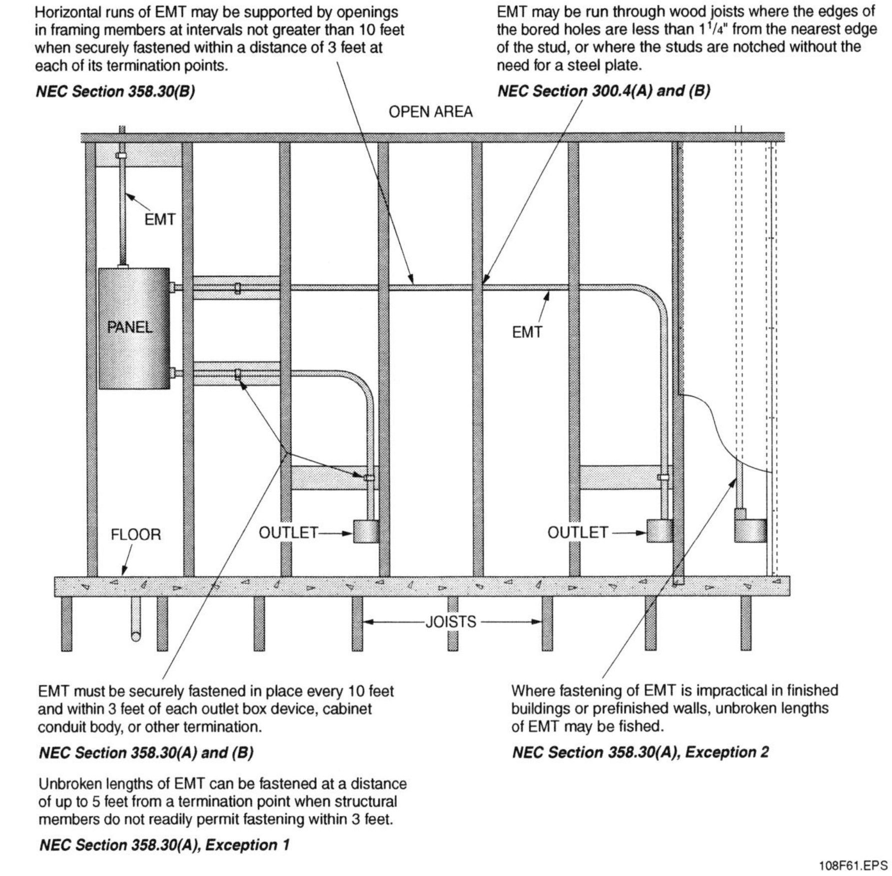

Horizontal runs of EMT may be supported by openings in framing members at intervals not greater than 10 feet when securely fastened within a distance of 3 feet at each of its termination points.
NEC Section 358.30(B)
EMT may be run through wood joists where the edges of the bored holes are less than 1 1/4" from the nearest edge of the stud, or where the studs are notched without the need for a steel plate.
NEC Section 300.4(A) and (B)
OPEN AREA
EMT
PANEL
EMT
FLOOR
OUTLET
OUTLET
JOISTS
EMT must be securely fastened in place every 10 feet and within 3 feet of each outlet box device, cabinet conduit body, or other termination.
NEC Section 358.30(A) and (B)
Unbroken lengths of EMT can be fastened at a distance of up to 5 feet from a termination point when structural members do not readily permit fastening within 3 feet.
NEC Section 358.30(A), Exception 1
Where fastening of EMT is impractical in finished buildings or prefinished walls, unbroken lengths of EMT may be fished.
NEC Section 358.30(A), Exception 2
108F61.EPS

Figure 61 ◈ Installing wire or conduit in a wood-frame building.

ports floor joists or other weight. An additional requirement is for the notch to be covered with a steel reinforcement bracket. This bracket aids in retaining the original strength of the timber.

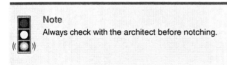

> **Note**
> Always check with the architect before notching.

16.4.0 Steel Environment

Electrical installations in buildings where steel beams are the structural framework are most often industrial buildings and warehouses. This type of construction is typical of the pre-engineered building where beams and other supports are pre-cut and pre-drilled so that erection of the building is fast and simple.

The interior of the building in most cases will be unfinished, and the wiring will be supported by the metal beams and purlings. Beams and purlings

Classroom

Stress NEC requirements.

Discuss the procedures for running conduit and installing boxes in steel beam construction.

RACEWAYS, BOXES, AND FITTINGS 297

Teaching Tip

Discuss the "Think About It."

Think About It

Putting It All Together

Think about the effort that goes into the design of a large industrial installation. If you were to design a large complex, such as the one shown here, where would you start and why?

108PO809.EPS

Audiovisual

Show Transparency 50 (Figure 62).

should not be drilled through; consequently, the conduit is supported from the metal beams by anchoring devices designed especially for that purpose. The supports attach to the beams or supports and have clamps to secure the conduit to the structure. All conduit runs should be plumb since they are exposed. Bends should be correct and have a neat and orderly appearance.

Since steel construction usually takes place in buildings where load handling and moving large and heavy items is common, rigid metal conduit is often required. If a large number of conduits are run along the same path, strut-type systems are used. These systems are sometimes referred to as *Unistrut* systems (Unistrut is a manufacturer of these systems). Another manufacturer of strut systems is *B-Line* systems. Both are very similar. These systems use a channel-type member that can support conduits from the ceiling by using threaded rod supports for the channel, as shown in *Figure 62.* Strut channel can also be secured to masonry walls to support vertical runs of conduit, wireways, and various types of boxes.

Instructor's Notes:

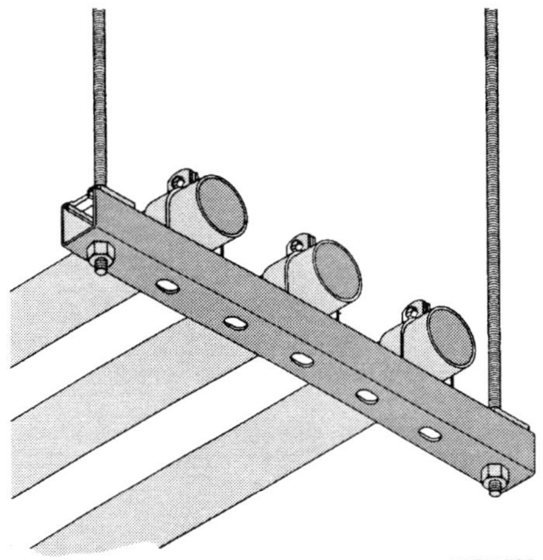

108F62.EPS

Figure 62 ◆ Steel strut system.

Summary

This module discussed the various types of raceways, boxes, and fittings, including their uses and procedures for installation. The primary purpose of raceways is to house electric wire used for power distribution, communication, or electronic signal transmission. Raceways provide protection to the wiring and even a means of identifying one type of wire from another when run adjacent to each other. This process requires proper planning to allow for current needs, future expansion, and a neat and orderly appearance.

Review Questions

1. _____ is the lightest duty and most widely used non-flexible metal conduit available for enclosing and protecting electrical wiring.
   a. Electrical metallic tubing
   b. Rigid metal conduit
   c. Aluminum conduit
   d. Plastic-coated RMC

2. Because the wall thickness of _____ is much less than that of rigid conduit, it is often referred to as *thinwall conduit.*
   a. intermediate metal conduit
   b. electrical metallic tubing
   c. rigid nonmetallic conduit
   d. galvanized rigid steel conduit

3. The type of conduit that provides the best physical protection for the wire inside is _____ .
   a. flexible metal conduit
   b. rigid metal conduit
   c. electrical metallic tubing
   d. intermediate metal conduit

4. Bending IMC is often easier than bending the same size and quantity of rigid metal conduit because IMC has a slightly _____ .
   a. larger outside diameter
   b. larger internal diameter
   c. smaller outside diameter
   d. smaller internal diameter

5. Flexible metal conduit _____ .
   a. is made from a single strip of steel or aluminum
   b. can be used in wet locations if lead-covered conductors are installed
   c. can be used in underground locations
   d. is available in sizes up to six inches in diameter

6. Flexible metal conduit cannot be used in _____ locations.
   a. dry
   b. wet
   c. underground
   d. outdoor

7. The requirements for boxes and fittings are found in _____ .
   a. *NEC Section 300.15*
   b. *NEC Section 348.10*
   c. *NEC Section 370.6*
   d. *NEC Section 390.14*

8. A Type LB conduit body has the cover on _____ .
   a. the left
   b. the right
   c. the back
   d. both sides

Classroom

Have the trainees complete the Review Questions and go over the answers prior to administering the Module Examination.

Examination

Administer the Module Examination. Record the results on Craft Training Report Form 200 and submit the results to the Training Program Sponsor.

Performance Testing

Administer the Performance Test and fill out Performance Profile Sheets for each trainee. If desired, trainee proficiency noted during laboratory sessions may be used to complete the Performance Test. Be sure to record the results on Craft Training Report Form 200 and submit the results to the Training Program Sponsor.

RACEWAYS, BOXES, AND FITTINGS 299

9. When conduit is joined to metal boxes, _____ protect the wires from the sharp edges of the conduit.
   a. washers
   b. locknuts
   c. bushings
   d. couplings

10. _____ are rigid rectangular raceways used for housing electric wires and cables.
   a. Troughs
   b. Gutters
   c. Pull boxes
   d. Conduits

11. Wireways must not be installed where they are subject to _____.
   a. physical damage
   b. corrosion and sunlight
   c. hazardous locations
   d. physical damage, corrosive vapors, and hazardous locations

12. Pancake raceways are a type of _____.
   a. rigid tubing
   b. flexible metal conduit
   c. surface metal raceway
   d. surface nonmetallic raceway

13. Surface metal raceways are designed primarily to protect conductors that run to devices _______ in a room.
   a. located near the walls and outlets
   b. located on the surface of the walls and ceilings
   c. not located near the walls and outlets
   d. not located near the walls and ceiling

14. When storing raceway, it is best to _____.
   a. lay it on the floor so that there is no chance that it will fall and injure someone
   b. store all raceway together regardless of the type of raceway material
   c. stand it on end so that it does not collect moisture
   d. cover it with a tarpaulin or other suitable covering to prevent contamination and corrosion

15. When transporting threaded metal conduit, it is best to _____.
   a. remove your gloves so that you do not contaminate the conduit
   b. remove the thread protection caps so that they do not get lost
   c. place the metal conduit inside a larger raceway to keep it from bending
   d. keep the thread protection caps on to prevent damage to the threads

Instructor's Notes:

300 CHAPTER 7

PROFILE IN SUCCESS

Matt Leitsinger, Beacon Electric Co.

Matt Leitsinger's apprentice training got off on the wrong foot, but he persevered and went on to win the National Craft Olympics in the fourth year of his apprenticeship. Following his apprenticeship, he achieved success as a journeyman, lead electrician, foreman, and instructor.

How did you decide to become an electrician?
It was a natural choice for me. My grandfather, father, brother, uncle, and cousin are all electricians.

What was your apprentice training like?
I started my training while I was still in high school. I would attend classes for half a day and work half a day. After I graduated, I would work a full day and go to class two nights a week for three hours each night. I was fortunate enough to have an excellent instructor and to learn from the NCCER Standardized Craft Training Program. One thing I remember especially is working eight or ten hours in the cold and then going to a nice, warm classroom.

I started off on the wrong foot because I tried to test out of the first year. I didn't pass, and it wound up costing me quite a bit of money. As it turned out, I learned a lot in my first year, so it worked out okay in the end.

The work experience during my apprenticeship was excellent training because it was demanding and complicated work on hospital and research facilities.

In the fourth year of my apprentice training, I qualified for the National Craft Olympics in Hawaii, and I came in first. My employer generously paid for both my wife and me to attend.

What positions have you held since you completed your apprentice training?
I have worked as a journeyman, a lead electrician, and a foreman. In my leadership positions, I have had the opportunity to work on all kinds of jobs, from small residential projects to multimillion-dollar commercial projects. This experience has helped me to become a well-rounded electrician. Currently, I work in our special projects division. I deal directly with clients and perform all kinds of work from simple maintenance to high-voltage terminations.

I have also had the opportunity to participate in training new electricians, both as an advisor to the local vocational school and an instructor at the vocational school and at my own company.

What would you say was the greatest single factor that contributed to your success?
I am highly motivated to continue learning. Also, I make it a point to show up for work every day and do my job. It sounds simple, but it's important that your employer be able to count on you. It really can have a damaging effect on a project if members of the team just don't show up for work. I guess you would call that having a sense of responsibility.

My company has been very supportive and my superiors have helped me at every step along my career path. A career in the electrical industry is not easy, but with the right attitude and training, the rewards will greatly outweigh the effort.

GLOSSARY

Trade Terms Introduced in This Module

Accessible: Able to be reached, as for service or repair.

Approved: Meeting the requirements of an appropriate regulatory agency.

Bonding wire: A wire used to make a continuous grounding path between equipment and ground.

Cable trays: Rigid structures used to support electrical conductors.

Conductors: Wires or cables used to carry electrical current.

Conduit: A round raceway, similar to pipe, that houses conductors.

Exposed location: Not permanently closed in by the structure or finish of a building; able to be installed or removed without damage to the structure.

Kick: A bend in a piece of conduit, usually less than 45°, made to change the direction of the conduit.

Raceways: Enclosed channels designed expressly for holding wires, cables, or busbars, with additional functions as permitted in the NEC.

Splice: Connection of two or more conductors.

Tap: Intermediate point on a main circuit where another wire is connected to supply electrical current to another circuit.

Trough: A long, narrow box used to house electrical connections that could be exposed to the environment.

Underwriters Laboratories (UL): An agency that evaluates and approves electrical components and equipment.

Wireways: Steel troughs designed to carry electrical wire and cable.

Instructor's Notes:

Additional Resources

This module is intended to present thorough resources for task training. The following reference works are suggested for further study. These are optional materials for continued education rather than for task training.

Benfield Conduit Bending Manual, Latest Edition. New York: McGraw-Hill Publishing Company.

National Electrical Code Handbook, Latest Edition. Quincy, MA: National Fire Protection Association.

RACEWAYS, BOXES, AND FITTINGS 303

ANSWER KEY

Answers to Review Questions

Answer	Section
1. a	3.2.1
2. b	3.2.1
3. b	3.2.2
4. b	3.2.7
5. a	3.2.10
6. b	3.2.9
7. c	4.2.2
8. c	6.0.0
9. a	Terms/9.0.0
10. d	9.2.0
11. c	9.4.1
12. d	9.4.1
13. b	9.4.1
14. d	11.0.0
15. d	12.0.0

Section 3.2.1 *Think About It—EMT Use*

The requirements governing the use of EMT are covered in *NEC Article 358.* EMT is widely used in both exposed and concealed locations with certain restrictions, as defined in *NEC Section 358.12.* EMT should not be used in installations where it may be subject to physical damage. For such installations, rigid metal conduit provides greater protection against mechanical injury to conductors.

Section 3.2.7 *Think About It—Applications of RNC*

The requirements governing the use of RNC are covered in *NEC Article 352.* RNC is corrosion-proof and may be placed underground as well as above grade, indoors and out. It is commonly used outdoors, especially for service-entrance conductor raceways and wiring to swimming pools and spas. Its use is restricted in certain hazardous locations and where it could be physically damaged or exposed to temperatures exceeding 122°F. It cannot be used to support fixtures.

Section 10.0.0 *Think About It—Cable Trays and Wireways*

Wireways are fully enclosed troughs, while cable trays are open to allow easy access to the cables. Wireways generally have a smaller capacity for holding conductors. Both wireways and cable trays are used to support large single conductors. Cable trays are also widely used to carry multiconductor cables as well as different types of telecommunications cables.

Section 16.4.0 *Think About It—Putting It All Together*

The design of a large industrial electrical installation starts with a thorough review of the construction drawings (blueprints) for the building and the specifications for the job. Once the job requirements are fully understood, the building and surrounding area should be surveyed to determine information important to both the design and the installation. Such information typically includes: the physical location, size, and type of construction for all equipment rooms and other building structures; the location of system grounding points and prime power inputs; the locations where equipment is to be installed; and the best paths for installing the various conduit runs, wireways, and/or cable trays.

CONTREN™ LEARNING SERIES — USER UPDATES

The NCCER makes every effort to keep these textbooks up-to-date and free of technical errors. We appreciate your help in this process. If you have an idea for improving this textbook, or if you find an error, a typographical mistake, or an inaccuracy in NCCER's Contren™ textbooks, please write us, using this form or a photocopy. Be sure to include the exact module number, page number, a detailed description, and the correction, if applicable. Your input will be brought to the attention of the Technical Review Committee. Thank you for your assistance.

Instructors – If you found that additional materials were necessary in order to teach this module effectively, please let us know so that we may include them in the Equipment/Materials list in the Instructor's Guide.

Write: Curriculum Revision and Development Department
National Center for Construction Education and Research
P.O. Box 141104, Gainesville, FL 32614-1104

Fax: 352-334-0932

E-mail: curriculum@nccer.org

Craft ______________________ Module Name ______________________

Copyright Date ____________ Module Number ____________ Page Number(s) ____________

Description

____

____

____

____

(Optional) Correction

____

____

____

(Optional) Your Name and Address

____

____

____

Chapter 8

Fasteners and Anchors

26103-02

Fasteners and Anchors
Annotated Instructor's Guide

Module 26103

MODULE OVERVIEW

This course introduces the electrical trainee to the applications and installation procedures for various types of fasteners and anchors.

PREREQUISITES

Please refer to the Course Map in the Trainee Module. Prior to training with this module, it is recommended that the trainee shall have successfully completed the following:

Core Curriculum; Residential Electrical I, Chapters 1 through 7.

LEARNING OBJECTIVES

Upon completion of this module, the trainee will be able to:

1. Identify and explain the use of threaded fasteners.
2. Identify and explain the use of non-threaded fasteners.
3. Identify and explain the use of anchors.
4. Demonstrate the correct applications for fasteners and anchors.
5. Install fasteners and anchors.

PERFORMANCE OBJECTIVES

Under supervision of the instructor, the trainee should be able to:

1. Install selected threaded fasteners.
2. Install blind (pop) rivets.
3. Install selected screws.
4. Install selected anchors.
5. Install selected toggle bolts.

NCCER STANDARDIZED CRAFT TRAINING PROGRAM

The National Center for Construction Education and Research (NCCER) provides a standardized national program of accredited craft training. Key features of the program include instructor certification, competency-based training, and performance testing. The program provides trainees, instructors, and companies with a standard form of recognition through a National Craft Training Registry. The program is described in full in the *Guidelines for Accreditation,* published by the NCCER. For more information on standardized craft training, contact the NCCER at P.O. Box 141104, Gainesville, FL 32614-1104, 352-334-0911, visit our Web site at www.nccer.org, or e-mail info@nccer.org.

HOW TO USE THIS ANNOTATED INSTRUCTOR'S GUIDE

Each page presents two sections of information. The larger section displays each page exactly as it appears in the Trainee Module. The narrow column ties suggested trainee and instructor actions to each page and provides icons to call your attention to material, safety, audiovisual, or testing requirements. The bottom of each page includes space for your notes.

If you see the Teaching Tip icon, that means there is a teaching tip associated with this section. Also refer to the suggested teaching tips at the end of the module.

SAFETY CONSIDERATIONS

Ensure that the trainees are equipped with appropriate personal protective equipment. Stress the importance of following all safety precautions and procedures when working with the tools associated with the installation of fasteners and anchors. Emphasize the importance of safety goggles/glasses.

PREPARATION

Before teaching this module, you should review the Module Outline, Learning and Performance Objectives, and the Materials and Equipment List. Be sure to allow ample time to prepare your own training or lesson plan and gather all required equipment and materials.

MATERIALS AND EQUIPMENT LIST

Materials:

Transparencies

Markers/chalk

Assorted threaded fasteners, including:
- Bolts
- Cap screws
- Studs
- Machine screws
- Nuts
- Washers

Assorted non-threaded fasteners, including:
- Retainers
- Keys
- Pins
- Blind (pop) rivets and rivet tool
- Tie wraps

Assorted special threaded fasteners, including:
- Eye bolts
- Anchor bolts
- J-bolts

Assorted screws, including:
- Wood screws
- Lag screws and shields
- Concrete/masonry screws
- Thread-forming (sheet metal) and thread-cutting screws
- Deck screws
- Drywall screws

Hammer-driven tools and related pin and stud fasteners

Powder-actuated tool, powder charges, and related pin and stud fasteners

Assorted mechanical anchors and associated anchor setting tool(s), including:
- Wedge
- Stud
- Sleeve
- One-piece
- Hammer-driven
- Drop-in
- Expansion shields
- Lead (caulk-in)
- Screw (fiber, lead, and plastic)
- Self-drilling
- Toggle bolts
- Sleeve-type
- Wallboard
- Metal drive-in

Concrete, masonry, wood

Module Examinations*

Performance Profile Sheets*

Equipment:

Overhead projector and screen

Whiteboard/chalkboard

Appropriate personal protective equipment

Mechanic's tool box with an assortment of wrenches, screwdrivers, etc.

Torque wrenches

Drills/drivers and assorted drill bits

*Located in the Test Booklet packaged with this Annotated Instructor's Guide.

ADDITIONAL RESOURCES

This module is intended to present thorough resources for task training. The following reference works are suggested for both instructors and motivated trainees interested in further study. These are optional materials for continued education rather than for task training.

American Electrician's Handbook, Latest Edition. New York, NY: McGraw-Hill.

The Sheet Metal Toolbox Manual, Latest Edition. Upper Saddle River, NJ: Prentice Hall.

NOTES

The designations "National Electrical Code," "NE Code," and "NEC," where used in this document, refer to the *National Electrical Code®*, which is a registered trademark of the National Fire Protection Association, Quincy, MA. All National Electrical Code (NEC) references in this module refer to the 2002 edition of the NEC.

If you feel that additional math instruction would be helpful, Prentice Hall offers a basic math textbook entitled *Fundamentals of Electrical and Mechanical Mathematics.* It covers the basic math requirements for electrical trainees and may be ordered by contacting Prentice Hall Customer Service at 1-800-922-0579.

TEACHING TIME FOR THIS MODULE

An outline for use in developing your lesson plan is presented below. Note that each Roman numeral in the outline equates to one session of instruction. Each session has a suggested time period of 2½ hours. This includes 10 minutes at the beginning of each session for administrative tasks and one 10-minute break during the session. Approximately 5 hours are suggested to cover *Fasteners and Anchors.* You will need to adjust the time required for hands-on activity and testing based on your class size and resources.

Topic	Planned Time
Session I. Threaded Fasteners; Non-Threaded Fasteners; Special Threaded Fasteners; Screws	
A. Threaded Fasteners	________
1. Thread Standards	________
a. Thread Series	________
b. Thread Classes	________
c. Thread Identification	________
d. Grade Markings	________
2. Bolt and Screw Types	________
a. Machine Screws	________
b. Machine Bolts	________
c. Cap Screws	________
d. Set Screws	________
e. Stud Bolts	________
3. Nuts	________
a. Jam Nuts	________
b. Castellated, Slotted, and Self-Locking Nuts	________
c. Acorn Nuts	________
d. Wing Nuts	________
4. Washers	________
a. Lock Washers	________
b. Flat and Fender Washers	________

5. Installing Fasteners
    a. Torque Tightening
    b. Installing Threaded Fasteners

B. Laboratory

Under instructor supervision, have the trainees practice installing selected threaded fasteners.

C. Non-Threaded Fasteners
1. Retainer Fasteners
2. Keys
3. Pin Fasteners
    a. Dowel Pins
    b. Taper and Spring Pins
    c. Cotter Pins
4. Blind/Pop Rivets
5. Tie Wraps

D. Laboratory

Under instructor supervision, have the trainees practice installing blind (pop) rivets.

E. Special Threaded Fasteners
1. Eye Bolts
2. Anchor Bolts
3. J-Bolts

F. Screws
1. Wood Screws
2. Lag Screws and Shields
3. Concrete/Masonry Screws
4. Thread-Forming and Thread-Cutting Screws
5. Deck Screws
6. Drywall Screws
7. Drive Screws

G. Laboratory

Under instructor supervision, have the trainees practice installing selected screws.

Session II. Hammer-Driven Pins and Studs; Powder-Actuated Tools and Fasteners; Mechanical Anchors; Epoxy Anchoring Systems; Module Examination and Performance Testing

A. Hammer-Driven Pins and Studs ____________

B. Powder-Actuated Tools and Fasteners ____________

C. Mechanical Anchors ____________

  1. One-Step Anchors ____________
    a. Wedge Anchors ____________
    b. Stud Bolt Anchors ____________
    c. Sleeve Anchors ____________
    d. One-Piece Anchors ____________
    e. Hammer-Set Anchors ____________
  2. Bolt Anchors ____________
    a. Drop-In Anchors ____________
    b. Single- and Double-Expansion Anchors ____________
    c. Lead (Caulk-In) Anchors ____________
  3. Screw Anchors ____________
  4. Self-Drilling Anchors ____________
  5. Guidelines for Drilling Anchor Holes in Hardened Concrete or Masonry ____________
  6. Hollow-Wall Anchors ____________
    a. Toggle Bolts ____________
    b. Sleeve-Type Wall Anchors ____________
    c. Wallboard Anchors ____________
    d. Metal Drive-In Anchors ____________

D. Laboratory ____________

Under instructor supervision, have the trainees practice installing selected anchors and toggle bolts.

E. Epoxy Anchoring Systems ____________

F. Module Examination ____________

  1. Trainees must score 70% or higher to receive recognition from the NCCER.
  2. Record the testing results on Craft Training Report Form 200 and submit the results to the Training Program Sponsor.

G. Performance Testing ____________

  1. Trainees must perform each task to the satisfaction of the instructor to receive recognition from the NCCER.
  2. Record the testing results on Craft Training Report Form 200 and submit the results to the Training Program Sponsor.

Chapter 8

Module 26103-02

Fasteners and Anchors

Course Map

This course map shows all of the modules in the first level of the Residential Electrical curriculum. The suggested training order begins at the bottom and proceeds up. Skill levels increase as you advance on the course map. The local Training Program Sponsor may adjust the training order.

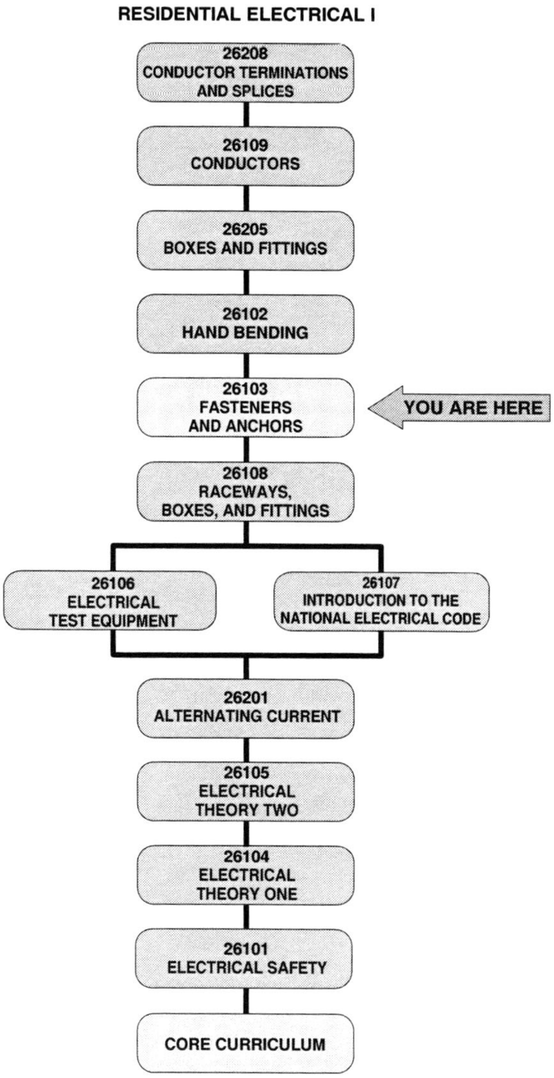

Homework

Assign reading of Module 26103.

MODULE 26103 CONTENTS

FASTENERS AND ANCHORS 311

Instructor's Notes:

Figures

Instructor's Notes:

Chapter 8

MODULE 26103

Fasteners and Anchors

Objectives

When you have completed this module, you will be able to do the following:

1. Identify and explain the use of threaded fasteners.
2. Identify and explain the use of non-threaded fasteners.
3. Identify and explain the use of anchors.
4. Demonstrate the correct applications for fasteners and anchors.
5. Install fasteners and anchors.

Prerequisites

Before you begin this module, it is recommended that you successfully complete the following: Core Curriculum; Residential Electrical I, Chapters 1 through 7.

Required Trainee Materials

1. Paper and pencil
2. Copy of the latest edition of the *National Electrical Code*
3. Appropriate personal protective equipment

1.0.0 ◆ INTRODUCTION

Fasteners are used to assemble and install many different types of equipment, parts, and materials. Fasteners include screws, bolts, nuts, pins, clamps, retainers, tie wraps, rivets, and **keys.** Fasteners are used extensively in the electrical craft. You need to be familiar with the many different types of fasteners in order to identify, select, and properly install the correct fastener for a specific application.

The two primary categories of fasteners are:

- Threaded fasteners
- Non-threaded fasteners

Within each of these two categories, there are numerous different types and sizes of fasteners. Each type of fastener is designed for a specific application. The kind of fastener used for a job may be listed in the project specifications, or you may have to select an appropriate fastener.

Failure of fasteners can result in a number of different problems. To perform quality electrical work, it is important to use the correct type and size of fastener for the particular job. It is equally important that the fastener be installed properly.

In this module, you will be introduced to fasteners commonly used in electrical work.

2.0.0 ◆ THREADED FASTENERS

Threaded fasteners are the most commonly used type of fastener. Many threaded fasteners are assembled with nuts and washers. The following sections describe standard threads used on threaded fasteners, as well as different types of bolts, screws, nuts, and washers. *Figure 1* shows several types of threaded fasteners.

Note: The designations "National Electrical Code," "NE Code," and "NEC," where used in this document, refer to the *National Electrical Code®*, which is a registered trademark of the National Fire Protection Association, Quincy, MA. *All National Electrical Code (NEC) references in this module refer to the 2002 edition of the NEC.*

Materials

Ensure you have everything required to teach the course. Check the Materials and Equipment List at the front of this Instructor's Guide.

Audiovisual

Show Transparency 1, Course Objectives.

Show Transparency 2, Performance Profile Tasks.

Classroom

Discuss threaded fasteners and their applications.

Audiovisual

Show Transparency 3 (Figure 1).

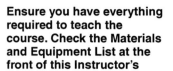

Teaching Tip

Discuss "What's Wrong with This Picture?" Explain that these meters and service panels are not supported by any fasteners.

Classroom

Explain that terms shown in bold (blue) are defined in the Glossary at the back of this module.

Describe Unified thread series standards.

Describe Unified thread class standards.

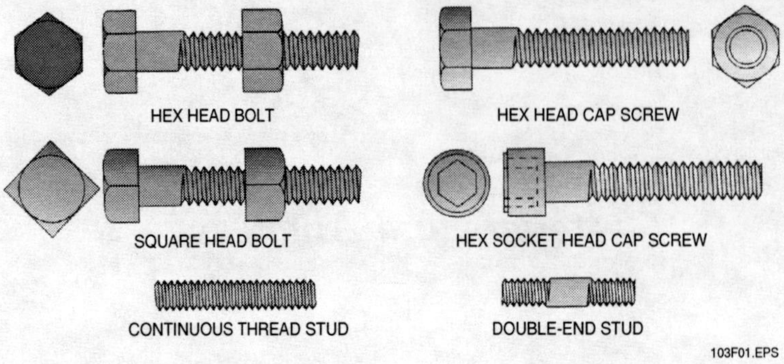

Figure 1 ◆ Threaded fasteners.

What's Wrong with This Picture?

103PO301.EPS

2.1.0 Thread Standards

There are many different types of threads used for manufacturing fasteners. The different types of threads are designed to be used for different jobs. Threads used on fasteners are manufactured to industry-established standards for uniformity. The most common **thread standard** is the Unified standard, sometimes referred to as the *American standard.* Unified standards are used to establish thread series and classes.

2.1.1 Thread Series

Unified standards are established for three series of threads, depending on the number of threads per inch for a certain diameter of fastener. These three series are:

- ***Unified National Coarse (UNC) thread***—Used for bolts, screws, nuts, and other general purposes. Fasteners with UNC threads are commonly used for rapid assembly or disassembly of parts and where corrosion or slight damage may occur.
- ***Unified National Fine (UNF) thread***—Used for bolts, screws, nuts, and other applications where a finer thread for a tighter fit is desired.
- ***Unified National Extra Fine (UNEF) thread***—Used on thin-walled tubes, nuts, ferrules, and couplings.

2.1.2 Thread Classes

The Unified standards also establish **thread classes.** Classes 1A, 2A, and 3A apply to external

Instructor's Notes:

316 CHAPTER 8

threads only. Classes 1B, 2B, and 3B apply to internal threads only. Thread classes are distinguished from each other by the amounts of **tolerance** provided. Classes 3A and 3B provide a minimum **clearance** and classes 1A and 1B provide a maximum clearance.

Classes 2A and 2B are the most commonly used. Classes 3A and 3B are used when close tolerances are needed. Classes 1A and 1B are used where quick and easy assembly is needed and a large tolerance is acceptable.

2.1.3 Thread Identification

Thread identification is done using a standard method. *Figure 2* shows how screw threads are designated for a common fastener.

- *Nominal size*—The **nominal size** is the approximate diameter of the fastener.
- *Number of threads per inch (TPI)*—The TPI is standard for all diameters.
- *Thread series symbol*—The Unified standard thread type (UNC, UNF, or UNEF).
- *Thread class symbol*—The closeness of fit between the bolt threads and nut threads.
- *Left-hand thread symbol*—Specified by the symbol LH. Unless threads are specified with the LH symbol, the threads are right-hand threads.

2.1.4 Grade Markings

Special markings on the head of a bolt or screw can be used to determine the quality of the fastener. The **Society of Automotive Engineers (SAE)** and the **American Society for Testing of Materials (ASTM)** have developed the standards for these markings. These grade or line markings for steel bolts and screws are shown in *Figure 3.*

THINK ABOUT IT

Thread Classes

External and internal thread classes differ widely with regard to tolerances. Why would you use Classes 3A or 3B for high-grade commercial products such as precision tools and machines? What household products would warrant the use of Classes 1A and 1B?

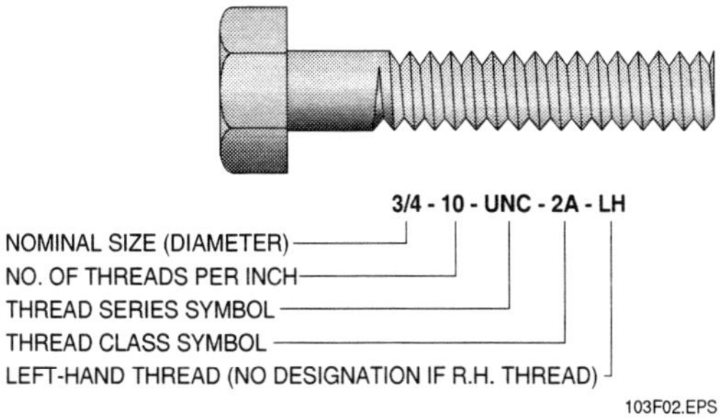

Figure 2 ◆ Screw thread designations.

THINK ABOUT IT

Screw Diameters

A screw's diameter and TPI for a given project depend on what condition(s)?

Classroom

Discuss thread identification and grade markings.

Audiovisual

Show Transparencies 4 and 5 (Figures 2 and 3).

Teaching Tip

Discuss the "Think About It." Precision tools require the close tolerances provided by Classes 3A/3B, while Classes 1A/1B provide for quick assembly. Household applications vary widely.

Teaching Tip

Discuss the "Think About It." The screw diameter and TPI depend on the strength required and the materials to be joined.

Demonstration

Display a variety of bolts and screws and demonstrate how to identify them using ASTM and SAE grade markings.

ASTM AND SAE GRADE MARKINGS FOR STEEL BOLTS & SCREWS

GRADE MARKING	SPECIFICATION	MATERIAL
	SAE-GRADE 0	STEEL
	SAE-GRADE 1 ASTM-A 307	LOW CARBON STEEL
	SAE-GRADE 2	LOW CARBON STEEL
	SAE-GRADE 3	MEDIUM CARBON STEEL, COLD WORKED
	SAE-GRADE 5	MEDIUM CARBON STEEL, QUENCHED AND TEMPERED
	ASTM-A 449	
	ASTM-A 325	MEDIUM CARBON STEEL, QUENCHED AND TEMPERED
	ASTM-A 354 GRADE BB	LOW ALLOY STEEL, QUENCHED AND TEMPERED
	ASTM-A 354 GRADE BC	LOW ALLOY STEEL, QUENCHED AND TEMPERED
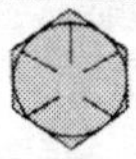	SAE-GRADE 7	MEDIUM CARBON ALLOY STEEL, QUENCHED AND TEMPERED ROLL THREADED AFTER HEAT TREATMENT
	SAE-GRADE 8	MEDIUM CARBON ALLOY STEEL, QUENCHED AND TEMPERED
	ASTM-A 354 GRADE BD	ALLOY STEEL, QUENCHED AND TEMPERED
	ASTM-A 490	ALLOY STEEL, QUENCHED AND TEMPERED

ASTM SPECIFICATIONS
- A 307 - LOW CARBON STEEL EXTERNALLY AND INTERNALLY THREADED STANDARD FASTENERS.
- A 325 - HIGH STRENGTH STEEL BOLTS FOR STRUCTURAL STEEL JOINTS, INCLUDING SUITABLE NUTS AND PLAIN HARDENED WASHERS.
- A 449 - QUENCHED AND TEMPERED STEEL BOLTS AND STUDS.
- A 354 - QUENCHED AND TEMPERED ALLOY STEEL BOLTS AND STUDS WITH SUITABLE NUTS.
- A 490 - HIGH STRENGTH ALLOY STEEL BOLTS FOR STRUCTURAL STEEL JOINTS, INCLUDING SUITABLE NUTS AND PLAIN HARDENED WASHERS.

SAE SPECIFICATION
- J 429 - MECHANICAL AND QUALITY REQUIREMENTS FOR THREADED FASTENERS.

103F03.EPS

Figure 3 ◈ Grade markings for steel bolts and screws.

Generally, the higher-quality steel fasteners have a greater number of marks on the head. If the head is unmarked, the fastener is usually considered to be made of mild steel (having low carbon content).

2.2.0 Bolt and Screw Types

Bolts and screws are made in many different sizes and shapes and from a variety of materials. They are usually identified by the head type or other

Instructor's Notes:

318 CHAPTER 8

THINK ABOUT IT

Installing Fasteners

Different types of fasteners require different installation techniques, but the same basic steps and safety precautions apply to most threaded fasteners. What are some of the basic steps and safety precautions you should be aware of?

special characteristics. The following sections describe several different types of bolts and screws.

2.2.1 Machine Screws

Machine screws are used for general assembly work. They come in a variety of types with slotted or recessed heads. Machine screws are generally available in diameters ranging from 0 (0.060") to ½" (0.500"). The length of machine screws typically varies from ⅛" to 3". Machine screws are also manufactured in metric sizes. *Figure 4* shows different types of machine screws.

As shown, the heads of machine screws are made in different shapes and with slots made to fit various kinds of manual and power tool screwdrivers. Flat-head screws are used in a countersunk hole and tightened so that the head is flush with the surface. Oval-head screws are also used in a countersunk hole in applications where a more decorative finish is desired. Pan and round-head screws are general-use fastening screws. Fillister, hex socket, and TORX® socket screws are typically used in confined space applications on machined assemblies that need a finished appearance. They are often installed in a recessed hole. Truss screws

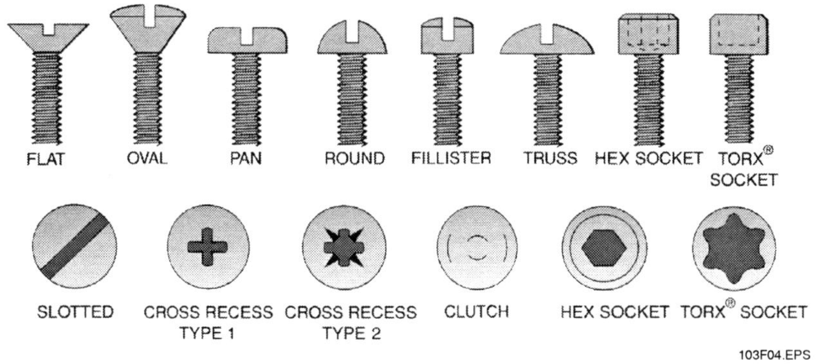

Figure 4 ◆ Machine screws.

THINK ABOUT IT

Using Machine Screws

The shape of a screw head is determined by where and how the screw will be used. Provide examples of where and how you would use the following types of screws:

- Flat-head screws
- Oval-head screws
- TORX® socket screws
- Truss screws

Teaching Tip

Discuss the "Think About It."

Classroom

Discuss machine screws and their applications.

Audiovisual

Show Transparency 6 (Figure 4).

Demonstration

Display a variety of machine screws and demonstrate how to identify them by head type.

Teaching Tip

Discuss the "Think About It." Answers will vary, but you should emphasize use in electrical applications.

Classroom

Discuss machine bolts, cap screws, set screws, and stud bolts. List applications for each type.

Audiovisual

Show Transparencies 7 through 10 (Figures 5 through 8).

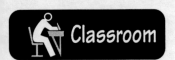

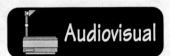

are a low-profile screw generally used without a washer. To prevent damage when tightening and removing machine screws (regardless of head type), make sure to use a screwdriver or power tool bit with the proper tip to drive them.

2.2.2 Machine Bolts

Machine bolts are generally used to assemble parts where close tolerances are not required. Machine bolts have square or hexagonal heads and are generally available in diameters ranging from ¼" to 3". The length of machine bolts typically varies from ½" to 30". Nuts used with machine bolts are similar in shape to the bolt heads. The nuts are usually purchased at the same time as the bolts. *Figure 5* shows two different types of machine bolts.

2.2.3 Cap Screws

Cap screws are often used on high-quality assemblies requiring a finished appearance. The cap screw passes through a clearance hole in one of the assembly parts and screws into a threaded hole in the other part. The clamping action occurs by tightening the cap screw.

Cap screws are made to close tolerances and are provided with a machined or semi-finished bearing surface under the head. They are normally made in coarse and fine thread series and in diameters from ¼" to 2". Lengths may range from ⅜" to 10". Metric sizes are also available. *Figure 6* shows typical cap screws.

2.2.4 Set Screws

Heat-treated steel is normally used to make set screws. Common uses of set screws include preventing pulleys from slipping on shafts, holding collars in place on shafts, and holding shafts in place. The head style and point style are typically used to classify set screws. *Figure 7* shows several set screw heads and point styles.

2.2.5 Stud Bolts

Stud bolts (*Figure 8*) are headless bolts that are threaded over the entire length of the bolt or for a length on both ends of the bolt. One end of the

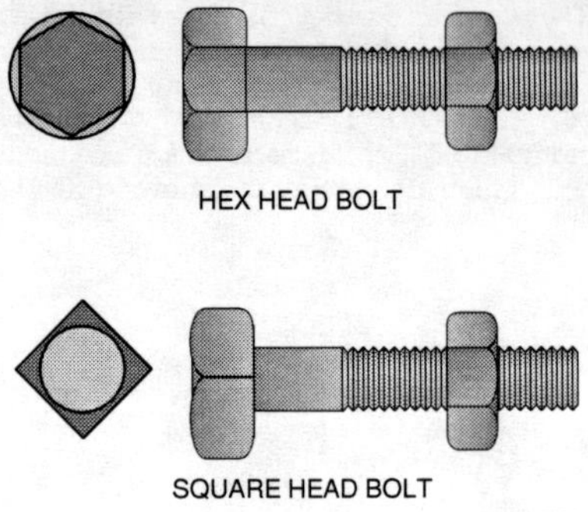

Figure 5 ◆ Machine bolts.

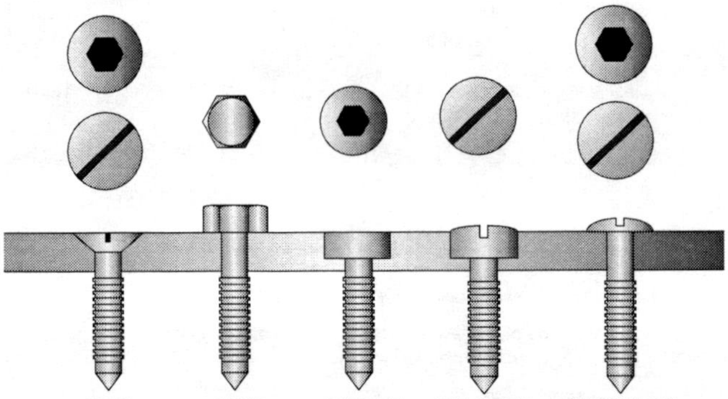

Figure 6 ◆ Cap screws.

Instructor's Notes:

320 CHAPTER 8

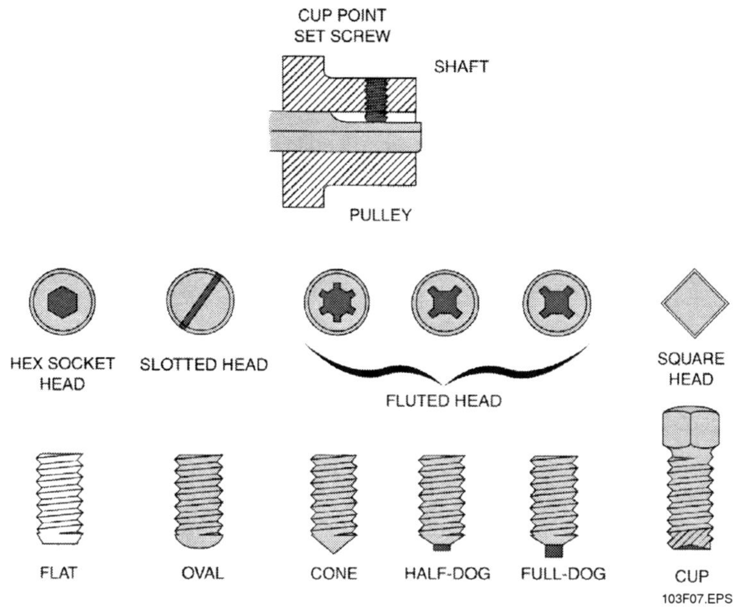

Figure 7 ◈ Set screws.

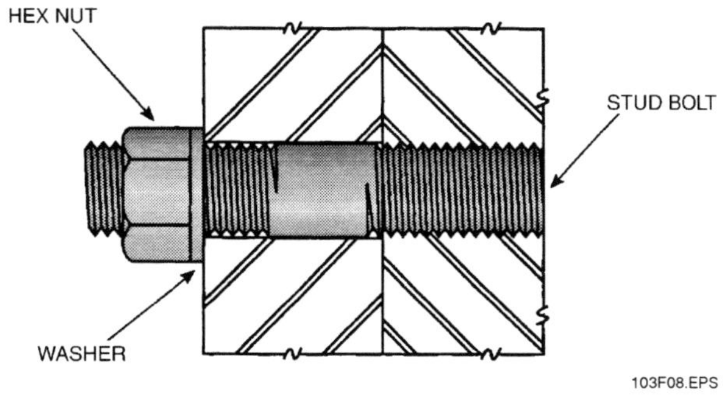

Figure 8 ◈ Stud bolt.

stud bolt is screwed into a tapped hole. The part to be clamped is placed over the remaining portion of the stud, and a nut and washer are screwed on to clamp the two parts together. Other stud bolts have machine-screw threads on one end and lag-screw threads on the other so that they can be screwed into wood.

Stud bolts are used for several purposes, including holding together inspection covers on equipment and bearing caps.

FASTENERS AND ANCHORS 321

Classroom

Discuss nut shapes and classifications.

Audiovisual

Show Transparencies 11 and 12 (Figures 9 and 10).

2.3.0 Nuts

Most nuts used with threaded fasteners are hexagonal or square. They are usually used with bolts having the same shape head. *Figure 9* shows several different types of nuts that are used with threaded fasteners.

Nuts are typically classified as regular, semi-finished, or finished. The only machining done on regular nuts is to the threads. In addition to the threads, semi-finished nuts are also machined on the bearing face. Machining the bearing face makes a truer surface for fitting the washer. The only difference between semi-finished and finished nuts is that finished nuts are made to closer tolerances.

The standard machine screw nut has a regular finish. Regular and semi-finished nuts are shown in *Figure 10*.

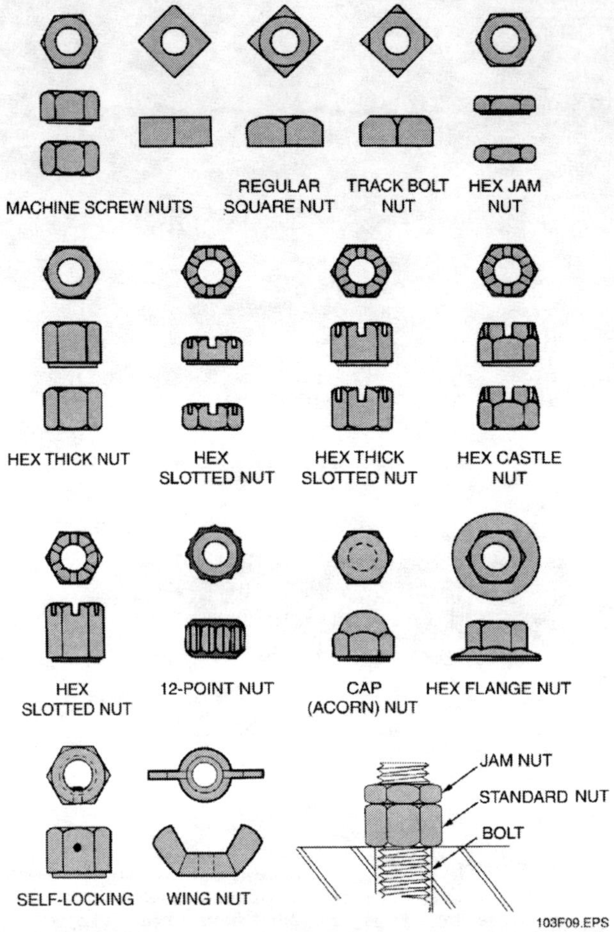

Figure 9 ◈ Nuts.

Instructor's Notes:

322 CHAPTER 8

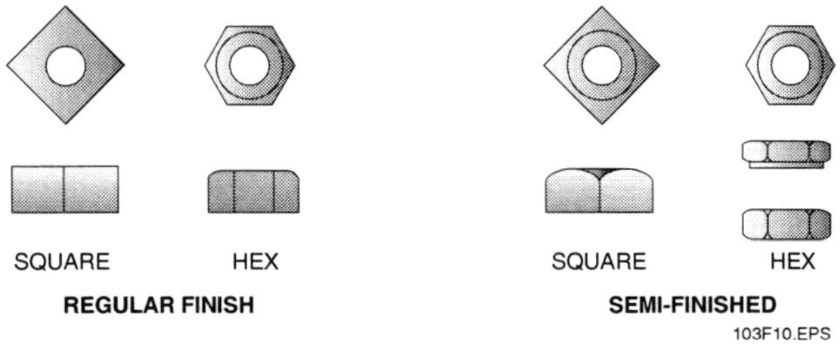

Figure 10 ◆ Nut finishes.

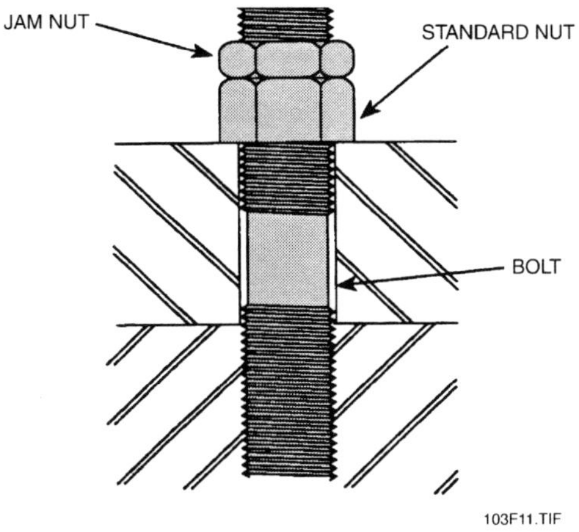

Figure 11 ◆ Jam nut.

2.3.1 Jam Nuts

A jam nut is used to lock a standard nut in place. A jam nut is a thin nut installed on top of the standard nut. *Figure 11* shows an example of a jam nut installation. Note that a regular nut can also be used as a jam nut.

2.3.2 Castellated, Slotted, and Self-Locking Nuts

Castellated (castle) and slotted nuts are slotted across the flat part of the nut. They are used with specially manufactured bolts in applications where little or no loosening of the fastener can be tolerated. After the nut has been tightened, a cotter pin is fitted in through a hole in the bolt and one set of slots in the nut. The cotter pin keeps the nut from loosening under working conditions.

Self-locking nuts are also used in many applications where loosening of the fastener cannot be tolerated. Self-locking nuts are designed with nylon inserts, or they are deliberately deformed in such a manner so they cannot work loose. An advantage of self-locking nuts is that no hole in the bolt is needed. *Figure 12* shows typical castellated, slotted, and self-locking nuts.

2.3.3 Acorn Nuts

When appearance is important or exposed, sharp thread edges on the fastener must be avoided, acorn (cap) nuts are used. The acorn nut tightens on the bolt and covers the ends of the threads. The tightening capability of an acorn nut is limited by the depth of the nut. *Figure 13* shows a typical acorn nut.

2.3.4 Wing Nuts

Wing nuts are designed to allow rapid loosening and tightening of the fastener without the need for a wrench. They are used in applications where limited **torque** is required and where frequent adjustments and service are necessary. *Figure 14* shows a typical wing nut.

> **Note**
> Wing nuts should be used for applications where hand tightening is sufficient.

Classroom

Discuss various types of nuts and their applications. Pass around examples for the trainees to examine.

Audiovisual

Show Transparencies 13 through 16 (Figures 11 through 14).

FASTENERS AND ANCHORS 323

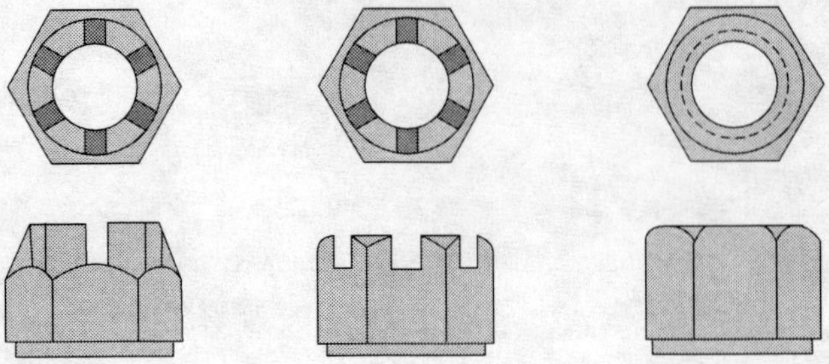

Figure 12 ◆ Castellated, slotted, and self-locking nuts.

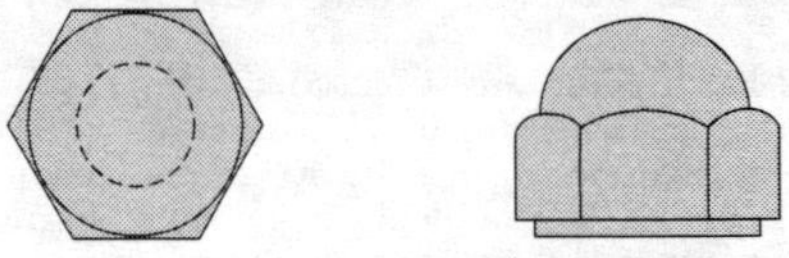

Figure 13 ◆ Acorn nut.

> **Note**
> The threads of the bolt or screw should have minimal clearance from the hole in the washer.

Classroom

Discuss various types of washers and their applications. Pass around examples for the trainees to examine.

Audiovisual

Show Transparency 17 (Figure 15).

2.4.0 Washers

There are several different types and sizes of washers. They fit over a bolt or screw to provide an enlarged surface for bolt heads and nuts. Washers also serve to distribute the fastener load over a larger area and to prevent marring of the surfaces. Standard washers are made in light, medium, heavy-duty, and extra heavy-duty series. *Figure 15* shows different types of washers.

2.4.1 Lock Washers

Lock washers are designed to keep bolts or nuts from working loose. There are various types of lock washers for different applications.

- *Split-ring*—Commonly used with bolts and cap screws.
- *External*—Used for the greatest resistance.
- *Internal*—Used with small screws.
- *Internal-external*—Used for oversized mounting holes.
- *Countersunk*—Used with flat or oval-head screws.

2.4.2 Flat and Fender Washers

Flat washers are used under bolts or nuts to spread the load over a larger area and protect the surface. Common flat washers are made to fit bolt or screw

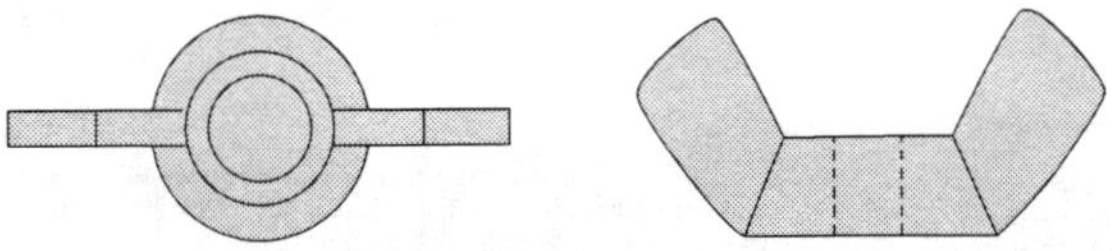

Figure 14 ◆ Wing nut.

Instructor's Notes:

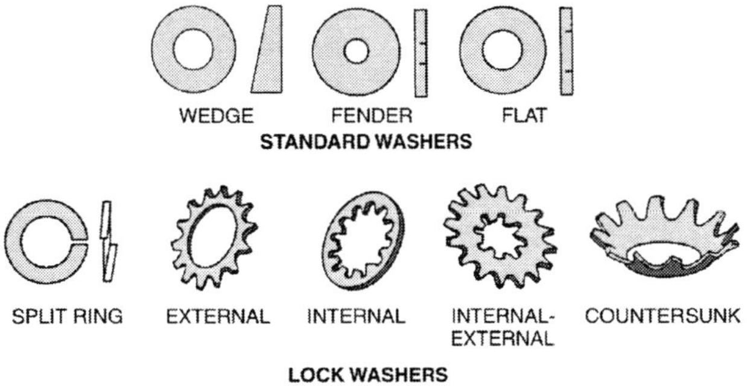

Figure 15 ◆ Washers.

> **INSIDE TRACK**
>
> ***Carrying Fasteners***
>
> A nail apron is handy for carrying a large amount of fasteners.

sizes ranging from No. 6 up to 1" with outside diameters ranging from ⅜" to 2", respectively.

Fender washers are wide-surfaced washers made to bridge oversized holes or other wide clearances to keep bolts or nuts from pulling through the material being fastened. They are flat washers that have a larger diameter and surface area than regular washers. They may also be thinner than a regular washer. Fender washers are typically made to fit bolt or screw sizes ranging from 3⁄16" to ½" with outside diameters ranging from ¾" to 2", respectively.

2.5.0 Installing Fasteners

Different types of fasteners require different installation techniques. However, all installations require knowing the proper installation methods, tightening sequence, and torque specifications for the type of fastener being used. Some bolts and nuts require that special safety wires or pins be installed to keep them from working loose.

Most fastener manufacturers provide charts that specify the size hole that should be drilled into the base material for use with each of their products (*Figure 16*). The charts typically show the proper size drill bit to use if it is necessary to first drill and tap holes for use with machine bolts, screws, or other threaded fasteners. They also show the proper size drill to use for drilling pilot holes used with metal and wood screws. (Various kinds of screws are described in detail later in this module.)

2.5.1 Torque Tightening

To properly tighten a threaded fastener, two primary factors must be considered:

- The strength of the fastener material
- The degree to which the fastener is tightened

A torque wrench is used to control the degree of tightness. The torque wrench measures how much a fastener is being tightened. *Torque* is the turning force applied to the fastener. Torque is normally expressed in **inch pounds (in. lbs.)** or **foot pounds (ft. lbs.).** A one-pound force applied to a wrench that is 1 foot long exerts 1 foot pound, or 12 inch pounds, of torque. The torque reading is shown on the indicator on the torque wrench as the fastener is being tightened. *Figure 17* shows two types of torque wrenches.

Classroom

Explain how to determine pilot hole sizes.

Audiovisual

Show Transparency 18 (Figure 16).

Classroom

Discuss torque.

FASTENERS AND ANCHORS 325

DRILL THIS SIZE HOLE		To Tap For This Size Bolt or Screw	For This Size Wood Screw Pilot in Hard Wood
Drill Size	Dec. Equiv.		
60	.0400		
59	.0410		
58	.0420		
57	.0430		
56	.0465	0 x 80	
3/64	.0469		
55	.0520		
54	.0550	1 x 56	No. 3
53	.0595	1 x 64-72	
1/16	.0625		
52	.0635		No. 4
51	.0670		
50	.0700	2 x 56-64	
49	.0730		No. 5
48	.0760		
5/64	.0781		
47	.0785	3 x 48	No. 6
46	.0810		
45	.0820	3 x 56	
44	.0860	4 x 36	No. 7
43	.0890	4 x 40	
42	.0935	4 x 48	
3/32	.0937		
41	.0960		
40	.0980	5 x 36	No. 8
39	.0995		
38	.1015	5 x 40	
37	.1040	5 x 44	No. 9
36	.1069		
7/64	.1094		
35	.1100	6 x 32	
34	.1110	6 x 36	
33	.1130	6 x 40	No. 10
32	.1160		
31	.1200		No. 11
1/8	.1250	7 x 36	
30	.1285	8 x 30	No. 12
29	.1360	8 x 32-36	
28	.1405	8 x 40	
27	.1440	9 x 30	
26	.1470	3/16 x 24	
25	.1495	10 x 24	No. 14
24	.1520		
23	.1540	10 x 28	
5/32	.1562		
22	.1570	10 x 30	
21	.1590	10 x 32	
20	.1610	3/16 x 32	
19	.1660		
18	.1695		No. 16
11/64	.1719		
17	.1730		
16	.1770	12 x 24	
15	.1800		
14	.1820	12 x 28	
13	.1850	12 x 32	No. 18
3/16	.1875		
12	.1890		

DRILL THIS SIZE HOLE		To Tap For This Size Bolt or Screw	For This Size Wood Screw Pilot in Hard Wood
Drill Size	Dec. Equiv.		
11	.1910		
10	.1935	15 x 20	
9	.1960		
8	.1990		
7	.2010	1/4 x 20	
13/64	.2031		
6	.2040		
5	.2055		
4	.2090	1/4 x 24	No. 20
3	.2130	1/4 x 28	
7/32	.2187	1/4 x 32	
2	.2210		
1	.2280		No. 24
A	.2340		
15/64	.2344		
B	.2380		
C	.2420		
D	.2460		
1/4	.2500		

DRILL THIS SIZE HOLE		To Tap For This Size Bolt or Screw
Drill Size	Dec. Equiv.	
E	.2500	
F	.2570	5/16 x 18
G	.2610	
17/64	.2656	5/16 x 18
H	.2660	
I	.2720	
J	.2770	5/16 x 24-32*
K	.2810	
9/32	.2812	5/16 x 24-32*
L	.2900	
M	.2950	
19/64	.2969	
N	.3020	
5/16	.3125	3/8" x 16-1/8" P
O	.3160	
P	.3230	
21/64	.3281	3/8 x 20-24
Q	.3332	
R	.3390	
11/32	.3437	
S	.3480	
T	.3580	
23/64	.3594	
U	.3680	
3/8	.3750	7/16 x 14
V	.3770	
W	.3860	
25/64	.3906	7/16 x 14
X	.3970	
Y	.4040	
13/32	.4062	
Z	.4130	
27/64	.4219	1/2 x 12-13
7/16	.4375	1/4" Pipe
29/64	.4531	1/2 x 20-24
15/32	.4687	1/2 x 27
31/64	.4844	9/16 x 12
1/2	.5000	

* All tap drill sizes are for 75% full thread except asterisked sizes which are 60% full thread.

103F16.EPS

Figure 16 ◆ Fastener hole guide chart.

Instructor's Notes:

326 CHAPTER 8

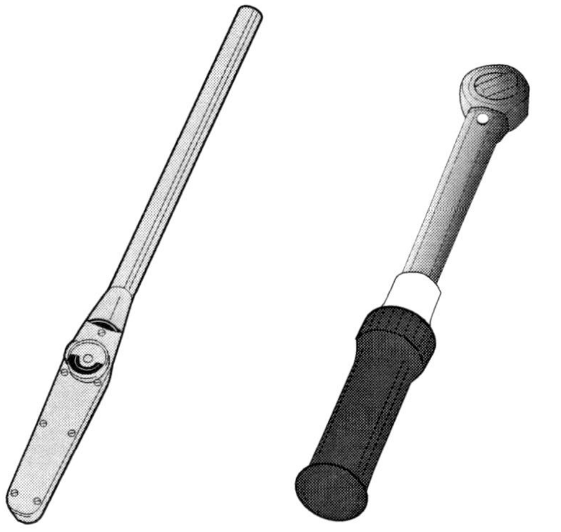

Figure 17 ◆ Torque wrenches.

Different types of bolts, nuts, and screws are torqued to different values depending on the application. Always check the project specifications and the manufacturer's manual to determine the proper torque for a particular type of fastener. *Figure 18* shows selected torque values for various graded steel bolts.

2.5.2 Installing Threaded Fasteners

The following general procedure can be used to install threaded fasteners in a variety of applications.

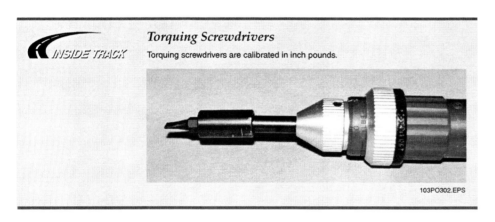

INSIDE TRACK

Torquing Screwdrivers

Torquing screwdrivers are calibrated in inch pounds.

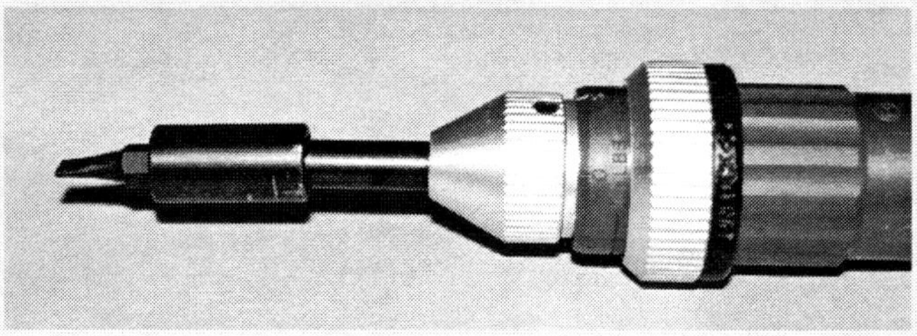

103PO302.EPS

Audiovisual

Show Transparencies 19 and 20 (Figures 17 and 18).

Demonstration

Demonstrate the use of torque wrenches and screwdrivers.

Teaching Tip

Discuss the "Think About It." Answers will vary (e.g., installing spark plugs, making connections inside switchgear, etc.).

THINK ABOUT IT

Torque Wrenches

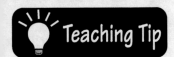

Think about situations in which the precise torque is important when installing a fastener. Do you commonly use an ordinary wrench in situations that dictate the use of a torque wrench?

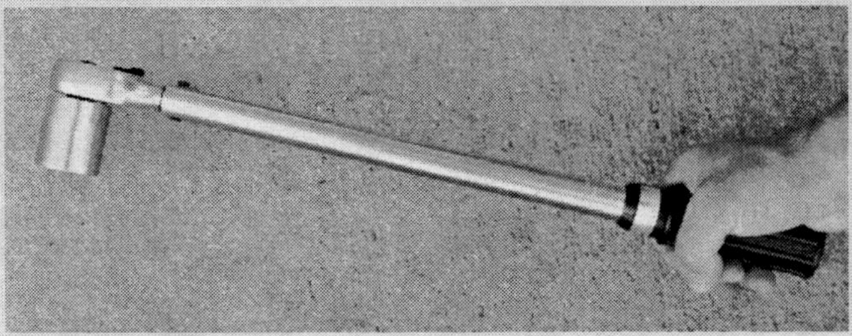

103PO303.EPS

TORQUE IN FOOT POUNDS

FASTENER DIAMETER	THREADS PER INCH	MILD STEEL	STAINLESS STEEL 18-8	ALLOY STEEL
1/4	20	4	6	8
5/16	18	8	11	16
3/8	16	12	18	24
7/16	14	20	32	40
1/2	13	30	43	60
5/8	11	60	92	120
3/4	10	100	128	200
7/8	9	160	180	320
1	8	245	285	490

SUGGESTED TORQUE VALUES FOR GRADED STEEL BOLTS

GRADE		SAE 1 OR 2	SAE 5	SAE 6	SAE 8
TENSILE STRENGTH		64,000 PSI	105,000 PSI	130,000 PSI	150,000 PSI
GRADE MARK		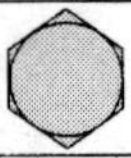	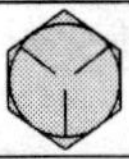	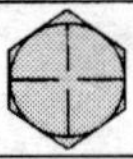	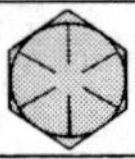
BOLT DIAMETER	THREADS PER INCH	FOOT POUNDS TORQUE			
1/4	20	5	7	10	10
5/16	18	9	14	19	22
3/8	16	15	25	34	37
7/16	14	24	40	55	60
1/2	13	37	60	85	92
9/16	12	53	88	120	132
5/8	11	74	120	169	180
3/4	10	120	200	280	296
7/8	9	190	302	440	473
1	8	282	466	660	714

103F18.EPS

Figure 18 ◆ Torque value chart.

Instructor's Notes:

Note
When installing threaded fasteners for a specific job, make sure to check all installation requirements.

WARNING!
Follow all safety precautions.

Step 1 Select the proper bolts or screws for the job.

Step 2 Check for damaged or dirty internal and external threads.

Step 3 Clean the bolt or screw threads. Do not lubricate the threads if a torque wrench is to be used to tighten the nuts.

Step 4 Insert the bolts through the predrilled holes and tighten the nuts by hand. Or, insert the screws through the holes and start the threads by hand.

Note
Turn the nuts or screws several turns by hand and check for cross threading.

Step 5 Following the proper tightening sequence, tighten the bolts or screws snugly.

Step 6 Check the torque specification. Following the proper tightening sequence, tighten each bolt, nut, or screw several times approaching the specified torque. Tighten to the final torque specification.

Step 7 If required to keep the bolts or nuts from working loose, install jam nuts, cotter pins, or safety wire. *Figure 19* shows fasteners with a safety wire installed.

3.0.0 ◆ NON-THREADED FASTENERS

Non-threaded fasteners have many uses in the electrical field. Different types of non-threaded fasteners include retainers, keys, pins, clamps, washers, rivets, and tie wraps.

3.1.0 Retainer Fasteners

Retainer fasteners, also called *retaining rings,* are used for both internal and external applications. Some retaining rings are seated in grooves in the fastener. Other types of retainer fasteners are self-locking and do not require a groove. To easily remove internal and external retainer rings without damaging the ring or the fastener, special pliers are used. *Figure 20* shows several types of retainer fasteners.

3.2.0 Keys

To prevent a gear or pulley from rotating on a shaft, keys are inserted. Half of the key fits into a

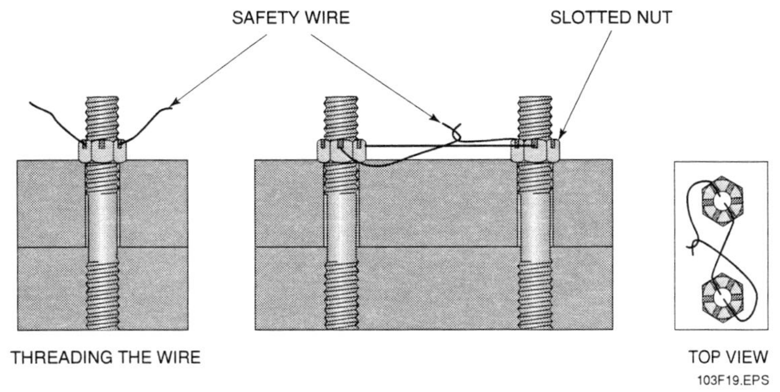

Figure 19 ◆ Safety-wired fasteners.

Demonstration

Demonstrate the procedure for installing threaded fasteners.

Audiovisual

Show Transparency 21 (Figure 19).

Laboratory

Have the trainees practice installing threaded fasteners. Note the proficiency of each trainee.

Classroom

Discuss retainer fasteners and their applications. Pass around examples for the trainees to examine.

Audiovisual

Show Transparency 22 (Figure 20).

FASTENERS AND ANCHORS 329

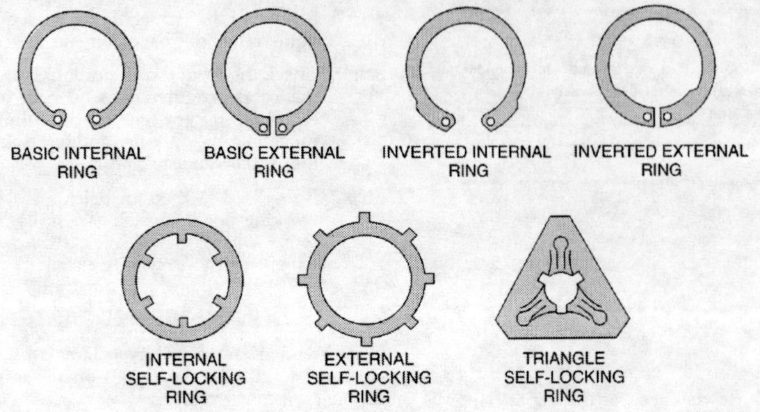

Figure 20 ◆ Retainer fasteners (rings).

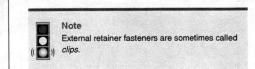

> **Note**
> External retainer fasteners are sometimes called *clips.*

keyseat on the shaft. The other half fits into a **keyway** in the hub of the gear or pulley. The key fastens the two parts together, stopping the gear or pulley from turning on the shaft. *Figure 21* shows several types of keys and keyways and their uses.

Some different types of keys include:

- *Square key*—Usually one-quarter of the shaft diameter. It may be slightly tapered on the top for easier fitting.
- *Pratt and Whitney key*—Similar to the square key, but rounded at both ends. It fits into a keyseat of the same shape.
- *Gib head key*—Interchangeable with the square key. The head design allows easy removal from the assembly.
- *Woodruff key*—Semicircular shape that fits into a keyseat of the same shape. The top of the key fits into the keyway of the mating part.

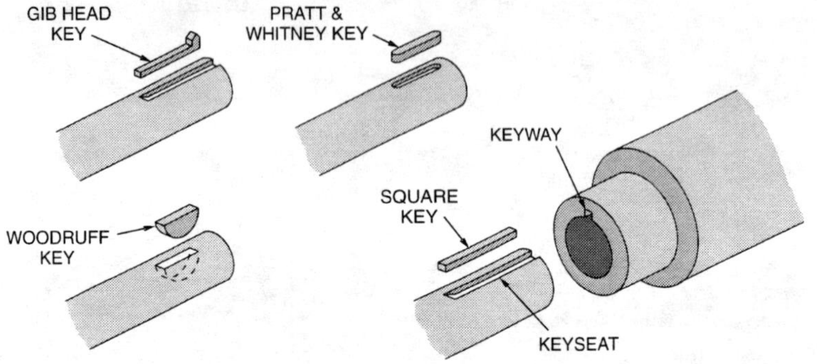

Figure 21 ◆ Keys and keyways.

Classroom

Discuss various types of keys and their applications. Pass around examples for the trainees to examine.

Audiovisual

Show Transparency 23 (Figure 21).

Instructor's Notes:

3.3.0 Pin Fasteners

Pin fasteners come in several types and sizes. They have a variety of applications. Common uses of pin fasteners include holding moving parts together, aligning mating parts, fastening hinges, holding gears and pulleys on shafts, and securing slotted nuts. *Figure 22* shows several pin fasteners.

3.3.1 Dowel Pins

Dowel pins fit into holes to position mating parts. They may also support a portion of the load placed on the parts. *Figure 23* shows an application of dowel pins used to position mating parts.

3.3.2 Taper and Spring Pins

Taper and spring pins are used to fasten gears, pulleys, and collars to a shaft. *Figure 24* shows how taper and spring pins are used to attach a component to a shaft. The groove in a spring pin allows it to compress against the walls in a spring-like fashion.

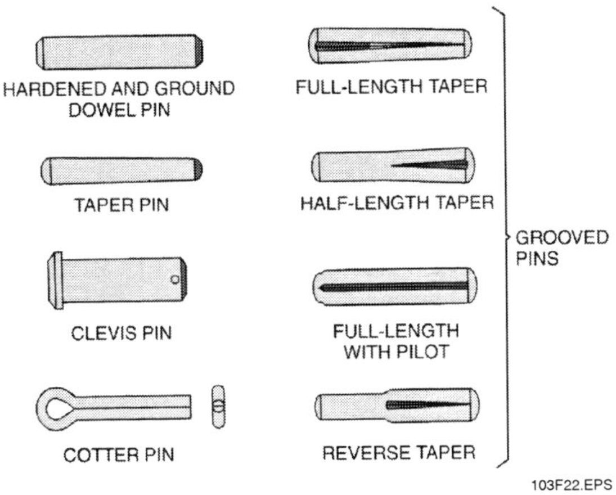

Figure 22 ◆ Pin fasteners.

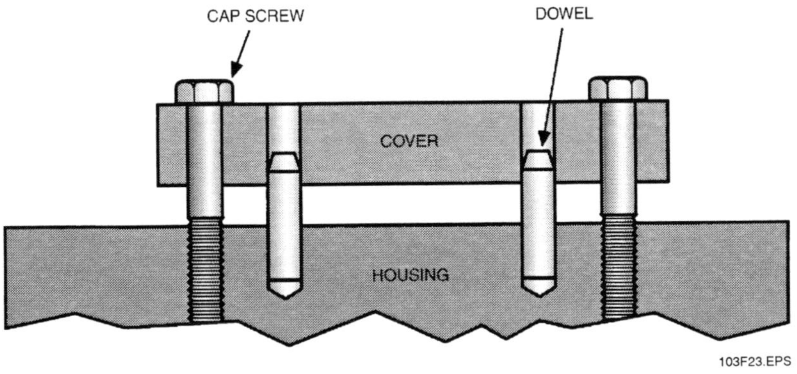

Figure 23 ◆ Dowel pins.

Classroom

Discuss various types of pin fasteners and their applications. Pass around examples for the trainees to examine.

Audiovisual

Show Transparencies 24 through 27 (Figures 22 through 25).

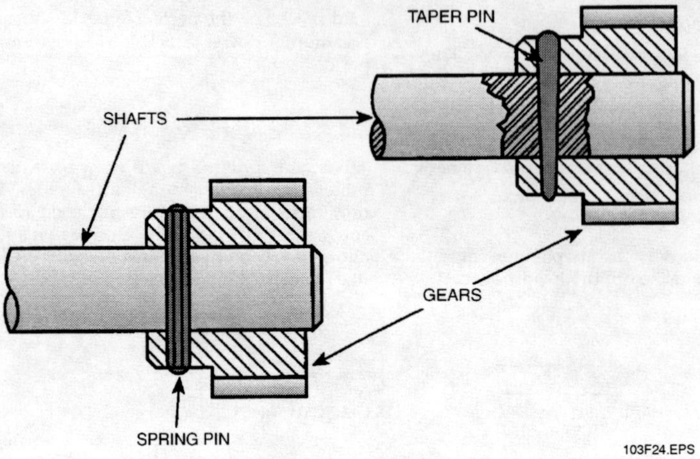

Figure 24 ◆ Taper and spring pins.

3.3.3 Cotter Pins

There are several different types of cotter pins used as a locking device for a variety of applications. Cotter pins are often inserted through a hole drilled crosswise through a shaft to prevent parts from slipping on or off the shaft. They are also used to keep slotted nuts from working loose. Standard cotter pins are general-use pins. When installed, the extended prong is normally bent back over the nut to provide the locking action. If it is ever removed, throw it away and replace it with a new one. The humped, cinch, and hitch-type cotter pins are self-locking pins. The humped and cinch type should also be thrown away and replaced with a new one if removed. The hitch pin, also called a *hair pin,* is a reusable pin made to be installed and removed quickly. *Figure 25* shows several common types of cotter pins.

Classroom

Discuss blind/pop rivets and their applications. Pass around examples for the trainees to examine.

Audiovisual

Show Transparencies 28 and 29 (Figures 26 and 27).

3.4.0 Blind/Pop Rivets

When only one side of a joint can be reached, blind rivets can be used to fasten the parts together. Some applications of blind rivets include fastening light to heavy gauge sheet metal, fiberglass, plastics, and belting. Blind rivets are made of a variety of materials and come in several sizes and lengths. They are installed using special riveting tools. *Figure 26* shows a typical blind rivet installation.

Blind rivets are installed through drilled or punched holes using a special blind (pop) rivet gun. *Figure 27* shows a typical rivet gun.

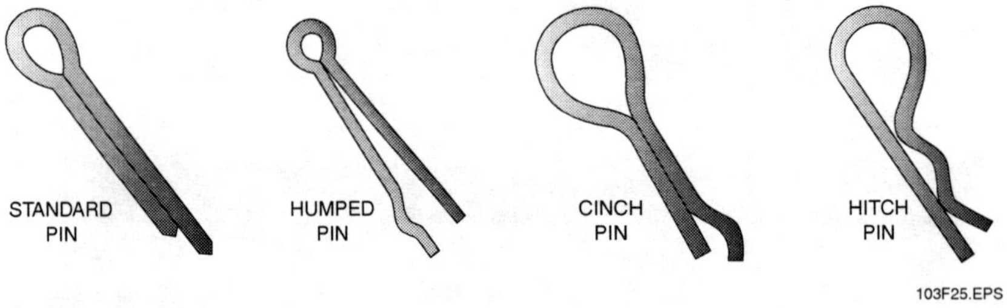

Figure 25 ◆ Cotter pins.

Instructor's Notes:

332 CHAPTER 8

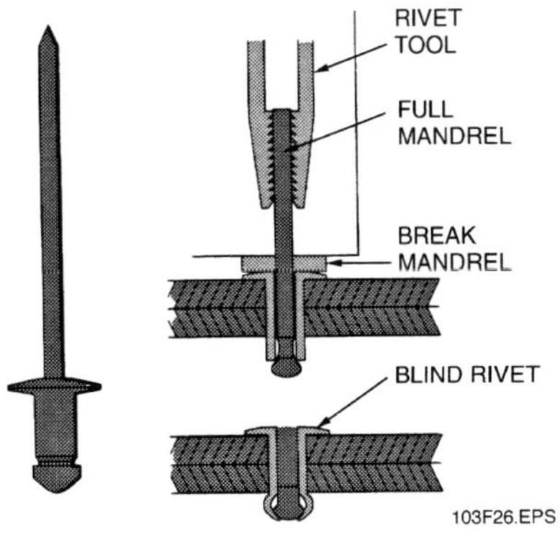

Figure 26 ◆ Blind rivet installation.

Use the following general procedure to install blind rivets.

> **WARNING!**
> Follow all safety precautions. Make sure to wear proper eye and face protection when riveting.

Step 1 Select the correct length and diameter of blind rivet to be used.

Step 2 Select the appropriate drill bit for the size of rivet being used.

Step 3 Drill a hole through both parts being connected.

Safety

Discuss the safety precautions involved with the use of a rivet gun.

Demonstration

Demonstrate the use of a rivet gun.

> INSIDE TRACK
> ### *Rivet Guns*
> Rivet guns are designed for blind riveting (i.e., when you can reach only one side of the material).

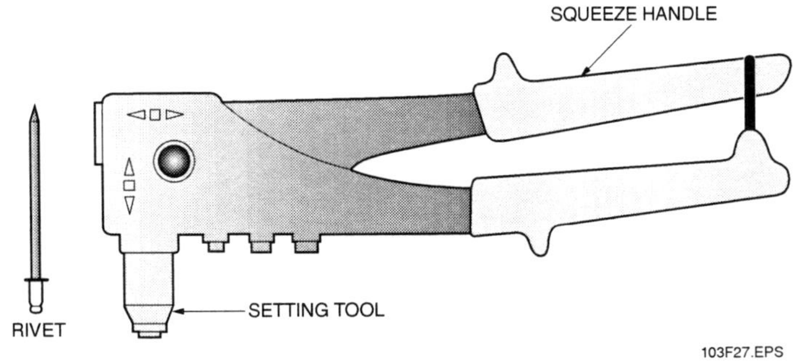

Figure 27 ◆ Rivet gun.

> INSIDE TRACK
> ### *Using a Rivet Gun*
> Before you attach materials using a rivet gun and possibly do something you'll regret later, make a spacing template. Hold the template up to your material to ensure that all rivets fall into the correct locations, then secure your pieces. When using the template for actual riveting, hold it in position with an awl or drift pin inserted into your predrilled rivet holes. Note that rivets may pull through certain materials or when joining two pieces of flexible material together when the hole is slightly larger than required.

Audiovisual

Show Transparencies 30 and 31 (Figures 28 and 29).

Laboratory

Under careful supervision, have the trainees practice installing blind/pop rivets. Note the proficiency of each trainee.

Classroom

Discuss various types of tie wraps and their applications. Pass around examples for the trainees to examine.

Audiovisual

Show Transparency 32 (Figure 30).

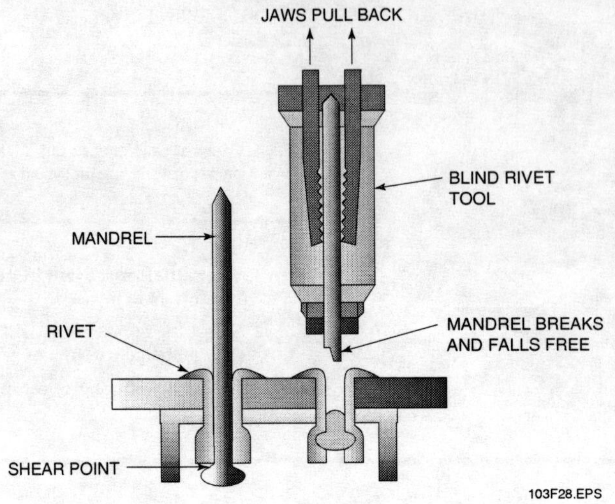

Figure 28 ◆ Joining parts.

Step 4 Inspect the rivet gun for any defects that might make it unsafe for use.

Step 5 Place the rivet mandrel into the proper size setting tool.

Step 6 Insert the rivet end into the predrilled hole.

Step 7 Install the rivet by squeezing the handle of the rivet gun, causing the jaws in the setting tool to grip the mandrel. The mandrel is pulled up, expanding the rivet until it breaks at the shear point. *Figure 28* shows the rivet and tool positioned for joining parts together.

Step 8 Inspect the rivet to make sure the pieces are firmly riveted together and that the rivet is properly installed. *Figure 29* shows a properly installed blind rivet.

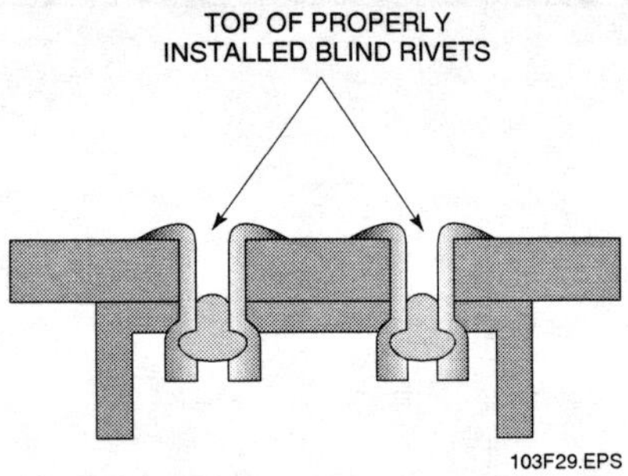

Figure 29 ◆ Properly installed blind rivets.

3.5.0 Tie Wraps

A tie wrap is a one-piece, self-locking cable tie, usually made of nylon, that is used to fasten a bundle of wires and cables together. Tie wraps can be quickly installed either manually or using a special installation tool. Black tie wraps resist ultraviolet light and are recommended for outdoor use.

Tie wraps are made in standard, cable strap and clamp, and identification configurations (*Figure 30*). All types function to clamp bundled wires or cables together. In addition, the cable strap and clamp has a molded mounting hole in the head used to secure the tie with a rivet, screw, or bolt after the tie wrap has been installed around the wires or cable. Identification tie wraps have a large flat area provided for imprinting or writing cable identification information. There is also a releasable version available. It is a non-permanent tie used for bundling wires or cables that may require frequent additions or deletions. Cable ties are made in various lengths ranging from about 3"

Instructor's Notes:

334 CHAPTER 8

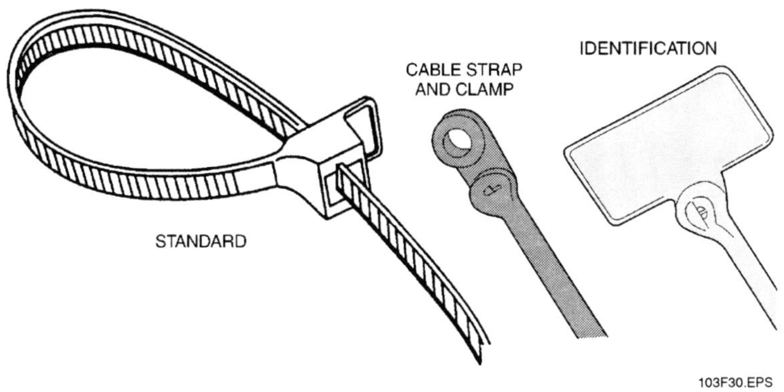

Figure 30 ◆ Tie wraps.

to 30", allowing them to be used for fastening wires and cables into bundles with diameters ranging from about ½" to 9", respectively. Tie wraps can also be attached to a variety of adhesive mounting bases made for that purpose.

4.0.0 ◆ SPECIAL THREADED FASTENERS

Special threaded fasteners consist of hardware manufactured in several shapes and sizes and designed to perform specific jobs. Certain types of nuts may be considered special threaded fasteners if they are designed especially for a particular application. In the electrical craft, special threaded fasteners are used on a number of different jobs.

The three types of special threaded fasteners described below are eye bolts, anchor bolts, and J-bolts.

4.1.0 Eye Bolts

Eye bolts get their name from the eye or loop at one end. The other end of an eye bolt is threaded. There are many types of eye bolts. The eye on some eye bolts is formed and welded, while the eye on other types is forged. Shoulder-forged eye bolts are commonly used as lifting devices and guides for wires, cables, and cords. *Figure 31* shows some typical eye bolts.

4.2.0 Anchor Bolts

An anchor bolt is used to fasten parts, machines, and equipment to concrete or masonry foundations, floors, and walls. There are several types of anchor bolts designed for different applications. *Figure 32* shows a type of anchor bolt for use in wet concrete. If the concrete has already hardened, expansion anchor bolts are used. These are covered later in this module.

One common method used to install anchor bolts in wet concrete involves making a wooden template to locate the anchor bolts. The template positions the anchor bolts so that they correspond to those in the equipment to be fastened.

4.3.0 J-Bolts

J-bolts get their name from the curve on one end that gives them a J shape. The other end of a J-bolt is threaded. There are many types of J-bolts. Some J-bolts are used to hold tubing bundles and include a plastic jacket to protect the tubing. Others are used to attach equipment to existing grating. Most J-bolts used in tubing racks are attached using two nuts. The upper nut allows for adjustment. The tubing bundle is clamped firmly, but not flattened. Both nuts are tightened against the tube track for positive holding. *Figure 33* shows a typical J-bolt.

5.0.0 ◆ SCREWS

Screws are made in a variety of shapes and sizes for different fastening jobs. The finish or coating used on a screw determines whether it is for interior or exterior use, corrosion resistant, etc. Screws of all types have heads with different shapes and slots similar to those previously described for machine

Classroom

Discuss eye bolts, anchor bolts, and J-bolts. Pass around examples for the trainees to examine.

Audiovisual

Show Transparencies 33 through 35 (Figures 31 through 33).

FASTENERS AND ANCHORS 335

INSIDE TRACK

Tie Wraps

Tie wraps are available in a wide variety of colors that can be used to color code different cable bundles.

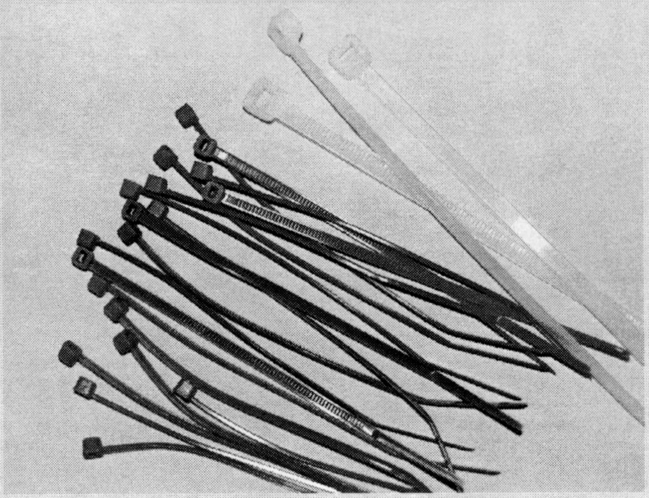

103PO304.EPS

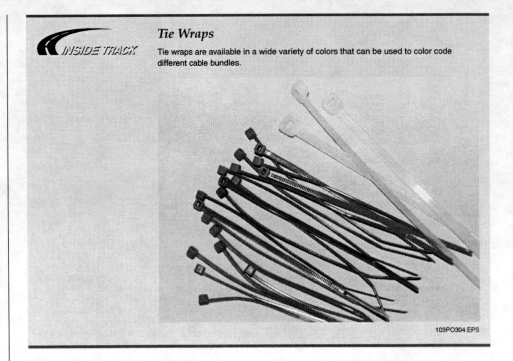

screws. Some have machine threads and are self-drilling. The size or diameter of a screw body or shank is given in gauge numbers ranging from No. 0 to No. 24, and in fractions of an inch for screws with diameters larger than ¼". The higher the gauge number, the larger the diameter of the shank. Screw lengths range from ¼" to 6", measured from the tip to the part of the head that is flush to the surface when driven in. When choosing a screw for an application, you must consider the type and thickness of the materials to be fastened, the size of the screw, the material it is made of, the

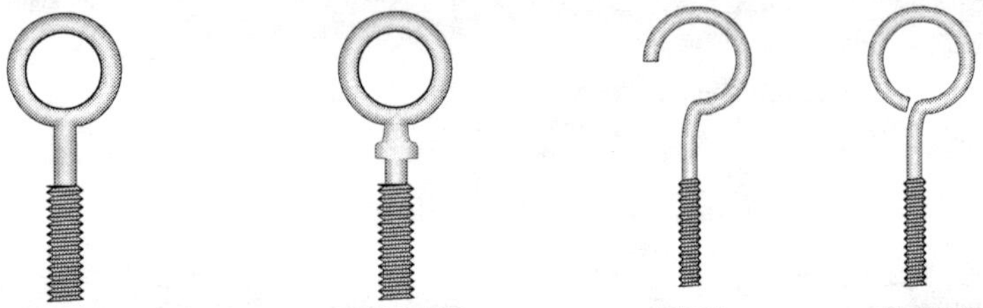

103F31.EPS

Figure 31 ◆ Eye bolts.

Instructor's Notes:

336 CHAPTER 8

shape of its head, and the type of driver. Because of the wide diversity in the types of screws and their application, always follow the manufacturer's recommendation to select the right screw for the job. To prevent damage to the screw head or the material being fastened, always use a screwdriver or power driver bit with the proper size and shape tip to fit the screw.

Some of the more common types of screws are:

- Wood screws
- Lag screws
- Masonry/concrete screws
- Thread-forming and thread-cutting screws
- Deck screws
- Drywall screws
- Drive screws

5.1.0 Wood Screws

Wood screws (*Figure 34*) are typically used to fasten boxes, panel enclosures, etc. to wood framing or structures where greater holding power is

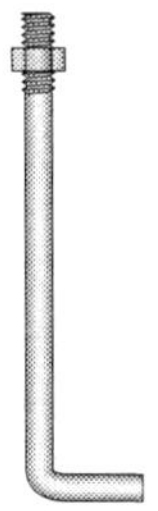

103F32.EPS

Figure 32 ◈ Anchor bolt.

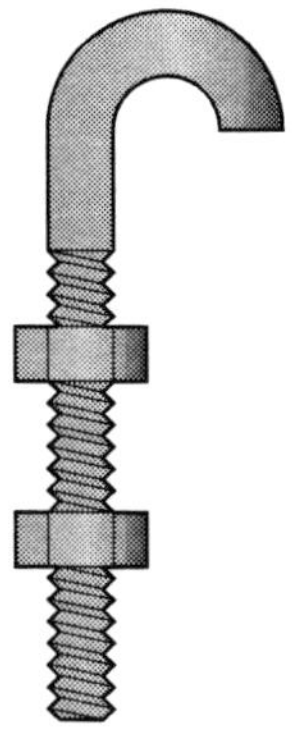

103F33.EPS

Figure 33 ◈ Typical J-bolt.

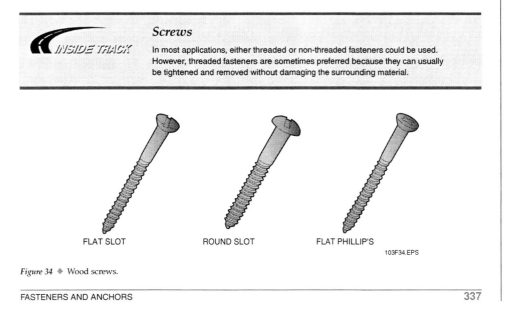

INSIDE TRACK

Screws

In most applications, either threaded or non-threaded fasteners could be used. However, threaded fasteners are sometimes preferred because they can usually be tightened and removed without damaging the surrounding material.

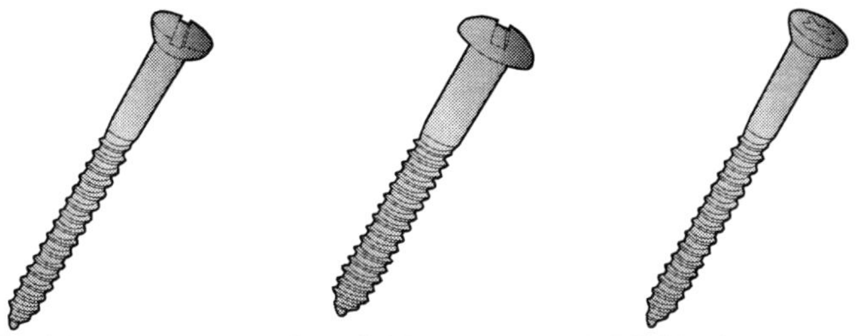

103F34.EPS

Figure 34 ◈ Wood screws.

Classroom

Discuss wood screws and their applications. Pass around examples for the trainees to examine.

Audiovisual

Show Transparency 36 (Figure 34).

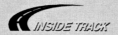

INSIDE TRACK

Driving Wood Screws

To maintain holding power, be careful not to drill your pilot hole too large. It's wise to drill a pilot hole deep enough to equal about two-thirds of the length of the threaded portion of the screw. Additionally, to lubricate screw threads, use soap, which makes the screw easier to drive.

Classroom

Discuss lag screws and their applications. Pass around examples for the trainees to examine.

Discuss concrete/masonry screws and their applications. Pass around examples for the trainees to examine.

Audiovisual

Show Transparency 37 (Figure 36).

needed than can be provided by nails. They are also used to fasten equipment to wood in applications where it may occasionally need to be unfastened and removed. Wood screws are commonly made in lengths from ¼" to 4", with shank gauge sizes ranging from 0 to 24. The shank size used is normally determined by the size hole provided in the box, panel, etc. to be fastened. When determining the length of a wood screw to use, a good rule of thumb is to select screws long enough to allow about ⅔ of the screw length to enter the piece of wood that is being gripped.

5.2.0 Lag Screws and Shields

Lag screws (*Figure 35*) or lag bolts are heavy-duty wood screws with square- or hex-shaped heads that provide greater holding power. Lag screws with diameters ranging between ¼" and ½" and lengths ranging from 1" to 6" are common. They are typically used to fasten heavy equipment to wood, but can also be used to fasten equipment to concrete when a lag shield is used.

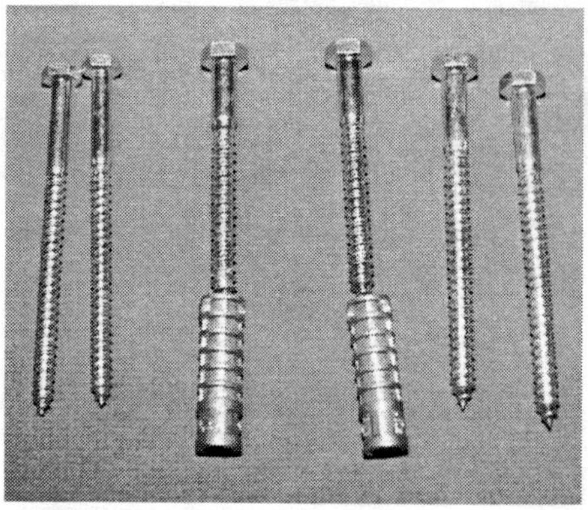

103F35.EPS

Figure 35 ◆ Lag screws and shields.

A lag shield is a lead tube that is split lengthwise but remains joined at one end. It is placed in a predrilled hole in the concrete. When a lag screw is screwed into the lag shield, the shield expands in the hole, firmly securing the lag screw. In hard masonry, short lag shields (typically 1" to 2" long) may be used to minimize drilling time. In soft or weak masonry, long lag shields (typically 1½" to 3" long) should be used to achieve maximum holding strength.

Make sure to use the proper length lag screw to achieve proper expansion. The length of the lag screw used should be equal to the thickness of the component being fastened plus the length of the lag shield. Also, drill the hole in the masonry to a depth approximately ½" longer than the shield being used. If the head of a lag screw rests directly on wood when installed, a flat washer should be placed under the head to prevent the head from digging into the wood as the lag screw is tightened down. Be sure to take the thickness of any washers used into account when selecting the length of the screw.

5.3.0 Concrete/Masonry Screws

Concrete/masonry screws (*Figure 36*), commonly called *self-threading anchors,* are used to fasten a device or fixture to concrete, block, or brick. No anchor is needed. To provide a matched tolerance anchor-

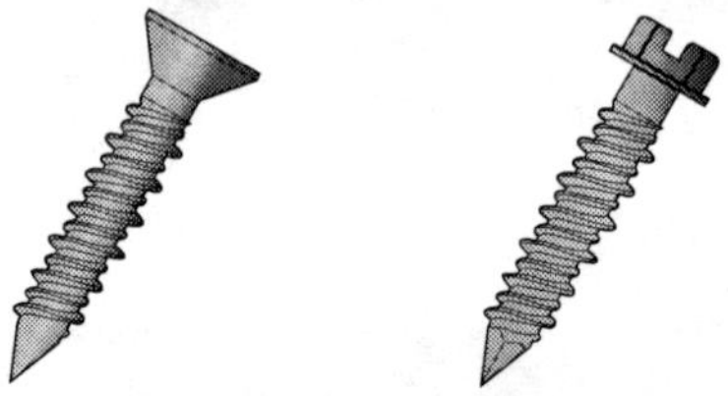

103F36.EPS

Figure 36 ◆ Concrete screws.

Instructor's Notes:

338 CHAPTER 8

ing system, the screws are installed using specially designed carbide drill bits and installation tools made for use with the screws. These tools are typically used with a standard rotary drill hammer. The installation tool, along with an appropriate drive socket or bit, is used to drive the screws directly into predrilled holes that have a diameter and depth specified by the screw manufacturer. When being driven into the concrete, the widely spaced threads on the screws cut into the walls of the hole to provide a tight friction fit. Most types of concrete/masonry screws can be removed and reinstalled to allow for shimming and leveling of the fastened device.

5.4.0 Thread-Forming and Thread-Cutting Screws

Thread-forming screws (*Figure 37*), commonly called *sheet metal screws,* are made of hard metal. They form a thread as they are driven into the work. This thread-forming action eliminates the need to tap a hole before installing the screw. To achieve proper holding, it is important to make sure to use the proper size bit when drilling pilot holes for thread-forming screws. The correct drill bit size used for a specific size screw is usually marked on the box containing the screws. Some types of thread-forming screws also drill their own holes, eliminating drilling, punching, and aligning parts. Thread-forming screws are primarily used to fasten light-gauge metal parts together. They are made in the same diameters and lengths as wood screws.

Hardened steel thread-cutting metal screws with blunt points and fine threads (*Figure 38*) are used to join heavy-gauge metals, metals of different gauges, and nonferrous metals. They are also used to fasten sheet metal to building structural members. These screws are made of hardened steel that is harder than the metal being tapped.

Figure 37 ◈ Thread-forming screws.

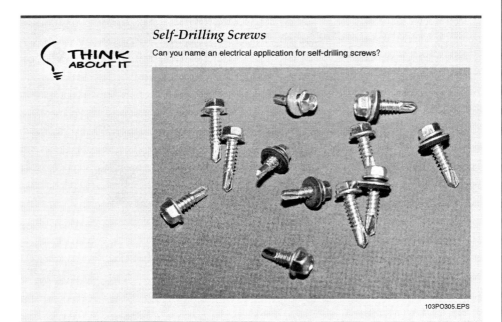

THINK ABOUT IT

Self-Drilling Screws

Can you name an electrical application for self-drilling screws?

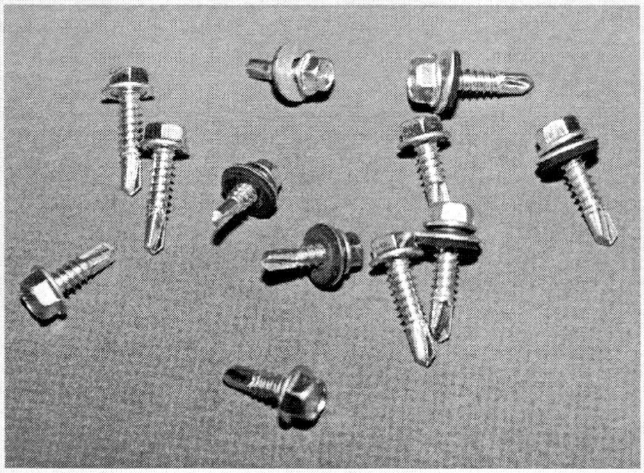

103PO305.EPS

Classroom

Discuss thread-forming and thread-cutting screws and their applications. Pass around examples for the trainees to examine.

Audiovisual

Show Transparencies 38 and 39 (Figures 37 and 38).

Teaching Tip

Discuss the "Think About It." Self-drilling screws are commonly used to mount boxes to metal studs.

FASTENERS AND ANCHORS 339

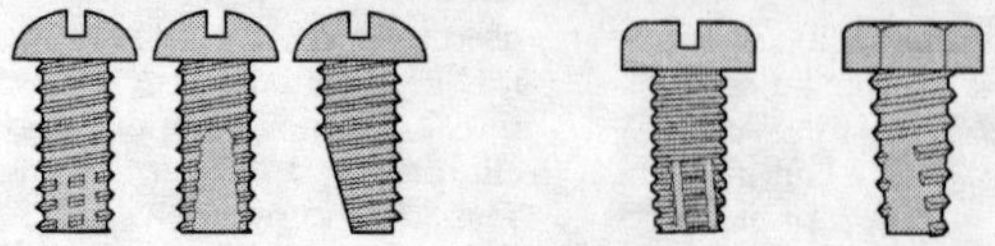

Figure 38 ◆ Thread-cutting screws.

Classroom

Discuss deck, drywall, and drive screws. Pass around examples for the trainees to examine.

Audiovisual

Show Transparencies 40 through 42 (Figures 39 through 41).

Demonstration

Demonstrate how to install selected screws.

Laboratory

Have the trainees practice installing selected screws. Note the proficiency of each trainee.

They cut threads by removing and cutting a portion of the metal as they are driven into a pilot hole and through the material.

5.5.0 Deck Screws

Deck screws (*Figure 39*) are made in a wide variety of shapes and sizes for different indoor and outdoor applications. Some are made to fasten pressure-treated and other types of wood decking to wood framing. Self-drilling types are made to fasten wood decking to different gauges of metal support structures. Similarly, other self-drilling kinds are made to fasten metal decking and sheeting to different gauges and types of metal structural support members. Because of their wide diversity, it is important to follow the manufacturer's recommendations for selection of the proper screw for a particular application. Many manufacturers make a stand-up installation tool used for driving their deck screws. Use of this tool eliminates angle driving, underdriven or overdriven screws, screw wobble, etc. It also reduces operator fatigue.

5.6.0 Drywall Screws

Drywall screws (*Figure 40*) are thin, self-drilling screws with bugle-shaped heads. Depending on the type of screw, it cuts through the wallboard and anchors itself into wood and/or metal studs, holding the wallboard tight to the stud. Coarse thread screws are normally used to fasten wallboard to wood studs. Fine thread and high-and-low thread types are generally used for fastening to metal studs. Some screws are made for use in either wood or metal. A Phillip's or Robertson drive head allows the drywall screw to be countersunk without tearing the surface of the wallboard.

5.7.0 Drive Screws

Drive screws do not require that the hole be tapped. They are installed by hammering the screw into a drilled or punched hole of the proper size. Drive screws are mostly used to fasten parts that will not be exposed to much pressure. A typical use of drive screws is to attach permanent name plates on electric motors and other types of equipment. *Figure 41* shows typical drive screws.

6.0.0 ◆ HAMMER-DRIVEN PINS AND STUDS

Hammer-driven pins or threaded studs (*Figure 42*) can be used to fasten wood or steel to concrete or block without the need to predrill holes. The

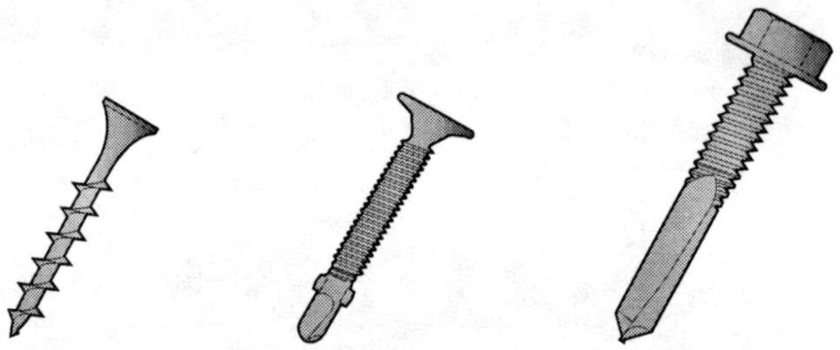

Figure 39 ◆ Typical deck screws.

Instructor's Notes:

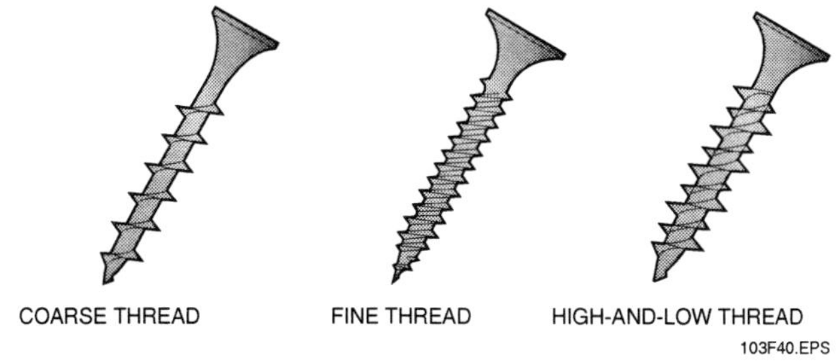

Figure 40 ◈ Drywall screws.

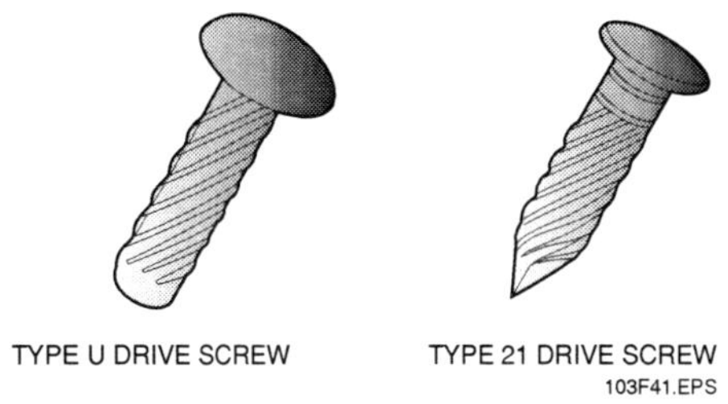

Figure 41 ◈ Drive screws.

pin or threaded stud is inserted into a hammer-driven tool designed for its use. The pin or stud is inserted in the tool point end out with the washer seated in the recess. The pin or stud is then positioned against the base material where it is to be fastened and the drive rod of the tool tapped lightly until the striker pin contacts the pin or stud. Following this, the tool's drive rod is struck using heavy blows with about a two-pound engineer's hammer. The force of the hammer blows is transmitted through the tool directly to the head of the fastener, causing it to be driven into the concrete or block. For best results, the drive pin or stud should be embedded a minimum of ½" in hard concrete to 1¼" in softer concrete block.

7.0.0 ◈ POWDER-ACTUATED TOOLS AND FASTENERS

Powder-actuated tools (*Figure 43*) can be used to drive a wide variety of specially designed pin and

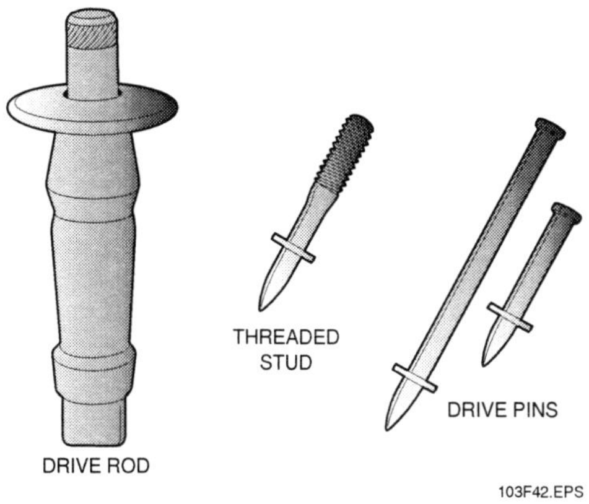

Figure 42 ◈ Hammer-driven pins and installation tool.

Classroom

Discuss hammer-driven pins and studs and their applications. Pass around examples for the trainees to examine.

Audiovisual

Show Transparency 43 (Figure 42).

Demonstration

Demonstrate how to install hammer-driven pins.

FASTENERS AND ANCHORS 341

![Classroom]

Discuss powder-actuated tools and their applications.

![Audiovisual]

Show Transparency 44 (Figure 43).

![Safety]

Discuss the safety precautions involved with the use of powder-actuated tools.

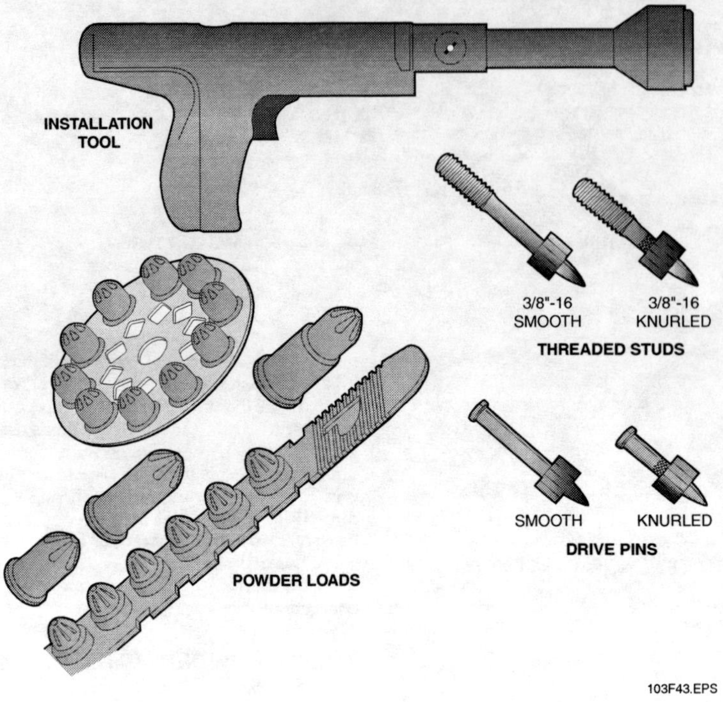

Figure 43 ◆ Powder-actuated installation tool and fasteners.

threaded stud-type fasteners into masonry and steel. These tools look and fire like a gun and use the force of a detonated gunpowder load (typically 22, 25, or 27 caliber) to drive the fastener into the material. The depth to which the pin or stud is driven is controlled by the density of the base material in which the pin or stud is being installed and by the power level or strength of the cased powder load.

Powder loads and their cases are designed for use with specific types and/or models of powder-actuated tools and are not interchangeable. Typically, powder loads are made in 12 increasing power or load levels used to achieve the proper penetration. The different power levels are identified by a color-code system and load case types. Note that different manufacturers may use different color codes to identify load strength. Power level 1 is the lowest power level while 12 is the highest. Higher number power levels are used when driving into hard materials or when a deeper penetration is needed. Powder loads are available as single-shot units for use with single-shot tools. They are also made in multi-shot strips or disks for semiautomatic tools.

> **WARNING!**
> Powder-actuated fastening tools are to be used only by trained and licensed operators and in accordance with the tool operator's manual. You must carry your license with you whenever you are using a powder-actuated tool.

OSHA Standard 29 CFR 1926.302(e) governs the use of powder-actuated tools and states that only those individuals who have been trained in the op-

Instructor's Notes:

eration of the particular powder-actuated tool in use be allowed to operate it. Authorized instructors available from the various powder-actuated tool manufacturers generally provide such training and licensing. Trained operators must take precautions to protect both themselves and others in the area when using a powder-actuated driver tool:

- Always use the tool in accordance with the published tool operation instructions. The instructions should be kept with the tool. Never attempt to override the safety features of the tool.
- Never place your hand or other body parts over the front muzzle end of the tool.
- Use only fasteners, powder loads, and tool parts specifically made for use with the tool. Use of other materials can cause improper and unsafe functioning of the tool.
- Operators and bystanders must wear eye and hearing protection along with hard hats. Other personal safety gear, as required, must also be used.
- Always post warning signs that state *Powder-Actuated Tool in Use* within 50 feet of the area where tools are used.
- Prior to using a tool, make sure it is unloaded and perform a proper function test. Check the functioning of the unloaded tool as described in the published tool operation instructions.
- Do not guess before fastening into any base material; always perform a center punch test.
- Always make a test firing into a suitable base material with the lowest power level recommended for the tool being used. If this does not set the fastener, try the next higher power level. Continue this procedure until the proper fastener penetration is obtained.
- Always point the tool away from operators or bystanders.
- Never use the tool in an explosive or flammable area.
- Never leave a loaded tool unattended. Do not load the tool until you are prepared to complete the fastening. Should you decide not to make a fastening after the tool has been loaded, always remove the powder load first, then the fastener. Always unload the tool before cleaning, servicing, or when changing parts, prior to work breaks, and when storing the tool.
- Always hold the tool perpendicular to the work surface and use the spall (chip or fragment) guard or stop spall whenever possible.
- Always follow the required spacing, edge distance, and base material thickness requirements.
- Never fire through an existing hole or into a weld area.
- In the event of a misfire, always hold the tool depressed against the work surface for at least 30 seconds. If the tool still does not fire, follow the published tool instructions. Never carelessly discard or throw unfired powder loads into a trash receptacle.
- Always store the powder loads and unloaded tool under lock and key.

8.0.0 ◈ MECHANICAL ANCHORS

Mechanical anchors are devices used to give fasteners a firm grip in a variety of materials, where the fasteners by themselves would otherwise have a tendency to pull out. Anchors can be classified in many ways by different manufacturers. In this module, anchors have been divided into five broad categories:

- One-step anchors
- Bolt anchors
- Screw anchors
- Self-drilling anchors
- Hollow-wall anchors

8.1.0 One-Step Anchors

One-step anchors are designed so that they can be installed through the mounting holes in the component to be fastened. This is because the anchor and the drilled hole into which it is installed have the same size diameter. They come in various diameters ranging from ¼" to 1¼" with lengths ranging from 1¾" to 12". Wedge, stud, sleeve, one-piece, screw, and nail anchors (*Figure 44*) are common types of one-step anchors.

8.1.1 Wedge Anchors

Wedge anchors are heavy-duty anchors supplied with nuts and washers. The drill bit size used to drill the hole is the same diameter as the anchor. The depth of the hole is not critical as long as the minimum length recommended by the manufacturer is drilled. After the hole is blown clean of dust and other material, the anchor is inserted into the hole and driven with a hammer far enough so that at least six threads are below the top surface of the component. Then, the component is fastened by tightening the anchor nut to expand the anchor and tighten it in the hole.

8.1.2 Stud Bolt Anchors

Stud bolt anchors are heavy-duty threaded anchors. Because this type of anchor is made to

Classroom

List the common types of mechanical anchors. Refer the trainees to *Appendix A* for an overview.

Discuss one-step anchors and their applications. Pass around examples for the trainees to examine.

Audiovisual

Show Transparency 45 (Figure 44).

FASTENERS AND ANCHORS 343

INSIDE TRACK

Powder-Actuated Fasteners

Powder-actuated fasteners can be used directly against the surface to be fastened or they can be used with special extension rods for overhead or remote fastening in tight locations.

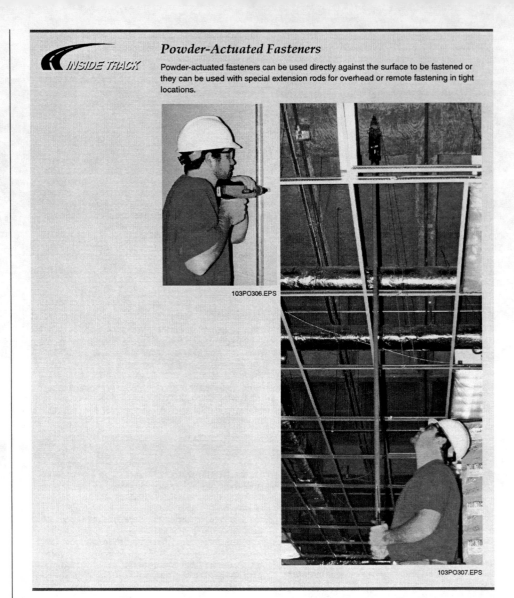

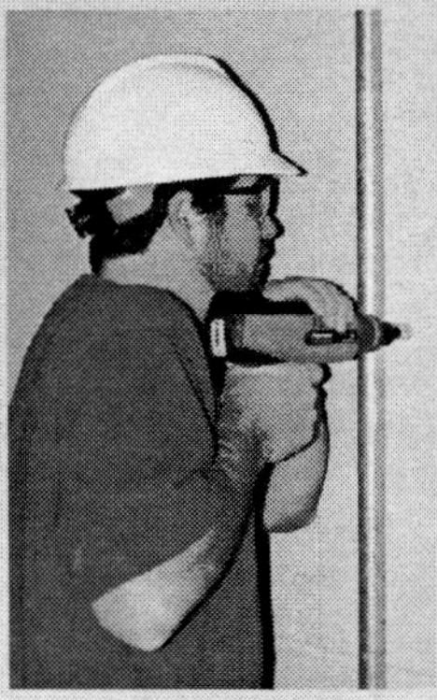

103PO306.EPS

103PO307.EPS

Instructor's Notes:

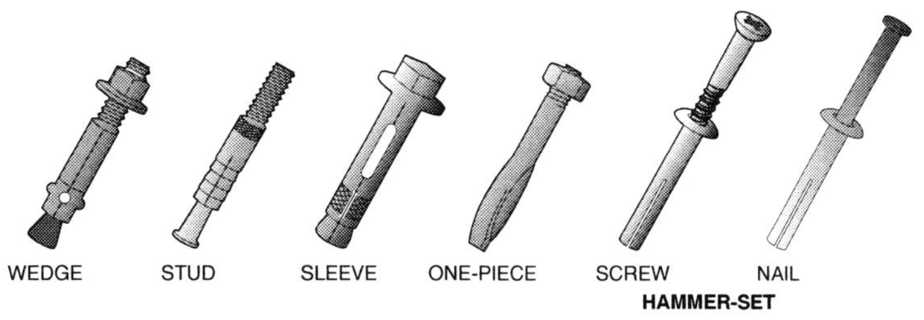

Figure 44 ◈ One-step anchors.

bottom in its mounting hole, it is a good choice to use when jacking or leveling of the fastened component is needed. The depth of the hole drilled in the masonry must be as specified by the manufacturer in order to achieve proper expansion. After the hole is blown clean of dust and other material, the anchor is inserted in the hole with the expander plug end down. Following this, the anchor is driven into the hole with a hammer (or setting tool) to expand the anchor and tighten it in the hole. The anchor is fully set when it can no longer be driven into the hole. The component is fastened using the correct size and thread bolt for use with the anchor stud.

8.1.3 Sleeve Anchors

Sleeve anchors are multi-purpose anchors. The depth of the anchor hole is not critical as long as the minimum length recommended by the manufacturer is drilled. After the hole is blown clean of dust and other material, the anchor is inserted into the hole and tapped until flush with the component. Then, the anchor nut or screw is tightened to expand the anchor and tighten it in the hole.

8.1.4 One-Piece Anchors

One-piece anchors are multi-purpose anchors. They work on the principle that as the anchor is driven into the hole, the spring force of the expansion mechanism is compressed and flexes to fit the size of the hole. Once set, it tries to regain its original shape. The depth of the hole drilled in the masonry must be at least ½" deeper than the required embedment. The proper depth is crucial. Overdrilling is as bad as underdrilling. After the hole is blown clean of dust and other material, the anchor is inserted through the component and driven with a hammer into the hole until the head is firmly seated against the component. It is important to make sure that the anchor is driven to the proper embedment depth. Note that manufacturers also make specially designed drivers and manual tools that are used instead of a hammer to drive one-piece anchors. These tools allow the anchors to be installed in confined spaces and help prevent damage to the component from stray hammer blows.

8.1.5 Hammer-Set Anchors

Hammer-set anchors are made for use in concrete and masonry. There are two types: nail and screw. An advantage of the screw-type anchors is that they are removable. Both types have a diameter the same size as the anchoring hole. For both types, the anchor hole must be drilled to the diameter of the anchor and to a depth of at least ¼" deeper than that required for embedment. After the hole is blown clean of dust and other material, the anchor is inserted into the hole through the mounting holes in the component to be fastened; then the screw or nail is driven into the anchor body to expand it. It is important to make sure that the head is seated firmly against the component and is at the proper embedment.

8.2.0 Bolt Anchors

Bolt anchors are designed to be installed flush with the surface of the base material. They are used in conjunction with threaded machine bolts or screws. In some types, they can be used with threaded rod. Drop-in, single and double expansion, and caulk-in anchors (*Figure 45*) are commonly used types of bolt anchors.

Classroom

Discuss bolt anchors and their applications. Pass around examples for the trainees to examine.

Audiovisual

Show Transparency 46 (Figure 45).

FASTENERS AND ANCHORS **345**

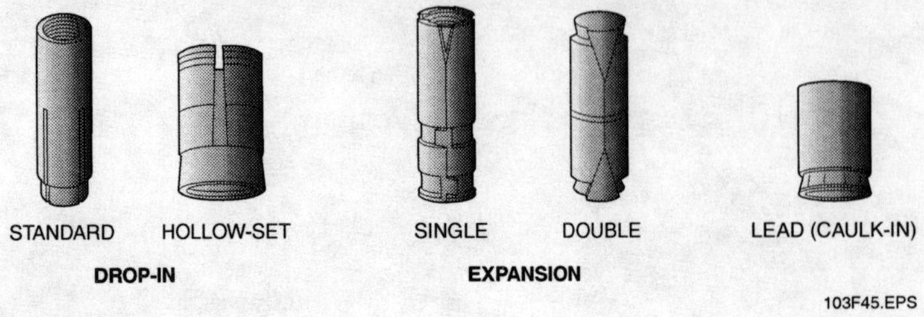

Figure 45 ◆ Bolt anchors.

8.2.1 Drop-In Anchors

Drop-in anchors are typically used as heavy-duty anchors. There are two types of drop-in anchors. The first type, made for use in solid concrete and masonry, has an internally threaded expansion anchor with a preassembled internal expander plug. The anchor hole must be drilled to the specific diameter and depth specified by the manufacturer. After the hole is blown clean of dust and other material, the anchor is inserted into the hole and tapped until it is flush with the surface. Following this, a setting tool supplied with the anchor is driven into the anchor to expand it. The component to be fastened is positioned in place and fastened by threading and tightening the correct size machine bolt or screw into the anchor.

The second type, called a *hollow set drop-in anchor,* is made for use in hollow concrete and masonry base materials. Hollow set drop-in anchors have a slotted, tapered expansion sleeve and a serrated expansion cone. They come in various lengths compatible with the outer wall thickness of most hollow base materials. They can also be used in solid concrete and masonry. The anchor hole must be drilled to the specific diameter specified by the manufacturer. When installed in hollow base materials, the hole is drilled into the cell or void. After the hole is blown clean of dust and other material, the anchor is inserted into the hole and tapped until it is flush with the surface. Following this, the component to be fastened is positioned in place; then the proper size machine bolt or screw is threaded into the anchor and tightened to expand the anchor in the hole.

8.2.2 Single- and Double-Expansion Anchors

Single- and double-expansion anchors are both made for use in concrete and other masonry. The double-expansion anchor is used mainly when fastening into concrete or masonry of questionable strength. For both types, the anchor hole must be drilled to the specific diameter and depth specified by the manufacturer. After the hole is blown clean of dust and other material, the anchor is inserted into the hole, threaded cone end first. It is then tapped until it is flush with the surface. Following this, the component to be fastened is positioned in place; then the proper size machine bolt or screw is threaded into the anchor and tightened to expand the anchor in the hole.

8.2.3 Lead (Caulk-In) Anchors

Lead (caulk-in) anchors are a cast-type anchor used in concrete and masonry. They consist of an internally threaded expander cone with a series of vertical internal ribs and a lead sleeve. The vertical internal ribs prevent the cone from turning in the sleeve as the anchor is tightened. The anchor hole must be drilled to the specific diameter and depth specified by the manufacturer. However, in weak or soft masonry, a slightly deeper hole can be drilled to countersink the anchor below the surface. After the hole is blown clean of dust and other material, the anchor is inserted into the hole, threaded cone end first. Following this, a setting tool supplied with the anchor is driven into the anchor to expand it. The component to be fastened is positioned in place and fastened by threading and tightening the correct size machine bolt or screw into the anchor.

8.3.0 Screw Anchors

Screw anchors are lighter-duty anchors made to be installed flush with the surface of the base material. They are used in conjunction with sheet metal, wood, or lag screws depending on the an-

Classroom

Discuss screw anchors and their applications. Pass around examples for the trainees to examine.

Instructor's Notes:

346 CHAPTER 8

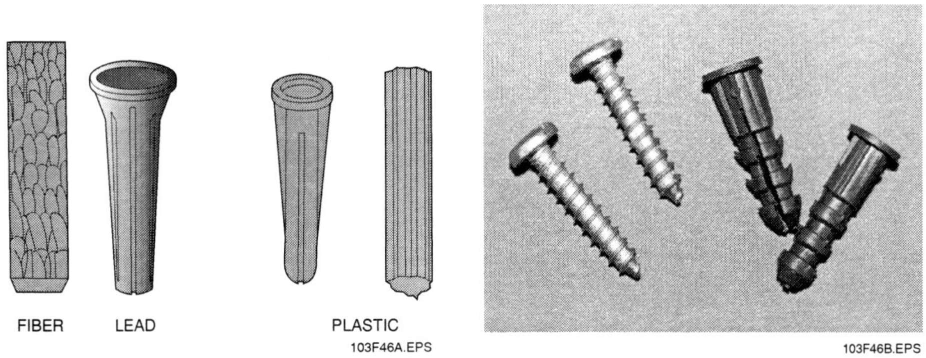

Figure 46 ◆ Screw anchors and screws.

chor type. Fiber, lead, and plastic anchors are common types of screw anchors (*Figure 46*). The lag shield anchor used with lag screws was described earlier in this module.

Fiber, lead, and plastic anchors are typically used in concrete and masonry. Plastic anchors are also commonly used in wallboard and similar base materials. The installation of all types is simple. The anchor hole must be drilled to the diameter specified by the manufacturer. The minimum depth of the hole must equal the anchor length. After the hole is blown clean of dust and other material, the anchor is inserted into the hole and tapped until it is flush with the surface. Following this, the component to be fastened is positioned in place; then the proper type and size screw is driven through the component mounting hole and into the anchor to expand the anchor in the hole.

8.4.0 Self-Drilling Anchors

Some anchors made for use in masonry are self-drilling anchors. *Figure 47* is typical of those in common use. This fastener has a cutting sleeve that is first used as a drill bit and later becomes the expandable fastener itself. A rotary hammer is used to drill the hole in the concrete using the anchor sleeve as the drill bit. After the hole is drilled, the anchor is pulled out and the hole cleaned. This is followed by inserting the anchor's expander plug into the cutting end of the sleeve. The anchor sleeve and expander plug are driven back into the hole with the rotary hammer until they are flush with the surface of the concrete. As the fastener is hammered down, it hits the bottom, where the tapered expander causes the fastener to expand and lock into the hole. The anchor is then snapped off at the shear point with a quick lateral movement of the hammer. The component to be fastened can then be attached to the anchor using the proper size bolt.

8.5.0 Guidelines for Drilling Anchor Holes in Hardened Concrete or Masonry

When selecting masonry anchors, regardless of the type, always take into consideration and follow the manufacturer's recommendations pertaining to hole diameter and depth, minimum embedment in concrete, maximum thickness of material to be fastened, and the pullout and shear load capacities.

When installing anchors and/or anchor bolts in hardened concrete, make sure the area where the equipment or component is to be fastened is smooth so that it will have solid footing. Uneven footing might cause the equipment to twist, warp, not tighten properly, or vibrate when in operation. Before starting, carefully inspect the rotary hammer or hammer drill and the drill bit(s) to ensure they are in good operating condition. Be sure to use the type of carbide-tipped masonry or percussion drill bits recommended by the drill/hammer or anchor manufacturer because these bits are made to take the higher impact of the masonry materials. Also, it is recommended that the drill or hammer tool depth gauge be set to the depth of the hole needed. The trick to using masonry drill bits is not to force them into the material by pushing down hard on the drill. Use a little pressure

Audiovisual

Show Transparency 47 (Figure 46A).

Classroom

Discuss self-drilling anchors and their applications. Pass around examples for the trainees to examine.

Audiovisual

Show Transparency 48 (Figure 47).

FASTENERS AND ANCHORS 347

Classroom

Review the procedure for installing anchors in concrete.

Audiovisual

Show Transparency 49 (Figure 48).

Safety

Emphasize the safety precautions necessary when drilling in concrete.

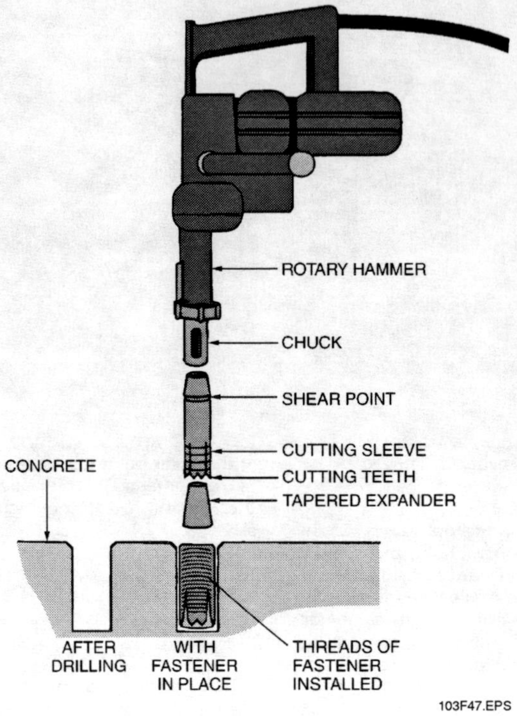

Figure 47 ◆ Self-drilling anchor.

and let the drill do the work. For large holes, start with a smaller bit, then change to a larger bit.

The methods for installing the different types of anchors in hardened concrete or masonry were briefly described in the sections above. Always install the selected anchors according to the manufacturer's directions. Here is an example of a typical procedure used to install many types of expansion anchors in hardened concrete or masonry. Refer to *Figure 48* as you study the procedure.

WARNING!

Drilling in concrete generates noise, dust, and flying particles. Always wear safety goggles, ear protectors, and gloves. Make sure other workers in the area also wear protective equipment.

Step 1 Drill the anchor bolt hole the same size as the anchor bolt. The hole must be deep enough for six threads of the bolt to be below the surface of the concrete (see *Figure 48, Step 1*). Clean out the hole using a squeeze bulb.

Step 2 Drive the anchor bolt into the hole using a hammer (*Figure 48, Step 2*). Protect the threads of the bolt with a nut that does not allow any threads to be exposed.

Step 3 Put a washer and nut on the bolt, and tighten the nut with a wrench until the anchor is secure in the concrete (*Figure 48, Step 3*).

8.6.0 Hollow-Wall Anchors

Hollow-wall anchors are used in hollow materials such as concrete plank, block, structural steel,

Instructor's Notes:

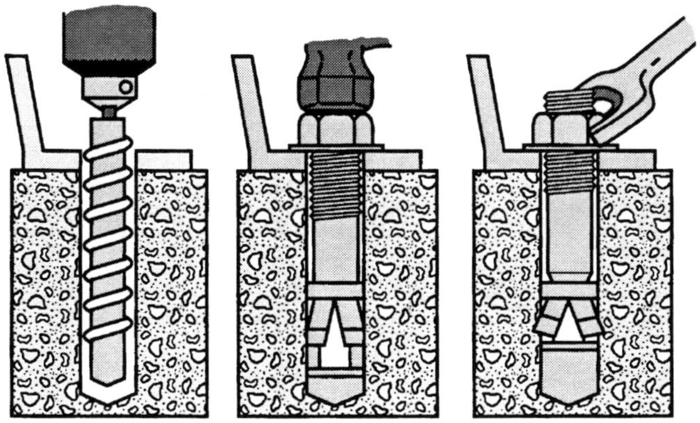

103F48.EPS

Figure 48 ◆ Installing an anchor bolt in hardened concrete.

INSIDE TRACK

Safety

Be sure to wear safety goggles whenever you tackle any fastening project, regardless of how small the job may seem. Remember, you can never replace lost eyesight.

wallboard, and plaster. Some types can also be used in solid materials. Toggle bolts, sleeve-type wall anchors, wallboard anchors, and metal drive-in anchors are common anchors used when fastening to hollow materials.

When installing anchors in hollow walls or ceilings, regardless of the type, always follow the manufacturer's recommendations pertaining to use, hole diameter, wall thickness, grip range (thickness of the anchoring material), and the pullout and shear load capacities.

8.6.1 Toggle Bolts

Toggle bolts (*Figure 49*) are used to fasten equipment, hangers, supports, and similar items into hollow surfaces such as walls and ceilings. They consist of a slotted bolt or screw and spring-loaded wings. When inserted through the item to be fastened, then through a predrilled hole in the wall or ceiling, the wings spring apart and provide a firm hold on the inside of the hollow wall or ceiling as the bolt is tightened. Note that the hole drilled in the wall or ceiling should be just large enough for the compressed wing-head to pass through. Once the toggle bolt is installed, be careful not to completely unscrew the bolt because the wings will fall off, making the fastener useless. Screw-actuated plastic toggle bolts are also made. These are similar to metal toggle bolts, but they come with a pointed screw and do not require as large a hole. Unlike the metal version, the plastic wings remain in place if the screw is removed.

Toggle bolts are used to fasten a part to hollow block, wallboard, plaster, panel, or tile. The following general procedure can be used to install toggle bolts.

WARNING!
Follow all safety precautions.

Step 1 Select the proper size drill bit or punch and toggle bolt for the job.

Classroom

Discuss various types of hollow-wall anchors and their applications. Pass around examples for the trainees to examine.

Safety

Emphasize that safety goggles must be worn when installing fasteners.

Audiovisual

Show Transparency 50 (Figure 49).

Demonstration

Demonstrate how to install toggle bolts.

FASTENERS AND ANCHORS 349

Teaching Tip

Discuss the "Think About It." Explain that over-tightening can result in damaged material and stripped threads.

Classroom

Discuss sleeve-type wall anchors.

Audiovisual

Show Transparency 51 (Figure 50).

Classroom

Discuss wallboard and metal drive-in anchors.

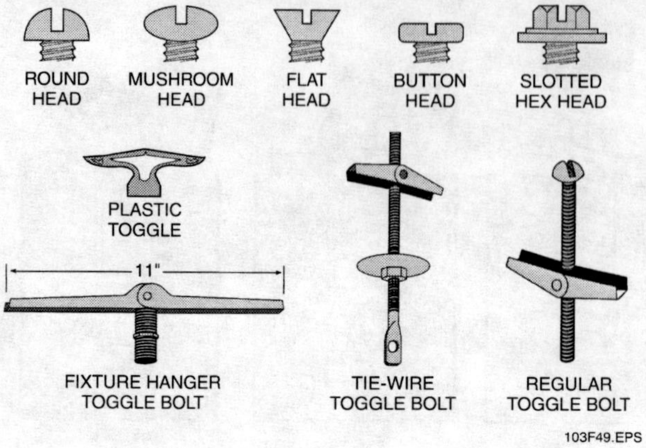

Figure 49 ◆ Toggle bolts.

THINK ABOUT IT

Toggle Bolts

What will happen if you fasten a toggle bolt too tightly?

Step 2 Check the toggle bolt for damaged or dirty threads or a malfunctioning wing mechanism.

Step 3 Drill a hole completely through the surface to which the part is to be fastened.

Step 4 Insert the toggle bolt through the opening in the item to be fastened.

Step 5 Screw the toggle wing onto the end of the toggle bolt, ensuring that the flat side of the toggle wing is facing the bolt head.

Step 6 Fold the wings completely back and push them through the drilled hole until the wings spring open.

Step 7 Pull back on the item to be fastened in order to hold the wings firmly against the inside surface to which the item is being attached.

Step 8 Tighten the toggle bolt with a screwdriver until it is snug.

8.6.2 Sleeve-Type Wall Anchors

Sleeve-type wall anchors (*Figure 50*) are suitable for use in concrete, block, plywood, wallboard, hollow tile, and similar materials. The two types made are standard and drive. The standard type is commonly used in walls and ceilings and is installed by drilling a mounting hole to the required diameter. The anchor is inserted into the hole and tapped until the gripper prongs embed in the base material. Following this, the anchor's screw is tightened to draw the anchor tight against the inside of the wall or ceiling. Note that the drive-type anchor is hammered into the material without the need for drilling a mounting hole. After the anchor is installed, the anchor screw is removed, the component being fastened is positioned in place, then the screw is reinstalled through the mounting hole in the component and into the anchor. The screw is tightened into the anchor to secure the component.

Instructor's Notes:

350 CHAPTER 8

Figure 50 ◆ Sleeve-type, wallboard, and metal drive-in anchors.

THINK ABOUT IT

Sleeve-Type Drive Anchors

What happens when you remove the screw when a sleeve-type drive anchor is in place?

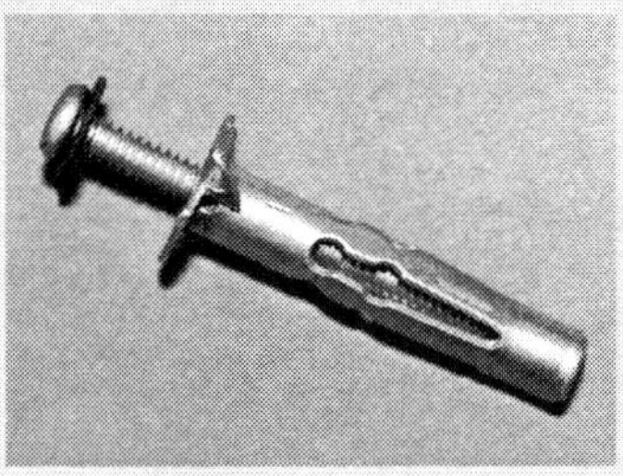

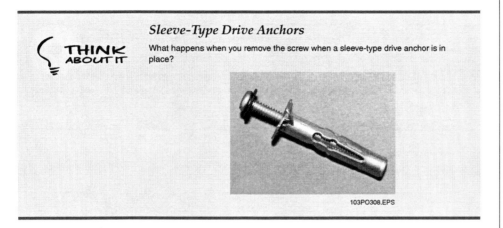

8.6.3 Wallboard Anchors

Wallboard anchors (*Figure 50*) are self-drilling medium- and light-duty anchors used for fastening in wallboard. The anchor is driven into the wall with a Phillip's head manual or cordless screwdriver until the head of the anchor is flush with the wall or ceiling surface. Following this, the component being fastened is positioned over the anchor, then secured with the proper size sheet metal screw driven into the anchor.

8.6.4 Metal Drive-In Anchors

Metal drive-in anchors (*Figure 50*) are used to fasten light to medium loads to wallboard. They have two pointed legs that stay together when the anchor is hammered into a wall and spread out against the inside of the wall when a No. 6 or 8 sheet metal screw is driven in.

9.0.0 ◆ EPOXY ANCHORING SYSTEMS

Epoxy resin compounds can be used to anchor threaded rods, dowels, and similar fasteners in solid concrete, hollow wall, and brick. For one manufacturer's product, a two-part epoxy is packaged in a two-chamber cartridge that keeps the resin and hardener ingredients separated until use. This cartridge is placed into a special tool similar to a caulking gun. When the gun handle is pumped, the epoxy resin and hardener components are mixed within the gun; then the epoxy is ejected from the gun nozzle.

To use the epoxy to install an anchor in solid concrete (*Figure 51*), a hole of the proper size is drilled in the concrete and cleaned using a nylon (not metal) brush. Following this, a small amount of epoxy is dispensed from the gun to make sure that the resin and hardener have mixed properly. This is indicated

Demonstration

Demonstrate the installation of various anchors.

Teaching Tip

Discuss the "Think About It." Explain that nothing happens; the anchor remains in place.

Laboratory

Have the trainees practice installing selected anchors and toggle bolts. Note the proficiency of each trainee.

Classroom

Discuss epoxy anchoring systems and their applications.

Audiovisual

Show Transparency 52 (Figure 51).

FASTENERS AND ANCHORS 351

Teaching Tip

Discuss the "Think About It." Explain that toggle bolts are designed for use in hollow materials, while expandable anchors are used in solid materials.

Teaching Tip

Discuss the "Think About It." Explain that toggle bolts would be the best choice in this application, but a heavy-duty expansion anchor could also be used.

THINK ABOUT IT

Nuts and Bolts

Expandable anchors and toggle bolts each do the same job: they fasten objects on hollow walls and from ceilings. However, each is installed differently. Why is this?

CASE HISTORY

Installation Requirements

In a college dormitory, battery-powered emergency lights were anchored to sheetrock hallway ceilings with sheetrock screws, with no additional support. These fixtures weigh 8–10 pounds each and might easily have fallen out of the ceiling, causing severe injury. When the situation was discovered, the contractor had to remove and replace dozens of fixtures.

The Bottom Line: Incorrect anchoring methods can be both costly and dangerous.

THINK ABOUT IT

Ceiling Installations

In the dormitory problem discussed above, which of the following fasteners could have been used to safely secure the emergency lights?

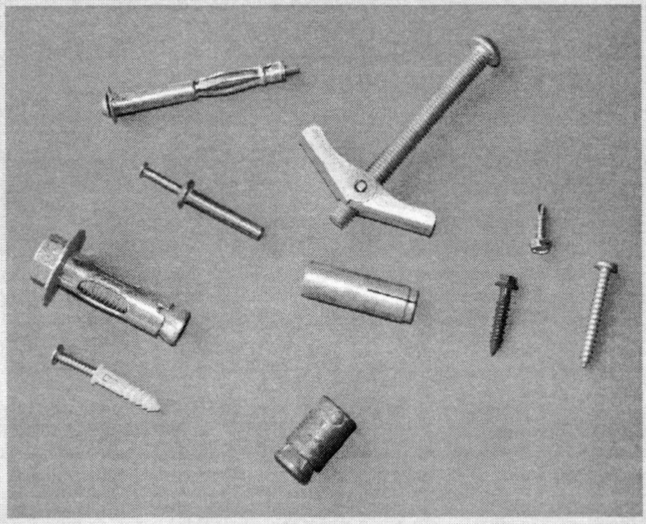

103PO309.EPS

Instructor's Notes:

Using Epoxy

Once mixed, epoxy has a limited working time. Therefore, mix exactly what you need and work quickly. After the working time is up, epoxy requires a specific curing time. Always give epoxy its recommended curing time; since epoxy is so strong and sets so quickly, you'll be tempted to stress the bond before it's fully cured.

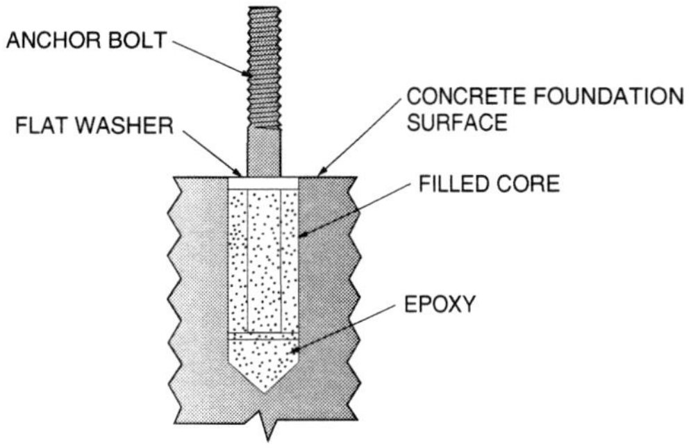

Figure 51 ◆ Fastener anchored in epoxy.

Use the Proper Tool for the Application

To avoid damaging fasteners, use the correct tool for the job. For example, don't use pliers to install bolts, and don't use a screwdriver that is too large or too small.

Putting It All Together

You are installing a surface-mounted panel. What fasteners would you use in each of the following types of wall construction?

- Sheetrock wall with wood studs
- Sheetrock wall with metal studs
- Concrete masonry units (hollow block)
- Poured concrete

Teaching Tip

Discuss the "Think About It."

by the epoxy being of a uniform color. The gun nozzle is then placed into the hole, and the epoxy is injected into the hole until half the depth of the hole is filled. Following this, the selected fastener is pushed into the hole with a slow twisting motion to make sure that the epoxy fills all voids and crevices, then is set to the required plumb (or level) position. After the recommended cure time for the epoxy has elapsed, the fastener nut can be tightened to secure the component or fixture in place.

The procedure for installing a fastener in a hollow wall or brick using epoxy is basically the same

FASTENERS AND ANCHORS 353

Classroom

Have the trainees complete the Review Questions and go over the answers prior to administering the Module Examination.

Examination

Administer the Module Examination. Record the results on Craft Training Report Form 200 and submit the results to the Training Program Sponsor.

Performance Testing

Administer the Performance Test and fill out Performance Profile Sheets for each trainee. If desired, trainee proficiency noted during laboratory sessions may be used to complete the Performance Test. Be sure to record the results on Craft Training Report Form 200 and submit the results to the Training Program Sponsor.

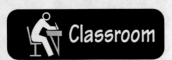

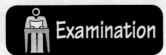

as described above. The difference is that the epoxy is first injected into an anchor screen to fill the screen, then the anchor screen is installed into the drilled hole. Use of the anchor screen is necessary to hold the epoxy intact in the hole until the anchor is inserted into the epoxy.

Summary

Fasteners and anchors are used for a variety of tasks in the electrical craft. In this module, you learned about various types of fasteners and anchors and their uses. Basic installation procedures were also included. Selecting the correct fastener or anchor for a particular job is required to perform high-quality work. In the electrical craft, it is important to be familiar with the correct terms used to describe fasteners and anchors. Using the proper technical terms helps avoid confusion and improper selection. Installation techniques for fasteners and anchors may vary depending on the job. Make sure to check the project specifications and manufacturer's information when installing any fastener or anchor.

New fasteners and anchors are being developed every day. Your local distributor/manufacturer is an excellent source of information.

Review Questions

1. The quality of some fasteners can be determined by the _____ on the head of a bolt or screw.
   a. number of grooves cut
   b. number of sides
   c. length of the lines
   d. grade markings

2. The purpose of a jam nut is to _____.
   a. hold a piece of material stationary
   b. lock a standard nut in place
   c. stop rotation of a machine quickly for safety reasons
   d. compress a lock washer in place

3. Washers are used to _____.
   a. distribute the load over a larger area
   b. attach an item to a hollow surface
   c. anchor materials that expand due to temperature changes
   d. allow the bolts to expand with temperature changes

4. Torque is normally expressed in _____.
   a. pounds per square inch (psi)
   b. gallons per minute (gpm)
   c. foot pounds (ft. lbs.)
   d. cubic feet per minute (cpm)

5. When tightening bolts, nuts, or screws, always use the proper _____.
   a. tightening sequence
   b. bolt pattern
   c. ratchet wrench
   d. lubrication

6. What type of fasteners are commonly used as lifting devices?
   a. Toggle bolts
   b. Anchor bolts
   c. J-bolts
   d. Eye bolts

7. When installing a panel in concrete using a lag screw, you would also use a _____.
   a. drop-in anchor
   b. lag shield
   c. caulk-in anchor
   d. double-expansion anchor

8. When fastening a device or fixture to concrete using concrete/masonry screws, the screws _____.
   a. need not be installed in predrilled holes because they are self-tapping
   b. are installed using specially designed carbide drill bits and installation tools made for use with the screws
   c. are installed by hammering them into a drilled or punched hole of the proper size
   d. are installed in a hole drilled with a standard masonry drill bit

9. When using powder-actuated tools and fasteners, _____.
   a. higher-numbered power level loads are used to drive into soft materials
   b. use only powder loads and fasteners that are specifically designed for use with the installation tool
   c. you may use powder loads interchangeably between different tools
   d. you may use standard fasteners

10. Wedge and sleeve anchors are classified as _____ anchors.
   a. bolt
   b. hollow-wall
   c. one-step
   d. screw

Instructor's Notes:

PROFILE IN SUCCESS

Gary Edgington, EdgeWel Electric, Inc.

Gary Edgington started his own electrical contracting business at the age of 24, and has developed it into a successful organization that employs dozens of electricians and is able to handle any type of electrical job. Gary credits part of his success to the mentors who helped him early in his career and taught him valuable lessons that have guided him ever since. Here is his success story as he told it:

I got into electrical work because I really wanted to work with my hands. I had an early childhood interest in electrical things that continued through high school.

I got into the trade in 1973 with no formal training. I started on-the-job training with Harry Kinsey, a seasoned electrical contractor who did commercial and residential work. He was really my first role model. He was a greatly admired individual with a very good reputation. I learned from him that if you're going to do a job, be the best at it, and take pride in your work.

Afterwards, I worked for Des Moines, Iowa-based AMF Corporation as an industrial electrician. During this time, I also attended college and studied industrial electronics because it gives you theory and basic electrical knowledge, which allows you to resolve problems better and to troubleshoot on your own. My role model at AMF was Mr. George McCoig. He was a retired Air Force electrician. He was the most highly respected electrician at AMF. He taught me that knowledge is important and that you don't have to know everything, you just have to know how to find the answers to everything. I was only 20 years old at the time, but I guess I had a knack for motor controls and so I found myself in the industrial field. I was assigned to George as his helper. George's job was to teach me electrical systems.

After George, I worked for a man named Bernard Sharples. He was an electrician from Manchester, England, and was second only in reputation to George at AMF. Bernard took me under his wing and he not only recognized my abilities, but helped me expand them. He reminded me of what I knew, not what I didn't know. Bernard was the type of person who brought out the best in other people.

I got my master's (contractor's) license in 1978 after being in the field 5 years. This gave me everything I really needed to be an electrical contractor. I was 24 years old. I started the company I own now with a partner, but after 3 years, I became sole owner and grew a business that started with two people to one with over 35 people. We have a 12,000 sq. ft. facility and do every kind of job. That's really our claim to fame; we do everything from single-family houses to industrial to commercial to telephone data work.

Today, most of my work entails estimating, design, and management. In 1994, I became involved with NCCER's standardized craft training. I already had 9 years worth of teaching experience behind me, so I got involved because of my commitment to training. I'm a person who believes that training costs nothing; in the long run, you get back much more than you invest in it.

FASTENERS AND ANCHORS 355

GLOSSARY

Trade Terms Introduced in This Module

American Society for Testing of Materials (ASTM): An organization that publishes specifications and standards relating to fasteners.

Clearance: The amount of space between the threads of bolts and their nuts.

Foot pounds (ft. lbs.): The normal method used for measuring the amount of torque being applied to bolts or nuts.

Inch pounds (in. lbs.): A method of measuring the amount of torque applied to small bolts or nuts that require measurement in smaller increments than foot pounds.

Key: A machined metal part that fits into a keyway and prevents parts such as gears or pulleys from rotating on a shaft.

Keyway: A machined slot in a shaft and on parts such as gears and pulleys that accepts a key.

Nominal size: A means of expressing the size of a bolt or screw. It is the approximate diameter of a bolt or screw.

Society of Automotive Engineers (SAE): An organization that publishes specifications and standards relating to fasteners.

Thread classes: Threads are distinguished by three classifications according to the amount of tolerance the threads provide between the bolt and nut.

Thread identification: Standard symbols used to identify threads.

Thread standards: An established set of standards for machining threads.

Tolerance: The amount of difference allowed from a standard.

Torque: The turning force applied to a fastener.

Unified National Coarse (UNC) thread: A standard type of coarse thread.

Unified National Extra Fine (UNEF) thread: A standard type of extra-fine thread.

Unified National Fine (UNF) thread: A standard type of fine thread.

Instructor's Notes:

APPENDIX

Mechanical Anchors and Their Uses

Mechanical Anchors and Their Uses

Anchor Type	Typically Used In	Use With Fastener	Typical Working Load Range*
	One-Step Anchors		
Wedge	Concrete **Stone	None	Light, medium, and heavy duty
Stud	Concrete **Stone, solid brick and block	None	Light, medium, and heavy duty
Sleeve	Concrete, solid brick and block **Stone, hollow brick and block	None	Light and medium duty
One-piece	Concrete, solid block **Stone, solid and hollow brick, hollow block	None	Light and medium duty
Hammer-set	Concrete, solid block **Stone, solid and hollow brick, hollow block	None	Light duty
	Bolt Anchors		
Drop-in	Concrete **Stone, solid brick	Machine screw or bolt	Light, medium, and heavy duty
Hollow-set drop-in	Concrete, solid brick and block **Stone, hollow brick and block	Machine screw or bolt	Light and medium duty
Single-expansion	Concrete, solid brick and block **Stone, hollow brick and block	Machine screw or bolt	Light and medium duty
Double-expansion	Concrete, solid brick and block **Stone, hollow brick and block	Machine screw or bolt	Light and medium duty
Lead (caulk-in)	Concrete, solid brick and block **Stone, hollow brick and block	Machine screw or bolt	Light and medium duty

FASTENERS AND ANCHORS 357

Anchor Type	Typically Used In	Use With Fastener	Typical Working Load Range*
Screw Anchors			
Lag shield	Concrete **Stone, solid and hollow brick and block	Lag screw	Light and medium duty
Fiber	Concrete, stone, solid brick and block **Hollow brick and block, wallboard	Wood, sheet metal, or lag screw	Light and medium duty
Lead	Concrete, solid brick and block **Hollow brick and block, wallboard	Wood or sheet metal screw	Light duty
Plastic	Concrete, stone, solid brick and block **Hollow brick and block, wallboard	Wood or sheet metal screw	Light duty
Hollow-Wall Anchors			
Toggle bolts	Concrete, plank, hollow block, wallboard, plywood/paneling	None	Light and medium duty
Plastic toggle bolts	Wallboard, plywood/paneling **Hollow block, structural tile	Wood or sheet metal screw	Light duty
Sleeve-type wall	Wallboard, plywood/paneling **Hollow block, structural tile	None	Light duty
Wallboard	Wallboard	Sheet metal screw	Light duty
Metal drive-in	Wallboard	Sheet metal screw	Light duty

*Anchor working loads given in the table are defined below. These are approximate loads only. Actual allowable loads depend on such factors as the anchor style and size, base material strength, spacing and edge distance, and the type of service load applied. Always consult the anchor manufacturer's product literature to determine the correct type of anchor and size to use for a specific application.

- Light duty—Less than 400 lbs.
- Medium duty—400 to 4,000 lbs.
- Heavy duty—Above 4,000 lbs.

**Indicates use may be suitable depending on the application.

Instructor's Notes:

Additional Resources

This module is intended to present thorough resources for task training. The following reference works are suggested for further study. These are optional materials for continued education rather than for task training.

American Electrician's Handbook, Latest Edition. New York: McGraw-Hill.

The Sheet Metal Toolbox Manual, Latest Edition. New York: Prentice Hall.

FASTENERS AND ANCHORS 359

ANSWER KEY

Answers to Review Questions

Answer	Section
1. d	2.1.4
2. b	2.3.1
3. a	2.4.0
4. c	2.5.1
5. a	2.5.2
6. d	4.1.0
7. b	5.2.0
8. b	5.3.0
9. b	7.0.0
10. c	8.1.0

Section 2.2.0 *Think About It—Installing Fasteners*

Trainees should be concerned with appearance, decorative finish, strength of fastener and material to be fastened, proper installation, amount of torque required, appropriate personal protective equipment (particularly safety goggles), and using the correct tool(s) for the job.

Section 9.0.0 *Think About It—Putting It All Together*

- *Sheetrock wall with wood studs*—Use wood screws if in studs; if not, use toggle bolts.
- *Sheetrock wall with metal studs*—Use sheet metal screws if in studs; if not, use toggle bolts.
- *Concrete masonry units (hollow block)*—May use hollow-wall anchors, expansion anchors, or powder-actuated fasteners.
- *Poured concrete*—Use expansion anchors or powder-actuated fasteners.

The NCCER makes every effort to keep these textbooks up-to-date and free of technical errors. We appreciate your help in this process. If you have an idea for improving this textbook, or if you find an error, a typographical mistake, or an inaccuracy in NCCER's Contren™ textbooks, please write us, using this form or a photocopy. Be sure to include the exact module number, page number, a detailed description, and the correction, if applicable. Your input will be brought to the attention of the Technical Review Committee. Thank you for your assistance.

Instructors – If you found that additional materials were necessary in order to teach this module effectively, please let us know so that we may include them in the Equipment/Materials list in the Instructor's Guide.

Write: Curriculum Revision and Development Department
National Center for Construction Education and Research
P.O. Box 141104, Gainesville, FL 32614-1104

Fax: 352-334-0932

E-mail: curriculum@nccer.org

Craft ______ Module Name ______

Copyright Date ______ Module Number ______ Page Number(s) ______

Description ______

(Optional) Correction ______

(Optional) Your Name and Address ______

Chapter 9

Hand Bending

26102-02

Hand Bending
Annotated Instructor's Guide

Module 26102

MODULE OVERVIEW

This course introduces the electrical trainee to the methods and procedures used in cutting, bending, and reaming conduit.

PREREQUISITES

Please refer to the Course Map in the Trainee Module. Prior to training with this module, it is recommended that the trainee shall have successfully completed the following modules:

Core Curriculum; Residential Electrical I, Chapters 1 through 8.

LEARNING OBJECTIVES

Upon completion of this module, the trainee will be able to:

1. Identify the methods of hand bending conduit.
2. Identify the various methods used to install conduit.
3. Use math formulas to determine conduit bends.
4. Make 90° bends, back-to-back bends, offsets, kicks, and saddle bends using a hand bender.
5. Cut, ream, and thread conduit.

PERFORMANCE OBJECTIVES

Under supervision of the instructor, the trainee should be able to:

1. Given a piece of EMT, complete the following using a hand bender, hacksaw, and reaming tool. No couplings are allowed, and there should be no kinks in the pipe.
   - Offset
   - Saddle
   - 90° bend

NCCER STANDARDIZED CRAFT TRAINING PROGRAM

The National Center for Construction Education and Research (NCCER) provides a standardized national program of accredited craft training. Key features of the program include instructor certification, competency-based training, and performance testing. The program provides trainees, instructors, and companies with a standard form of recognition through a National Craft Training Registry. The program is described in full in the *Guidelines for Accreditation*, published by the NCCER. For more information on standardized craft training, contact the NCCER at P.O. Box 141104, Gainesville, FL 32614-1104, 352-334-0911, visit our Web site at www.nccer.org, or e-mail info@nccer.org.

HOW TO USE THIS ANNOTATED INSTRUCTOR'S GUIDE

Each page presents two sections of information. The larger section displays each page exactly as it appears in the Trainee Module. The narrow column ties suggested trainee and instructor actions to each page and provides icons to call your attention to material, safety, audiovisual, or testing requirements. The bottom of each page includes space for your notes.

If you see the Teaching Tip icon, that means there is a teaching tip associated with this section.

SAFETY CONSIDERATIONS

Ensure that the trainees are equipped with appropriate personal protective equipment. Stress the importance of avoiding contact with sharp edges when handling conduit.

PREPARATION

Before teaching this module, you should review the Module Outline, Learning and Performance Objectives, and the Materials and Equipment List. Be sure to allow ample time to prepare your own training or lesson plan and gather all required equipment and materials.

MATERIALS AND EQUIPMENT LIST

Materials:

Transparencies

Markers/chalk

Pencils/scratch paper

Various pieces of conduit

One piece of EMT per trainee

PVC pieces

PVC cement

Cutting oil

Shop towels

No. 10 or No. 12 solid wire

Copy of the latest edition of the *National Electrical Code®*

OSHA Electrical Safety Guidelines (pocket guide)

Module Examinations*

Performance Profile Sheets*

Equipment:

Overhead projector and screen

Whiteboard/chalkboard

Appropriate personal protective equipment

Tape measure

Hickey

Hand bender

Calculator

Hacksaw

Pipe vise

Pipe cutter

Reamer

Hand-operated threader

Sandbox or drip pan

Torpedo level

*Located in the Test Booklet packaged with this Annotated Instructor's Guide.

ADDITIONAL RESOURCES

This module is intended to present thorough resources for task training. The following reference works are suggested for both instructors and motivated trainees interested in further study. These are optional materials for continued education rather than for task training.

Benfield Conduit Bending Manual, Latest Edition. New York, NY: McGraw-Hill Publishing Company.

National Electrical Code Handbook, Latest Edition. Quincy, MA: National Fire Protection Association.

Tom Henry's Conduit Bending Package (includes video, book, and bending chart). Winter Park, FL: Code Electrical Classes, Inc.

NOTES

The designations "National Electrical Code," "NE Code," and "NEC," where used in this document, refer to the *National Electrical Code®*, which is a registered trademark of the National Fire Protection Association, Quincy, MA. All National Electrical Code (NEC) references in this module refer to the 2002 edition of the NEC.

If you feel that additional math instruction would be helpful, Prentice Hall offers a basic math textbook entitled *Fundamentals of Electrical and Mechanical Mathematics.* It covers the basic math requirements for electrical trainees and may be ordered by contacting Prentice Hall Customer Service at 1-800-922-0579.

TEACHING TIME FOR THIS MODULE

An outline for use in developing your lesson plan is presented below. Note that each Roman numeral in the outline equates to one session of instruction. Each session has a suggested time period of 2½ hours. This includes 10 minutes at the beginning of each session for administrative tasks and one 10-minute break during the session. Approximately 7½ hours are suggested to cover *Hand Bending.* You will need to adjust the time required for hands-on activity and testing based on your class size and resources.

Topic	Planned Time
Session I. Introduction to Hand Bending; Laboratory; Gain; Back-to-Back Bends	
A. Introduction to Hand Bending	________
1. Equipment	________
2. Geometry Required to Make a Bend	________
3. Making a 90° Bend	________
B. Laboratory	________
Under instructor supervision, have the trainees practice making 90° bends.	________
C. Gain	________
D. Back-to-Back 90° Bends	________
Session II. Offset Bends; Laboratory; Saddle Bends; Laboratory	
A. Making an Offset	________
B. Laboratory	________
Under instructor supervision, have the trainees practice making offset bends.	________
C. Parallel Offsets	________
D. Saddle Bends	________
E. Laboratory	________
Under instructor supervision, have the trainees practice making saddle bends.	________
F. Four-Bend Saddles	________
Session III. Cutting and Reaming Conduit; Laboratory; Threading Conduit; Cutting and Joining PVC Conduit; Module Examination and Performance Testing	
A. Cutting and Reaming Conduit	________
1. Using a Hacksaw	________
2. Using a Pipe Cutter	________
3. Reaming Conduit	________
B. Laboratory	________
Under instructor supervision, have the trainees practice cutting and reaming conduit.	________
C. Threading Conduit	________
D. Cutting and Joining PVC Conduit	________

F. Module Examination

1. Trainees must score 70% or higher to receive recognition from the NCCER.
2. Record the testing results on Craft Training Report Form 200 and submit the results to the Training Program Sponsor.

G. Performance Testing

1. Trainees must perform each task to the satisfaction of the instructor to receive recognition from the NCCER.
2. Record the testing results on Craft Training Report Form 200 and submit the results to the Training Program Sponsor.

Chapter 9

Module 26102-02

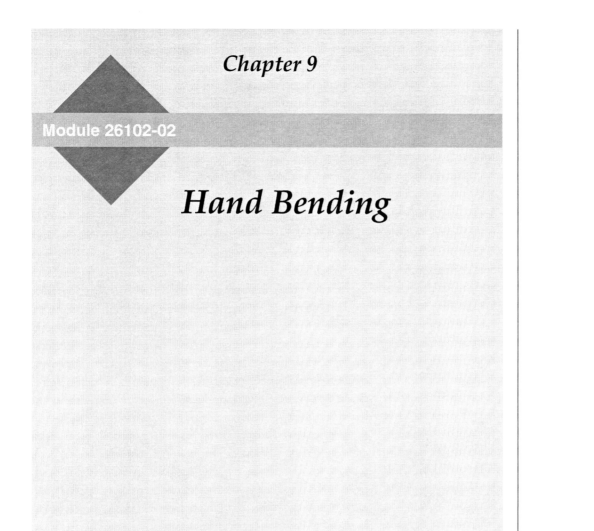

Hand Bending

Course Map

This course map shows all of the modules in the first level of the Residential Electrical curriculum. The suggested training order begins at the bottom and proceeds up. Skill levels increase as you advance on the course map. The local Training Program Sponsor may adjust the training order.

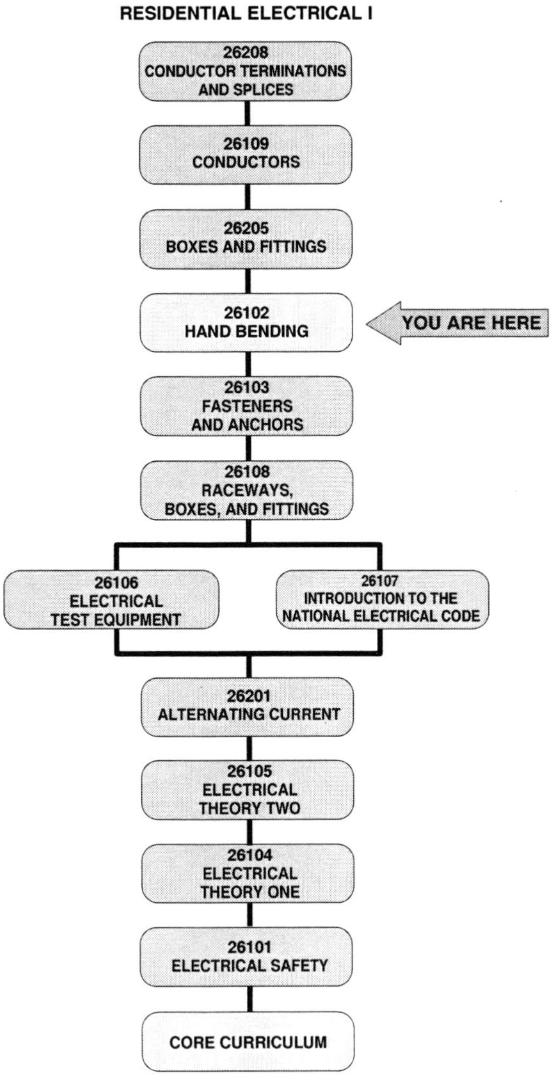

Homework

Assign reading of Module 26102.

MODULE 26102 CONTENTS

HAND BENDING 367

Figures

Tables

Instructor's Notes:

Chapter 9

MODULE 26102

Hand Bending

Objectives

When you have completed this module, you will be able to do the following:

1. Identify the methods of hand bending conduit.
2. Identify the various methods used to install conduit.
3. Use math formulas to determine conduit bends.
4. Make 90° bends, back-to-back bends, offsets, kicks, and saddle bends using a hand bender.
5. Cut, ream, and thread conduit.

Prerequisites

Before you begin this module, it is recommended that you successfully complete the following: Core Curriculum; Residential Electrical I, Chapters 1 through 8.

Required Trainee Materials

1. Paper and pencil
2. Copy of the latest edition of the *National Electrical Code*
3. Appropriate personal protective equipment

1.0.0 ◆ INTRODUCTION

The art of conduit bending is dependent upon the skills of the electrician and requires a working knowledge of basic terms and proven procedures. Practice, knowledge, and training will help you gain the skills necessary for proper conduit bending and installation. You will be able to practice conduit bending in the lab and in the field under the supervision of experienced coworkers. In this module, the techniques for using hand-operated and step conduit benders such as the hand bender and the hickey will be covered. The process of hand bending conduit, and cutting, reaming, and threading conduit will also be explained.

2.0.0 ◆ HAND BENDING EQUIPMENT

Figure 1 shows a hand bender. Hand benders are convenient to use on the job because they are portable and no electrical power is required. Hand benders have a shape that supports the walls of the conduit being bent.

These benders are used to make various bends in smaller-size conduit (½" to 1¼"). Most hand benders are sized to bend rigid conduit and electrical metallic tubing (EMT) of corresponding sizes. For example, a single hand bender can bend either ¾" EMT or ½" rigid conduit. The next larger size of hand bender will bend either 1" EMT or ¾" rigid conduit. This is because the corresponding sizes of conduit have nearly equal outside diameters.

The first step in making a good bend is familiarizing yourself with the bender. The manufacturer of the bender will typically provide documentation indicating starting points, distance between **offsets,** **gains,** and other important values associated with

Note: The designations "National Electrical Code," "NE Code," and "NEC," where used in this document, refer to the *National Electrical Code®*, which is a registered trademark of the National Fire Protection Association, Quincy, MA. *All National Electrical Code (NEC) references in this module refer to the 2002 edition of the NEC.*

Materials

Ensure you have everything required to teach the course. Check the Materials and Equipment List at the front of this Instructor's Guide.

Audiovisual

Show Transparency 1, Course Objectives.

Show Transparency 2, Performance Profile Tasks.

Classroom

Discuss hand benders and their applications.

Explain that terms shown in bold (blue) are defined in the Glossary at the back of this module.

Demonstration

Demonstrate how to mock up a conduit run using a piece of solid wire.

Classroom

Point out the NEC requirements pertaining to conduit bends.

Explain the difference between a hand bender and a hickey.

Audiovisual

Show Transparency 3 (Figure 2).

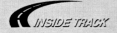

> **INSIDE TRACK**
>
> ### *Working with Conduit*
>
> Unprotected electrical cable is susceptible to moisture and physical damage; therefore, protect the wiring with conduit.

> **INSIDE TRACK**
>
> ### *Bending Conduit*
>
> A good way to practice bending conduit is to use a piece of No. 10 or No. 12 solid wire and bend it to resemble the bends you need. This gives you some perspective on how to bend the conduit and it will also help you to anticipate any problems with the bends.

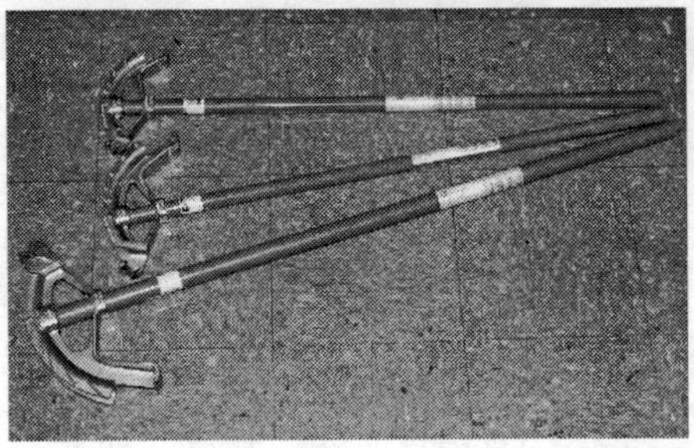

102F01.EPS

Figure 1 ◆ Hand benders.

that particular bender. There is no substitute for taking the time to review this information. It will make the job go faster and result in better bends.

> **CAUTION**
>
> When making bends, be sure you have a firm grip on the handle to avoid slippage and possible injury.

When performing a bend, it is important to keep the conduit on a stable, firm, flat surface for the entire duration of the bend. Hand benders are designed to have force applied using one foot and the hands. See *Figure 2.* It is important to use constant foot pressure as well as force on the handle to achieve uniform bends. Allowing the conduit to rise up or performing the bend on soft ground can result in distorting the conduit outside the bender.

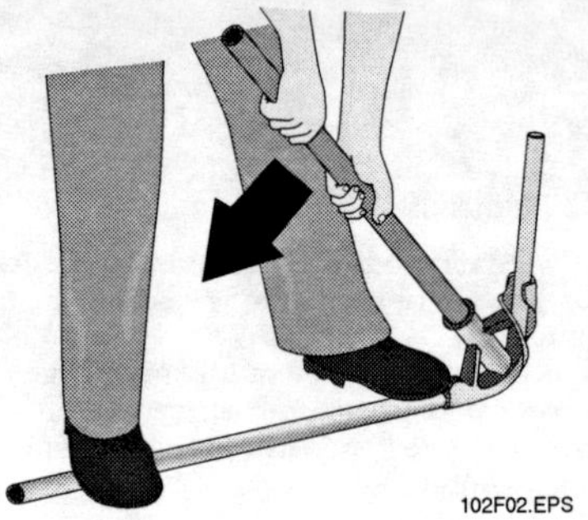

102F02.EPS

Figure 2 ◆ Pushing down on the bender to complete the bend.

> **Note**
>
> Bends should be made in accordance with the guidelines of ***NEC Article 342*** (intermediate metal conduit or IMC), ***Article 344*** (rigid metal conduit or RMC), ***Article 352*** (rigid nonmetallic conduit or RNC), or ***Article 358*** (electrical metallic tubing or EMT).

A hickey should not be confused with a hand bender. The hickey, which is used for RMC and IMC only, functions quite differently. See *Figure 3.*

When you use a hickey to bend conduit, you are forming the bend as well as the radius. When using a hickey, be careful not to flatten or kink the conduit. Hickeys should only be used with RMC and IMC because very little support is given to the walls of the conduit being bent. A hickey is a

Instructor's Notes:

INSIDE TRACK

Proper Bends

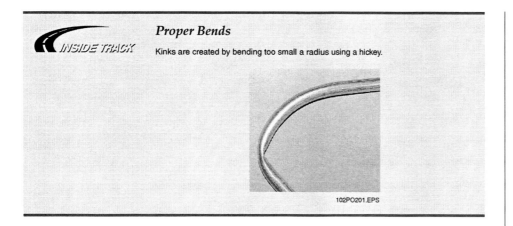

Kinks are created by bending too small a radius using a hickey.

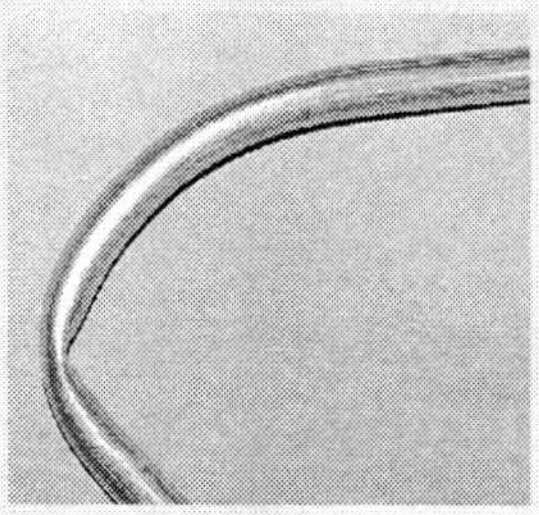

102PO201.EPS

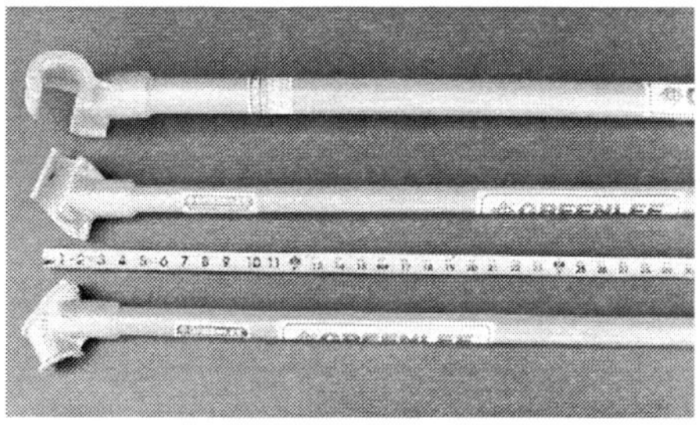

102F03.EPS

Figure 3 ◆ Hickeys.

102F04.EPS

Figure 4 ◆ Typical PVC heating units.

segment bending device. First, a small bend of about 10° is made. Then, the hickey is moved to a new position and another small bend is made. This process is continued until the bend is completed. A hickey can be used for conduit **stub-ups** in slabs and decks.

PVC conduit is bent using a heating unit (*Figure 4*). The PVC must be rotated regularly while it is in the heater so that it heats evenly. Once heated, the PVC is removed, and the bending is performed by hand. Some units use an electric heating element, while others use liquid propane (LP). After bending, a damp sponge or cloth is often used so that the PVC sets up faster.

When bending PVC that is 2" or larger in diameter, there is a risk of wrinkling or flattening the bend. A plug set eliminates this problem (*Figure 5*). A plug is inserted into each end of the piece of PVC being bent. Then, a hand pump is used to pressurize the conduit before bending it. The pressure is about 3 to 5 psi.

> **CAUTION**
> Avoid contact with the case of the heating unit; it can become very hot and cause burns. Also, to avoid a fire hazard, ensure that the unit is cool before storage. If using an LP unit, keep a fire extinguisher nearby.

> **Note**
> The plugs must remain in place until the pipe is cooled and set.

Classroom

Describe how to bend PVC using a heating unit.

Safety

Discuss the precautions that must be taken when using PVC heating units.

HAND BENDING 371

Audiovisual

Show Transparency 4 (Figure 6).

Teaching Tip

Discuss the "What's Wrong with This Picture?" This PVC was overheated and burned, creating a kink.

Classroom

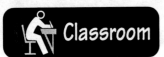

Discuss geometric terms relating to:

- **Right triangles**
- **Circles**

Refer the trainees to *Appendix A* in the back of this module.

102F05.EPS

Figure 5 ◆ Typical plug set.

What's Wrong with This Picture?

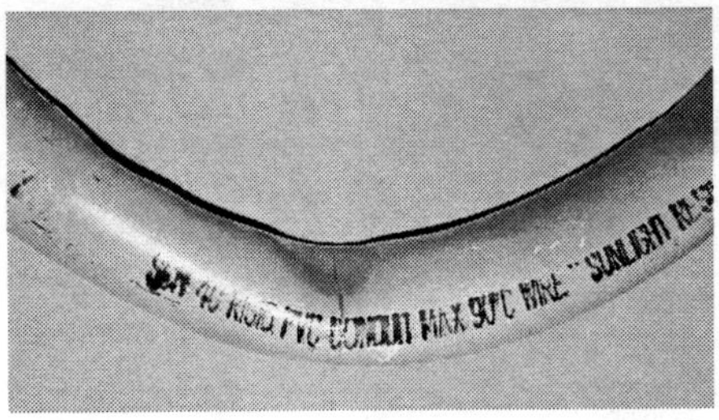

102PO202.EPS

2.1.0 Geometry Required to Make a Bend

Bending conduit requires that you use some basic geometry. You may already be familiar with most of the concepts needed; however, here is a review of the concepts directly related to this task. A right triangle is defined as any triangle with a 90° angle. The side directly opposite the 90° angle is called the *hypotenuse*, and the side on which the triangle sits is the *base*. The vertical side is called the *height*. On the job, you will apply the relationships in a right triangle when making an offset bend. The offset forms the hypotenuse of a right triangle (*Figure 6*).

> **Note**
> There are reference tables for sizing offset bends based on these relationships (see *Appendix A*).

A circle is defined as a closed curved line whose points are all the same distance from its center. The distance from the center point to the edge of the circle is called the *radius*. The length from one edge of the circle to the other edge through the center point is the *diameter*. The distance around the circle is called the *circumference*. A circle can be divided into four equal quadrants. Each quadrant accounts for 90°, making a total of 360°. When you make a **90° bend,** you will use ¼ of a circle, or one quadrant. Concentric circles are circles that have a common center but different radii. The concept of concentric circles can be applied to **concentric bends** in conduit. The angle of each bend is 90°.

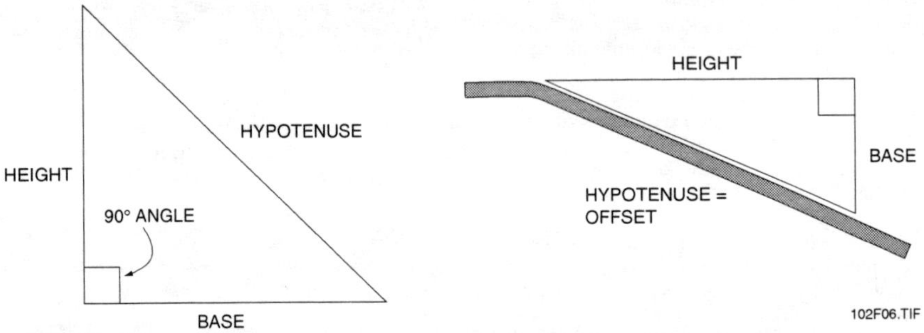

Figure 6 ◆ Right triangle and offset bend.

Instructor's Notes:

372 CHAPTER 9

INSIDE TRACK

Practical Bending

When making offset bends of 45° to step conduit up to another level, square floor tiles make a convenient grid to gauge the distance and angles.

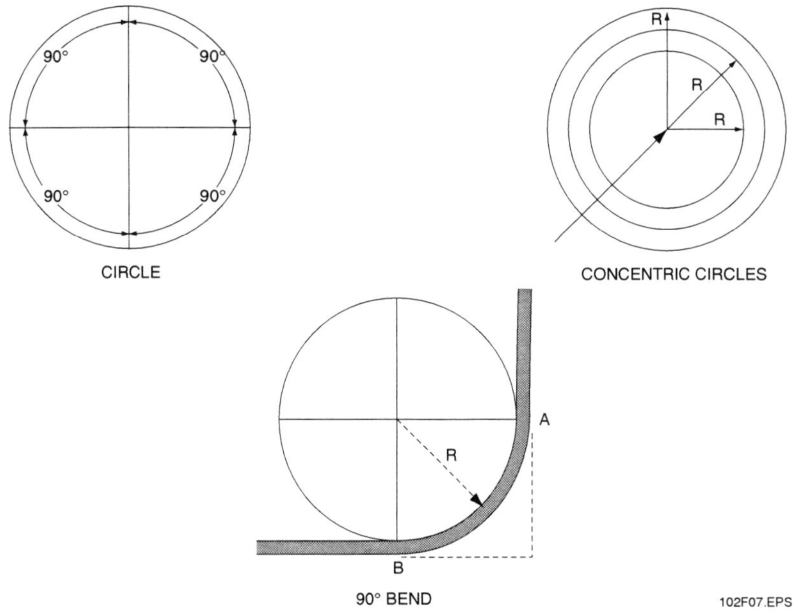

Figure 7 ◈ Circles and 90° bends.

Such bends have the same center point, but the radius of each is different. See *Figure 7.*

To calculate the circumference of a circle, use the following formula:

$$C = \pi \times D \text{ or } C = \pi D$$

In this formula, C = circumference, D = diameter, and $\pi = 3.14$. Another way of stating the formula for circumference is $C = 2\pi R$, where R equals the radius or ½ the diameter.

To figure the arc of a quadrant use:

$$\text{Length of arc} = (.25)\ 2\pi R = 1.57R$$

For this formula, the arc of a quadrant equals ¼ the circumference of the circle or 1.57 times the radius.

A bending radius table is included in *Appendix B.*

2.2.0 Making a 90° Bend

The 90° stub bend is probably the most basic bend of all. The stub bend is used much of the time, regardless of the type of conduit being installed. Before beginning to make the bend, you need to know two measurements:

- Desired **rise** or stub-up
- Take-up distance of the bender

The desired rise is the height of the stub-up. The *take-up* is the amount of conduit the bender will use to form the bend. Take-up distances are usually listed in the manufacturer's instruction manual. Typical bender take-up distances are shown in *Table 1.*

Audiovisual

Show Transparency 5 (Figure 7).

Classroom

Explain how to calculate circumference and arc length.

Refer the trainees to *Appendix B* in the back of this module.

Table 1 Typical Bender Take-Up Distances

EMT	Rigid/IMC	Take-Up
½"	—	5"
¾"	½"	6"
1"	¾"	8"
1¼"	1"	11"

Once you have determined the take-up, subtract it from the stub-up height. Mark that distance on the conduit (all the way around) at that distance from the end. The mark will indicate the point at which you will begin to bend the conduit. Line up the starting point on the conduit with the starting point on the bender. Most benders have a mark, like an arrow, to indicate this point. *Figure 8* shows the take-up required to achieve an 18" stub-up on a piece of ½" EMT.

Once you have lined up the bender, use one foot to hold the conduit steady. Keep your heel on the floor for balance. Apply pressure on the bender foot pedal with your other foot. Make sure you hold the bender handle level, as far up as possible, to get maximum leverage. Then, bend the conduit in one smooth motion, pulling as steadily as possible. Avoid overstretching.

> **Note**
> When bending conduit using the take-up method, always place the bender on the conduit and make the bend facing the hook of the conduit from which the measurements were taken.

After finishing the bend, check to make sure you have the correct angle and measurement. Use the following steps to check a 90° bend:

Step 1 With the back of the bend on the floor, measure to the end of the conduit stub-up to make sure it is the right length.

INSIDE TRACK

Matching Bends in Parallel Runs

Suppose you are running 1" rigid conduit along with a 2" rigid conduit in a rack and you come to a 90° bend. If you used a 1" shoe, the radius would not match that of the 2" conduit bend. To match the 2" 90° bend, take your 1" conduit and put it in the 2" shoe of your bender, then bend as usual. This 1" 90° bend will now have the same radius as the 2" 90° bend. This trick will only work on rigid conduit. If done with EMT, it will flatten the pipe.

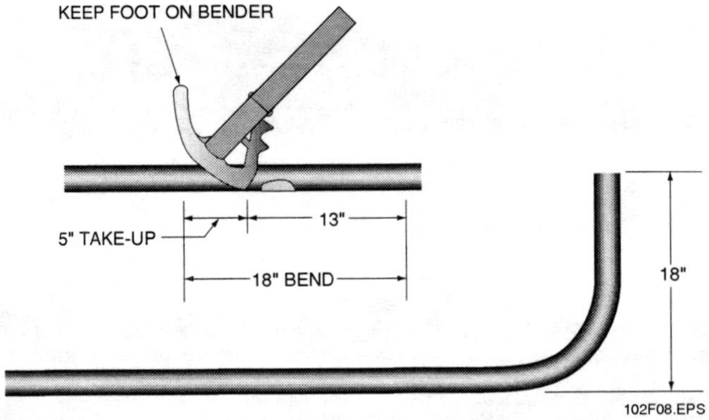

Figure 8 ◆ Bending an 18-inch stub-up.

Audiovisual

Show Transparency 6 (Figure 8).

Demonstration

Demonstrate how to make and check a 90° one-shot bend.

Laboratory

Have the trainees practice making and checking a 90° one-shot bend. Note the proficiency of each trainee.

Instructor's Notes:

374 CHAPTER 9

INSIDE TRACK

Take-Up Method

When bending conduit using the take-up method, always place the bender on the conduit and make the bend facing the end of the conduit from which the measurements were taken. It helps to make a narrow mark with a soft lead pencil or marker completely around the conduit to ensure straight bends.

INSIDE TRACK

Checking Vertical Rise

Use a torpedo level to check for plumb on a vertical rise.

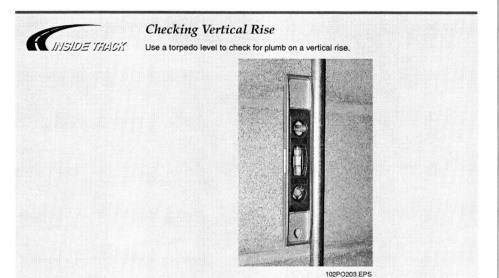

102PO203.EPS

Step 2 Check the 90° angle of the bend with a square or at the angle formed by the floor and a wall. A torpedo level may also be used.

> **Note**
> If you overbend a conduit slightly past the desired angle, you can use the bender to bend the conduit back to the correct angle.

The above procedure will produce a 90° *one-shot bend.* That means that it took a single bend to form the conduit bend. A **segment bend** is any bend that is formed by a series of bends of a few degrees each, rather than a single one-shot bend. A shot is actually one bend in a segment bend. Segment or sweep bends must conform to the provisions of the NEC.

2.3.0 Gain

The gain is the distance saved by the arc of a 90° bend. Knowing the gain can help you to precut, ream, and prethread both ends of the conduit before you bend it. This will make your work go more quickly because it is easier to work with conduit while it is straight. *Figure 9* shows that the overall **developed length** of a piece of conduit with a 90° bend is less than the sum of the horizontal and vertical distances when measured square to the corner. This is shown by the following equation:

$$\text{Developed length} = (A + B) - \text{gain}$$

Demonstration

Demonstrate how to check for plumb using a torpedo level.

Audiovisual

Show Transparency 7 (Figure 9).

Teaching Tip

Discuss the "Think About It." Explain that smooth bends help to prevent conductor damage.

Discuss the "Think About It." Explain that take-up represents the length needed while gain is the distance saved.

THINK ABOUT IT

Smooth Bends

Why are smooth bends so important?

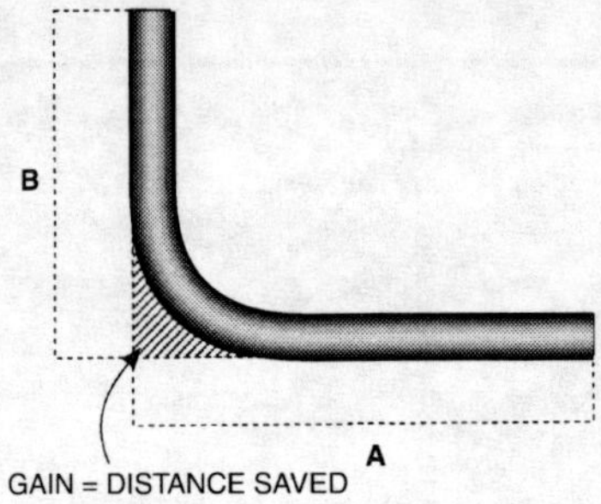

Conduit Size	NEC Radius	90° Gain
½"	4"	2⅝"
¾"	5"	3¼"
1"	6"	4"
1¼"	8"	5⅝"

TYPICAL GAIN TABLE

102F09.TIF

Figure 9 ◆ Gain.

THINK ABOUT IT

Gain

What is the difference between the gain and the take-up of a bend?

An example of a manufacturer's gain table is also shown in *Figure 9*. These tables are used to determine the gain for a certain size conduit.

2.4.0 Back-to-Back 90° Bends

A **back-to-back bend** consists of two 90° bends made on the same piece of conduit and placed back-to-back (*Figure 10*).

To make a back-to-back bend, make the first bend (labeled *X* in *Figure 10*) in the usual manner. To make the second bend, measure the required distance between the bends from the back of the first bend. This distance is labeled *L* in the figure. Reverse the bender on the conduit, as shown in *Figure 10*. Place the bender's back-to-back indicating mark at point *Y* on the conduit. Note that outside measurements from point *X* to point *Y* are used. Holding the bender in the reverse position and properly aligned, apply foot pressure and complete the second bend.

2.5.0 Making an Offset

Many situations require that the conduit be bent so that it can pass over objects such as beams and other conduits, or enter meter cabinets and junction boxes. Bends used for this purpose are called *offsets (kicks)*. To produce an offset, two equal bends of less than 90° are required, a specified distance apart, as shown in *Figure 11*.

Offsets are a trade-off between space and the effort it will take to pull the wire. The larger the degree of bend, the harder it will be to pull the wire. The smaller the degree of bend, the easier it will be to pull the wire. Use the shallowest degree of bend that will still allow the conduit to bypass the obstruction and fit in the given space.

When conduit is offset, some of the conduit length is used. If the offset is made into the area, an allowance must be made for this shrinkage. If the offset angle is away from the obstruction, the shrinkage can be ignored.

Audiovisual

Show Transparency 8 (Figure 10A).

Demonstration

Demonstrate how to make a back-to-back bend.

Instructor's Notes:

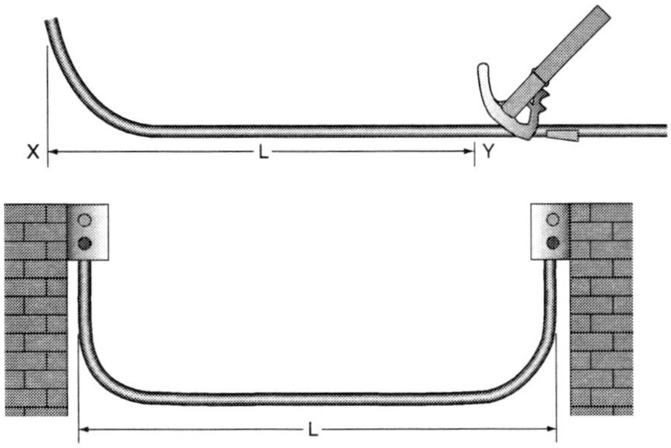

102F10A.EPS

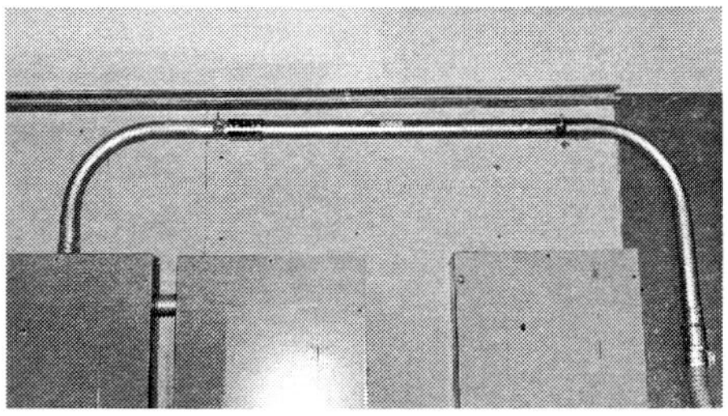

102F10B.EPS

Figure 10 ◈ Back-to-back bends.

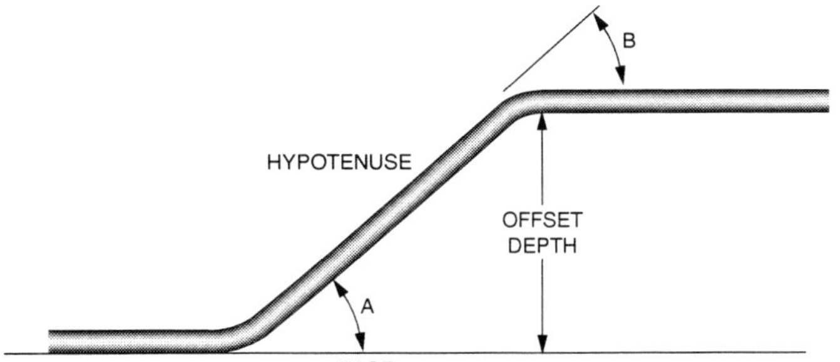

102F11A.EPS

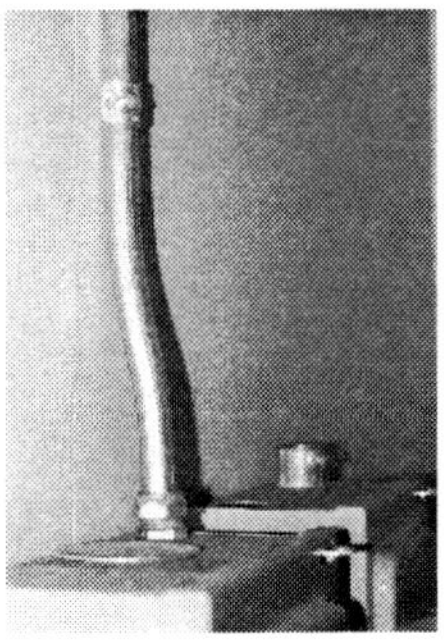

102F11B.EPS

Figure 11 ◈ Offsets.

Audiovisual

Show Transparency 9 (Figure 11A).

HAND BENDING 377

Demonstration

Demonstrate how to make an offset bend.

Laboratory

Have the trainees practice making offset bends. Note the proficiency of each trainee.

Audiovisual

Show Transparencies 10 through 12 (Figures 12 through 14).

Teaching Tip

Discuss the "Think About It."

6" Offset:
Distance between bends = 12"
Shrink = 1½"

10" Offset:
Distance between bends = 14"
Shrink = 3¾"

Table 2 shows the amount of shrinkage per inch of rise for common offset angles.

The formula for figuring the distance between bends is as follows:

Distance between bends =
depth of offset × multiplier

The distance between the offset bends can generally be found in the manufacturer's documentation for the bender. *Table 3* shows the distance between bends for the most common offset angles.

Calculations related to offsets are derived from the branch of mathematics known as *trigonometry,* which deals with triangles. The multipliers shown in *Table 2* represent the *cosecant* (COS) of the related offset angle. The multiplier is determined by dividing the depth of the offset by the hypotenuse of the triangle created by the offset (*Figure 11*).

Basic trigonometry (trig) functions are briefly covered in *Appendix A*. As you will see in the next section, the *tangent* (TAN) of the offset angle is also used in calculating parallel offsets. Understanding trig functions will help you understand how offsets are determined. If you have a scientific calculator and understand these functions, you can calculate offset angles when you know the dimensions of the triangle created by the offset and the obstacle.

Table 2 Shrinkage Calculation

Offset Angle	Multiplier	Shrinkage (per inch of rise)
10° × 10°	6	1⁄16"
22½° × 22½°	2.6	3⁄16"
30° × 30°	2	¼"
45° × 45°	1.4	⅜"
60° × 60°	1.2	½"

2.6.0 Parallel Offsets

Often, multiple pieces of conduit must be bent around a common obstruction. In this case, parallel offsets are made. Since the bends are laid out along a common radius, an adjustment must be made to ensure that the ends do not come out uneven, as shown in *Figure 12*.

The center of the first bend of the innermost conduit is found first, as shown in *Figure 13*. Each successive conduit must have its centerline moved farther away from the end of the pipe, as shown in *Figure 14*. The amount to add is calculated as follows:

Amount added = center-to-center spacing
× tangent (TAN) of ½ offset angle

Table 3 Common Offset Factors (in Inches)

	22½°		30°		45°		60°	
Offset Depth	Between Bends	Shrinkage	Between Bends	Shrinkage	Between Bends	Shrinkage	Between Bends	Shrinkage
2	5¼	⅜	—	—	—	—	—	—
3	7¾	9⁄16	6	¾	—	—	—	—
4	10½	¾	8	1	—	—	—	—
5	13	15⁄16	10	1¼	7	1⅞	—	—
6	15½	1⅛	12	1½	8½	2¼	7¼	3
7	18¼	15⁄16	14	1¾	9¾	2⅝	8⅜	3½
8	20¾	1½	16	2	11¼	3	9⅝	4
9	23½	1¾	18	2¼	12½	3⅜	10⅞	4½
10	26	1⅞	20	2½	14	3¾	12	5

THINK ABOUT IT

Calculating Shrinkage

You're making a 30° by 30° offset to clear a 6" obstruction. What will be the distance between bends? What will be the developed length shrink? Make the same calculations for a 10" offset with 45° bends.

Instructor's Notes:

Equal Angles

THINK ABOUT IT

Why is it important that the angles be identical when making an offset bend?

Tangents can be found using the trig tables provided in *Appendix A.*

For example, *Figure 15* shows three pipes laid out as parallel and offset. The angle of the offset is 30°. The center-to-center spacing is 3". The start of the innermost pipe's first bend is 12".

The starting point of the second pipe will be:

12" + [center-to-center spacing × TAN (½ offset angle)]

12" + (3" × TAN 15°) = 12" + (3" × .2679) = 12" + .8037"

This is approximately $12\frac{13}{16}$".

The starting point for the outermost pipe is:

$12\frac{13}{16}$" + $\frac{13}{16}$" = $13\frac{5}{8}$"

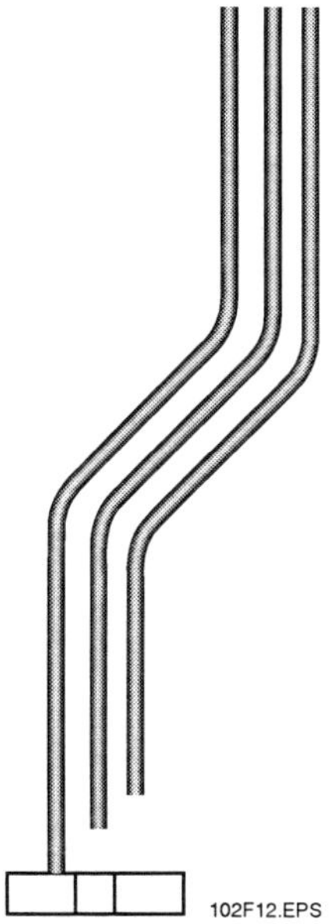

Figure 12 ◈ Incorrect parallel offsets.

2.7.0 Saddle Bends

A saddle bend is used to go around obstructions. *Figure 16* illustrates an example of a saddle bend that is required to clear a pipe obstruction. Making a saddle bend will cause the center of the saddle to shorten $\frac{3}{16}$" for every inch of saddle depth (see *Table 4*). For example, if the pipe diameter is 2 inches, this would cause a $\frac{3}{8}$" shortening of the conduit on each side of the bend.

When making saddle bends, the following steps should apply:

Step 1 Locate the center mark A on the conduit by using the size of the obstruction (i.e., pipe diameter) and calculate the shrink rate

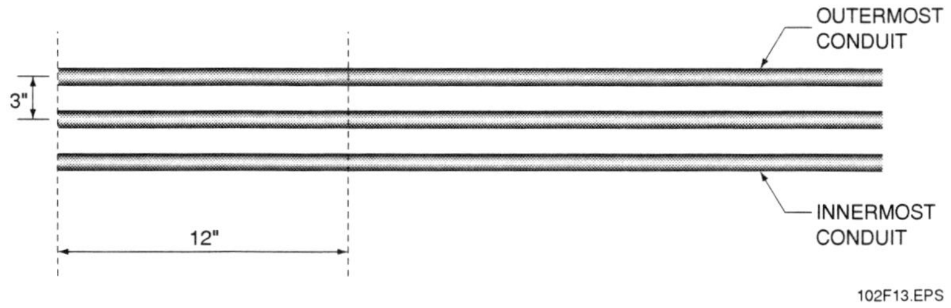

Figure 13 ◈ Center of first bend.

Teaching Tip

Discuss the "Think About It." Explain that uniform bends are necessary to ensure that the run remains parallel.

Audiovisual

Show Transparencies 13 and 14 (Figures 15 and 16).

Demonstration

Demonstrate how to make a saddle bend.

Laboratory

Have the trainees practice making saddle bends. Note the proficiency of each trainee.

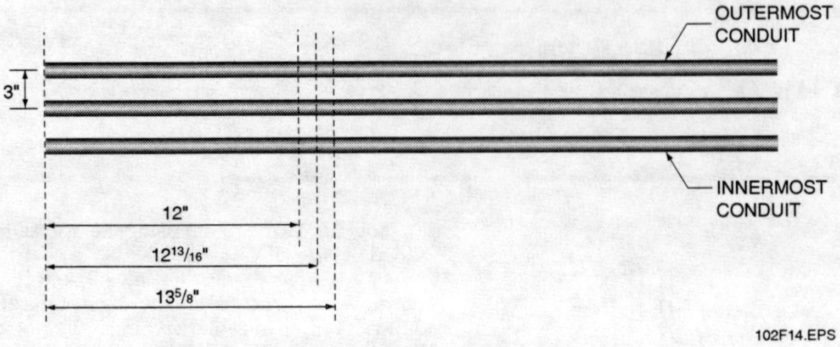

Figure 14 ◆ Successive centerlines.

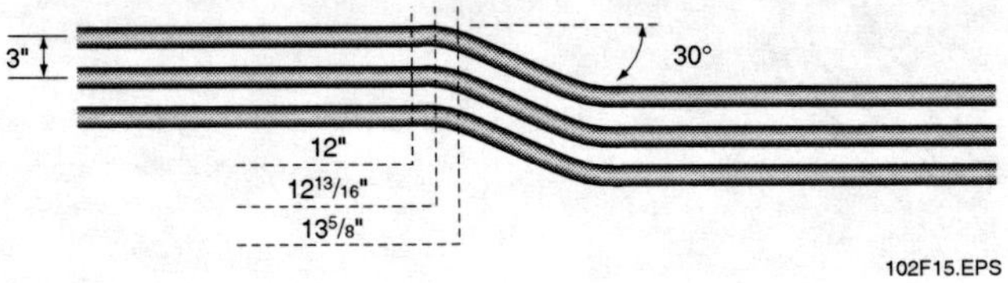

Figure 15 ◆ Parallel offset pipes.

Teaching Tip

Discuss the "Think About It." Starting point of second pipe = 13⅝".

THINK ABOUT IT

Calculating Parallel Offsets

You're making parallel offsets of 45° and the lengths of conduit are spaced 4" center to center. If the offset starts 12" down the pipe, what is the starting point for the bend on the second pipe?

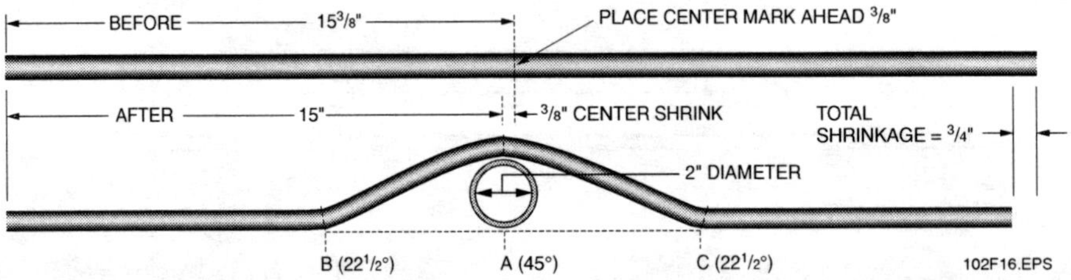

Figure 16 ◆ Saddle measurement.

Instructor's Notes:

Table 4 Shrinkage Chart for Saddle Bends with a 45° Center Bend and Two 22½° Bends

Obstruction Depth	Shrinkage Amount (Move Center Mark Forward)	Make Outside Marks from *New* Center Mark
1	3/16"	2½"
2	3/8"	5"
3	9/16"	7½"
4	3/4"	10"
5	15/16"	12½"
6	1⅛"	15"
For each additional inch, add	3/16"	2½"

of the obstruction (for example, if the pipe diameter is 2 inches, 3/8" of conduit will be lost on each side of the bend for a total shrinkage of 3/4"). This figure will be added to the measurement from the end of the conduit to the centerline of the obstruction (for example, if the distance measured from the conduit end and the obstruction centerline was 15", the distance to A would be 15⅜").

Step 2 Locate marks B and C on the conduit by measuring 2½" for every 1" of saddle depth *from* the A mark (i.e., for the saddle depth of 2 inches, the B mark would be 5" before the A mark and the C mark would be 5" after the A mark). See *Figure 17.*

Step 3 Refer to *Figure 18* and make a 45° bend at point A, make a 22½° bend at point B, and make a 22½° bend at point C. (Be sure to check the manufacturer's specifications.)

2.8.0 Four-Bend Saddles

Four-bend saddles can be difficult. The reason is that four bends must be aligned exactly on the same plane. Extra time spent laying it out and performing the bends will pay off in not having to scrap the whole piece and start over.

Figure 19 illustrates that the four-bend saddle is really two offsets formed back-to-back. Working left to right, the procedure for forming this saddle is as follows:

Step 1 Determine the height of the offset.

Step 2 Determine the correct spacing for the first offset and mark the conduit.

Step 3 Bend the first offset.

Audiovisual

Show Transparencies 15 through 17 (Figures 17 through 19).

Demonstration

Demonstrate how to make a four-bend saddle.

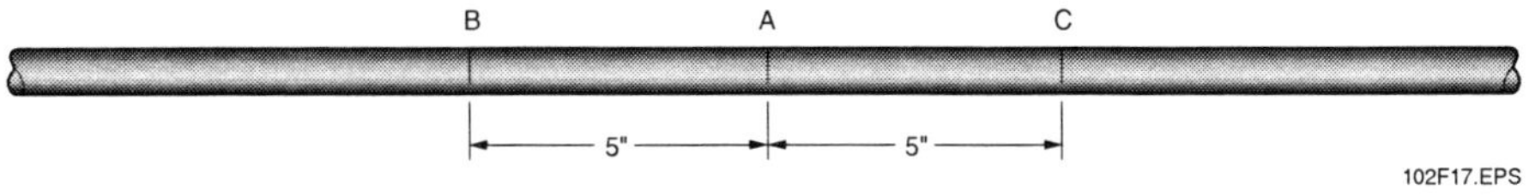

Figure 17 ◈ Measurement locations.

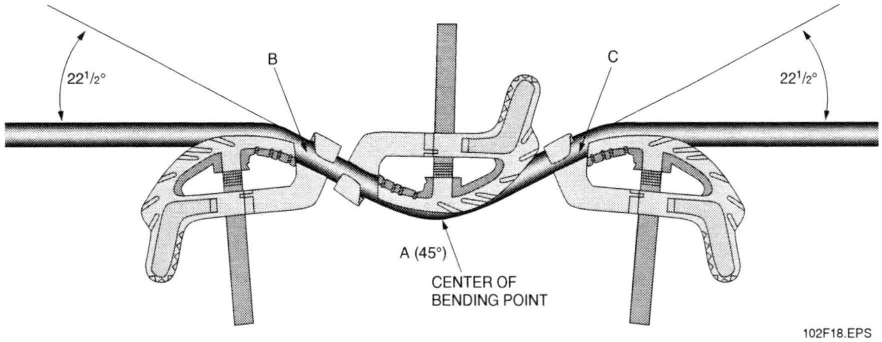

Figure 18 ◈ Location of bends.

HAND BENDING 381

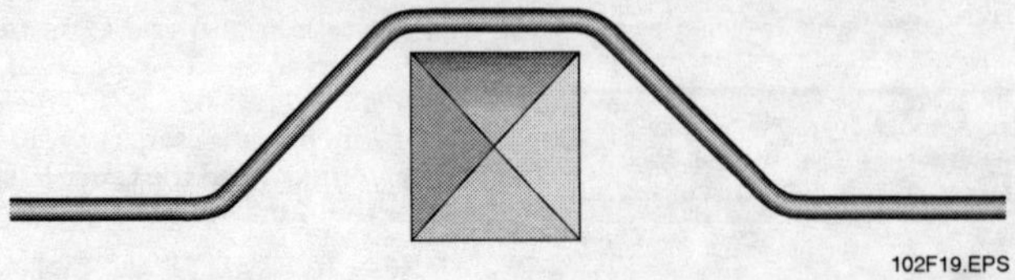

Figure 19 ◆ Typical four-bend saddle.

Audiovisual

Show Transparencies 18 through 20 (Figures 20 through 22).

Step 4 Mark the start point for the second offset at the trailing edge of the obstruction.

Step 5 Mark the spacing for the second offset.

Step 6 Bend the second offset.

Using *Figure 20* as an example, a four-bend saddle using ½" EMT is laid out as follows:

- Height of the box = 6"
- Width of the box = 8"
- Distance to the obstruction = 36"

Two 30° offsets will be used to form the saddle. It is created as follows:

Step 1 See *Figure 21.* Working from left to right, calculate the start point for the first bend. The distance to the obstruction is 36", the offset is 6", and the 30° multiplier from *Table 2* is 2.00:

Distance to the obstruction − (offset × constant for the angle) + shrinkage = distance to the first bend

36" − (6" × 2.00) + 1½" = 25½"

Step 2 Determine where the second bend will end to ensure the conduit clears the obstruction. See *Figure 22.*

Distance to the first bend + distance to second bend + shrinkage = total length of the first offset

25½" + 12" + 1½" = 39"

Step 3 Determine the start point of the second offset. The width of the box is 8"; therefore, the start point of the second offset should be 8" beyond the end of the first offset:

8" + 39" = 47"

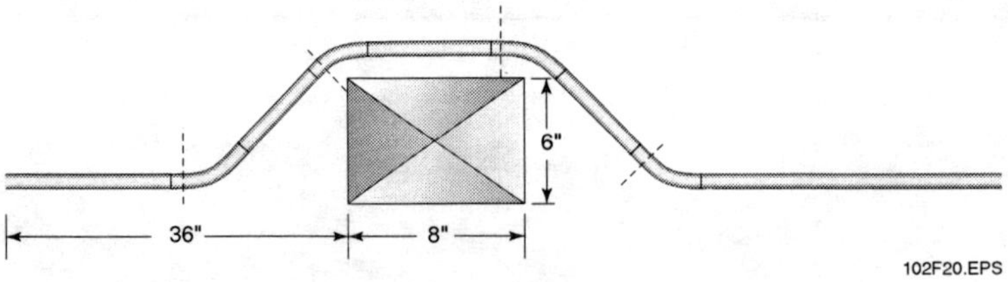

Figure 20 ◆ Four-bend saddle.

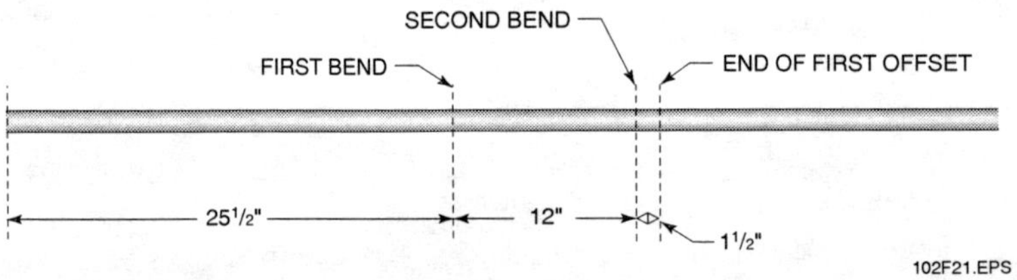

Figure 21 ◆ Four-bend saddle measurements.

Instructor's Notes:

382 CHAPTER 9

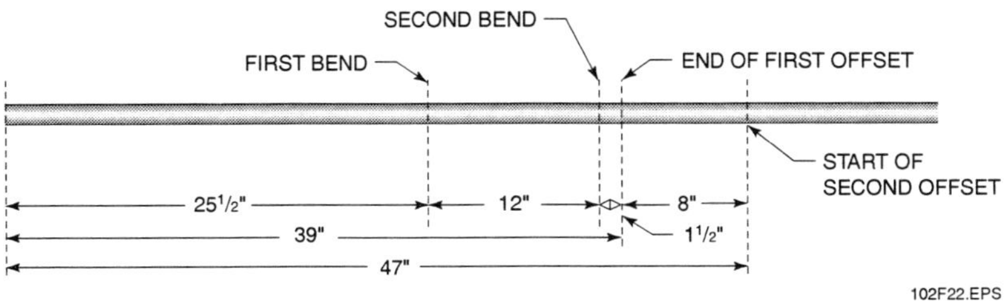

Figure 22 ◆ Bend and offset measurements.

INSIDE TRACK

Planning Bends

The more bends you make, the more difficult it is to pull the wires through the conduit. Therefore, plan your bends in advance, avoid sharp bends, and make as few bends as possible. The NEC allows the bends in a single run of conduit to total 360°; however, 360° is not as much as you might think. For example, if you bend the conduit 90° for two corners of a room, with two 45° offsets where the conduit connects to a panelboard and junction box, you've used up your 360°.

Classroom

Emphasize the importance of careful planning to ensure proper bends.

INSIDE TRACK

Using Your Bender Head to Secure Conduit

To secure conduit while cutting, insert the conduit into one of the holes in the bender head, brace your knee against the bender to secure it, then proceed to cut the conduit.

Step 4 Determine the spacing for the second offset. Since the first and second offsets have the same rise and angle, the distance between bends will be the same, or 12".

3.0.0 ◆ CUTTING, REAMING, AND THREADING CONDUIT

RMC, IMC, and EMT are available in standard 10-foot lengths. When installing conduit, it is cut to fit the job requirements.

3.1.0 Hacksaw Method of Cutting Conduit

Conduit is normally cut using a hacksaw. To cut conduit with a hacksaw:

Step 1 Inspect the blade of the hacksaw and replace it, if needed. A blade with 18, 24, or 32 cutting teeth per inch is recommended for conduit. Use a higher tooth count for EMT and a lower tooth count for rigid conduit and IMC. If the blade needs to be replaced, point the teeth toward the front of the saw when installing the new blade.

Step 2 Secure the conduit in a pipe vise.

Step 3 Rest the middle of the hacksaw blade on the conduit where the cut is to be made. Position the saw so the end of the blade is pointing slightly down and the handle is pointing slightly up. Push forward gently until the cut is started. Make even strokes until the cut is finished.

CAUTION

To avoid bruising your knuckles on the newly cut pipe, use gentle strokes for the final cut.

Demonstration

Demonstrate how to use a hacksaw to cut conduit.

Laboratory

Have the trainees practice cutting conduit using a hacksaw. Note the proficiency of each trainee.

Audiovisual

Show Transparency 21 (Figure 23).

Demonstration

Demonstrate how to use a pipe cutter.

Audiovisual

Show Transparency 22 (Figure 24A).

Step 4 Check the cut. The end of the conduit should be straight and smooth. *Figure 23* shows correct and incorrect cuts. Ream the conduit.

3.2.0 Pipe Cutter Method

A pipe cutter can also be used to cut RMC and IMC. To use a pipe cutter:

Step 1 Secure the conduit in a pipe vise and mark a place for the cut.

Step 2 Open the cutter and place it over the conduit with the cutter wheel on the mark.

Step 3 Tighten the cutter by rotating the screw handle.

> **CAUTION**
> Do not overtighten the cutter. Overtightening can break the cutter wheel and distort the wall of the conduit.

Step 4 Rotate the cutter counterclockwise to start the cut. *Figure 24* shows the proper way to rotate the cutter.

Step 5 Tighten the cutter handle ¼ turn for each full turn around the conduit. Again, make sure that you do not overtighten it.

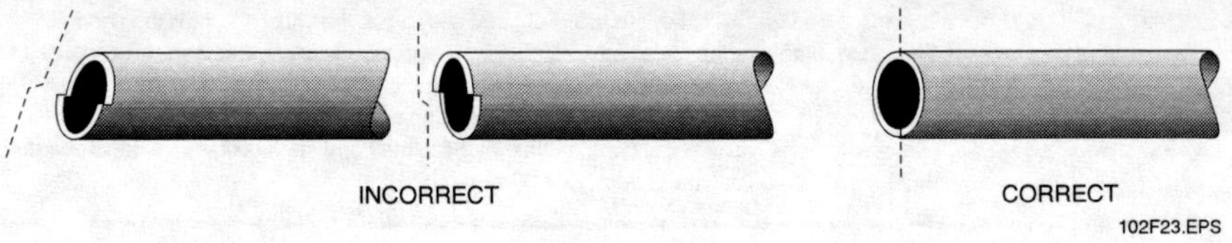

Figure 23 ◆ Conduit ends after cutting.

> INSIDE TRACK
> ***Using Tape***
> Use a piece of tape as a guide for marking your cutting lines around the conduit. This ensures a straight cut.

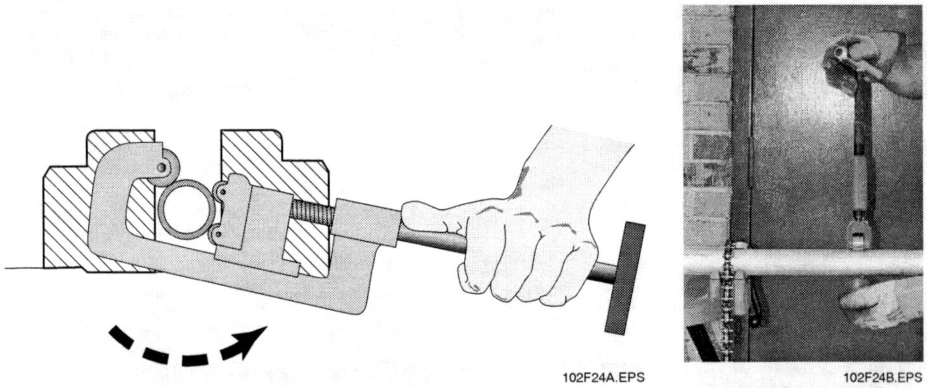

Figure 24 ◆ Cutter rotation.

Instructor's Notes:

Step 6 Add a few drops of cutting oil to the groove and continue cutting. Avoid skin contact with the oil.

Step 7 When the cut is almost finished, stop cutting and snap the conduit to finish the cut. This reduces the ridge that can be formed on the inside of the conduit.

Step 8 Clean the conduit and cutter with a shop towel rag.

Step 9 Ream the conduit.

3.3.0 Reaming Conduit

When the conduit is cut, the inside edge is sharp. This edge will damage the insulation of the wire when it is pulled through. To avoid this damage, the inside edge must be smoothed or *reamed* using a reamer (*Figure 25*).

To ream the inside edge of a piece of conduit using a hand reamer, proceed as follows:

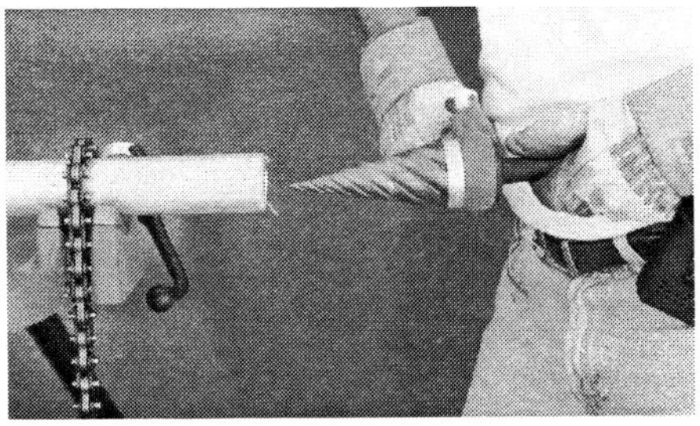

102F25.EPS

Figure 25 ◈ Rigid conduit reamer.

Step 1 Place the conduit in a pipe vise.

Step 2 Insert the reamer tip in the conduit.

Step 3 Apply light forward pressure and start rotating the reamer. *Figure 26* shows the

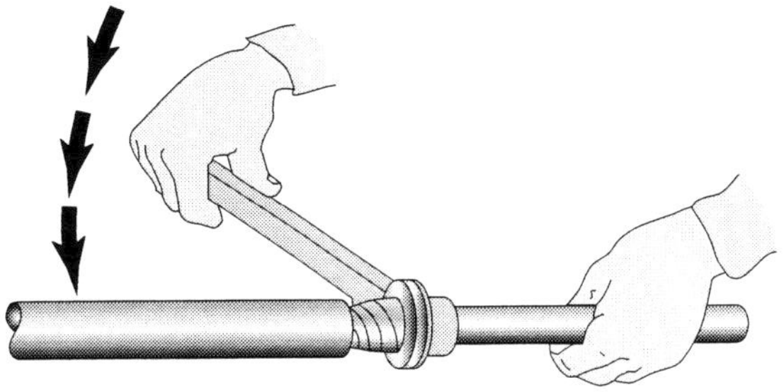

102F26A.EPS

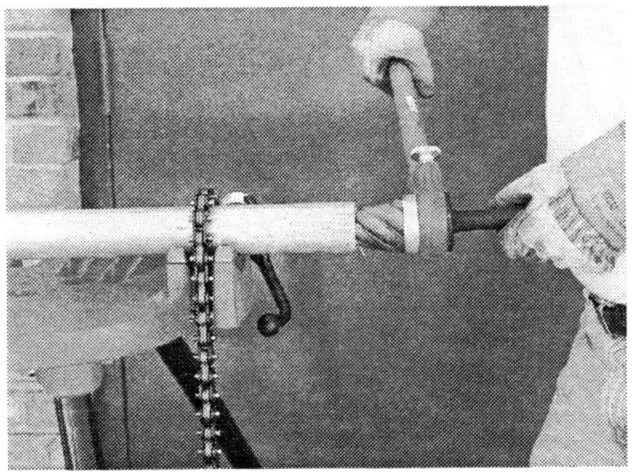

102F26B.EPS

Figure 26 ◈ Reamer rotation.

Audiovisual

Show Transparency 23 (Figure 26A).

Demonstration

Demonstrate how to ream conduit.

Laboratory

Have the trainees practice reaming conduit. Note the proficiency of each trainee.

Safety

Emphasize that burrs are sharp and can cause cuts if gloves are not worn.

Demonstration

Demonstrate the use of a ratchet threader.

proper way to rotate the reamer. It should be rotated using a downward motion. The reamer can be damaged if you rotate it in the wrong direction. The reamer should bite as soon as you apply the proper pressure.

Step 4 Remove the reamer by pulling back on it while continuing to rotate it. Check the progress and then reinsert the reamer. Rotate the reamer until the inside edge is smooth. You should stop when all burrs have been removed.

> **Note**
> If a conduit reamer is not available, use a half-round file (the tang of the file must have a handle attached). EMT may be reamed using the nose of diagonal cutters or small hand reamers.

3.4.0 Threading Conduit

After conduit is cut and reamed, it is usually threaded so it can be properly joined. Only RMC and IMC have walls thick enough for threading.

The tool used to cut threads in conduit is called a *die*. Conduit dies are made to cut a taper of ¾ inch per foot. The number of threads per inch varies from 8 to 18, depending upon the diameter of the conduit. A thread gauge is used to measure how many threads per inch are cut.

The threading dies are contained in a die head. The die head can be used with a hand-operated ratchet threader (*Figure 27*) or with a portable power drive.

To thread conduit using a hand-operated threader, proceed as follows:

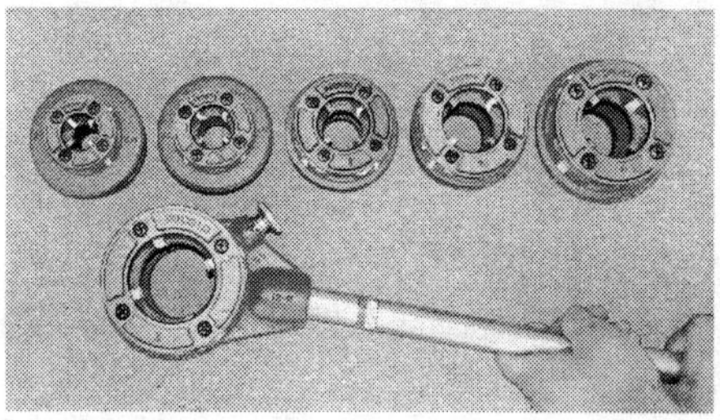

102F27.EPS

Figure 27 ◆ Hand-operated ratchet threader.

Step 1 Insert the conduit in a pipe vise. Make sure the vise is fastened to a strong surface. Place supports, if necessary, to help secure the conduit.

Step 2 Determine the correct die and head. Inspect the die for damage such as broken teeth. Never use a damaged die.

Step 3 Insert the die securely in the head. Make sure the proper die is in the appropriately numbered slot on the head.

Step 4 Determine the correct thread length to cut for the conduit size used (match the manufacturer's thread length).

Step 5 Lubricate the die with cutting oil at the beginning and throughout the threading operation. Avoid skin contact with the oil.

Step 6 Cut threads to the proper length. Make sure that the conduit enters the tapered side of the die. Apply pressure and start turning the head. You should back off the head each ¼ turn to clear away chips.

Step 7 Remove the die when the proper cut is made. Threads should be cut only to the length of the die. Overcutting will leave the threads exposed to corrosion.

Step 8 Inspect the threads to make sure they are clean, sharp, and properly made. Use a thread gauge to measure the threads. The finished end should allow for a wrench-tight fit with one or two threads exposed.

> **Note**
> The conduit should be reamed again after threading to remove any burrs and edges. Cutting oil must be swabbed from the inside and outside of the conduit. Use a sandbox or drip pan under the threader to collect drips and shavings.

Die heads can also be used with portable power drives. You will follow the same steps when using a portable power drive. Threading machines are often used on larger conduit and where frequent threading is required. Threading machines hold and rotate the conduit while the die is fed onto the conduit for cutting. When using a threading machine, make sure you secure the legs properly and follow the manufacturer's instructions.

Instructor's Notes:

3.5.0 Cutting and Joining PVC Conduit

PVC conduit may be easily cut with a fine-tooth handsaw. To ensure square cuts, a miter box or similar device is recommended for cutting 2" and larger PVC. You can deburr the cut ends using a pocket knife. Smaller diameter PVC conduit, up to 1½", may be cut using a PVC cutter.

Use the following steps to join PVC conduit sections or attachments to plastic boxes:

Step 1 Wipe all the contacting surfaces clean and dry.

Step 2 Apply a coat of cement (a brush or aerosol can is recommended) to the male end to be attached.

Step 3 Press the conduit and fitting together and rotate about a half-turn to evenly distribute the cement.

Forming PVC in the field requires a special tool called a *hot box* or other specialized methods. PVC may not be threaded when it is used for electrical applications.

Note

Cementing the PVC must be done quickly. The aerosol spray cans of cement or the cement/brush combination are usually provided by the PVC manufacturer. Make sure you use the recommended cement.

CAUTION

Solvents and cements used with PVC are hazardous. Wear gloves and follow the product instructions.

Demonstration

Demonstrate how to cut and join PVC conduit.

Safety

Emphasize the precautions required when working with PVC solvents and cements.

INSIDE TRACK

Oiling the Threader

For smoother operation, oil the threader often while threading the conduit.

102PO204.EPS

INSIDE TRACK

Threading Conduit

The key to threading conduit is to start with a square cut. If you don't get it right the first time, the fitting won't thread properly.

Demonstration

Demonstrate how to cut a piece of PVC using a nylon string.

Teaching Tip

Discuss the "Think About It." Emphasize that poor bending techniques can result in pulling problems, damaged conductors, and a sloppy appearance.

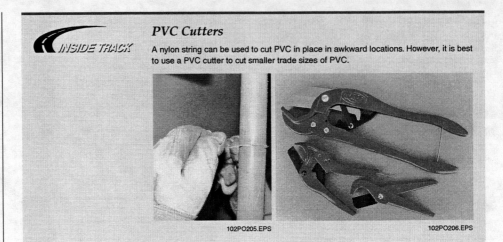

INSIDE TRACK

PVC Cutters

A nylon string can be used to cut PVC in place in awkward locations. However, it is best to use a PVC cutter to cut smaller trade sizes of PVC.

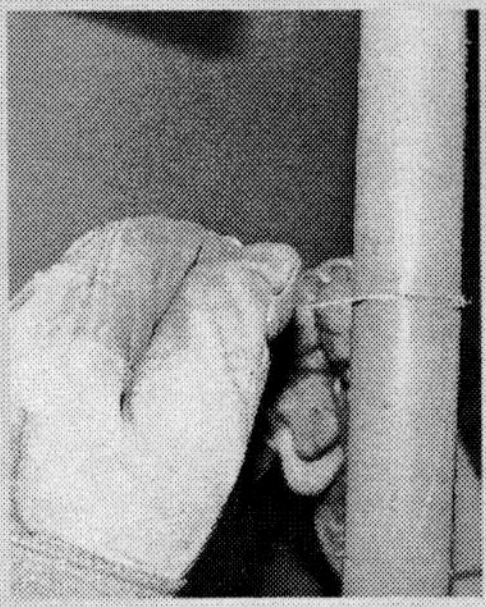

102PO205.EPS

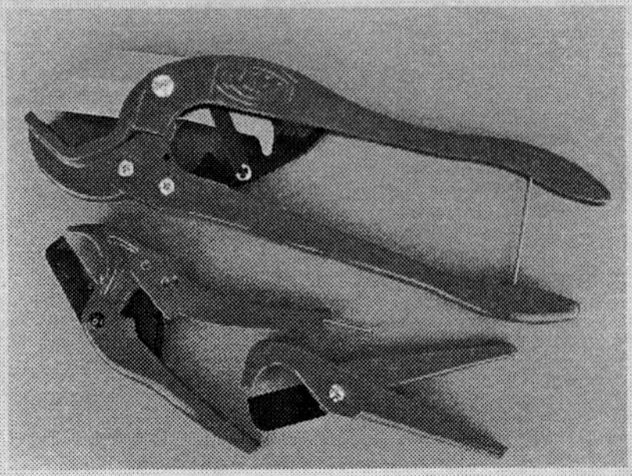

102PO206.EPS

THINK ABOUT IT

Putting It All Together

This module has stressed the precision necessary for creating accurate and uniform bends. Why is this important? What practical problems can result from sloppy or inaccurate bends?

Summary

You must choose a conduit bender to suit the kind of conduit being installed and the type of bend to be made. Some knowledge of the geometry of right triangles and circles needs to be mastered to make the necessary calculations. You must be able to calculate, lay out, and perform bending operations on a single run of conduit and also on two or more parallel runs of conduit. At times, data tables for the figures may be consulted for the calculations. All work must conform to the requirements of the NEC.

Instructor's Notes:

388 CHAPTER 9

Review Questions

1. The field bending of PVC requires a _____.
   a. hickey
   b. heating unit
   c. segmented bender
   d. one-shot bender
2. After bending PVC, the bend can be set by using _____.
   a. a damp sponge or cloth
   b. dry ice
   c. ice-cold water
   d. a blow dryer
3. A hickey is used for bending _____.
   a. RMC
   b. EMT and IMC
   c. PVC conduit
   d. RMC, EMT, IMC, and PVC conduit
4. A plug set is typically used to prevent _____ PVC when bending it.
   a. overpressurizing
   b. flattening
   c. corroding
   d. cutting
5. What is the key to accurate bending with a hand bender?
   a. Correct size and length of handle
   b. Constant foot pressure on the back piece
   c. Using only the correct brand of bender
   d. Correct inverting of the conduit bender
6. In a right triangle, the side directly opposite the 90° angle is called the _____.
   a. right side
   b. hypotenuse
   c. altitude
   d. base
7. The formula for calculating the circumference of a circle is _____.
   a. $\pi \times R^2$
   b. $2\pi \times R^2$
   c. $\pi \times D$
   d. $2\pi \times D$
8. Prior to making a 90° bend, what two measurements must be known?
   a. Length of conduit and size of conduit
   b. Desired rise and length of conduit
   c. Size of bender and size of conduit
   d. Stub-up distance and take-up distance
9. A back-to-back bend is _____.
   a. a two-shot 90° bend
   b. two 90° bends made back-to-back
   c. an offset with four bends
   d. a segmented bend
10. To ensure the conduit enters straight into the junction box, a(n) _____ may be required.
    a. back-to-back bend
    b. saddle bend
    c. offset
    d. take-up
11. To prevent the ends of the conduit from being staggered, what additional information must be used when making parallel offset bends?
    a. Center-to-center spacing and tangent of ½ the offset angle
    b. Length of conduit and size of conduit
    c. Stub-up distance and take-up distance
    d. Offset angle and length of conduit
12. When making a saddle bend, the center of the saddle will cause the conduit to shrink _____ for every inch of saddle depth.
    a. 3/8"
    b. 3/16"
    c. 3/4"
    d. 3/32"
13. When using a pipe cutter, always rotate the cutter _____ to start the cut.
    a. in a clockwise direction
    b. with the grain
    c. in a counterclockwise direction
    d. against the grain
14. You would use _____ to smooth the sharp inside edge of metal conduit after it has been cut.
    a. a flat file
    b. rough sandpaper
    c. a reamer
    d. a pocket knife
15. What tool is used to cut threads in RMC or IMC?
    a. A thread gauge
    b. A cutter
    c. A tap
    d. A die

Classroom

Have the trainees complete the Review Questions and go over the answers prior to administering the Module Examination.

Examination

Administer the Module Examination. Record the results on Craft Training Report Form 200 and submit the results to the Training Program Sponsor.

Performance Testing

Administer the Performance Test and fill out Performance Profile Sheets for each trainee. If desired, trainee proficiency noted during laboratory sessions may be used to complete the Performance Test. Be sure to record the results on Craft Training Report Form 200 and submit the results to the Training Program Sponsor.

HAND BENDING 389

PROFILE IN SUCCESS

Timothy Ely, Beacon Electric Company

Tim Ely is a man who believes in giving something back to the industry that nurtured his successful career. Despite working in a demanding executive position, he serves on many industry committees and was instrumental in the development of the NCCER Electrical Program.

What made you decide to become an electrician?
During my last two years of high school, I worked for a do-it-all construction company. We laid concrete, installed roofs, hung drywall, installed plumbing, and did electrical work. I liked the electrical work the best.

How did you learn the trade?
I learned through on-the-job training, hard work, and studying on my own. I had good teachers who were patient with me and took the time to help me succeed.

What kinds of jobs did you hold on the way to your current position?
I started out wiring houses and did that for the first two years. Then I switched over to commercial and industrial work, and worked as an apprentice in that area for two more years before becoming a journeyman. From there, I served as a lead electrician, then foreman, then finally city superintendent, before being promoted to my current job as general superintendent.

What factor or factors have contributed most to your success?
Hard work helps a lot. I also try to bring a positive attitude to work with me every day. My family and friends have supported me throughout my career.

What does a general superintendent do in your company?
In my job, I have responsibility for all the job sites, as well as the warehouse and service trucks. I also have responsibility for employee hiring, safety training, job planning and scheduling, quality control, and licensing. I personally hold 26 different state and city licenses, and I firmly believe that getting the training to obtain your licenses and then doing the in-service training to keep your licenses current are important factors in an electrician's success. For example, an electrical contractor can bid on jobs in a wide geographical area. Electricians working for that contractor can work on projects in different cities, even different states. Every place you go will require you to have a valid license.

What advice would you give to someone entering the electrical trade?
Work hard, treat people with respect, and keep an open mind. Be careful how you deal with people. Someone you offend today may wind up being your boss or a potential customer tomorrow.

Instructor's Notes:

GLOSSARY

Trade Terms Introduced in This Module

90° bend: A bend that changes the direction of the conduit by 90°.

Back-to-back bend: Any bend formed by two 90° bends with a straight section of conduit between the bends.

Concentric bends: Making 90° bends in two or more parallel runs of conduit and increasing the radius of each conduit from the inside of the run toward the outside.

Developed length: The actual length of the conduit that will be bent.

Gain: Because a conduit bends in a radius and not at right angles, the length of conduit needed for a bend will not equal the total determined length. Gain is the distance saved by the arc of a 90° bend.

Offsets: An offset (kick) is two bends placed in a piece of conduit to change elevation to go over or under obstructions or for proper entry into boxes, cabinets, etc.

Rise: The length of the bent section of conduit measured from the bottom, centerline, or top of the straight section to the end of the bent section.

Segment bend: A large bend formed by multiple short bends or *shots.*

Stub-up: Another name for the rise in a section of conduit. Also, a term used for conduit penetrating a slab or the ground.

APPENDIX A

Using Trigonometry to Determine Offset Angles and Multipliers

You do not have to be a mathematician to use trigonometry. Understanding the basic trig functions and how to use them can help you calculate unknown distances or angles. Assume that the right triangle below represents a conduit offset. If you know the length of one side and the angle, you can calculate the length of the other sides, or if you know the length of any two of the sides of the triangle, you can then find the offset angle using one or more of these trig functions. You can use a trig table such as that shown on the following pages or a scientific calculator to determine the offset angle. For example, if the cosecant of angle A is 2.6, the trig table tells you that the offset angle is 22½°.

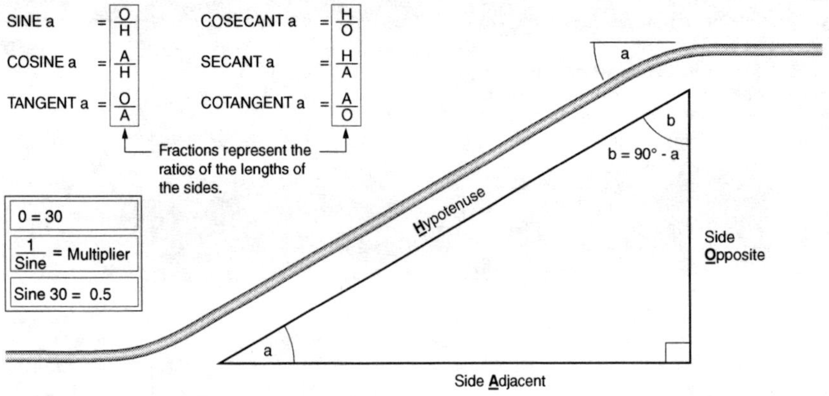

To determine the multiplier for the distance between bends in an offset:

1. Determine the angle of the offset: 30°
2. Find the sine of the angle: 0.5
3. Find the inverse (reciprocal) of the sine: $\frac{1}{0.5} = 2$. This is also listed in trig tables as the cosecant of the angle.
4. This number multiplied by the height of the offset gives the hypotenuse of the triangle, which is equal to the distance between bends.

102APX01.EPS

Instructor's Notes:

392 CHAPTER 9

TRIGONOMETRY TABLE

Angle	Sine	Cosine	Tangent	Cotangent	Cosecant
1°	.0175	.9998	.0175	57.3	57.3065
2°	.0349	.9994	.0349	28.6	28.6532
3°	.0523	.9986	.0524	19.1	19.1058
4°	.0698	.9976	.0699	14.3	14.3348
5°	.0872	.9962	.0875	11.4	11.4731
6°	.1045	.9945	.1051	9.51	9.5666
7°	.1219	.9925	.1228	8.14	8.2054
8°	.1392	.9903	.1405	7.12	7.1854
9°	.1564	.9877	.1584	6.31	6.3926
10°	.1736	.9848	.1763	5.67	5.7587
11°	.1908	.9816	.1944	5.14	5.2408
12°	.2079	.9781	.2126	4.70	4.8097
13°	.2250	.9744	.2309	4.33	4.4454
14°	.2419	.9703	.2493	4.01	4.1335
15°	.2588	.9659	.2679	3.73	3.8636
16°	.2756	.9613	.2867	3.49	3.5915
17°	.2924	.9563	.3057	3.27	3.4203
18°	.3090	.9511	.3249	3.08	3.2360
19°	.3256	.9455	.3443	2.90	3.0715
20°	.3420	.9397	.3640	2.75	2.9238
21°	.3584	.9336	.3839	2.61	2.7904
22°	.3744	.9272	.4040	2.48	2.6694
23°	.3907	.9205	.4245	2.36	2.5593
24°	.4067	.9135	.4452	2.25	2.4585
25°	.4226	.9063	.4663	2.14	2.3661
26°	.4384	.8988	.4877	2.05	2.2811
27°	.4540	.8910	.5095	1.96	2.2026
28°	.4695	.8829	.5317	1.88	2.1300
29°	.4848	.8746	.5543	1.80	2.0626
30°	.5000	.8660	.5774	1.73	2.0000
31°	.5150	.8572	.6009	1.66	1.9415
32°	.5299	.8480	.6249	1.60	1.8870
33°	.5446	.8387	.6494	1.54	1.8360
34°	.5592	.8290	.6745	1.48	1.7883
35°	.5736	.8192	.7002	1.43	1.7434
36°	.5878	.8090	.7265	1.38	1.7012
37°	.6018	.7986	.7536	1.33	1.6616
38°	.6157	.7880	.7813	1.28	1.6242
39°	.6293	.7771	.8098	1.23	1.5890
40°	.6428	.7660	.8391	1.19	1.5557
41°	.6561	.7547	.8693	1.15	1.5242
42°	.6691	.7431	.9004	1.11	1.4944
43°	.6820	.7314	.9325	1.07	1.4662
44°	.6947	.7193	.9657	1.04	1.4395
45°	.7071	.7071	1.0000	1.00	1.4142

102APX02.TIF

HAND BENDING 393

TRIGONOMETRY TABLE (continued)

Angle	Sine	Cosine	Tangent	Cotangent	Cosecant
46°	.7193	.6947	1.035	.966	1.4395
47°	.7314	.6820	1.0724	.933	1.3673
48°	.7431	.6691	1.1106	.900	1.3456
49°	.7547	.6561	1.1504	.869	1.3250
50°	.7660	.6428	1.1918	.839	1.3054
51°	.7771	.6293	1.2349	.810	1.2867
52°	.7880	.6157	1.2799	.781	1.2690
53°	.7986	.6018	1.3270	.754	1.2521
54°	.8090	.5878	1.3764	.727	1.2360
55°	.8192	.5736	1.4281	.700	1.2207
56°	.8290	.5592	1.4826	.675	1.2062
57°	.8387	.5446	1.5399	.649	1.1923
58°	.8480	.5299	1.6003	.625	1.1791
59°	.8572	.5150	1.6643	.601	1.1666
60°	.8660	.5000	1.7321	.577	1.1547
61°	.8746	.4848	1.8040	.554	1.1433
62°	.8829	.4695	1.8807	.532	1.1325
63°	.8910	.4540	1.9626	.510	1.1223
64°	.8988	.4384	2.0503	.488	1.1126
65°	.9063	.4226	2.1445	.466	1.1033
66°	.9135	.4067	2.2460	.445	1.0946
67°	.9205	.3907	2.3559	.424	1.0863
68°	.9272	.3746	2.4751	.404	1.0785
69°	.9336	.3584	2.6051	.384	1.0711
70°	.9397	.3420	2.7475	.364	1.0641
71°	.9455	.3256	2.9042	.344	1.0576
72°	.9511	.3090	3.0777	.325	1.0514
73°	.9563	.2924	3.2709	.306	1.0456
74°	.9613	.2756	3.4874	.287	1.0402
75°	.9659	.2588	3.7321	.268	1.0352
76°	.9703	.2419	4.0108	.249	1.0306
77°	.9744	.2250	4.3315	.231	1.0263
78°	.9781	.2079	4.7046	.213	1.0223
79°	.9816	.1908	5.1446	.194	1.0187
80°	.9848	.1736	5.6713	.176	1.0154
81°	.9877	.1564	6.3138	.158	1.0124
82°	.9903	.1392	7.1154	.141	1.0098
83°	.9925	.1219	8.1443	.123	1.0075
84°	.9945	.1045	9.5144	.105	1.0055
85°	.9962	.0872	11.4300	.088	1.0038
86°	.9976	.0698	14.3010	.070	1.0024
87°	.9986	.0523	19.0810	.052	1.0013
88°	.9994	.0349	28.6360	.035	1.0006
89°	.9998	.0175	57.2900	.018	1.0001
90°	1.0000	.0000	00	.000	1.0000

102APX03.TIF

Instructor's Notes:

394 CHAPTER 9

APPENDIX B

Bending Radius Table

BENDING RADIUS TABLE

RADIUS (INCHES)	RADIUS INCREMENTS (INCHES)									
	0	**1**	**2**	**3**	**4**	**5**	**6**	**7**	**8**	**9**
0	0.00	1.57	3.14	4.71	6.28	7.85	.942	10.99	12.56	14.13
10	15.70	17.27	18.84	20.41	21.98	23.85	25.12	26.69	28.26	29.83
20	31.40	32.97	34.54	36.11	37.68	39.25	40.82	42.39	43.96	45.83
30	47.10	48.67	50.24	51.81	53.38	54.95	56.52	58.09	59.66	61.23
40	62.80	64.37	65.94	67.50	69.03	70.65	72.22	73.79	75.36	76.93
50	87.50	80.07	81.64	83.21	84.78	86.35	87.92	89.49	91.06	92.63
60	94.20	95.77	97.34	98.91	100.48	102.05	103.62	105.19	106.76	108.33
70	109.90	111.47	113.04	114.61	116.18	117.75	119.32	120.89	122.46	124.03
80	125.60	127.17	128.74	130.31	131.88	133.45	135.02	136.59	138.16	139.73
90	141.30	142.87	144.44	146.01	147.58	149.15	150.72			

Developed length for following angles use fraction of 90° chart.

For	15°	22-1/2°	30°	45°	60°	67-1/2°	75°	90°
Take	1/6	1/4	1/3	1/2	2/3	3/4	5/6	See Chart

For any other degrees: Developed length = .01744 x radius x degrees.

102APX04.TIF

HAND BENDING 395

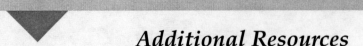

Additional Resources

This module is intended to present thorough resources for task training. The following reference works are suggested for further study. These are optional materials for continued education rather than for task training.

Benfield Conduit Bending Manual, Latest Edition. New York, NY: McGraw-Hill Publishing Company.

National Electrical Code Handbook, Latest Edition. Quincy, MA: National Fire Protection Association.

Tom Henry's Conduit Bending Package (includes video, book, and bending chart). Winter Park, FL: Code Electrical Classes, Inc.

Instructor's Notes:

396 CHAPTER 9

Answers to Review Questions

Answer	Section
1. b	2.0.0
2. a	2.0.0
3. a	2.0.0
4. b	2.0.0
5. b	2.0.0
6. b	2.1.0
7. c	2.1.0
8. d	2.2.0
9. b	2.4.0
10. c	2.5.0
11. a	2.6.0
12. b	2.7.0
13. c	3.2.0
14. c	3.3.0
15. d	3.4.0

The NCCER makes every effort to keep these textbooks up-to-date and free of technical errors. We appreciate your help in this process. If you have an idea for improving this textbook, or if you find an error, a typographical mistake, or an inaccuracy in NCCER's Contren™ textbooks, please write us, using this form or a photocopy. Be sure to include the exact module number, page number, a detailed description, and the correction, if applicable. Your input will be brought to the attention of the Technical Review Committee. Thank you for your assistance.

Instructors – If you found that additional materials were necessary in order to teach this module effectively, please let us know so that we may include them in the Equipment/Materials list in the Instructor's Guide.

Write: Curriculum Revision and Development Department
National Center for Construction Education and Research
P.O. Box 141104, Gainesville, FL 32614-1104

Fax: 352-334-0932

E-mail: curriculum@nccer.org

Craft _______________ Module Name _______________

Copyright Date _______________ Module Number _______________ Page Number(s) _______________

Description _______________

____

____

____

(Optional) Correction _______________

____

____

(Optional) Your Name and Address _______________

____

____

Chapter 10

Boxes and Fittings

26205-03

MODULE OVERVIEW

This course introduces the electrical trainee to the methods and procedures used in the selection and installation of outlet boxes and fittings.

PREREQUISITES

Please refer to the Course Map in the Trainee Module. Prior to training with this module, it is recommended that the trainee shall have successfully completed the following modules:

Core Curriculum; Residential Electrical I, Chapters 1 through 9

OBJECTIVES

Upon completion of this module, the trainee will be able to:

1. Describe the different types of nonmetallic and metallic boxes.
2. Understand the NEC requirements for box fill.
3. Calculate the required box size for any number and size of conductors.
4. Explain the NEC regulations for volume required per conductor in outlet boxes.
5. Properly locate, install, and support boxes of all types.
6. Describe the NEC regulations governing pull and junction boxes.
7. Explain the radius rule when installing conductors in pull boxes.
8. Understand the NEC requirements for boxes supporting lighting fixtures.
9. Describe the purpose of conduit bodies and Type FS boxes.
10. Install the different types of fittings used in conjunction with boxes.
11. Describe the installation rules for installing boxes and fittings in hazardous areas.
12. Explain how boxes and fittings are selected and installed.
13. Describe the various types of box supports.

PERFORMANCE TASKS

Under supervision of the instructor, the trainee should be able to:

1. Install selected conduit bodies.
2. Install selected fittings in a raceway system.
3. Calculate the outlet box capacities for various wiring configurations.

NCCER STANDARDIZED CRAFT TRAINING PROGRAM

The National Center for Construction Education and Research (NCCER) provides a standardized national program of accredited craft training. Key features of the program include instructor certification, competency-based training, and performance testing. The program provides trainees, instructors, and companies with a standard form of recognition through a National Craft Training Registry. The program is described in full in the *Guidelines for Accreditation*, published by the NCCER. For more information on standardized craft training, contact the NCCER by writing us at P.O. Box 141104, Gainesville, FL 32614-1104, calling 352-334-0911, or emailing info@nccer.org. More information can be found at our Web site, www.nccer.org.

HOW TO USE THIS ANNOTATED INSTRUCTOR'S GUIDE

Each page presents two sections of information. The larger section displays each page exactly as it appears in the Trainee Module. The narrow column ties suggested trainee and instructor actions to each page and provides icons to call your attention to material, safety, audiovisual, or testing requirements. The bottom of each page includes space for your notes.

Teaching Tip If you see the Teaching Tip icon, that means there is a teaching tip associated with this section. Also refer to any suggested teaching tips at the end of the module.

SAFETY CONSIDERATIONS

Ensure that the trainees are equipped with appropriate personal protective equipment. Stress the importance of following the proper safety precautions and procedures when installing outlet boxes and fittings.

PREPARATION

Before teaching this module, you should review the Module Outline, Objectives, Performance Tasks, and the Materials and Equipment List. Be sure to allow ample time to prepare your own training or lesson plan and gather all required equipment and materials.

MATERIALS AND EQUIPMENT LIST

Materials:

Transparencies

Markers/chalk

Copy of the latest edition of the *National Electrical Code®*

Examples of different types of metallic and nonmetallic boxes, device covers, and extension rings

Examples of pull and junction boxes

Example of FS and FD boxes

Examples of different types of conduit bodies, pulling elbows, and entrance ells

Examples of different types of boxes used in hazardous locations

Sealing fittings

Examples of fittings, including:
- EMT
- Rigid
- Aluminum
- IMC
- Locknuts and bushings

Module Examinations*

Performance Profile Sheets*

Equipment:

Overhead projector and screen

Whiteboard/chalkboard

Appropriate personal protective equipment

*Located in the Test Booklet packaged with this Annotated Instructor's Guide.

ADDITIONAL RESOURCES

This module is intended to present thorough resources for task training. The following reference works are suggested for both instructors and motivated trainees interested in further study. These are optional materials for continued education rather than for task training.

American Electricians' Handbook, Latest Edition. New York, NY: McGraw-Hill Publishing Company.

National Electrical Code Handbook, Latest Edition. Quincy, MA: National Fire Protection Association.

NOTES

The designations "National Electrical Code," "NE Code," and "NEC," where used in this document, refer to the *National Electrical Code®*, which is a registered trademark of the National Fire Protection Association, Quincy, MA. All National Electrical Code (NEC) references in this module refer to the 2002 edition of the NEC.

If you feel that additional math instruction would be helpful, Prentice Hall offers a basic math textbook entitled *Fundamentals of Electrical and Mechanical Mathematics*. It covers the basic math requirements for electrical trainees and may be ordered by contacting Prentice Hall Customer Service at 1-800-922-0579.

TEACHING TIME FOR THIS MODULE

An outline for use in developing your lesson plan is presented below. Note that each Roman numeral in the outline equates to one session of instruction. Each session has a suggested time period of 2½ hours. This includes 10 minutes at the beginning of each session for administrative tasks and one 10-minute break during the session. Approximately 10 hours are suggested to cover *Boxes and Fittings*. You will need to adjust the time required for hands-on activity and testing based on your class size and resources.

Topic	Planned Time
Session I. Introduction to Outlet Boxes and Fittings; Types of Boxes; Sizing Outlet Boxes	
A. Introduction to Outlet Boxes and Fittings	______
B. Types of Boxes	______
1. Octagon and Round Boxes	______
2. Square Boxes	______
3. Device Boxes	______
4. Concrete Boxes	______
5. Boxes for Damp and Wet Locations	______
C. Sizing Outlet Boxes	______
Session II. Pull and Junction Boxes; Conduit Bodies; Laboratory	
A. Pull and Junction Boxes	______
1. Sizing Pull and Junction Boxes	______
B. Conduit Bodies	______
1. Type C Conduit Body	______
2. Type LB Conduit Body	______
3. Type T Conduit Body	______
4. Type X Conduit Body	______
5. FS and FD Boxes	______
6. Pulling Elbows	______
7. Entrance Ell (SLB)	______
C. Laboratory	______
Under instructor supervision, have the trainees practice installing selected conduit bodies.	
Session III. Outlet Boxes in Hazardous Locations; Fittings; Laboratory	
A. Outlet Boxes in Hazardous Locations	______
1. Class I Locations	______
2. Class II Locations	______
3. Class III Locations	______
4. Seal-Off Fittings	______

B. Fittings

1. EMT Fittings
2. Rigid, Aluminum, and IMC Fittings
3. Locknuts and Bushings

C. Laboratory

Under instructor supervision, have the trainees practice installing fittings in a raceway system.

Session IV. Module Examination and Performance Testing

A. Review

B. Module Examination

1. Trainees must score 70% or higher to receive recognition from the NCCER.
2. Record the testing results on Craft Training Report Form 200 and submit the results to the Training Program Sponsor.

C. Performance Testing

1. Trainees must perform each task to the satisfaction of the instructor to receive recognition from the NCCER.
2. Record the testing results on Craft Training Report Form 200 and submit the results to the Training Program Sponsor.

Chapter 10

Module 26205-03

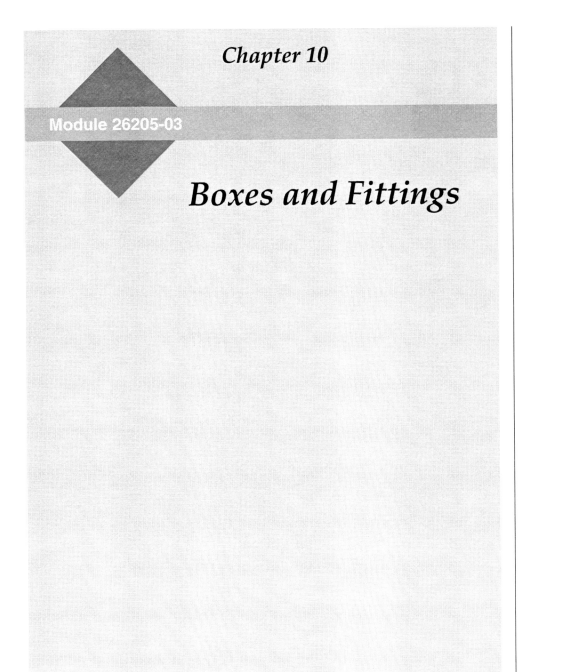

Boxes and Fittings

Course Map

This course map shows all of the modules in the first level of the Residential Electrical curriculum. The suggested training order begins at the bottom and proceeds up. Skill levels increase as you advance on the course map. The local Training Program Sponsor may adjust the training order.

RESIDENTIAL ELECTRICAL I

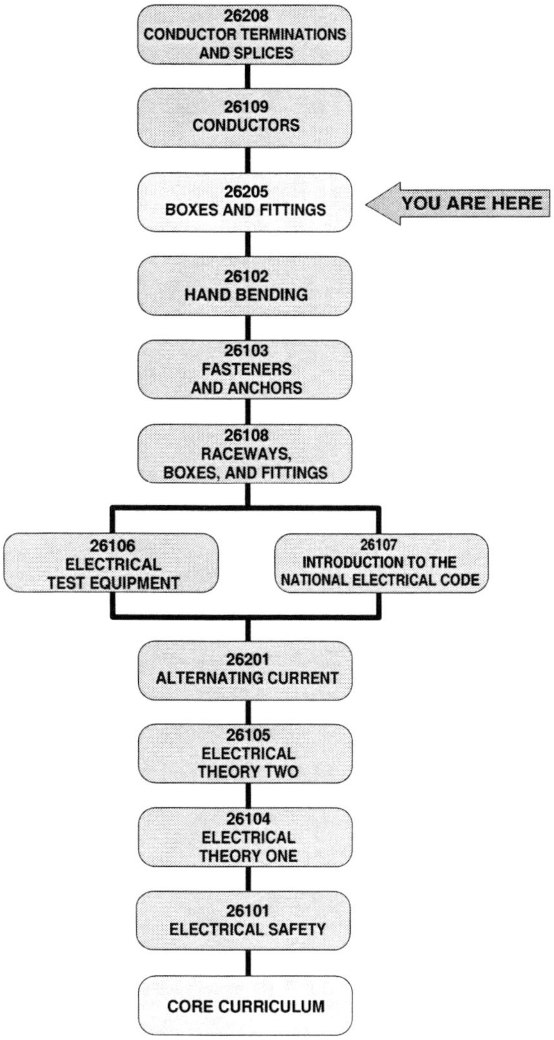

Homework

Assign reading of Module 26205.

MODULE 26205 CONTENTS

Figures

Instructor's Notes:

404 CHAPTER 10

MODULE 26205

Boxes and Fittings

Objectives

When you have completed this module, you will be able to do the following:

1. Describe the different types of nonmetallic and metallic boxes.
2. Understand the NEC requirements for box fill.
3. Calculate the required box size for any number and size of conductors.
4. Explain the NEC regulations for volume required per conductor in outlet boxes.
5. Properly locate, install, and support boxes of all types.
6. Describe the NEC regulations governing pull and junction boxes.
7. Explain the radius rule when installing conductors in pull boxes.
8. Understand the NEC requirements for boxes supporting lighting fixtures.
9. Describe the purpose of conduit bodies and Type FS boxes.
10. Install the different types of fittings used in conjunction with boxes.
11. Describe the installation rules for installing boxes and fittings in hazardous areas.
12. Explain how boxes and fittings are selected and installed.
13. Describe the various types of box supports.

Prerequisites

Before you begin this module, it is recommended that you successfully complete the following: Core Curriculum; Residential Electrical I, Chapters 1 through 9.

Required Trainee Materials

1. Pencil and paper
2. Appropriate personal protective equipment
3. Copy of the latest edition of the *National Electrical Code®*

1.0.0 ◆ INTRODUCTION TO OUTLET BOXES AND FITTINGS

On every job, boxes are required for outlets, switches, **pull boxes,** and **junction boxes.** All of these must be sized, installed, and supported to meet current NEC requirements. Since the NEC limits the number of conductors allowed in each outlet or switch box according to its size, you must install boxes that are large enough to accommodate the number of conductors that must be spliced in the box or fed through it. Therefore, a knowledge of the various types of boxes and the volume of each is essential.

Besides being able to calculate the required box sizes, you must also know how to select the proper type of box for any given application. For example, a metallic box used in concrete deck pours are different from those used as device boxes in residential or commercial buildings. Boxes used for the support of lighting fixtures or for securing devices in outdoor installations will be different from the two types just mentioned.

Note: The designations "National Electrical Code," "NE Code," and "NEC," where used in this document, refer to the *National Electrical Code®*, which is a registered trademark of the National Fire Protection Association, Quincy, MA. *All National Electrical Code (NEC) references in this module refer to the 2002 edition of the NEC.*

Materials

Ensure that you have everything required to teach the course. Check the Materials and Equipment List at the front of this Instructor's Guide.

Classroom

Review the goals of the course.

Explain that terms shown in bold (blue) are defined in the Glossary at the back of this module.

Audiovisual

Show Transparency 1, Course Objectives.

Show Transparency 2, Performance Tasks.

Classroom

Describe some of the applications of boxes and fittings.

Special Considerations

INSIDE TRACK

Before you can effectively plan and route conductors to their termination points and then select and install the correct types and sizes of boxes, you will need to have the following information:

- Length and number of conductors
- Conductor ampacity
- Allowances for voltage drops
- Type(s) of hazard(s) in the environment

Classroom

Discuss conduit caps and their uses.

Describe the categories of outlet boxes.

Boxes for use in certain hazardous locations will further differ in construction; many must be rated as being **explosion-proof.**

You must also know what fittings are available for terminating the various wiring methods in these boxes.

Electrical drawings rarely indicate the exact types of **outlet boxes** to be used in a given area, with the possible exception of boxes used in hazardous locations. Electricians who lack practical on-the-job experience may not always choose the best box for a given application. The use of improper boxes and other ill-adapted materials will cause excessive time to be taken on the job. For example, outlet boxes for use with Type AC cable should contain built-in clamps; many times boxes will be ordered with knockouts only. This latter case requires the use of additional **connectors,** which may only take a few additional seconds per connection, but when these are added up over the period of a large project, much additional time is wasted. Because the cost of labor is an expensive item, any excess labor required will more than offset any savings gained from the use of inadequate materials.

Excess labor is often consumed due to obstructions that enter conduit installations during the general construction work. Most of these obstructions can be avoided by plugging all conduit terminals with capped bushings or some similar means of protection (*Figure 1*). Care should also be taken to thoroughly tighten all fittings, couplings, and so on. In the case of outlet boxes installed in poured concrete structures, painting the inside of the boxes with grease and using capped bushings will greatly aid in keeping concrete out of the raceway system. If any concrete should enter the box, the grease will prevent it from adhering to the metal box.

This module is designed to cover boxes and fittings in depth, based on the applicable NEC regulations. A knowledge of this material should enable you to approach your work in an efficient manner.

205F01.EPS

Figure 1 ◆ Conduit caps.

2.0.0 ◆ TYPES OF BOXES

Outlet boxes normally fall into three categories:

- Pressed steel boxes with knockouts of various sizes for raceway or cable entrances
- Cast iron, aluminum, or brass boxes with threaded hubs of various sizes and locations for raceway entrances
- Nonmetallic boxes

Pressed steel boxes also fall into two categories:

- Boxes with conduit, electric metallic tubing, and cable
- Boxes designed for use with specific types of surface metal moldings

Outlet boxes vary in size and shape depending upon their use, the size of the raceway, the number of conductors entering the box, the type of

Instructor's Notes:

Identifying Boxes

INSIDE TRACK

After installing all the boxes in a complex installation, you can save yourself much time and trouble later on if you label each box location according to type. Using a can of spray paint, paint small dots on the floor below each box. Use one dot for a plug receptacle, two dots for a switch box, and three dots for a wall light. After the drywall installers are finished, you'll be able to easily identify any wall box.

building construction, the atmospheric conditions of the area, and special requirements.

Outlet box covers are usually required to adapt the box to the particular use it is to serve. For example, a 4" square box is adapted to one-gang or two-gang switches or receptacles by the use of either one-gang or two-gang flush device covers. A one-gang cast hub box can be adapted to provide a vapor-proof switch or a vapor-proof receptacle cover.

Special outlet box hangers are available to facilitate their installation, particularly in frame building construction.

The types of enclosures used as outlet and device boxes for the support of fixtures or for securing devices, such as switches, receptacles, or other equipment, on the same yoke or strap are available in various sizes and shapes. These enclosures may be used in the one-gang, two-gang, three-gang, or four-gang types. Ceiling outlet boxes are available in various shapes. Where a square metal box is used, it may be used for the support of a fixture or device, depending on the type of raised cover used. Device boxes are those that are usually installed to enclose receptacles and switches.

Boxes installed for the support of a lighting fixture are required to be specifically designed for the purpose. Most device boxes are typically not designed or listed for use to support lighting fixtures. The use of a device box to support fixtures is addressed in *NEC Section 314.27(A), Exception.*

A floor box that is listed specifically for installation in a floor is required where receptacles or junction boxes are installed in a floor. Listed floor boxes are provided with covers and gaskets to exclude surface water and cleaning compounds.

A box used at fan outlets is not permitted to be used as the sole support for ceiling (paddle) fans, unless it is specifically listed for the application as the sole means of support. Where a listed ceiling fan does not exceed 35 pounds in weight, with or without its accessories, it is permitted to be supported by outlet boxes identified for such use. The support of this box must be made so that it is rigidly supported from a structural member of the building. A paddle fan box and its related accessories are shown in *Figure 2.*

2.1.0 Octagon and Round Boxes

Metallic octagon boxes are available with knockouts for use with either conduit (using locknuts and bushings) or cable box connectors. They are also available with both Type AC and NM cable clamps. The standard width of octagon boxes is four inches, with depths available in 1¼", 1½", or 2⅛". Extension rings are also available for increasing the depth.

Round boxes are available in the same dimensions, but *NEC Section 314.2* prevents using such boxes where conduit or connectors—requiring locknuts and bushings—are connected to the side of the box. The conduit or box connector must terminate in the top of such boxes. Round boxes with cable clamps, however, are permitted for use with Type AC and NM cables that may terminate in either the side or top of the box.

Nonmetallic round boxes are permitted only with open wiring on insulators, concealed knob-and-tube wiring, Type NM cable, and nonmetallic raceways.

Locating Boxes on Walls

INSIDE TRACK

The Universal Building Code (UBC) requires boxes on opposite sides of walls between occupancies to be separated by at least 24" in order to maintain the fire rating. (See ***NEC Section 300.21.***)

Classroom

Mention the special requirements of paddle fan boxes.

Audiovisual

Show Transparency 3 (Figure 2).

Classroom

Discuss the applications of octagon and round boxes. Pass around examples for the trainees to examine.

Note the NEC requirements pertaining to these boxes.

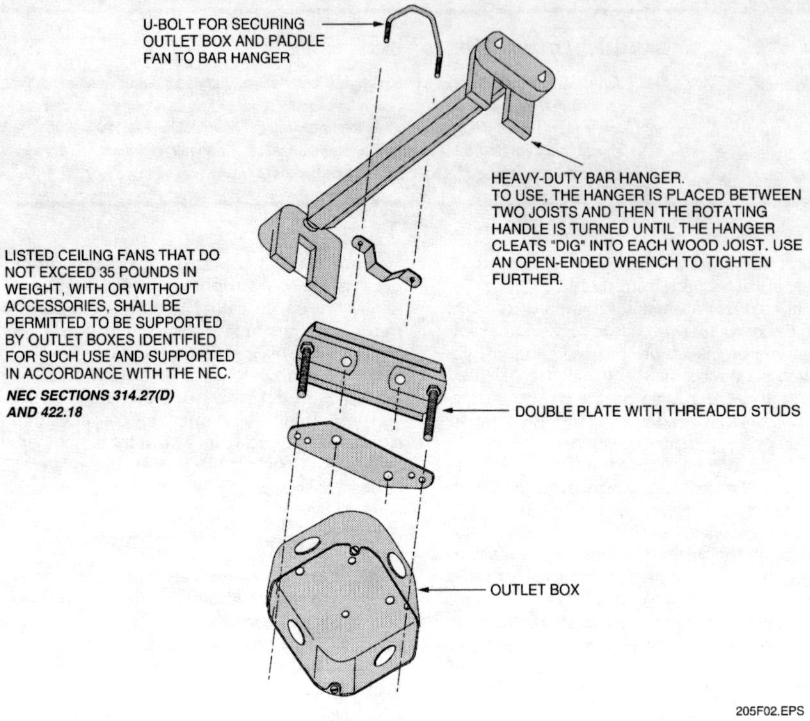
U-BOLT FOR SECURING OUTLET BOX AND PADDLE FAN TO BAR HANGER
HEAVY-DUTY BAR HANGER. TO USE, THE HANGER IS PLACED BETWEEN TWO JOISTS AND THEN THE ROTATING HANDLE IS TURNED UNTIL THE HANGER CLEATS "DIG" INTO EACH WOOD JOIST. USE AN OPEN-ENDED WRENCH TO TIGHTEN FURTHER.
LISTED CEILING FANS THAT DO NOT EXCEED 35 POUNDS IN WEIGHT, WITH OR WITHOUT ACCESSORIES, SHALL BE PERMITTED TO BE SUPPORTED BY OUTLET BOXES IDENTIFIED FOR SUCH USE AND SUPPORTED IN ACCORDANCE WITH THE NEC.
NEC SECTIONS 314.27(D) AND 422.18
DOUBLE PLATE WITH THREADED STUDS
OUTLET BOX
205F02.EPS

Figure 2 ◆ Typical paddle fan box for installation after the ceiling is installed.

Audiovisual

Show Transparencies 4 and 5 (Figures 3 and 5).

Classroom

Discuss the applications of square boxes and pass around examples for the trainees to examine.

Figure 3 shows typical metallic octagon boxes and an octagon extension ring. *Figure 3(A)* shows a box with concentric knockouts for conduit or box connectors, while *Figure 3(B)* shows a box that utilizes cable clamps. *Figure 3(C)* shows the extension ring. *Figure 4* shows a nonmetallic round box with a bar hanger for mounting between studs or joists.

Octagon and round boxes are used mostly for ceiling and wall-mounted incandescent lighting fixtures. However, covers are available for octagon boxes that will support receptacles and switches. Blank covers are also available when the box is used as a junction box.

Figure 5 shows a cross section of a $3\frac{1}{2}$" round shallow box used for supporting lighting fixtures that have integral wire termination space. *NEC Section 314.24* requires that this and all boxes have a minimum depth of $\frac{1}{2}$". Boxes intended to enclose flush devices must not have a depth of less than $\frac{15}{16}$".

2.2.0 Square Boxes

Square boxes are available in 4" and $4\frac{11}{16}$" square sizes. Both types are available in depths of $1\frac{1}{4}$", $1\frac{1}{2}$", and $2\frac{1}{8}$". Extension rings are also available to further increase the depth. These boxes are available with or without mounting brackets (for nailing to wooden structural members). Boxes designed for use with cable have either Type AC or NM clamps for securing the cable at the box entrance points. Either type of square box may be used with a single or two-gang device ring (e.g., plaster ring or tile ring) for mounting receptacles or switches. A ring with a round opening is also available for mounting lighting fixtures. Blank covers are available when the boxes are used as junction boxes.

Square boxes have been dubbed **nineteen hundred boxes** by some electricians because one of the early manufacturers gave its square boxes a

Instructor's Notes:

408 CHAPTER 10

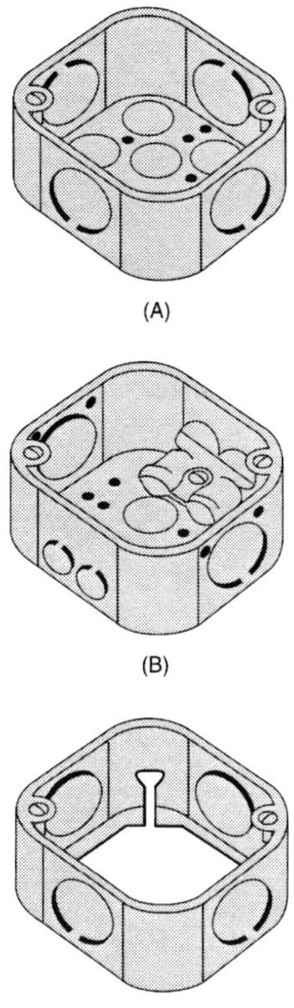

Figure 3 ◆ Typical octagon boxes and octagon extension ring.

model number of 1900. *Figure 6* shows a square box with a rectangular extension ring that is used to bring the box to the masonry surface.

Square boxes are used for mounting two wiring devices such as junction boxes, and when the number of wires warrant more capacity than is available in other types of boxes.

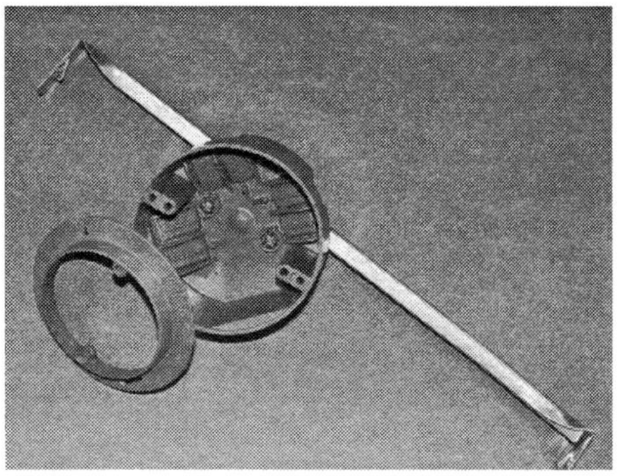

Figure 4 ◆ Nonmetallic round box with mounting bracket.

2.3.0 Device Boxes

Device boxes are designed for flush mounting mainly in residential and some commercial applications. They are available with or without cable clamps and brackets for mounting to wooden structural members. This type of box is also available with plaster ears for installation in finished wall partitions. Several types of device boxes are shown in *Figure 7*.

A special single-gang box with rounded edges is also used and is called a **handy box** (also known as a *utility box*). Such boxes are available in depths of 1½", 1⅞", and 2⅛". Care must be exercised when using these boxes as their limited volume restricts the number of conductors permitted in the box.

2.4.0 Concrete Boxes

Special boxes designed for use in concrete pours are used in flat-slab construction jobs. These boxes consist of a sleeve with external ears and a plate which is attached after the sleeve is nailed to the deck.

INSIDE TRACK

Use of Round Boxes Is Limited

Round boxes are not to be used in cases where conduit must enter from the side of the box because it is difficult to make a good connection with a locknut or bushing on a rounded surface.

Classroom

Discuss the applications of device boxes. Pass around examples for the trainees to examine.

Audiovisual

Show Transparency 6 (Figure 7).

Classroom

Discuss concrete boxes. Pass around an example for the trainees to examine.

BOXES AND FITTINGS **409**

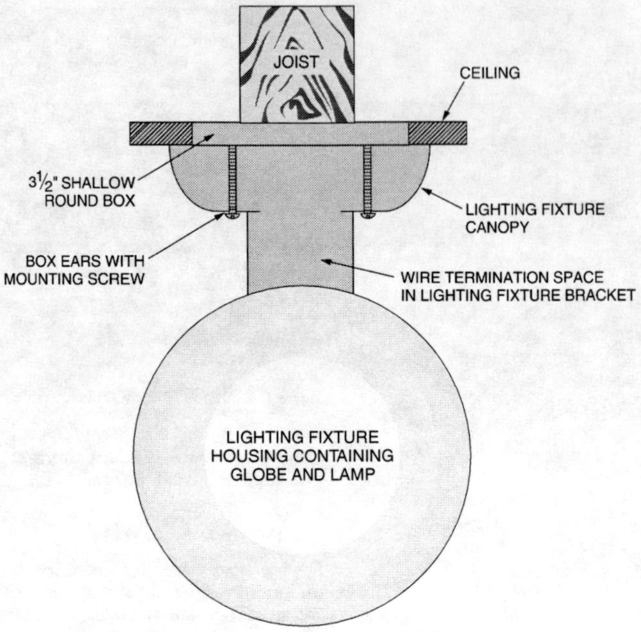

Figure 5 ◆ Shallow round box used for mounting a lighting fixture.

Audiovisual

Show Transparency 7 (Figure 8).

Classroom

Note the special requirements for boxes installed in damp and wet locations.

Figure 6 ◆ Square box with extension ring.

Concrete boxes are manufactured in different heights and care should be taken to use boxes of sufficient height to allow the knockouts to come well above the reinforcing rods. This eliminates the need for offsets in the conduit where it enters a box. *Figure 8* shows a practical application of a concrete box.

2.5.0 Boxes for Damp and Wet Locations

In damp or wet locations, boxes and fittings must be placed or equipped to prevent moisture or water from entering and accumulating within the box or fitting. It is recommended that approved boxes of nonconductive material be used with nonmetallic sheathed cable or approved nonmetallic conduit when the cable or conduit is used in locations where there is likely to be occasional moisture present. Boxes installed in wet locations must be approved for the purpose per *NEC Section 314.15(A)*.

Instructor's Notes:

410 CHAPTER 10

Inside Track: Knockouts

When completing building renovations, it may be difficult to remove a knockout from a previously installed box without dislodging the box. One way to do it is to drill into the knockout and partially insert a self-tapping screw. Then use diagonal or side-cut pliers to pull the knockout from the box.

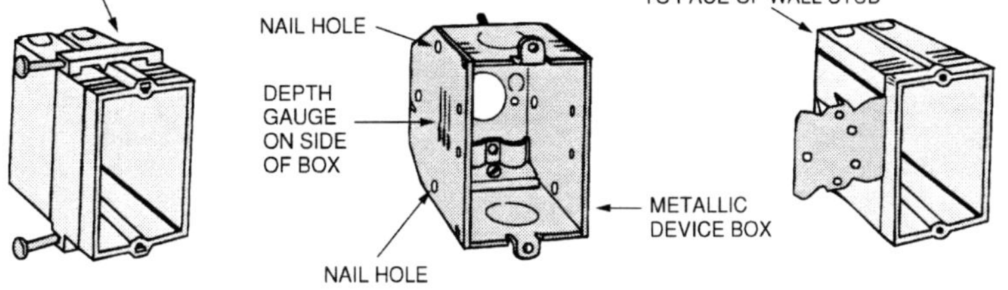

Figure 7 ◈ Typical device boxes.

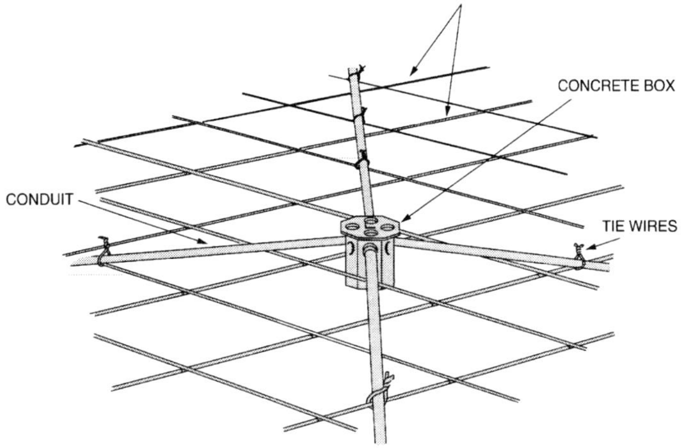

Figure 8 ◈ Practical application of concrete boxes.

BOXES AND FITTINGS 411

Classroom

Point out that *NEC Section 314.16* provides the requirements for sizing outlet boxes.

Discuss the devices that reduce box capacity and those that add to box capacity.

Audiovisual

Show Transparencies 8 and 9 (Figures 9 and 10).

A wet location is any location subject to saturation with water or other liquids, such as locations exposed to weather or water, washrooms, garages, and interiors that might be hosed down. Underground installations or those in concrete slabs or masonry in direct contact with the earth must be considered to be wet locations. **Raintight, waterproof,** or **watertight** equipment (including fittings) may satisfy the requirements for **weatherproof** equipment. Boxes with threaded conduit hubs and gasketed covers will normally prevent water from entering the box except for condensation within the box.

A damp location is a location subject to some degree of moisture. Such locations include partially protected outdoor locations—such as under canopies, marquees, and roofed open porches. It also includes interior locations subject to moderate degrees of moisture—such as some basements, some barns, and cold storage warehouses.

Weatherproof covers for outdoor receptacles must be chosen with care. If the receptacle feeds a permanently-connected load (such as a lighting fixture), the entire enclosure must be weatherproof with the plug inserted. If the receptacle is used only with portable tools or other portable equipment, the enclosure must be weatherproof with the cover closed and the cover must be self-closing. *NEC Section 406.8* covers installation of receptacles in damp or wet locations.

3.0.0 ◆ SIZING OUTLET BOXES

In general, the maximum number of conductors permitted in standard outlet boxes is listed in *NEC Table 314.16(A)*. These figures apply where no fittings or devices such as fixture studs, cable clamps, switches, or receptacles are contained in the box and where no grounding conductors are part of the wiring within the box. Obviously, in all modern residential wiring systems there will be one or more of these items contained in the outlet box. Therefore, where one or more of the above-mentioned items are present, the number of conductors is reduced by one less than that shown in the table for each type of fitting and by two for each device strap. For example, a deduction of two conductors must be made for each strap containing a device such as a switch or duplex receptacle; a further deduction of one conductor shall be made for one or more grounding conductors entering the box. For example, a 3" × 2" × 3½" box is listed in the table as containing a maximum number of eight No. 12 wires. If the box contains cable clamps and a duplex receptacle, three wires will have to be deducted from the total of eight—providing for only five No. 12 wires. If a ground wire is used, only four No. 12 wires may be used, which might be the case when a three-wire cable with ground is used to feed a three-way wall switch.

A pictorial definition of stipulated conditions as they apply to *NEC Section 314.16* is shown in the following illustrations. *Figure 9* illustrates an assortment of raised covers and outlet box extensions. These components, when combined with the appropriate outlet boxes, serve to increase the usable work space. Each type is marked with its cubic inch capacity, which may be added to the figures in *NEC Table 314.16(A)* to calculate the increased number of conductors allowed.

Figure 10 shows typical wiring configurations, which must be counted as conductors when calculating the total capacity of outlet boxes. A wire passing through the box without a splice or tap is counted as one conductor. Therefore, a cable containing two wires that passes in and out of an outlet box with a splice or tap is counted as two conductors. However, wires that enter a box and are either spliced or connected to a terminal, and then exit again, are counted as two conductors. In the case of two cables that each have two wires, the total conductors charged will be four. Wires that enter and terminate in the same box are charged as individual conductors and in this case, the total charge would be two conductors.

INSIDE TRACK

Outdoor Boxes

Outdoor wiring must be able to resist the entry of water. Outdoor boxes are either *drip tight*, which means sealed against falling water from above, or *watertight*, which means sealed against water from any direction. Drip-tight boxes simply have lids that deflect rain; they are not waterproof. Watertight boxes are sealed with gaskets to prevent the entry of water from any angle.

Instructor's Notes:

412 CHAPTER 10

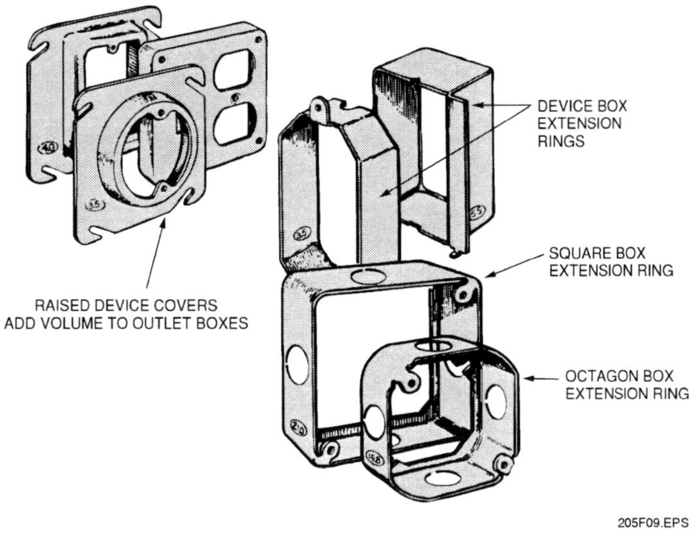

Figure 9 ◈ Devices or components that add to outlet box capacity.

Remember, when one or more grounding wires enter the box and are joined, a deduction on only one is required, regardless of their number.

Further components that require deduction adjustments from those specified in *NEC Table 314.16(A)* include fixture studs, hickeys, and fixture stud extensions [*NEC Section 314.16(B)(3)*]. One conductor must be deducted from the total for each type of fitting used. Two conductors must be deducted for each strap-mounted device, such as duplex receptacles and wall switches; a deduction of one conductor is made when one or more internally mounted cable clamps are used [*NEC Section 314.16(B)(2) and (4)*].

Figure 11 shows components that may be used in outlet boxes without affecting the total number of conductors. Such items include grounding clips and screws, wire nuts, and cable connectors when the latter are inserted through knockout holes in the outlet box and secured with locknuts. Prewired fixture wires are not counted against the total number of allowable conductors in an outlet box; neither are conductors originating and ending in the box.

INSIDE TRACK

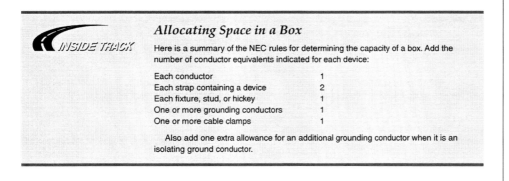

Allocating Space in a Box

Here is a summary of the NEC rules for determining the capacity of a box. Add the number of conductor equivalents indicated for each device:

Each conductor	1
Each strap containing a device	2
Each fixture, stud, or hickey	1
One or more grounding conductors	1
One or more cable clamps	1

Also add one extra allowance for an additional grounding conductor when it is an isolating ground conductor.

Classroom

List the devices that do not affect box capacity.

Audiovisual

Show Transparency 10 (Figure 11).

BOXES AND FITTINGS 413

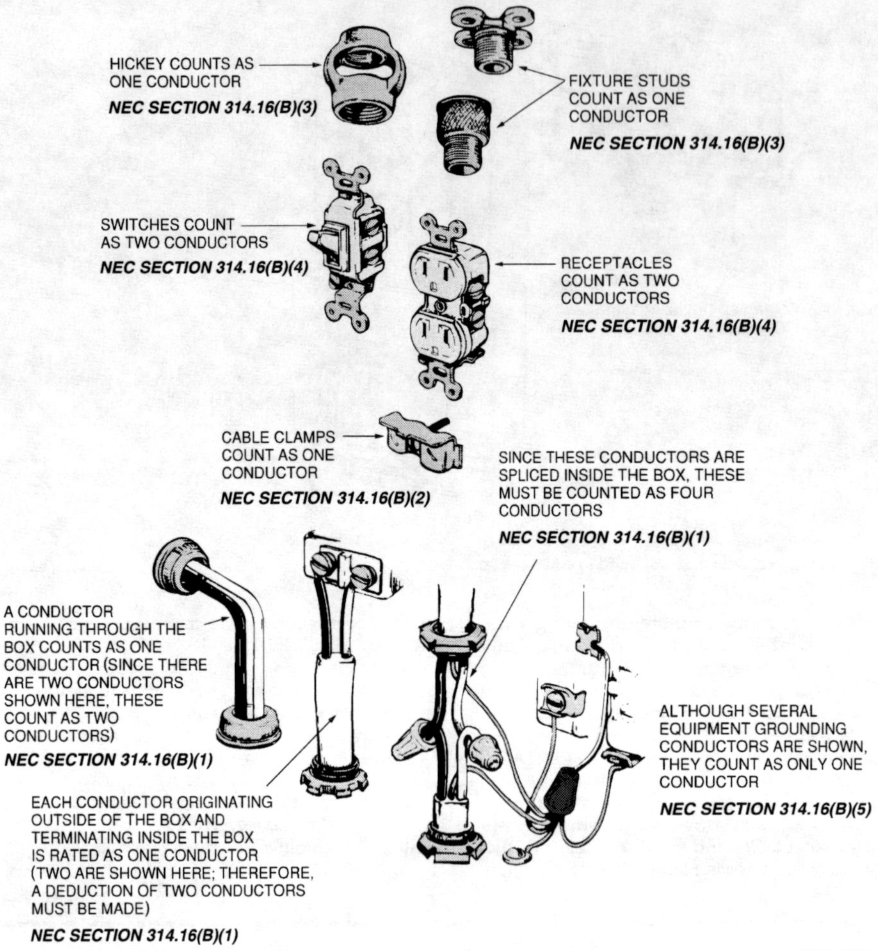
HICKEY COUNTS AS ONE CONDUCTOR
NEC SECTION 314.16(B)(3)
FIXTURE STUDS COUNT AS ONE CONDUCTOR
NEC SECTION 314.16(B)(3)
SWITCHES COUNT AS TWO CONDUCTORS
NEC SECTION 314.16(B)(4)
RECEPTACLES COUNT AS TWO CONDUCTORS
NEC SECTION 314.16(B)(4)
CABLE CLAMPS COUNT AS ONE CONDUCTOR
NEC SECTION 314.16(B)(2)
SINCE THESE CONDUCTORS ARE SPLICED INSIDE THE BOX, THESE MUST BE COUNTED AS FOUR CONDUCTORS
NEC SECTION 314.16(B)(1)
A CONDUCTOR RUNNING THROUGH THE BOX COUNTS AS ONE CONDUCTOR (SINCE THERE ARE TWO CONDUCTORS SHOWN HERE, THESE COUNT AS TWO CONDUCTORS)
NEC SECTION 314.16(B)(1)
ALTHOUGH SEVERAL EQUIPMENT GROUNDING CONDUCTORS ARE SHOWN, THEY COUNT AS ONLY ONE CONDUCTOR
NEC SECTION 314.16(B)(5)
EACH CONDUCTOR ORIGINATING OUTSIDE OF THE BOX AND TERMINATING INSIDE THE BOX IS RATED AS ONE CONDUCTOR (TWO ARE SHOWN HERE; THEREFORE, A DEDUCTION OF TWO CONDUCTORS MUST BE MADE)
NEC SECTION 314.16(B)(1)
205F10.EPS

Figure 10 ◆ Devices and components that require deductions in outlet box capacity.

Classroom

Demonstrate the procedure for sizing an outlet box.

Have the trainees practice sizing outlet boxes.

To better understand how outlet boxes are sized, we will take two No. 12 AWG conductors installed in ½" EMT and terminating into a metallic outlet box containing one duplex receptacle. What size outlet box will meet NEC requirements?

The first step is to count the total number of conductors and equivalents that will be used in the box (*NEC Section 314.16*).

Step 1 Calculate the total number of conductors and their equivalents:

One receptacle = 2
Two #12 conductors = 2
Total #12 conductors = 4

Instructor's Notes:

414 CHAPTER 10

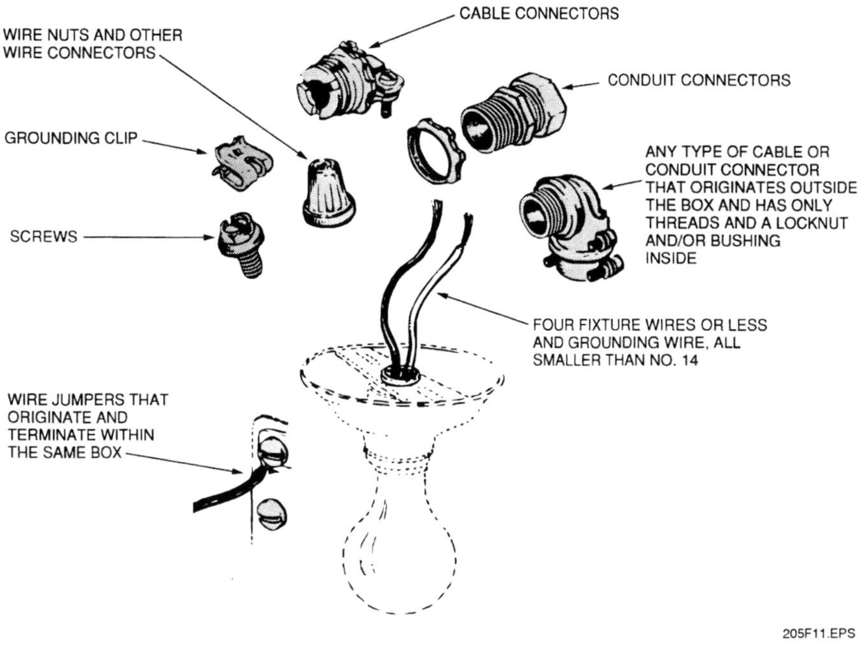

Figure 11 ◈ Items that may be disregarded when calculating outlet box capacity.

Step 2 Determine the amount of space required for each conductor. *NEC Table 314.16(B)* gives the box volume required for each conductor:

No. 12 AWG = 2.25 cubic inches

Step 3 Calculate the outlet box space required by multiplying the number of cubic inches required for each conductor by the number of conductors found in Step 1 above.

$4 \times 2.25 = 9.00$ cubic inches

Step 4 Once you have determined the required box capacity, again refer to *NEC Table 314.16(A)* and note that a 3" × 2" × 2" box comes closest to our requirements. This box size is rated for 10.0 cubic inches.

For another example, if four No. 12 conductors enter the box, two additional No. 12 conductors must be added to our previous count for a total of six conductors.

$6 \times 2.25 = 13.5$ cubic inches

Again, refer to *NEC Table 314.16(A)* and note that a 3" × 2" × 2¾" device box with a rated capacity of 14.0 cubic inches is the closest device box that meets NEC requirements. Of course, any box with a larger capacity is permitted.

THINK ABOUT IT

Oversized Devices

Some devices such as GFCI receptacles and dimmer switches are larger than the standard boxes. How does the NEC deal with this issue?

Classroom

Discuss the "Think About It."

Teaching Tip

When calculating box fill, a double volume allowance in accordance with *NEC Table 314.16(B)* is used for devices such as GFCI receptacles and dimmer switches, based on the largest conductor connected to the device.

Metal Boxes Must Be Grounded

INSIDE TRACK

Metal boxes are good conductors. Therefore, when metal boxes are used, they must be grounded to the circuit grounding system.

Classroom

Discuss the applications of pull and junction boxes. Pass around examples for the trainees to examine.

Discuss the NEC requirements for sizing pull and junction boxes.

Emphasize the importance of providing sufficient pull boxes.

4.0.0 ◆ PULL AND JUNCTION BOXES

Pull and junction boxes are provided in an electrical installation to facilitate the installation of conductors, or to provide a junction point for the connection of conductors, or both. In some instances, the location and size of pull boxes are designated on the drawings. In most cases, however, the electricians on the job will have to determine the proper number, location, and sizes of pull or junction boxes to facilitate conductor installation.

Pull boxes should be as large as possible. Workers need space within the box for both hands and in the case of the larger wire sizes, workers will need room for their arms to feed the wire. *NEC Section 314.28* specifies that pull and junction boxes must provide adequate space and dimensions for the installation of conductors. For raceways containing conductors of No. 4 or larger, and for cables containing conductors of No. 4 or larger, the minimum dimensions of pull or junction boxes installed in a raceway or cable run shall comply with the following:

- In straight pulls, the length of the box shall not be less than eight times the trade diameter of the largest raceway.
- Where angle or U pulls are made, the distance between each raceway entry inside the box and the opposite wall of the box shall not be less than six times the trade diameter of the largest raceway in a row. This distance shall be increased for additional entries by the amount of the sum of the diameter of all other raceway entries in the same row on the same wall of the box. Each row shall be calculated individually, and the single row that provides the maximum distance shall be used.
- The distance between raceway entries enclosing the same conductor shall not be less than six times the trade diameter of the largest raceway.
- When transposing cable size into raceway size, the minimum trade size raceway required for the number and size of conductors in the cable shall be used.

Long runs of conductors should not be made in one pull. Pull boxes, installed at convenient intervals, will relieve much of the strain on the conductors. The length of the pull, in many cases, is left to the judgment of the workers or their supervisor, and the condition under which the work is installed.

The installation of pull boxes may seem to cause a great deal of extra work and trouble, but they save a considerable amount of time and hard work when pulling conductors. Properly placed, they eliminate many bends and elbows and do away with the necessity of fishing from both ends of a conduit run.

If possible, pull boxes should be installed in a location that allows electricians to work easily and conveniently. For example, in an installation where the conduit comes up a corner of a wall and changes direction at the ceiling, a pull box that is installed too high will force the electrician to stand on a ladder when feeding conductors, and will allow no room for supporting the weight of the wire loop or for the cable-pulling tools.

Unless the contract drawings or project engineer state otherwise, it is just as easy for the pull boxes to be placed at a convenient height that allows workers to stand on the floor with sufficient room for both wire loop and tools.

In some electrical installations, a number of junction boxes must be installed to route the conduit in the shortest, most economical way. The NEC requires all junction boxes to be readily accessible. This means that a person must be able to get to the conductors inside the box without removing plaster, wall covering, or any other part of the building.

Junction boxes detract from the decorative scheme of a building. Therefore, where such boxes will be used in areas open to the public or in other areas where the boxes will be unattractive, they should be installed above suspended ceilings, in closets, or at least in corners of the room or area.

Junction boxes or pull boxes must be securely fastened in place on walls or ceilings or adequately suspended.

While certain sizes of factory-constructed boxes are available with concentric knockouts, in many instances it will be necessary to have them custom built to meet the job requirements. When it is not

Instructor's Notes:

Using Pull Boxes

INSIDE TRACK

Pull boxes make it easier to install conductors. They can also be installed to avoid having more than 360° worth of bends in a single run. (Remember, if a pull box is used, it is considered the end of the run for the purposes of the NEC 360° rule.)

possible to accurately anticipate the raceway entrance requirements, it will be necessary to cut the required knockouts on the job using hydraulic knockout cutters to keep labor to a minimum.

In the case of large pull boxes and troughs, shop drawings should be prepared prior to the construction of these items with all required knockouts accurately indicated in relation to the conduit run requirements.

4.1.0 Sizing Pull and Junction Boxes

Figure 12 shows a junction box with several runs of conduit. Since this is a straight pull, and 4" conduit is the largest size in the group, the minimum length required for the box can be determined by the following calculation:

Trade size of conduit x 8 [per ***NEC Section 314.28(A)(1)***] = minimum length of box

or:

$$4" \times 8 = 32"$$

Therefore, this particular pull box must be at least 32" in length. The width of the box, however, need only be of sufficient size to enable locknuts and bushings to be installed on all the conduits or connectors entering the enclosure.

Junction or pull boxes in which the conductors are pulled at an angle (*Figure 13*) must have a distance of not less than six times the trade diameter of the largest conduit [*NEC Section 314.28(A)(2)*]. The distance must be increased for additional conduit entries by the amount of the sum of the diameter of all other conduits entering the box on the same side. The distance between raceway entries enclosing the same conductors must not be less than six times the trade diameter of the largest conduit.

Since the 4" conduit is the largest size in this case:

$$L_1 = 6 \times 4" + (3 + 2) = 29"$$

Since the same conduit runs are located on the adjacent wall of the box, L_2 is calculated in the same way; therefore, L_2 = 29".

The distance (D) = 6 × 4" or 24". This is the minimum distance permitted between conduit entries enclosing the same conductor.

The depth of the box need only be of sufficient size to permit locknuts and bushings to be properly installed. In this case, a 6" deep box would suffice.

5.0.0 ◆ CONDUIT BODIES

Conduit bodies, also called *condulets,* are defined in *NEC Article 100* as a separate portion of a conduit or tubing system that provides access through a removable cover to the interior of the system at a junction of two or more sections of the system or at a terminal point of the system.

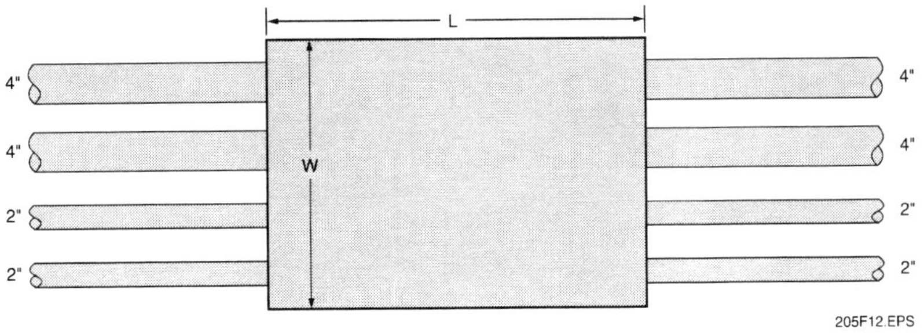

Figure 12 ◆ Pull box with two 4" and two 2" conduit runs.

Classroom

Demonstrate the procedure for sizing pull and junction boxes.

Audiovisual

Show Transparencies 11 and 12 (Figures 12 and 13).

Classroom

Discuss conduit bodies (condulets) and their applications. Pass around examples for the trainees to examine.

Classroom

Explain how to identify a Type L conduit body.

Audiovisual

Show Transparency 13 (Figure 14).

Classroom

Discuss the applications of various conduit bodies and note applicable NEC requirements:
- **Type C**
- **Type LB**
- **Type T**
- **Type X**

INSIDE TRACK

Label Junction Boxes

It's a good idea to label every junction box plate with the circuit number, the panel it came from, and its destination. The next person to service the installation will be grateful for this extra help.

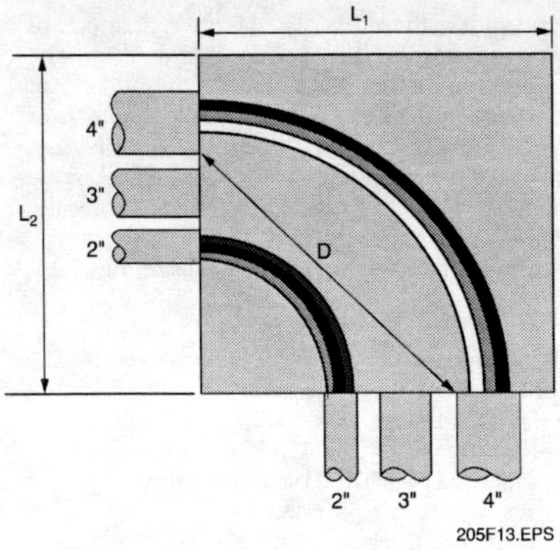

Figure 13 ◆ Pull box with conduit runs entering at right angles.

Conduit bodies are usually used with RMC and IMC. The cost of conduit bodies, because they are cast, is significantly higher than the stamped steel boxes. Splicing in conduit bodies is typically not recommended; however, it is permitted under certain conditions as specified in ***NEC Section 314.16(C)(2)***.

As an electrical trainee, you will hear such terms as *LL, LR,* as well as other letters to distinguish between the various types of conduit bodies. To identify certain conduit bodies, an old trick of the trade is to hold the conduit body like a pistol (*Figure 14*). When doing so, if the oval-shaped opening of the conduit body is to your left, it is called an *LL*—the first *L* stands for *elbow* and the second *L* stands for *left*. If the opening is on your right, it is called an *LR*—the *R* stands for *right*. If the opening is facing upward, this type of conduit body is called an *LB*—the *B* stands for *back*. If there is an opening on both sides, it is called an *LRL*—for both *left* and *right*. The other popular shapes are named for their letter look-alikes, that is, *T* and *X*. The only exception is the C conduit body. Let us take a closer look at each of these.

5.1.0 Type C Conduit Body

A Type C conduit body (*Figure 15*) may be used to provide a pull point in a long conduit run or a conduit run that has bends totaling more than 360° (see ***NEC Sections 342.26, 344.26, and 358.26***). In this application, the Type C conduit body is used as a pull point.

5.2.0 Type LB Conduit Body

A Type L conduit body (*Figure 16*) is used as a pulling point for conduit that requires a 90° change in direction. (Again, the letter *L* is short for *elbow*.) To use a Type L conduit body, the cover is removed, the wire is pulled out and coiled on the ground (or floor), and then it is reinserted into the other opening and pulled. Type L conduit bodies are available with the cover on the back (Type LB),

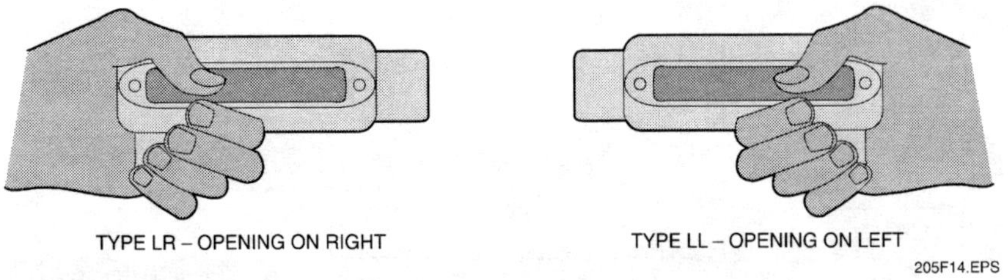

Figure 14 ◆ Identifying conduit bodies.

Instructor's Notes:

205F15.EPS

Figure 15 ◆ Type C conduit body.

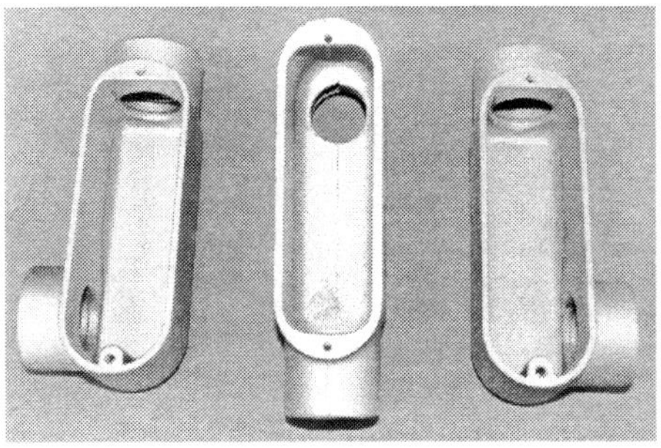

205F16.EPS

Figure 16 ◆ Type L conduit bodies.

on the sides (Type LL and Type LR), or on both sides (Type LRL). The cover and gasket for conduit body fittings must be ordered separately; do not assume that these parts come with the conduit body when it is ordered.

5.3.0 Type T Conduit Body

A Type T conduit body, also known as a *tee,* is used to provide a junction point for three intersecting conduits (*Figure 17*). Tees are used extensively in rigid conduit systems. The cost of a tee conduit body is more than twice that of a standard 4" square box with a cover. Therefore, the use of Type T conduit bodies with EMT is limited. According to ***NEC Section 314.16(C)(2),*** conductor splicing is permitted in a Type T conduit body if the necessary conditions are satisfied.

5.4.0 Type X Conduit Body

A Type X conduit body is used to provide a junction point for four intersecting conduits. The removable cover provides access to the interior of the X so that wire pulling and splicing may be performed (see *Figure 18*).

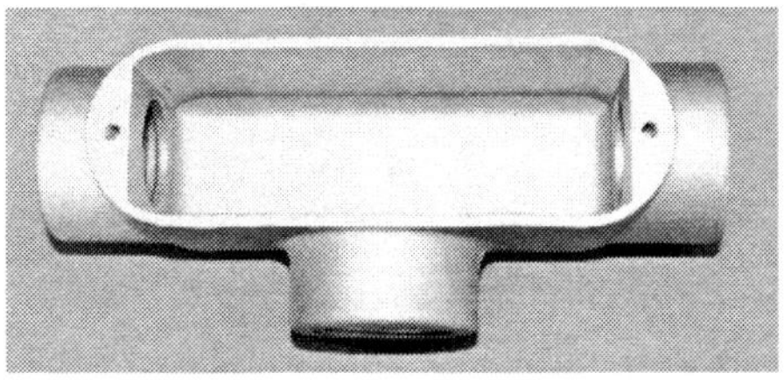

205F17.EPS

Figure 17 ◆ Type T conduit body.

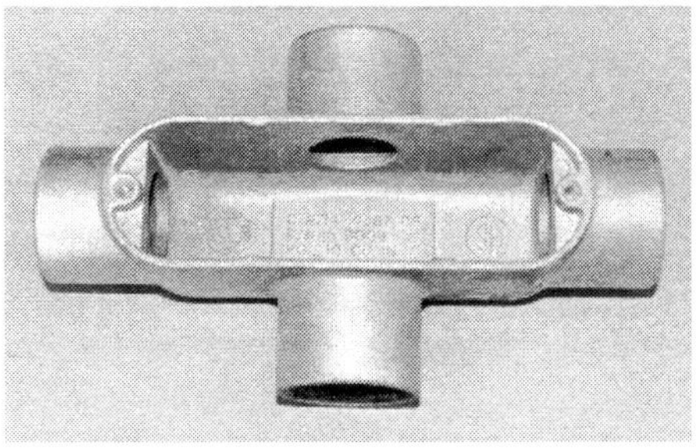

205F18.EPS

Figure 18 ◆ Type X conduit body.

5.5.0 FS and FD Boxes

FS boxes are cast boxes available in single-gang, two-gang, and three-gang configurations. They are sized to permit the installation of switches and receptacles. Covers for switches and receptacles are available for FS boxes that have formed openings much like switch and receptacle plates. FD boxes are similar to FS boxes. The letter *D* in the *FD* box indicates it is a deeper box ($2\frac{1}{2}$" deep versus $1\frac{5}{8}$" deep for an FS box). Neither FS nor FD boxes are considered by the NEC to be conduit bodies. FS and FD boxes may be used in environments defined by NEMA 1 (dry, clean environments); NEMA 3R (outdoor, wet environments); and NEMA 12 (dusty, oily environments). NEMA stands for the National Electrical Manufacturers' Association.

Engineers specify and electricians install FS and FD boxes for a reason. Never alter these boxes by drilling mounting holes in them. Most are provided with cast-in mounting eyes for this purpose. Mount these boxes only as recommended by the manufacturer.

Demonstration

Demonstrate how to install a conduit body.

Laboratory

Have the trainees practice installing conduit bodies. Note the proficiency of each trainee.

Classroom

Discuss the applications of FS and FD boxes. Note that they are typically used in industrial applications.

Classroom

Discuss the applications of pulling and entrance elbows. Pass around examples for the trainees to examine.

Audiovisual

Show Transparency 14 (Figure 19).

Classroom

Explain that special boxes are used in hazardous locations.

INSIDE TRACK

FS and FD Boxes

FS and FD boxes are precision-molded boxes with a cover and gasket designed to provide a tight seal. Because they cost significantly more than standard boxes, they are typically used only in industrial environments where the boxes are subjected to moisture, dirt, dust, and corrosion.

5.6.0 Pulling Elbows

Pulling elbows are used exclusively for pulling wire at a corner point of a conduit run. The volume of a pulling elbow is too low to permit splicing wire. See *Figure 19(A)*.

5.7.0 Entrance Ell (SLB)

An entrance ell, or SLB, is built with an offset so that it may be attached directly to the surface that is to have a conduit penetration. A cover on the back of the SLB permits wire to be pulled out and reinserted into the conduit that penetrates the support surface. See *Figure 19(B)*.

6.0.0 ◆ OUTLET BOXES IN HAZARDOUS LOCATIONS

Any area in which the atmosphere or a material in the area is such that the arcing of operating electrical contacts, components, and equipment may cause an explosion or fire is considered a hazardous location. In all such cases, explosion-proof equipment, raceways, and fittings are used to provide an explosion-proof wiring system, including the outlet boxes.

Hazardous locations have been classified in the NEC into certain class locations. Various atmospheric groups have been established (on the basis of the explosive character of the atmosphere) for the testing and approval of equipment for use in the various groups.

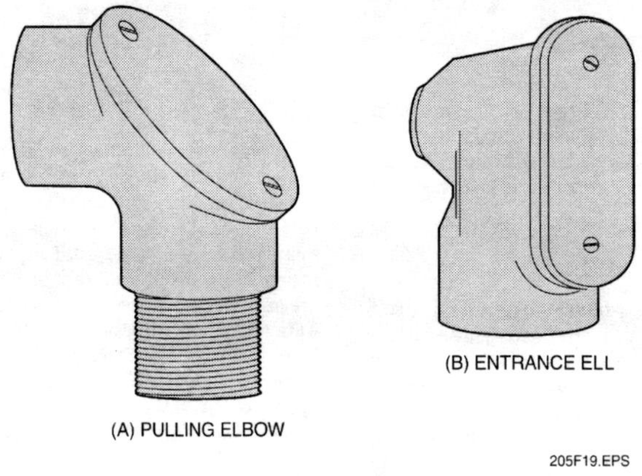

Figure 19 ◆ Elbows.

INSIDE TRACK

Making Splices in an Entrance Ell (SLB)

Never make splices in an entrance ell (SLB). Splices can be made in a conduit body that has the cubic inches marked, but never in an SLB.

Instructor's Notes:

6.1.0 Class I Locations

Those locations in which flammable gases or vapors may be present in the air in quantities sufficient to produce explosive or ignitable mixtures are classified as Class I locations. Examples of such locations are interiors of spray paint booths where volatile, flammable solvents are used, inadequately-ventilated pump rooms where flammable gas is pumped, and drying rooms for the evaporation of flammable solvents.

Classroom

Discuss the three classes of hazardous locations as defined by the NEC and list examples of each type.

6.2.0 Class II Locations

Class II locations are those that are hazardous because of the presence of combustible dust. Class II, Division 1 locations are areas where combustible dust under normal operating conditions may be present in the air in quantities sufficient to produce explosive or ignitable mixtures; examples are working areas of grain handling and storage plants and rooms containing grinders or pulverizers. Class II, Division 2 locations are areas where dangerous concentrations of suspended dust are not likely, but where dust accumulations might form.

6.3.0 Class III Locations

These locations are those areas that are hazardous because of the presence of easily ignitable fibers or flyings, but such fibers and flyings are not likely to be suspended in the air in these locations in quantities sufficient to produce ignitable mixtures. Such locations usually include some parts of rayon, cotton, and textile mills; clothing manufacturing plants; and woodworking plants.

The wide assortment of explosion-proof equipment now available makes it possible to provide adequate electrical installations under any of these hazardous conditions; however, you must be thoroughly familiar with all NEC requirements. You must also know what fittings are available, how to install them properly, and where and when to use them.

The usual construction documents (drawings and specifications) for a hazardous area are drawn the same as the layout of an electrical system for a non-hazardous area—the only distinction is a note on the drawings stating that the wiring in this particular area must conform to the NEC requirements for hazardous locations. The drafter or designer will sometimes add the letters *EXP* or *XP* next to all the symbols of the outlets that are to be explosion-proof. A few large-scale detail drawings may also be present. However, few engineers or drafters detail their drawings for hazardous areas sufficiently for the electricians to proceed with the installation without additional study and layout work on the job site. Therefore, you must be familiar with the layout and installation procedures before attempting such an installation. For other than very simple installations, it may be advisable to make rough, detailed wiring layouts of the proposed installation, even if they are merely sketches on the original working drawings.

When an electrical system in a hazardous location is designed or installed, the type of building structure and finish must be considered. If the building is under construction, this information may be obtained from the architectural drawings and specifications. If the installation is made in an existing building, a preliminary job site investigation is often necessary. The location of the explosion-proof outlets, whether concealed or exposed, and the class of hazardous locations should appear in the electrical drawings and specifications. If such information is not provided, the contractor or electrician will have to determine this information from the architect, owner, or local inspection authority.

Explain that the requirements for installations in hazardous locations are listed in *NEC Article 500.*

In general, rigid metallic conduit is required for all hazardous locations, except for special flexible terminations and as otherwise permitted in the NEC. The conduit should be threaded with a standard conduit butting die that provides ¾" taper per foot. The conduit should be made up wrench-tight in order to minimize sparking in the event that fault current flows through the conduit system [*NEC Section 500.8(D)*]. Where it is impractical to make a threaded joint tight, a bonding jumper should be used. All boxes, fittings, and joints must be threaded for connection to the conduit system and must be of an approved, explosion-proof type. Threaded joints must be made up with at least five threads fully engaged. Where it becomes necessary to employ flexible connectors at motor or fixture terminals, flexible fittings approved for the particular class location must be used.

6.4.0 Seal-Off Fittings

Seal-off fittings are required in conduit systems to minimize the passage of gases and vapors and prevent the passage of flames from one portion of the electrical installation to another through the

Explain the purpose of seal-off fittings.

BOXES AND FITTINGS 421

Hazardous Locations Summarized

INSIDE TRACK

This table provides an overview of how hazards are organized by class and division. Refer to ***NEC Article 500*** for more detail.

Class	Division	Hazard	Example
I	1	Flammable gas or vapors exist under normal conditions	Spray paint facilities where flammable solvents are used; rooms where flammable gas is pumped
	2	Flammable gas or vapors exist but are normally contained	Areas adjacent to Division 1 facilities; areas where flammable gases or vapors are handled in a closed system; similar areas where proper ventilation is provided
II	1	High concentrations of airborne combustible dust	Grain handling or storage plants
	2	Combustible dust accumulation on surfaces	Same as Division 1
III	1	Easily ignitable fibers or airborne particles released through a manufacturing process	Textile mills; woodworking plants
	2	Ignitable fiber storage	Same as Division 1

Classroom

Discuss the NEC requirements for sealing fittings.

Audiovisual

Show Transparency 15 (Figure 20A).

Demonstration

Demonstrate how to install a sealing fitting.

conduit. For Class I, Division 1 locations, *NEC Section 501.5(A)(1)* states that in each conduit run entering an explosion-proof enclosure containing switches, circuit breakers, fuses, relays, resistors, or other apparatus that may produce arcs, sparks, or high temperatures, seals shall be placed as close as practicable and in no case more than 18" from such enclosures. There shall be no junction box or similar enclosure in the conduit run between the sealing fitting and the apparatus enclosure.

Conduit seals are also required in each conduit run of 2" or larger entering the enclosure or fitting housing terminals, splices, or taps, and within 18" of such enclosures or fittings.

In each conduit run leaving the Class I, Division 1 hazardous area, the sealing fitting may be located on either side of the boundary of the hazardous area within 10' of the boundary, but shall be so designed and installed that any gases or vapors that may enter the conduit system, within the Division 1 hazardous area, will not enter or be communicated to the conduit beyond the seal. There shall be no union, coupling, box, or fitting, except approved explosion-proof reducers at the sealing fitting, in the conduit between the sealing fitting and the point at which the conduit leaves the area.

Sealing compound must be approved for the purpose, must not be affected by the surrounding atmosphere or liquids, and must not have a melting point of less than 200°F (93°C). Most sealing compound kits contain a powder in a polyethylene bag within an outer container. To mix, remove the bag of powder, fill the outside container with water up to the marked line on the container, pour in the powder, and mix.

To pack the seal-off, remove the threaded plug or plugs from the fitting and insert the fiber supplied with the packing kit. Tamp the fiber between the wires and the hub before pouring the sealing compound into the fitting. Then, pour in the sealing cement and reset the threaded plug tightly. The fiber packing prevents the sealing compound (in the liquid state) from entering the conduit lines.

The seal-off (sealing) fittings shown in *Figure 20* are typical of those used. *Figure 20(A)* is for vertical mounting and is provided with a threaded, plugged opening into which the sealing cement is poured. The seal-off in *Figure 20(B)* is for either horizontal or vertical runs, and has an additional plugged opening in the lower hub to facilitate packing fiber around the conductors in order to form a dam for the sealing cement.

Instructor's Notes:

422 CHAPTER 10

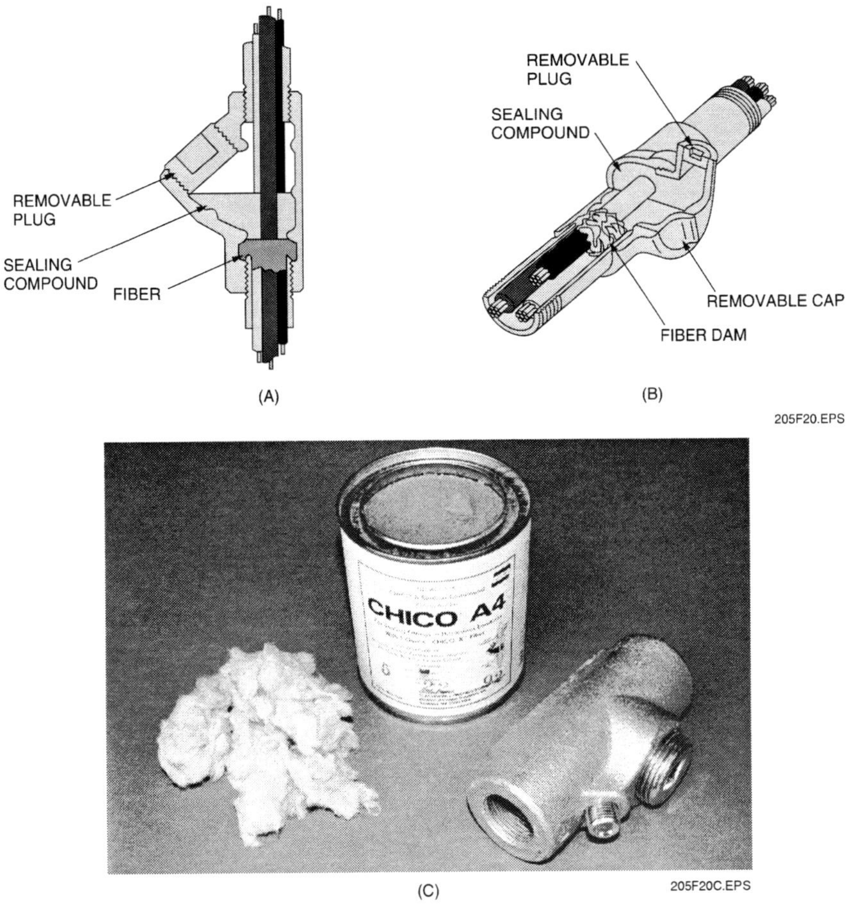

Figure 20 ◈ Typical seal-off fittings.

Choosing the correct type of seal-off compound is very important. Always use a type that is compatible with the type of conductor insulation and is recommended by the manufacturer of the seal-off itself. There are certain types on the market that are designed for certain applications; make sure you use the right one.

Most other explosion-proof fittings are provided with threaded hubs for securing the conduit. Typical fittings include switch and junction boxes, conduit bodies, union end connectors, flexible couplings, explosion-proof lighting fixtures, receptacles, and panelboard and motor starter enclosures. A practical representation of these and other fittings appears in *Figure 21.*

In certain hazardous locations, flexible couplings are used to connect vibrating machines, such as motors. Such a connection is shown in *Figure 22.*

7.0.0 ◈ FITTINGS

Certain fittings are required in every raceway system for joining runs of conduit and also when the

Classroom

Discuss the applications of explosion-proof fittings.

Audiovisual

Show Transparencies 16 and 17 (Figures 21 and 22).

BOXES AND FITTINGS 423

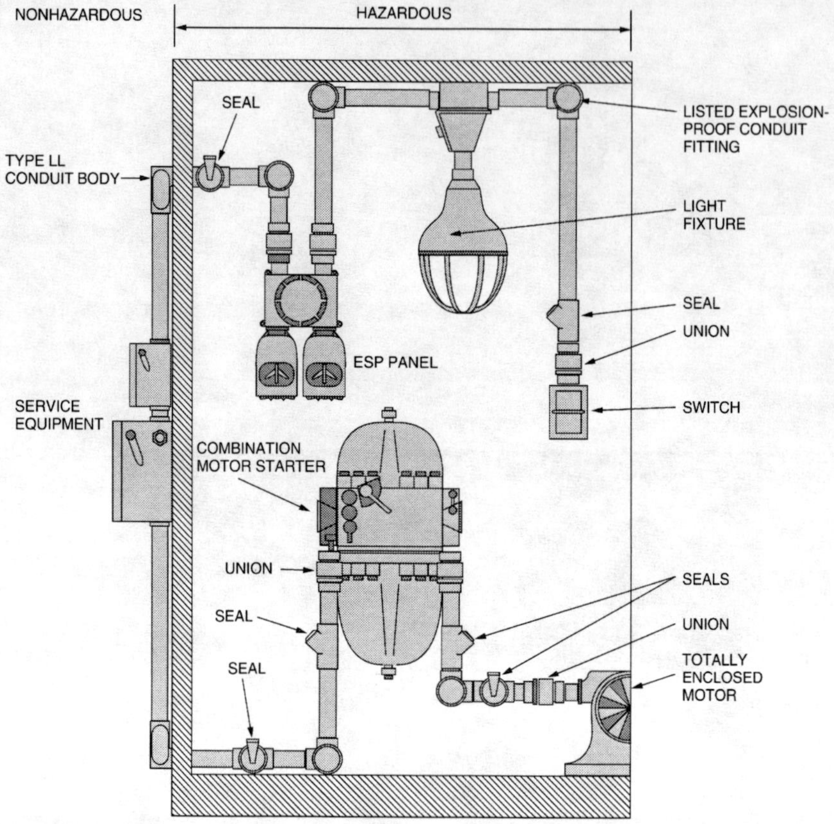

Figure 21 ◈ Practical representation of explosion-proof fittings in Class I, Division 1.

Classroom

Discuss the special fittings used with EMT. Pass around examples for the trainees to examine.

Audiovisual

Show Transparency 18 (Figure 23).

raceway terminates in an outlet box or other enclosure. Most metallic raceways qualify as an equipment grounding conductor provided they are tightly connected at each joint and termination point to provide a continuous grounding path.

7.1.0 EMT Fittings

Because EMT or *thinwall* is too thin for threads, special fittings must be used. For wet or damp locations, compression fittings such as those shown in *Figure 23* are used. This type of fitting contains compression rings made of plastic or other soft material that forms a watertight seal.

EMT fittings for dry locations can be either the setscrew type or the compression type. To use the setscrew type, the reamed ends of the EMT are inserted into the sleeve and the setscrews are tightened with a screwdriver to secure them and the conduit in place. Various types of setscrew couplings are shown in *Figure 24.*

EMT also requires connectors at each termination point, and EMT connectors are available to match the couplings described previously, that is, compression and setscrew types. They are similar to the couplings except that one end of the connector is threaded to accept a locknut and bushing.

Instructor's Notes:

424 CHAPTER 10

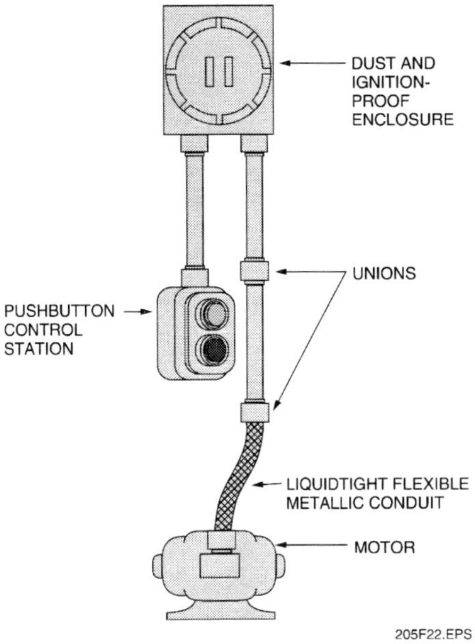

Figure 22 ◆ Power diagram for Class II, Division 2 installation with flexible conduit to motor.

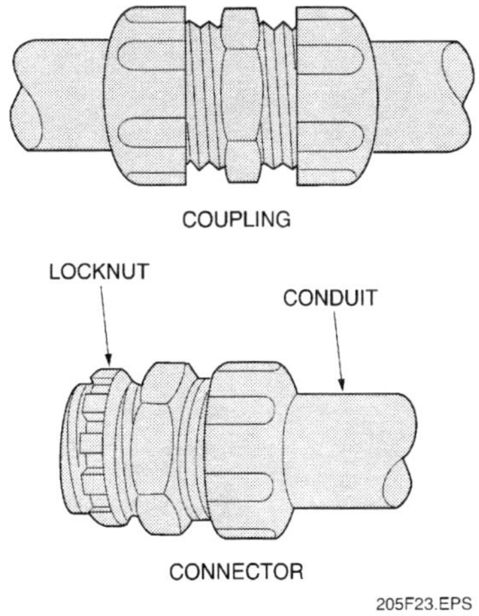

Figure 23 ◆ EMT compression fittings.

7.2.0 Rigid, Aluminum, and IMC Fittings

Rigid metal conduit, aluminum conduit, and intermediate metal conduit all have sufficient wall thicknesses to permit threading. Consequently, all three types may be joined with threaded couplings (*Figure 25*) and when any of these types terminate into an outlet box or other enclosure, double locknuts are used to secure the conduit to the box opening. Running threads are not permitted for connection at couplings.

Sometimes rigid conduit or tubing must be connected to flexible metal conduit for connection to electric motors and other machinery that may vibrate during operation. Combination couplings (*Figure 26*) are used to make the transition. When using combination couplings, be sure the flexible conduit is pushed as far as possible into the coupling. This covers the sharp edges of the conduit to protect the conductors from damage.

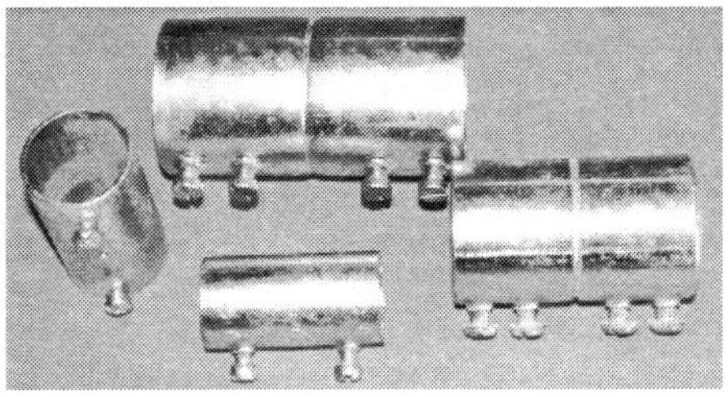

Figure 24 ◆ Setscrew fittings.

Figure 25 ◆ Rigid metal conduit with coupling.

Discuss the threaded fittings used with rigid, aluminum, and intermediate metal conduit. Pass around examples for the trainees to examine.

Show Transparency 19 (Figure 26).

Teaching Tip

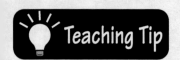

Discuss the "Think About It." Have the trainees examine these components and compare them to standard components.

THINK ABOUT IT

Components Used in Hazardous Locations

Photo (A) shows a special motor starter designed for use in a hazardous location. Photo (B) shows the same starter with the cover removed. Photo (C) shows an explosion-proof seal. Examine these components. How do they differ from standard components?

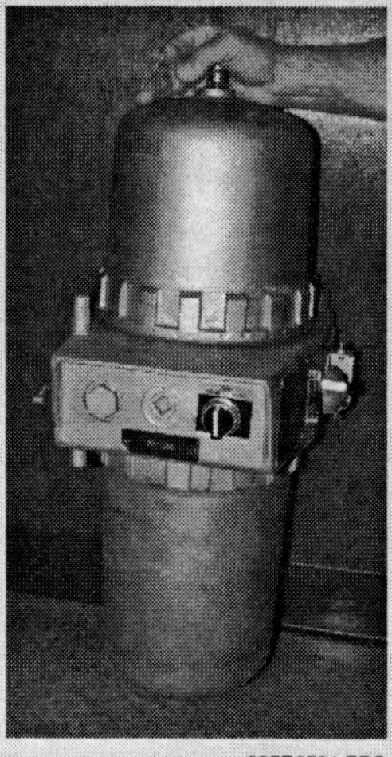

205P0501.EPS

(A)

205P0502.EPS

(B)

205P0503.EPS

(C)

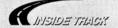

INSIDE TRACK

Explosionproof Flex

This photo shows a special type of metal conduit for use in hazardous locations. Note that all conduit of this type comes with factory-made fittings for optimum protection under extreme conditions.

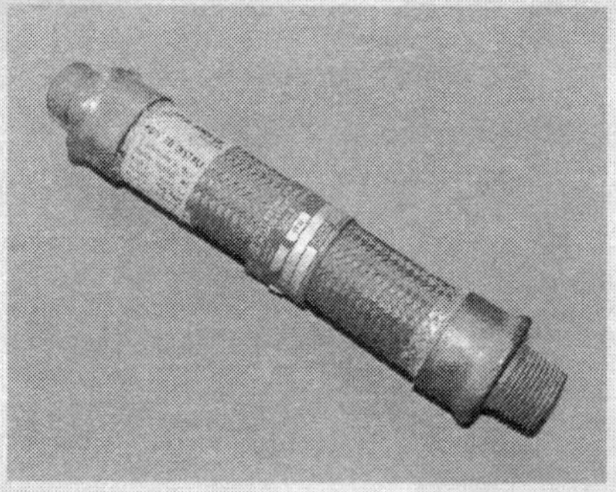

205P0504.EPS

Instructor's Notes:

426 CHAPTER 10

Setscrew and Compression Fittings

THINK ABOUT IT

This picture shows both setscrew and compression fittings. Which type provides a better connection? Why?

205P0505.EPS

Threadless couplings and connectors may also be used under certain conditions with rigid, IMC, and aluminum conduit. When used, they must be made up wrench-tight and where buried in masonry or concrete, they must be concrete-tight. Where installed in wet locations, they must be rainproof. This type of coupling is not permitted in most hazardous locations.

Other types of fittings used with raceway systems are shown in *Figure 27.*

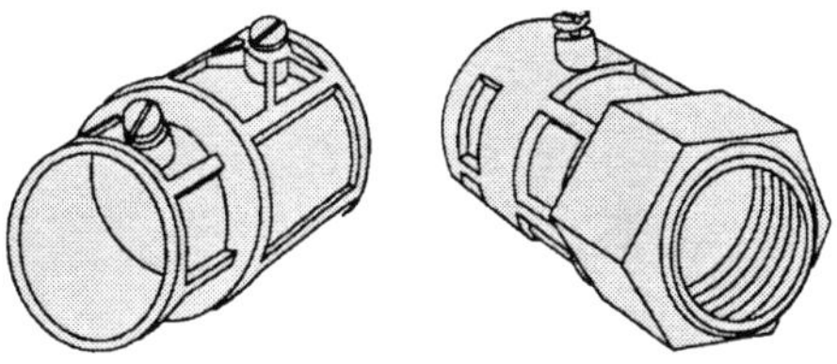

Figure 26 ◈ Combination couplings.

7.3.0 Locknuts and Bushings

In general, locknuts are used on the inside and outside walls of outlet boxes or other enclosures to which threaded conduit is connected. When conduit connectors are used, such as EMT connectors, only one locknut is required on the threads that protrude inside the box. A grounding locknut may be needed if bonding jumpers are used inside the box or enclosure. Special sealing locknuts are also available for use in wet locations. Locknuts are shown in *Figure 28.*

Bushings protect the wires from the sharp edges of the conduit or connector. Bushings are usually made of plastic, fiber, or metal. Some metal bushings have a grounding screw to permit an equipment or bonding jumper wire to be installed. Several types of bushings are shown in *Figure 29.*

An insulating bushing is installed on the threaded end of conduit that enters a sheet metal enclosure. The purpose of the bushing is to protect the conductors from being damaged by the sharp edges of the threaded conduit end. Any ungrounded conductor, No. 4 AWG or larger, that enters a raceway, box, or enclosure must be protected with an insulating bushing, as required in ***NEC Sections 300.4(F), 312.6(C), and 314.17(D).***

A grounding insulated bushing has provisions for protecting conductors and also has provisions

Teaching Tip

Discuss the "Think About It." Explain that setscrew fittings are held in place only by the pressure of the setscrews, while compression fittings are threaded to provide a much tighter connection.

Audiovisual

Show Transparency 20 (Figure 27).

Classroom

Discuss the purpose of locknuts. Pass around examples for the trainees to examine.

Discuss the use of bushings. Pass around examples for the trainees to examine.

Audiovisual

Show Transparency 21 (Figure 28).

Show Transparency 22 (Figure 29).

Demonstration

Demonstrate how to install various fittings in a raceway system.

Laboratory

Have the trainees practice installing various fittings in a raceway system. Note the proficiency of each trainee.

Classroom

Explain the difference between grounding insulating bushings and standard insulating bushings.

Demonstration

Demonstrate the use of a knockout punch.

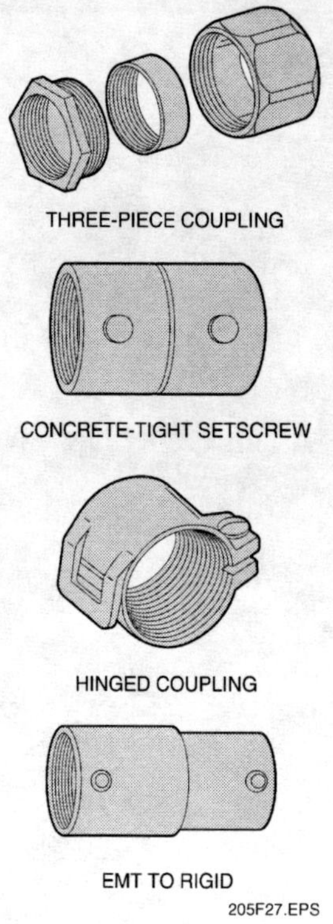

Figure 27 ◆ Metal conduit couplings.

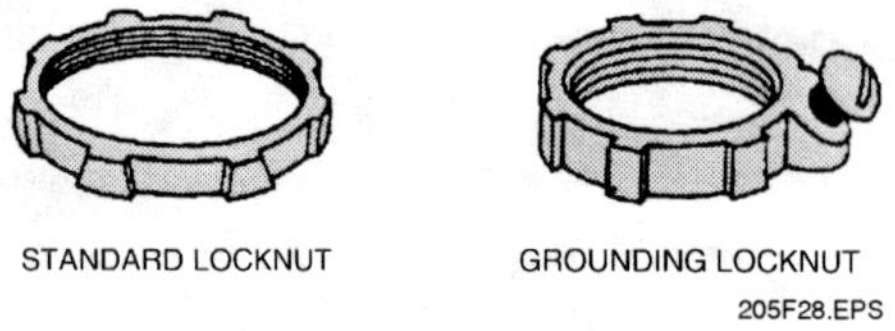

Figure 28 ◆ Common types of locknuts.

for the connection of a ground wire. The ground wire, once connected to the grounding bushing, may be connected to the box to which the conduit is connected. See *Figure 30.*

An opening must be provided in the outlet box or enclosure for the entrance of conduit and connectors when raceway systems terminate. Most boxes and enclosures are provided with an adequate number of concentric knockouts. However, some may not have precut knockouts or the ones that are available may not be in the required location. In these cases, a knockout punch must be used to make a hole for the conduit connection. A hand-operated knockout punch is shown in *Figure 31.*

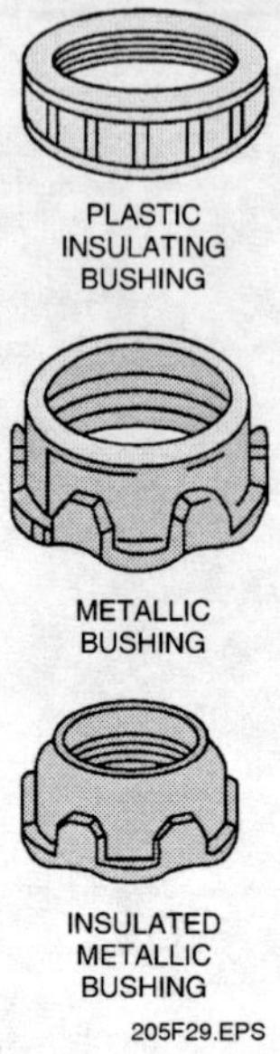

Figure 29 ◆ Typical bushings used at termination points.

To use the knockout punch, the center of the hole is located in the box or enclosure and marked with a center punch. A pilot hole is then drilled to accept the threaded drive bolt of the knockout punch. The punch is separated from the drive screw, the screw is then placed through the pilot hole with the die on one side of the box wall, and the punch is screwed onto the drive screw on the opposite side of the wall. The punch is then aligned and screwed onto the drive screw hand-tight, in which case the punch should lightly bite into the wall of the enclosure. A wrench is then used to tighten the drive nut until the punch is drawn through the enclosure wall, making a neat circular opening.

Where many such openings must be made, or when knockouts for the larger sizes of conduit must be cut, contractors normally furnish workers with power knockout tools to facilitate the operation.

Instructor's Notes:

428 CHAPTER 10

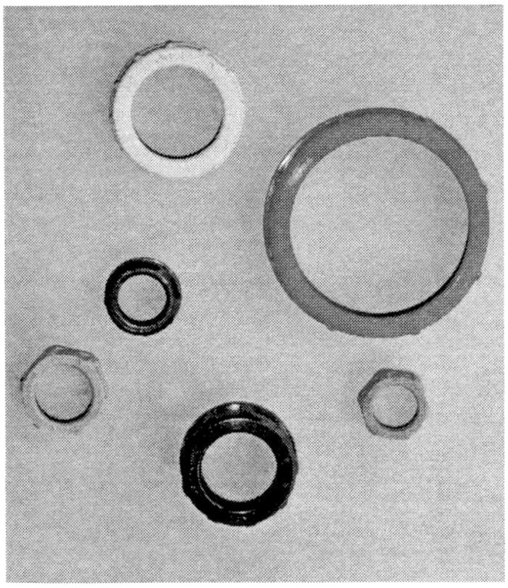
205F30A.EPS

(A)

205F30B.EPS

(B)

Figure 30 ◈ (A) Regular insulating bushings and (B) grounding insulating bushings.

205F31.EPS

Figure 31 ◈ Knockout punch kit.

THINK ABOUT IT

Putting It All Together

Turn off the power in one area of your home and then remove some of the switch and receptacle plates. Examine the wiring inside each box. Is the box adequately sized for the number of wires and devices?

Teaching Tip

Discuss the "Think About It." Have trainees report their findings to the rest of the class and discuss the options for correcting any deficiencies found.

BOXES AND FITTINGS 429

Classroom

Have the trainees complete the Review Questions and go over the answers prior to administering the Module Examination.

Examination

Administer the Module Examination. Record the results on Craft Training Report Form 200 and submit the results to the Training Program Sponsor.

Performance Testing

Administer the Performance Test and fill out Performance Profile Sheets for each trainee. If desired, trainee proficiency noted during laboratory sessions may be used to complete the Performance Test. Be sure to record the results on Craft Training Report Form 200 and submit the results to the Training Program Sponsor.

Summary

Electricians work with boxes and fittings almost every day on every project. Consequently, you must have a thorough knowledge of the types available and their applications. Portions of *NEC Chapters 3 and 4* cover the installation of boxes and fittings, so refer to appropriate sections in these chapters whenever a question arises.

One of the best ways to learn about boxes and fittings is to study manufacturers' catalogs. You will find a wealth of information in each of these, including detailed instructions (in many of them) on installation techniques. Some of these catalogs also provide simplified NEC explanations on the use of the manufacturer's products.

Review Questions

1. The maximum weight allowed by the NEC when ceiling fans are mounted directly to an approved outlet box is _____ pounds.
a. 25
b. 35
c. 45
d. 55

2. Square outlet boxes are available in 4" and _____ sizes.
a. $4\frac{11}{16}$"
b. 5"
c. $5\frac{1}{4}$"
d. 6"

3. Concrete boxes are normally used _____.
a. in residential wiring to house wiring devices
b. to feed lay-in troffer lighting fixtures
c. in flat-slab construction projects
d. in airport runway lighting

4. Deduct _____ conductor(s) for each strap-mounted device in a device box.
a. one
b. two
c. three
d. four

5. Using _____ adds to the capacity of an outlet box.
a. fixture studs
b. wire nuts
c. strap-mounted devices
d. raised device covers

6. When calculating the pull box size for straight pulls, the length of the box must not be less than _____ times the trade diameter of the largest raceway.
a. two
b. four
c. six
d. eight

7. If the largest trade diameter of a raceway entering a pull box is 3", and it is a straight pull, the minimum size box allowed is _____.
a. 20"
b. 24"
c. 30"
d. 36"

8. A _____ conduit body has openings on four different sides plus an access opening.
a. Type X
b. Type C
c. Type T
d. Type LL

9. Which of the following best describes how Type FS boxes should be installed?
a. Holes should be drilled in back of the box for mounting screws.
b. Holes should be drilled on the sides of the box only for mounting screws.
c. No holes should be drilled in the box for mounting.
d. Holes may be drilled only in existing installations.

10. The purpose of using seals or seal-off fittings in hazardous areas is to prevent _____.
a. other conductors from being pulled into the raceway
b. the passage of gases or vapors
c. water from entering the raceway system
d. the conductors from being removed from the raceway system

Instructor's Notes:

430 CHAPTER 10

GLOSSARY

Trade Terms Introduced in This Module

Conduit body: A separate portion of a conduit or tubing system that provides access through a removable cover (or covers) to the interior of the system at a junction of two or more sections of the system or at a terminal point of the system. Boxes such as FS and FD or larger cast or sheet metal boxes are not classified as conduit bodies.

Connector: Device used to physically connect conduit or cable to an outlet box, cabinet, or other enclosure.

Explosion-proof: Designed and constructed to withstand an internal explosion without creating an external explosion or fire.

Handy box: Single-gang outlet box used for surface mounting to enclose receptacles or wall switches on concrete or concrete block construction of industrial and commercial buildings; nongangable; also made for recessed mounting; also known as a *utility box.*

Junction box: An enclosure where one or more raceways or cables enter, and in which electrical conductors can be, or are, spliced.

Nineteen hundred box: A term commonly used to refer to any two-gang, four-inch-square outlet box.

Outlet box: A metallic or nonmetallic box installed in an electrical wiring system from which current is taken to supply some apparatus or device.

Pull box: A sheet metal box-like enclosure used in conduit runs to facilitate the pulling of cables from point to point in long runs, or to provide for the installation of conduit support bushings needed to support the weight of long riser cables, or to provide for turns in multiple-conduit runs.

Raintight: Constructed or protected so that exposure to a beating rain will not result in the entrance of water under specified test conditions.

Sealing compound: The material poured into an electrical fitting (seal-off) to seal and minimize the passage of vapors.

Waterproof: Constructed so that moisture will not interfere with successful operation.

Watertight: Constructed so that water will not enter the enclosure under specified test conditions.

Weatherproof: Constructed or protected so that exposure to the weather will not interfere with successful operation.

Additional Resources

This module is intended to present thorough resources for task training. The following reference works are suggested for further study. These are optional materials for continued education rather than for task training.

American Electricians' Handbook, Latest Edition. New York: Croft, McGraw-Hill.

National Electrical Code Handbook, Latest Edition. Quincy, MA: National Fire Protection Association.

Instructor's Notes:

ANSWER KEY

Answers to Review Questions

Answer	Section
1. b	2.0.0
2. a	2.2.0
3. c	2.4.0
4. b	3.0.0
5. d	3.0.0
6. d	4.1.0
7. b	4.1.0
8. a	5.4.0
9. c	5.5.0
10. b	6.4.0

CONTREN™ LEARNING SERIES — USER UPDATES

The NCCER makes every effort to keep these textbooks up-to-date and free of technical errors. We appreciate your help in this process. If you have an idea for improving this textbook, or if you find an error, a typographical mistake, or an inaccuracy in NCCER's Contren™ textbooks, please write us, using this form or a photocopy. Be sure to include the exact module number, page number, a detailed description, and the correction, if applicable. Your input will be brought to the attention of the Technical Review Committee. Thank you for your assistance.

Instructors – If you found that additional materials were necessary in order to teach this module effectively, please let us know so that we may include them in the Equipment/Materials list in the Instructor's Guide.

Write: Curriculum Revision and Development Department
National Center for Construction Education and Research
P.O. Box 141104, Gainesville, FL 32614-1104

Fax: 352-334-0932

E-mail: curriculum@nccer.org

Craft ______ Module Name ______

Copyright Date ______ Module Number ______ Page Number(s) ______

Description ______

(Optional) Correction ______

(Optional) Your Name and Address ______

Chapter 11

Conductors

26109-02

MODULE OVERVIEW

This course introduces the electrical trainee to the various types of conductors, explains how conductors are rated by the NEC, and discusses the different methods used for pulling conductors through conduit runs.

PREREQUISITES

Please refer to the Course Map in the Trainee Module. Prior to training with this module, it is recommended that the trainee shall have successfully completed the following:

Core Curriculum; Residential Electrical I, Chapters 1 through 10.

LEARNING OBJECTIVES

Upon completion of this module, the trainee will be able to:

1. Explain the various sizes and gauges of wire in accordance with American Wire Gauge standards.
2. Identify insulation and jacket types according to conditions and applications.
3. Describe voltage ratings of conductors and cables.
4. Read and identify markings on conductors and cables.
5. Use the tables in the NEC to determine the ampacity of a conductor.
6. State the purpose of stranded wire.
7. State the purpose of compressed conductors.
8. Describe the different materials from which conductors are made.
9. Describe the different types of conductor insulation.
10. Describe the color coding of insulation.
11. Describe instrumentation control wiring.
12. Describe the equipment required for pulling wire through conduit.
13. Describe the procedure for pulling wire through conduit.
14. Install conductors in conduit.
15. Pull conductors in a conduit system.

PERFORMANCE OBJECTIVES

Under supervision of the instructor, the trainee should be able to:

1. Perform a manual single cable pull:
   - Select the proper pulling rope for the pull.
   - Attach the pulling rope to the cable.
   - Attach the pulling rope to the puller.
   - Pull the cable through the conduit.
2. Perform the feed:
   - Make a pulling head attachment.
   - Apply lubricant to the cable as it is being pulled.
   - Seal the end of the cable after the pull is complete.

NCCER STANDARDIZED CRAFT TRAINING PROGRAM

The National Center for Construction Education and Research (NCCER) provides a standardized national program of accredited craft training. Key features of the program include instructor certification, competency-based training, and performance testing. The program provides trainees, instructors, and companies with a standard form of recognition through a National Craft Training Registry. The program is described in full in the *Guidelines for Accreditation*, published by the NCCER. For more information on standardized craft training, contact the NCCER at P.O. Box 141104, Gainesville, FL 32614-1104, 352-334-0911, visit our Web site at www.nccer.org, or e-mail info@nccer.org.

HOW TO USE THIS ANNOTATED INSTRUCTOR'S GUIDE

Each page presents two sections of information. The larger section displays each page exactly as it appears in the Trainee Module. The narrow column ties suggested trainee and instructor actions to each page and provides icons to call your attention to material, safety, audiovisual, or testing requirements. The bottom of each page includes space for your notes.

Teaching Tip

If you see the Teaching Tip icon, that means there is a teaching tip associated with this section. Also refer to the suggested teaching tips at the end of the module.

SAFETY CONSIDERATIONS

Ensure that the trainees are equipped with appropriate personal protective equipment. Emphasize the use of the proper safety precautions and procedures when working with the materials and tools used to install conductors.

PREPARATION

Before teaching this module, you should review the Module Outline, Learning and Performance Objectives, and the Materials and Equipment List. Be sure to allow ample time to prepare your own training or lesson plan and gather all required equipment and materials.

MATERIALS AND EQUIPMENT LIST

Materials:

Transparencies

Markers/chalk

Copy of the latest edition of the *National Electrical Code®*

Variety of solid wire conductors

Samples of stranded conductors

Samples of cable, including:
- Type NM
- Type NMC
- Type SE
- Type UF
- Type NMS
- Type MV
- High-voltage shielded
- Type FC
- Type FCC
- Type TC
- Type USE

Instrumentation control wiring

Module Examinations*

Performance Profile Sheets*

Equipment:

Overhead projector and screen

Whiteboard/chalkboard

Appropriate personal protective equipment

Fish tapes

Power fishing system

Basket grip

Wire grips

Manual wire puller

Power puller

Pull lines

Reel cart

Electrician's hand tools

Access to a conduit run

*Located in the Test Booklet packaged with this Annotated Instructor's Guide.

ADDITIONAL RESOURCES

This module is intended to present thorough resources for task training. The following reference is suggested for both instructors and motivated trainees interested in further study. This is optional material for continued education rather than for task training.

National Electrical Code Handbook, Latest Edition. Quincy, MA: National Fire Protection Association.

NOTES

The designations "National Electrical Code," "NE Code," and "NEC," where used in this document, refer to the *National Electrical Code*®, which is a registered trademark of the National Fire Protection Association, Quincy, MA. All National Electrical Code (NEC) references in this module refer to the 2002 edition of the NEC.

If you feel that additional math instruction would be helpful, Prentice Hall offers a basic math textbook entitled *Fundamentals of Electrical and Mechanical Mathematics.* It covers the basic math requirements for electrical trainees and may be ordered by contacting Prentice Hall Customer Service at 1-800-922-0579.

TEACHING TIME FOR THIS MODULE

An outline for use in developing your lesson plan is presented below. Note that each Roman numeral in the outline equates to one session of instruction. Each session has a suggested time period of 2½ hours. This includes 10 minutes at the beginning of each session for administrative tasks and one 10-minute break during the session. Approximately 15 hours are suggested to cover *Conductors.* You will need to adjust the time required for hands-on activity and testing based on your class size and resources.

Topic	Planned Time
Session I. Ampacity; Underground Installations; Wire Size; Conductor Material	
A. Ampacity	________
1. NEC Ampacity Tables	________
B. Underground Installations	________
1. Direct Burial	________
C. Wire Size	________
1. AWG System	________
2. Stranding	________
3. Compressed Conductors	________
4. Circular Mils	________
D. Conductor Material	________
1. Conductivity	________
2. Cost	________
3. Availability	________
4. Workability	________
Session II. Conductor Insulation	
A. Thermoplastic	________
B. Letter Coding	________
C. Color Coding	________
D. Wire Ratings	________
Session III. Fixture Wires; Cables; Instrumentation Control Wiring	
A. Fixture Wires	________
1. Heating Cables	________
B. Cables	________
1. Cable Markings	________
2. Nonmetallic-Sheathed Cable	________
3. Type UF Cable	________
4. Type NMS Cable	________
5. Type MV Cable	________
6. High-Voltage Shielded Cable	________
7. Channel Wire Assemblies	________

8. Flat Conductor Cable
9. Type TC Cable
10. Type USE Cable

C. Instrumentation Control Wiring
   1. Shields
   2. Grounding
   3. Jackets

Session IV. Fish Tape; Wire Grips; Pull Lines; Safety Precautions; Pulling Equipment

A. Fish Tape
   1. Power Conduit Fishing Systems
   2. Connecting Wire to a String Line

B. Wire Grips

C. Pull Lines

D. Safety Precautions

E. Pulling Equipment

Session V. Feeding Conductors into Conduit; Conductor Lubrication; Conductor Termination

A. Feeding Conductors into Conduit

B. Conductor Lubrication

C. Conductor Termination

Session VI. Laboratory; Module Examination and Performance Testing

A. Laboratory

Under instructor supervision, have the trainees practice performing a manual cable pull, including selecting and attaching the pulling rope, making a pulling head attachment, applying lubricant to the cable as it is being pulled, and sealing the end of the cable after the pull is complete.

B. Module Examination
   1. Trainees must score 70% or higher to receive recognition from the NCCER.
   2. Record the testing results on Craft Training Report Form 200 and submit the results to the Training Program Sponsor.

C. Performance Testing
   1. Trainees must perform each task to the satisfaction of the instructor to receive recognition from the NCCER.
   2. Record the testing results on Craft Training Report Form 200 and submit the results to the Training Program Sponsor.

Chapter 11

Module 26109-02

Conductors

Course Map

This course map shows all of the modules in the first level of the Residential Electrical curriculum. The suggested training order begins at the bottom and proceeds up. Skill levels increase as you advance on the course map. The training order may be adjusted by the local Training Program Sponsor.

RESIDENTIAL ELECTRICAL I

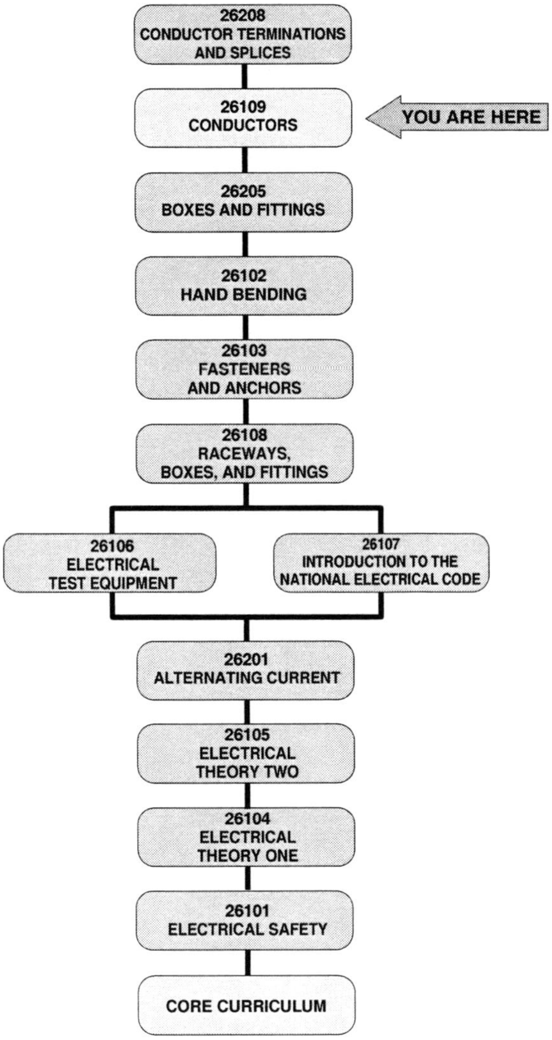

Homework

Assign reading of Module 26109.

MODULE 26109 CONTENTS

Instructor's Notes:

Figures

CONDUCTORS 441

Chapter 11

MODULE 26109

Conductors

Objectives

When you have completed this module, you will be able to do the following:

1. Explain the various sizes and gauges of wire in accordance with American Wire Gauge standards.
2. Identify insulation and jacket types according to conditions and applications.
3. Describe voltage ratings of conductors and cables.
4. Read and identify markings on conductors and cables.
5. Use the tables in the NEC to determine the ampacity of a conductor.
6. State the purpose of stranded wire.
7. State the purpose of compressed conductors.
8. Describe the different materials from which conductors are made.
9. Describe the different types of conductor insulation.
10. Describe the color coding of insulation.
11. Describe instrumentation control wiring.
12. Describe the equipment required for pulling wire through conduit.
13. Describe the procedure for pulling wire through conduit.
14. Install conductors in conduit.
15. Pull conductors in a conduit system.

Prerequisites

Before you begin this module, it is recommended that you successfully complete the following: Core Curriculum; Residential Electrical I, Chapters 1 through 10.

Required Trainee Materials

1. Pencil and paper
2. Copy of the latest edition of the *National Electrical Code*
3. Appropriate personal protective equipment

1.0.0 ◆ INTRODUCTION

As an electrician, you will be required to select the proper wire and/or cable for a job. You will also be required to pull this wire or cable through conduit runs in order to terminate it. This module will examine the different types of conductors and conductor insulation. It will also examine how these conductors are rated and classified by the NEC and the different methods used for pulling these conductors through conduit runs.

Note: The designations "National Electrical Code," "NE Code," and "NEC," where used in this document, refer to the *National Electrical Code®*, which is a registered trademark of the National Fire Protection Association, Quincy, MA. *All National Electrical Code (NEC) references in this module refer to the 2002 edition of the NEC.*

Materials

Ensure you have everything required to teach the course. Check the Materials and Equipment List at the front of this Instructor's Guide.

Audiovisual

Show Transparency 1, Course Objectives.

Show Transparency 2, Performance Profile Tasks.

Classroom

Discuss the importance of understanding how to properly identify, select, and pull conductors.

Classroom

Explain that items shown in bold (blue) are defined in the Glossary at the back of this module.

Discuss the NEC classifications of conductor insulation.

Explain how to use the NEC Ampacity Tables.

Discuss underground installation requirements.

2.0.0 ◆ CONDUCTORS AND INSULATION

The term *conductor* is used in two ways. It is used to describe the current-carrying portion of a wire or cable, and it is used to describe the wire or cable composed of a current-carrying portion and an outer covering (insulation). In this module, the term *conductor,* if not specified otherwise, will be used to describe the wire assembly, which includes the insulation and the current-carrying portion of the wire. Conductors are uniquely identified by size and insulation material. Size refers to the physical size of the current-carrying portion of the conductor.

NEC Table 310.13 presents application and construction data on the wide range of 600-volt insulated, individual conductors recognized by the NEC, with the appropriate letter designation used to identify each type of insulated conductor. Important data that should be noted in this table are as follows:

- The designation for one thousand circular mils is *kcmil,* which has been substituted for the long-time designation *MCM* in this table and throughout the NEC.
- Type MI (mineral insulated) cable may have either a copper or an alloy steel sheath.
- Type RHW-2 is a conductor insulation made of moisture- and heat-resistant rubber with a 90°C (194°F) rating for use in dry and wet locations.
- Type XHHW-2 is a moisture- and heat-resistant crosslinked synthetic polymer with a 90°C (194°F) rating for use in dry and wet locations.
- The suffix *LS* designates a conductor insulation as low-smoke producing and flame retardant. For example, Type THHN/LS is a THHN conductor with a limited smoke-producing characteristic.
- Type THHW is a moisture- and heat-resistant insulation rated at 75°C (167°F) for wet locations and 90°C (194°F) for dry locations. This is similar to THWN and THHN without the outer nylon covering but with thicker insulation.
- All insulations using asbestos have been deleted from *NEC Table 310.13* because they are no longer made.

2.1.0 Ampacity

Ampacity is the current in amperes a conductor can carry continuously under the conditions of use without exceeding its temperature rating. The ampacity of conductors for given conditions of use are listed in ***NEC Tables 310.16 through 310.19.***

2.1.1 NEC Ampacity Tables

NEC Table 310.16 covers conductors rated up to 2,000 volts where not more than three conductors are installed in a raceway or cable or are directly buried in the earth, based on an ambient temperature of 30°C (86°F).

NEC Table 310.17 covers both copper conductors and aluminum or copper-clad aluminum conductors up to 2,000 volts where conductors are used as single conductors in free air, based on an ambient temperature of 30°C (86°F).

NEC Tables 310.18 and 19 apply to conductors rated at 150°C to 250°C (302°F to 482°F), used either in raceway or cable or as single conductors in free air, based on an ambient temperature of 40°C (104°F).

Example:

Determine the ampacity of a No. 12 Cu (copper) THW conductor.

Solution:

25 amps (from ***NEC Table 310.16***).

2.2.0 Underground

Any conductor used in a wet location (see definition under *Wet Location* in ***NEC Article 100***) must be designated as suitable for wet locations. Any conduit run underground is assumed to be subject to water infiltration and is, therefore, in a wet location, requiring the use of only the listed conductor types.

2.2.1 Direct Burial

Direct burial conductors should be trench-laid without crossovers; slightly snaked to allow for

Teaching Tip

Discuss the "Think About It."

THINK ABOUT IT

Overheating

What happens when insulation is overheated? What affects the ampacity of an insulated conductor?

444 CHAPTER 11

Instructor's Notes:

possible earth settlement, movement, or heaving due to frost action; and have cushions and covers of sand or screened fill to protect conductors against sharp objects in trenches or backfill.

2.3.0 Wire Size

Wire sizes are expressed in gauge numbers. The standard system of wire sizes in the United States is the American Wire Gauge (AWG) system.

2.3.1 AWG System

The AWG system uses numbers to identify the different sizes of wire and cable (*Figure 1*). The larger the number, the smaller the cross-sectional area of the wire. The larger the cross-sectional area of the current-carrying portion of a conductor, the higher the amount of current the wire can conduct. The AWG numbers range from 50 to 1; then 0, 00, 000, and 0000 (one aught [1/0], two aught [2/0], three aught [3/0], and four aught [4/0]). Any wire larger than 0000 is identified by its area in circular mils. Wire sizes smaller than No. 18 AWG are usually solid, but may be stranded in some cases. Wire sizes of No. 6 AWG or larger are stranded.

For wire sizes larger than No. 16 AWG, the wire size is marked on the insulation (*Figure 2*).

NEC Chapter 9, Table 8 has descriptive information on wire sizes. Again, note that all wires smaller than No. 6 are available as solid or stranded. Wire sizes of No. 6 or larger are shown only as stranded. Solid wire larger than No. 6 is manufactured; however, the NEC only permits the use of solid wire in a raceway for sizes smaller than No. 8 (***NEC Section 310.3***).

2.3.2 Stranding

According to ***NEC Chapter 9, Table 8,*** wire sizes No. 18 to No. 2 have seven strands; wire sizes No. 1 to No. 4/0 have 19 strands; and wire sizes between 250 kcmil and 500 kcmil have 37 strands. The purpose of stranding is to increase the flexibility of the wire. Terminating solid wire sizes larger than No. 8 in pull boxes, disconnect switches, and panels would not only be very difficult, but might also result in damage to equipment and wire insulation.

Pulling solid wire in conduit around bends could pose a major problem and cause damage to equipment for wire sizes larger than No. 8. The

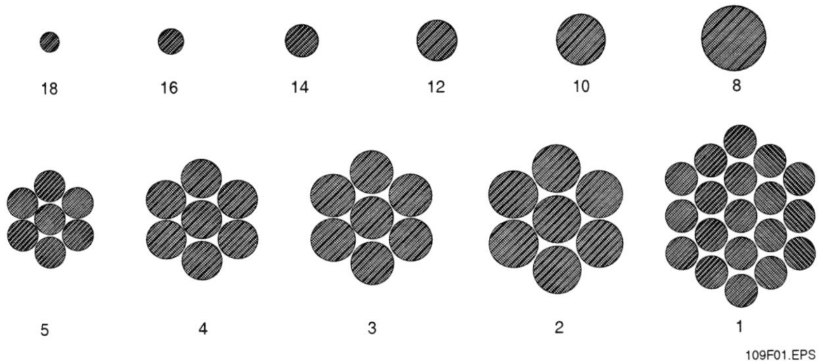

Figure 1 ◆ Comparison of wire sizes (enlarged) from No. 18 to No. 1 AWG.

THINK ABOUT IT

Wire Size

Why is wire size a critical factor in a wiring system? Other than load, what other factors may dictate wire size? What can happen when a wire isn't properly sized for the load?

Audiovisual

Show Transparencies 3 and 4 (Figures 1 and 2). Discuss the AWG wire sizing system.

Classroom

Discuss wire stranding.

Teaching Tip

Discuss the "Think About It."

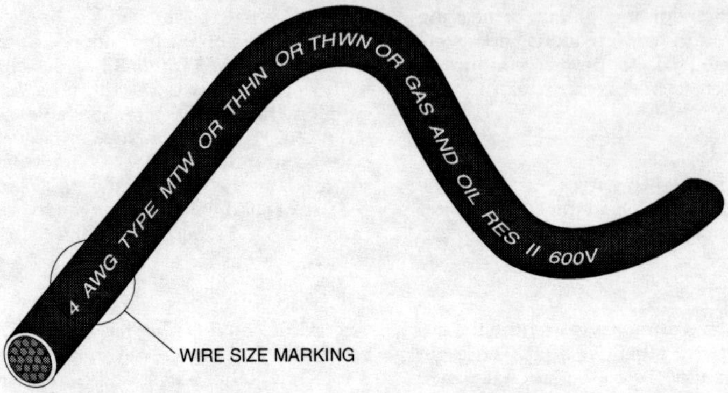

Figure 2 ◆ Wire size marking.

Audiovisual

Show Transparency 5 (Figure 3).

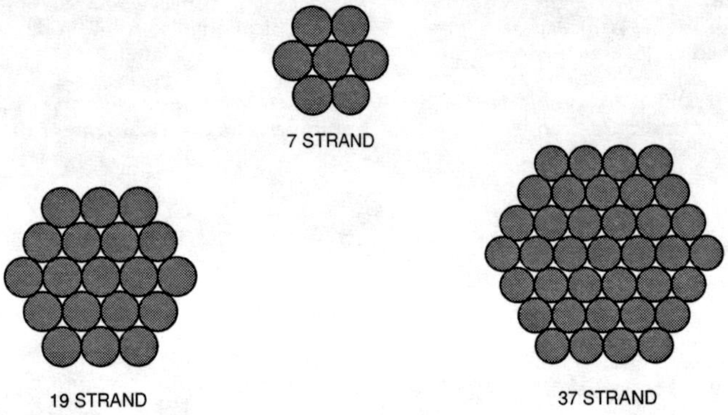

Figure 3 ◆ Strand configurations.

Classroom

Discuss the advantages of using compressed conductors.

Audiovisual

Show Transparency 6 (Figure 4).

Classroom

Define *circular mil.*

reason for choosing 7, 19, and 37 strands for stranded conductors is that it is necessary to provide a flexible, almost round conductor. In order for a conductor to be flexible, individual strands must not be too large. *Figure 3* shows how these conductors are configured.

2.3.3 Compressed Conductors

A relatively new entry in the scheme of conductors is compressed aluminum conductors *(Figure 4)*. Compressed aluminum conductors are those that have been compressed to reduce the air space between the strands.

The purpose of a compressed conductor is to reduce the overall diameter of the cable so that it may be installed in a conduit that is smaller than that required for standard conductors of the same wire size. Compressed conductors are especially useful when increasing the ampacity of an existing service.

2.3.4 Circular Mils

A circular mil is a circle that has a diameter of 1 mil. A mil is 0.001 inch. When a wire size is 250 kcmil, the cross-sectional area of the current-

Instructor's Notes:

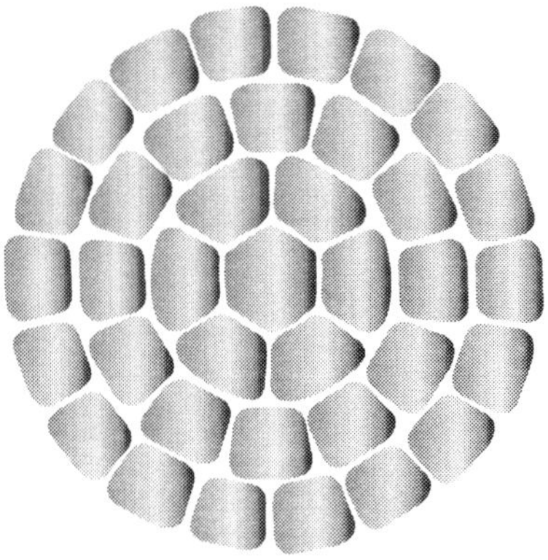

109F04.EPS

Figure 4 ◆ Compressed conductor.

carrying portion of the wire is the same as 250,000 circles having a diameter of 0.001 inch. This may seem to be a rather clumsy way of sizing wire at first; however, the alternative would be to size the wire as a function of its cross-sectional area expressed in square inches.

According to ***NEC Chapter 9, Table 8***, the cross-sectional area of a 250 kcmil conductor is 0.260 square inch.

If a conductor is to be sized by cross-sectional area, it is much easier to express the wire size in circular mils (or thousands of circular mils) than in square inches.

2.4.0 Conductor Material

The most common conductor material is copper. Copper is used because of its excellent conductivity (low resistance), ease of use, and value. The value of a material as an ingredient for wire is determined by several factors, including conductivity, cost, availability, and workability.

2.4.1 Conductivity

Conductivity is a word that describes the ease (or difficulty) of travel presented to an electric current by a conductor. If a conductor has a low resistance, it has a high conductivity. Silver is one of the best conductors since it has very low resistance and high conductivity. Copper has high conductivity and a lower price than silver. Aluminum, another material with good conductivity, is also a good choice for conductor material. The conductivity of aluminum is approximately two-thirds that of copper.

2.4.2 Cost

Cost is always an issue that contributes to the selection of a material to be used for a given application. Often, a material that has low cost may be selected as a conductor material even though it has physical properties that are inferior to the more expensive material. Such is the case in the selection of copper over platinum. Here, the cost of platinum is very high, and very little thinking is required to determine that copper is a better choice. The choice between copper and aluminum is often more difficult to make.

2.4.3 Availability

The availability of some material is often a concern when selecting components for a job. As applied to wire, the mining industry often controls the availability of raw materials, which could produce shortages of some material. The availability of a substance such as copper or aluminum affects the price of the finished product (copper or aluminum wire).

2.4.4 Workability

It is a good idea to select a material that requires less expense for tools and is easier to work with. Aluminum conductors are lighter than copper conductors of the same size. They are also much

INSIDE TRACK

Terminating Aluminum Wire

Care must be taken to use listed connectors when terminating aluminum wire. All aluminum connections also require the use of anti-oxidizing compound. Some connectors are precoated with compound; others require the addition of it. Be sure to check the connectors before beginning the installation.

Classroom

Discuss the factors involved in the selection of conductor material. Cover:

- **Conductivity**
- **Cost**
- **Availability**
- **Workability**

Stress the importance of using listed connectors with aluminum wire.

Teaching Tip

Discuss the "Think About It."

Classroom

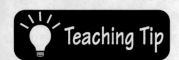

Provide an overview of the history of conductor insulation.

List the most commonly used types of thermoplastic insulation.

Discuss letter coding of insulation. Refer the trainees to the NEC Ampacity Tables.

Conductor Insulation

THINK ABOUT IT

What are the functions of conductor insulation? Under what conditions does the NEC allow uninsulated conductors?

more flexible than copper conductors and, in general, are easier to work with. However, terminating aluminum conductors often requires special tools and treatment of termination surfaces with an anti-oxidation material. Splicing and terminating aluminum conductors often requires a higher degree of training on the part of the electrician than do similar efforts with copper wire. This is partly due to the fact that aluminum expands and contracts with heat more than copper.

2.5.0 Conductor Insulation

The first attempt to insulate wire was made in the early 1800s during the development of the telegraph. This insulation was designed to provide physical protection rather than electrical protection. Electrical insulation was not an important issue because the telegraph operated at low-voltage DC. This early form of insulation was a substance composed of tarred hemp or cotton fiber and shellac and was used primarily for weatherproofing long-distance distribution lines to mines, industrial sites, and railroads.

Some early electrical distribution systems utilized the knob-and-tube technique of installing wire. The wire was often bare and was pulled between and wrapped around ceramic knobs that were affixed to the building structure. When it was necessary to pull wire through structural members, it was pulled through ceramic tubes. The structural member (usually wood) was drilled, the tube was pressed into the hole, and the wire was pulled through the hole in the tube. As dangerous as this may appear, older homes still exist that have knob-and-tube wiring that was installed in the early 1900s and is still operational. Knob-and-tube wiring was revised to use insulated conductors and was in use up to 1957 in some areas.

The grounded or neutral conductor in overhead services may be bare. Furthermore, the concentric grounded conductor in Type SE cable may be bare when used as a service-entrance cable. However, all current-carrying conductors (including the grounded conductor) must be insulated when used on the inside of buildings, or after the first overcurrent protection device.

2.5.1 Thermoplastic

Thermoplastic is a popular and effective insulation material used on conductors for the present-day market. The following thermoplastics are widely used as insulation materials:

- *Polyvinyl chloride (PVC)*—The base material used for the manufacture of TW and THW insulation.
- *Polyethylene (PE)*—An excellent weatherproofing material used primarily for insulation of control and communications wiring. It is not used for high-voltage conductors (those exceeding 5,000 volts).
- *Cross-linked polyethylene (XLP)*—An improved PE with superior heat- and moisture-resistant qualities. Used for THHN, THWN, and XHHW wiring as well as most high-voltage cables.
- *Nylon*—Primarily used as jacketing material. THHN building wire has an outer coating of nylon.
- *Teflon®*—A high-temperature insulation. Widely used for telephone wiring in a plenum (where other insulated conductors require conduit routing).

2.5.2 Letter Coding

Conductor insulation as applied to building wire is coded by letters. The letters generally, but not always, indicate the type of insulation or its environmental rating. The types of conductor insulation described in this module will be those indicated at the top of ***NEC Table 310.16.*** The various insulation designations are:

Letter	Description
B	Braid
E	Ethylene or Entrance
F	Fluorinated or Feeder
H	Heat-Rated or Flame-Retardant
N	Nylon
P	Propylene
R	Rubber
S	Silicon or Synthetic

Instructor's Notes:

T	Thermoplastic
U	Underground
W	Weather-Rated
X	Cross-Linked Polyethylene
Z	Modified Ethylene Tetrafluorethylene
TW	Weather-Rated Thermoplastic (60°C/140°F)
FEP	Fluorinated Ethylene Propylene
FEPB	Fluorinated Ethylene Propylene with Glass Braid
MI	Mineral Insulation
MTW	Moisture, Heat, and Oil-Resistant Thermoplastic
PFA	Perfluoroalkoxy
RH	Heat-Rated Rubber (75°C/167°F)
RHH	Flame-Retardant Heat-Rated Rubber
RHW	Weather-Rated, Heat-Rated Rubber (75°C/167°F)
SA	Silicon
SIS	Synthetic Heat-Resistant
TBS	Thermoplastic Braided Silicon
TFE	Extended Polytetrafluoroethylene
THHN	Heat-Resistant Thermoplastic
THHW	Moisture and Heat-Resistant Thermoplastic
THW	Moisture and Heat-Resistant Thermoset
THWN	Weather-Rated, Heat-Rated Thermoplastic with Nylon Cover
UF	Underground Feeder
USE	Underground Service Entrance
XHH	Thermoset
XHHW	Heat-Rated, Flame-Retardant, Weather-Rated Thermoset
ZW	Weather-Rated Modified Ethylene Tetrafluorethylene

2.5.3 Color Coding

A color code is used to help identify wires by the color of the insulation. This makes it easier to install and properly connect the wires. A typical color code is as follows:

- *Two-conductor cable*—One white wire, one black wire, and a grounding wire (usually bare)
- *Three-conductor cable*—One white, one black, one red, and a grounding wire
- *Four-conductor cable*—Same as three-conductor cable plus fourth wire (blue)
- *Five-conductor cable*—Same as four-conductor cable plus fifth wire (yellow)

The grounding conductor may be bare, green, or green with a yellow stripe. Color codes are shown in *Figure 5*.

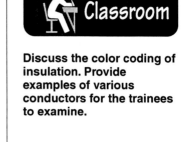

THINK ABOUT IT

Insulation Types

Use ***NEC Table 310.13*** to identify two types of insulation that are suitable for use in wet locations.

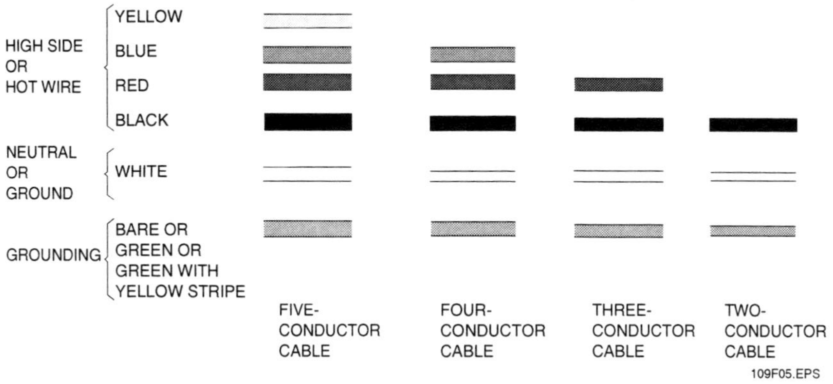

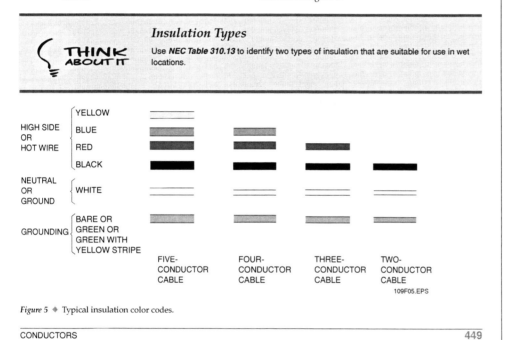

Figure 5 ◈ Typical insulation color codes.

Classroom

Discuss the color coding of insulation. Provide examples of various conductors for the trainees to examine.

Audiovisual

Show Transparency 7 (Figure 5).

Teaching Tip

Discuss the "Think About It."

CONDUCTORS 449

Classroom

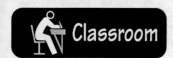

Discuss temperature ratings and correction factors.

Discuss the NEC requirements for fixture wires.

Audiovisual

Show Transparency 8 (Figure 6).

Teaching Tip

Discuss the "Think About It."

Safety

Stress the importance of verifying all conductors using a voltmeter.

The NEC does not require color coding of ungrounded conductors in many cases. The ungrounded conductors may be any color with the exception of white, grey, or green; however, it is a good practice to color code conductors as described here. In fact, many construction specifications require color coding. Furthermore, on a four-wire, delta-connected secondary where the midpoint of one phase is grounded to supply lighting and similar loads, the phase conductor having the higher voltage to ground must be identified by an outer finish that is orange in color, by tagging, or by other effective means. Such identification must be placed at each point where a connection is made if the grounded conductor is also present. In most cases, orange tape is used at all termination points when such a condition exists.

2.5.4 Wire Ratings

Conductor selection is based largely on the temperature rating of the wire. This requirement is extremely important and is the basis of safe operation for insulated conductors. As shown in ***NEC Table 310.13,*** conductors have various ratings (60°C, 75°C, 90°C, etc.). Since ***NEC Tables 310.16 through 310.19*** are based on an assumed ambient temperature of 30°C (86°F), conductor ampacities are based on the ambient temperature plus the heat (I^2R) produced by the conductor while carrying current. Therefore, the type of insulation used on the conductor is the first consideration in determining the maximum permitted conductor ampacity.

For example, a No. 3/0 THW copper conductor for use in a raceway has an ampacity of 200 according to ***NEC Table 310.16.*** In a 30°C ambient temperature, the conductor is subjected to this temperature when it carries no current. Since a THW-insulated conductor is rated at 75°C, this leaves 45°C (75 − 30) for increased temperature due to current flow. If the ambient temperature exceeds 30°C, the conductor maximum load-current rating must be reduced proportionally (see *Correction Factors* at the bottom of ***NEC Table 310.16***) so that the total temperature (ambient plus conductor temperature rise due to current flow) will not exceed the temperature rating of the conductor insulation (60°C, 75°C, etc.). For the same reason, the allowable ampacity must be reduced when more than three conductors are contained in a raceway or cable. See ***NEC Section 310.15(B).***

Using the ampacity tables—An important step in circuit design is the selection of the type of conductor to be used (TW, THW, THWN, RHH, THHN, XHHW, etc.). The various types of conductors are covered in ***NEC Article 310,*** and the ampacities of conductors are given in ***NEC Tables 310.16 through 310.19*** for the varying conditions of use (e.g., in a raceway, in open air, at normal or higher-than-normal ambient temperatures). Conductors must be used in accordance with the data in these tables and notes.

2.6.0 Fixture Wires

Fixture wire is used for the interior wiring of fixtures and for wiring fixtures to a power source. Guidelines concerning fixture wire are given in ***NEC Article 402.*** The list of approved types of fixture wire is given in ***NEC Table 402.3.*** *Figure 6* shows one example of fixture wire. The wires are composed of insulated conductors with or without an outer jacket. The conductors range in size from No. 18 to No. 10 AWG.

THINK ABOUT IT

Color Coding Ungrounded Conductors

Although the NEC does not require the use of color-coded ungrounded current-carrying conductors, why might it be a good idea to use them anyway?

INSIDE TRACK

Color Coding

Color designations are good indicators, but never trust your life to them. Always protect yourself by testing circuits with a voltmeter.

Instructor's Notes:

450 CHAPTER 11

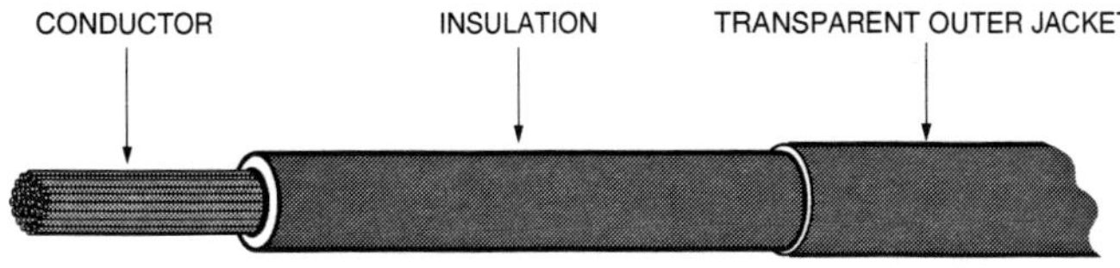

Figure 6 ◈ Fixture wire.

The decision of which fixture wire to use depends primarily upon the operating temperature that is expected within the fixture. Therefore, it is the character of the insulation that will determine the wire selected. For instance, fixture wires insulated with perfluoroalkoxy (PAF) or extruded polytetrafluorethylene (PTF) would be selected if the operating temperature of the fixture is expected to reach a maximum of 482°F. This is the highest operating temperature allowed for any fixture wire.

As indicated by *NEC Section 402.3,* fixture wires are suitable for service at 600 volts unless otherwise specified in *NEC Table 402.3.* The allowed ampacities of fixture wire are given in *NEC Table 402.5.*

Although the primary use for fixture wire is the internal wiring of fixtures, several of the wires listed in *NEC Table 402.3* may be used for wiring remote-control, signaling, or power-limited circuits in accordance with *NEC Section 725.15.* Fixture wires may never be used as substitutes for branch circuit conductors.

2.6.1 Heating Cables

Heating cables are another important group of special wires that are commonly used in the building industry. The purpose of these cables is to produce heat when energized. There are three general categories of heating cables:

- Interior space-heating cables
- De-icing and snow-melting cables
- Pipeline and vessel heating cables

These heating cables consist of a long insulated wire, cabled heating wire, or resistance wire, which produces heat in the wire when it is connected to a power source. The ends of the resistance wire are connected to non-heating lead wires, which extend to a thermostat or junction box.

Guidelines pertaining to heating cable used for interior space heating are given in *NEC Article 424.* The NEC considers the term *heating equipment* to include any factory-manufactured heating devices such as unit heaters, boilers, and local heating systems, as well as heating cables. Heating cables are specifically covered in *NEC Article 424, Part V.*

Heating cables meant for installation in ceilings are available from the factory in unit lengths from 75 to 1,800 feet, capacities ranging from 200 to 5,000 watts, and voltage ratings of 120, 208, and 240 volts. The cables are usually rated at 2¾ watts per linear foot. The nonheating lead wires are at least seven feet in length and are color coded to indicate the voltage rating. The insulation on the heating wire is designed to be resistant to high temperatures, water absorption, aging, and chemical action.

Heating cables meant for installation in floors have capacities that vary with the cable dimensions and heating wire resistance. They are typically rated at voltages from 120 to 600 volts. There are two popular floor heating cables. In the first type, the resistance wire is covered with a sheath of polyvinyl chloride (PVC). In the second, the resistance wire is encased in a mineral insulation and then covered with a copper sheath. This construction is identical to that of Type MI (mineral-insulated, metal-sheathed cable). In contrast to ceiling heating cables, which are available in specific unit lengths, these heating cables are sold in random lengths, which must be cut and properly terminated at the job site according to the job's heating requirements.

NEC Article 424, Part V specifies guidelines for installing floor and ceiling heating cables; it should be read carefully. Generally, ceiling heating cables are stapled to the ceiling with at least 1½ inches between adjacent runs. According to the NEC, the cables must not come within eight inches of recessed lighting fixtures or within six inches of other metallic materials in the ceiling. Floor heating cables are usually embedded in concrete, with at least a one-inch space between adjacent runs of cable. Spacing must also be maintained between the cable and other metallic

Classroom

Discuss the three classes of heating cables.

Have the trainees locate the NEC requirements for each type of heating cable.

CONDUCTORS 451

Classroom

Discuss cable classifications and markings.

Audiovisual

Show Transparency 9 (Figure 7A).

Classroom

Discuss Types NM and NMC cable and pass around examples for the trainees to examine.

Audiovisual

Show Transparency 10 (Figure 8A).

> **INSIDE TRACK**
>
> ***Cable Selection***
>
> There are two factors to be considered when determining the type of cable to be used for a specific application: the type of conductor insulation and the cable jacket. Both must be appropriate for the application.

objects in the floor, unless a grounded metal-clad cable is used.

Guidelines pertaining to de-icing and snow-melting cables are given in *NEC Article 426.* These cables are similar in construction to interior floor heating cables in that they consist of resistance wires insulated with a sheath of PVC or with mineral insulation and a copper sheath. Most de-icing and snow-melting cables are installed by embedding them in concrete or asphalt. As with floor heating cables, these cables are rated at various voltages and must be cut to the desired length.

Guidelines pertaining to pipeline and vessel heating cables are given in *NEC Article 427.* These cables are generally available in two types of construction. The cable can either be insulated resistance wire with nonheating lead wires attached to each end, or it can be in the form of a flat heating tape. The tape consists of a single piece of heating wire doubled over so that the halves of the wire run parallel to each other. The wire is encased in a plastic enclosure, thus producing an assembly that resembles a tape. The nonheating lead wires are attached to the open end of the tape.

The obvious purpose of pipeline and vessel heating cables is to prevent the fluid in the pipeline or vessel from falling below a specified minimum temperature. Flat heating tapes are often used in buildings under construction or not completely enclosed. They can be used on water pipes to keep the water from freezing in cold weather.

2.7.0 Cables

Cables are two or more insulated wires and may contain a grounding wire covered by an outer jacket or sheath. Cable is usually classified by the type of covering it has, either nonmetallic (i.e., plastic) or metallic, also called *armored cable.*

Cable may also be classified according to where it can be used (see *NEC Table 400.4*). Because water is such a good conductor of electricity, moisture on conductors can cause power loss or short circuits. For this reason, cables are classified for either dry, damp, or wet locations. Cables can also be classified regarding exposure to sunlight and rough use.

2.7.1 Cable Markings

All cables are marked to show important properties and uses. Cable markings show the wire size, number of conductors, cable type, and voltage rating. In addition, a marking may be included to signify approved service or applications. This information is printed on nonmetallic cable (*Figure 7*). On metallic cable, marking information is usually included on a tag.

2.7.2 Nonmetallic-Sheathed Cable

Nonmetallic-sheathed cable (Type NM and Type NMC) is widely used for branch circuits and feeders in residential and commercial systems. See *Figure 8* on page 9.12. Both types are commonly called *Romex,* even though the cable manufacturer only calls Type NM cable Romex. Guidelines for the use of nonmetallic-sheathed cable are given in *NEC Article 334.* This cable consists of two or three insulated conductors and one bare conductor enclosed in a nonmetallic sheath. The conductors may be wrapped individually with paper, and the spaces between the conductors may be filled with jute, paper, or other material to protect the conductors and help the cable keep its shape. The sheath covering both Type NM cable and Type NMC cable is flame-retardant and moisture-resistant. The sheath covering Type NMC cable has the additional characteristics of being fungus- and corrosion-resistant.

NEC Article 334 lists the allowed and prohibited uses for Type NM cable and Type NMC cable. Both are allowed to be installed in either exposed or concealed work. The primary difference in their applied uses is that Type NM cable is suitable for dry locations only, whereas Type NMC is permitted for dry, moist, damp, or corrosive locations. Since you will probably work with these two cables often, read *NEC Article 334* carefully.

Guidelines for the use of Type SE (service-entrance) cable are given in *NEC Article 338.* The NEC contains no specifications for the construc-

452 CHAPTER 11

Instructor's Notes:

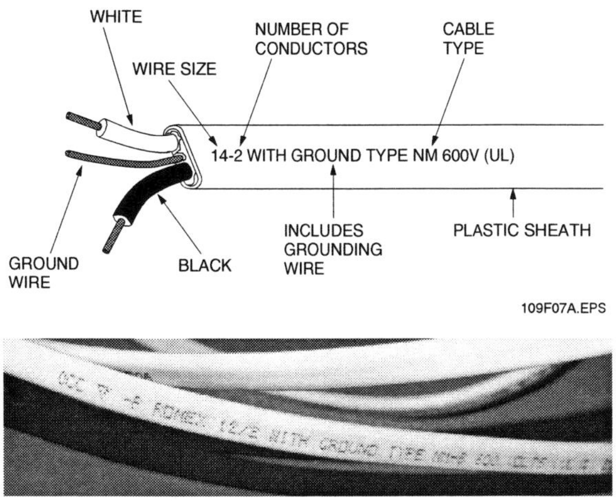

Figure 7 ◈ Nonmetallic cable markings.

tion of this cable; it is left to UL to determine what types of cable should be approved for this purpose. Currently, service-entrance cable is labeled in sizes No. 12 AWG and larger for copper, and No. 10 AWG and larger for aluminum or copper-clad aluminum, with Types RH, RHW, RHH, or XHHW conductors. If the type designation for the conductor is marked on the outside surface of the cable, the temperature rating of the cable corresponds to the rating of the individual conductor. When this marking does not appear, the temperature rating of the cable is 75°C (167°F). Type SE cable is for above-ground installation.

When used as a service-entrance cable, Type SE must be installed as specified in ***NEC Article 230.*** Service-entrance cable may also be used as feeder and branch circuit cable. Guidelines for the use of service-entrance cable are given in ***NEC Section 338.10.***

2.7.3 Type UF Cable

Guidelines for the use of Type UF (underground feeder and branch circuit) cable are given in ***NEC Article 340.*** Type UF cable is very similar in appearance, construction, and use to Type NMC cable. The main difference between these two cables is that Type UF cable is suitable for direct burial, whereas Type NMC cable is not.

2.7.4 Type NMS Cable

Refer to ***NEC Sections 334.10 and 334.12*** for the applications of Type NMS cable. Type NMS cable is a form of nonmetallic-sheathed cable that contains a factory assembly of power, communications, and signaling conductors enclosed within a moisture-resistant, flame-retardant sheath.

2.7.5 Type MV Cable

Type MV (medium-voltage) cable is covered in ***NEC Article 328.*** It consists of one or more insulated conductors encased in an outer jacket. This cable is suitable for use with voltages ranging from 2,001 to 35,000 volts. It may be installed in wet and dry locations and may be buried directly in the earth.

2.7.6 High-Voltage Shielded Cable

Shielding of high-voltage cables protects the conductor assembly against surface discharge or burning due to corona discharge in ionized air, which can be destructive to the insulation and jacketing.

Electrostatic shielding of cables makes use of both nonmetallic and metallic materials (*Figures 9* and *10* on pages 455 and 456).

Classroom

Discuss Type SE cable and pass around an example for the trainees to examine.

Discuss Type UF cable and pass around an example for the trainees to examine.

Discuss Type NMS cable and pass around an example for the trainees to examine.

Discuss Type MV cable and pass around an example for the trainees to examine.

Discuss high-voltage shielded cable and pass around an example for the trainees to examine.

Audiovisual

Show Transparencies 11 and 12 (Figures 9A and 10).

CONDUCTORS 453

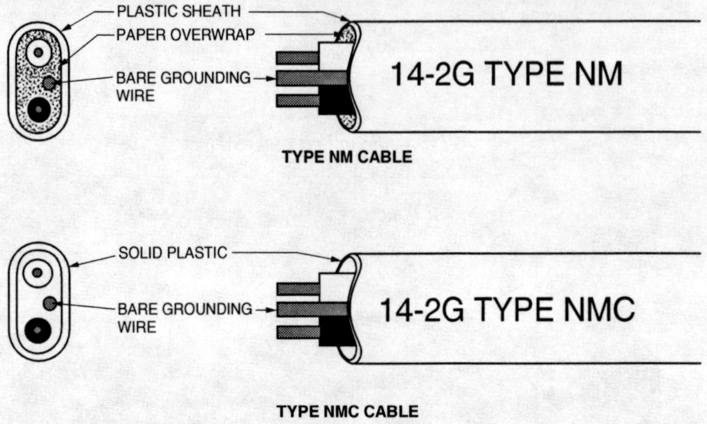

109F08A .EPS

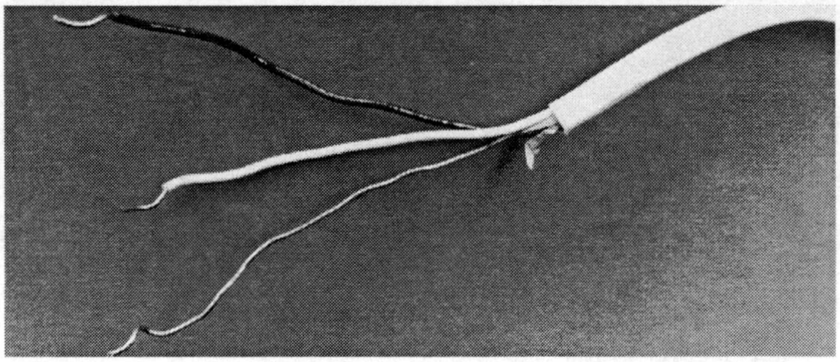

109F08B.EPS

Figure 8 ◆ Nonmetallic-sheathed cable.

Classroom

Discuss Type FC wiring systems and their components. Pass around example components for the trainees to examine.

Audiovisual

Show Transparencies 13 and 14 (Figures 11 and 12).

2.7.7 Channel Wire Assemblies

Channel wire assemblies (Type FC) comprise an entire wiring system, which includes the cable, cable supports, splicers, circuit taps, fixture hangers, and fittings (*Figure 11* on page 457). Guidelines for the use of this system are given in ***NEC Article 322.*** Type FC cable is a flat cable assembly with three or four parallel No. 10 special stranded copper conductors. The assembly is installed in an approved U-channel surface metal raceway with one side open. Tap devices can be inserted anywhere along the run. Connections from the tap devices to the flat cable assembly are made by pin-type contacts when the tap devices are fastened in place. The pin-type contacts penetrate the insulation of the cable assembly and contact the multi-stranded conductors in a matched phase sequence. These taps can then be wired to lighting fixtures or power outlets (*Figure 12* on page 458).

As indicated in ***NEC Section 322.10,*** this wiring system is suitable for branch circuits that only supply small appliances and lights. This system is suitable for exposed wiring only and may not be concealed within the building structure. It is ideal for quick branch circuit wiring at field installations.

2.7.8 Flat Conductor Cable

Type FCC (flat conductor) cable comprises an entire branch wiring system similar in many

Instructor's Notes:

454 CHAPTER 11

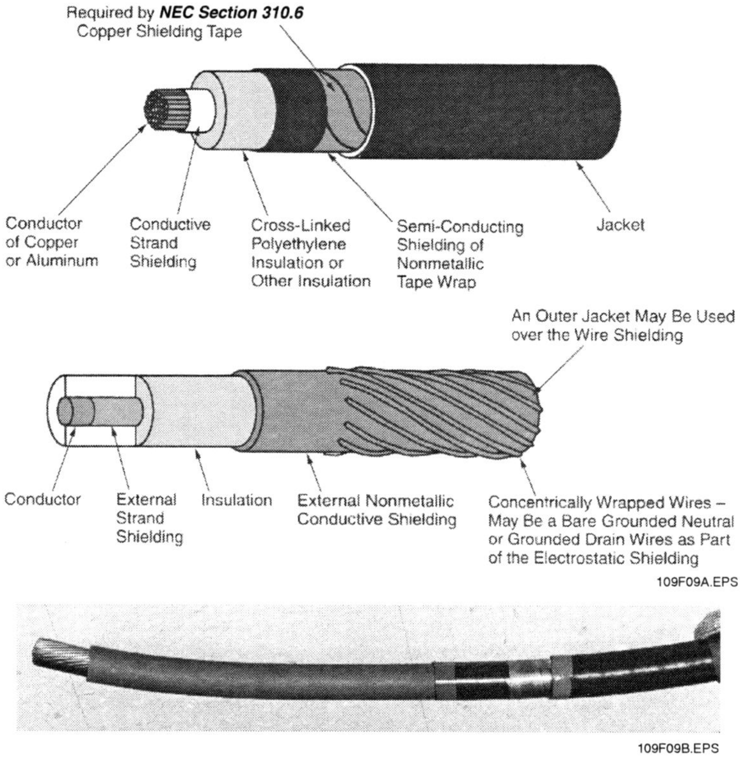

Figure 9 ◆ Metallic shielding.

respects to Type FC flat conductor assemblies. Guidelines for the use of this system are given in *NEC Article 324.* Type FCC cable consists of three to five flat conductors placed edge-to-edge, separated, and enclosed in a moisture-resistant and flame-retardant insulating assembly. Accessories include cable connectors, terminators, power source adapters, and receptacles.

This wiring system has been designed to supply floor outlets in office areas and other commercial and institutional interiors. It is meant to be run under carpets so that no floor drilling is required. This system is also suitable for wall mounting. As indicated in *NEC Article 324,* telephone and other communications circuits may share the same enclosure as Type FCC flat cable. The main advantage of the system is its ease of installation. It is the ideal wiring system for use when remodeling or expanding existing office facilities.

2.7.9 Type TC Cable

Guidelines for the use of Type TC (power and control tray) cable are given in *NEC Article 336.* Type TC cable consists of two or more insulated conductors twisted together, with or without associated bare or fully insulated grounding conductors, and covered with a nonmetallic jacket. The cables are rated at 600 volts. The cable is listed in conductor sizes No. 18 AWG to 2,000 kcmil copper or No. 12 AWG to 2,000 kcmil aluminum or copper-clad aluminum (*Figure 13* on page 458).

As the *T* in the letter designator indicates, this cable is tray cable. It can be used in cable trays and raceways. It may also be buried directly if the

Classroom

Discuss Types FC and FCC cable. Pass around examples for the trainees to examine.

Discuss Type TC cable. Pass around an example for the trainees to examine.

CONDUCTORS 455

Classroom

Discuss Type USE cable. Pass around an example for the trainees to examine.

Audiovisual

Show Transparency 15 (Figure 14).

Classroom

Discuss instrumentation control wiring. Pass around examples for the trainees to examine.

Audiovisual

Show Transparency 16 (Figure 15A).

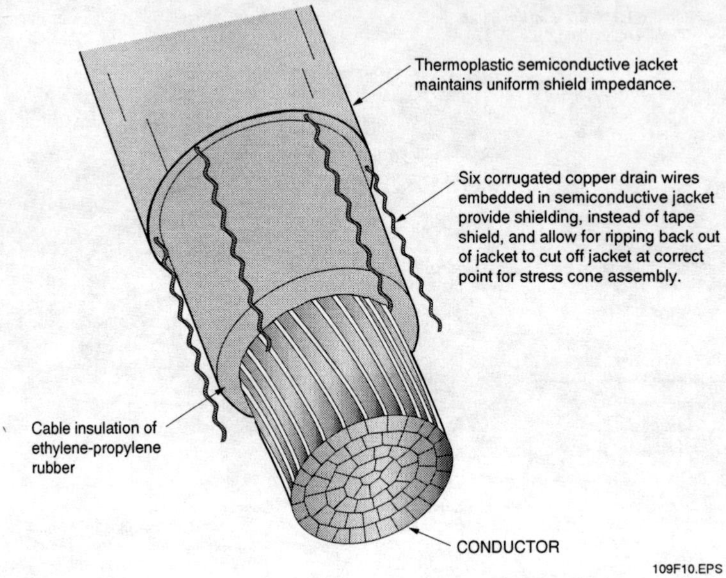

Thermoplastic semiconductive jacket maintains uniform shield impedance.
Six corrugated copper drain wires embedded in semiconductive jacket provide shielding, instead of tape shield, and allow for ripping back out of jacket to cut off jacket at correct point for stress cone assembly.
Cable insulation of ethylene-propylene rubber
CONDUCTOR
109F10.EPS

Figure 10 ◆ Nonmetallic shielding.

sheathing material is suitable for this use. Type TC cable is also good for use in sunlight when indicated by the cable markings.

2.7.10 Type USE Cable

Type USE cable is for underground installation including burial directly in the earth. Type USE cable in sizes No. 4/0 AWG and smaller with all conductors insulated is suitable for all of the underground uses for which Type UF cable is permitted by the NEC.

Type USE cable may consist of either single conductors or a multi-conductor assembly provided with a moisture-resistant covering, but it is not required to have a flame-retardant covering. This type of cable may have a bare copper conductor cabled with the assembly. Furthermore, Type USE single, parallel, or cabled conductor assemblies recognized for underground use may have a bare copper concentric conductor applied. These constructions do not require an outer overall covering. Guidelines for the use of Type USE cable are specified in ***NEC Article 338***. See *Figure 14* on page 459.

When used as a service-entrance cable, Type USE cable must be installed as specified in ***NEC Article 230***. Take the time to read ***NEC Article 230*** to ensure proper installation. Type USE service-entrance cable may also be used as feeder and branch circuit cable. Guidelines for this use of service-entrance cable are given in ***NEC Section 338.10.***

2.8.0 Instrumentation Control Wiring

Instrumentation control wiring links the field-sensing, controlling, printout, and operating devices that form an electronic instrumentation control system. The style and size of instrumentation control wiring must be matched to a specific job.

Instrumentation control wiring usually has two or more insulated conductor wires. These wires may also have a shield and a ground wire. An outer layer called the *jacket* protects the wiring (*Figure 15* on page 460). Instrumentation conductor wires come in pairs. The number of pairs in a multi-conductor cable depends on the size of the wire used. A multi-pair cable may have as many as 60 pairs of conductor wires.

2.8.1 Shields

Shields are provided on instrumentation control wiring to protect the electrical signals traveling

Instructor's Notes:

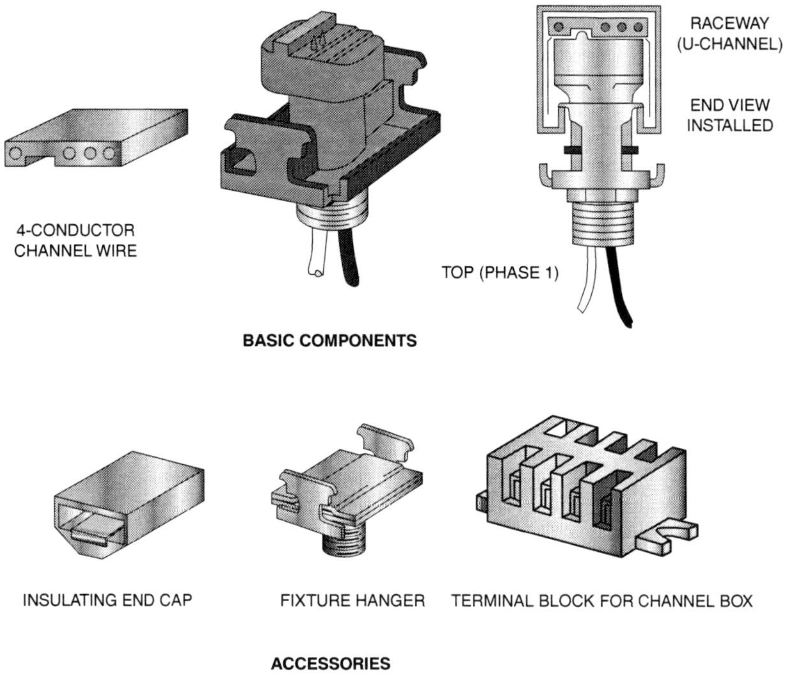

Figure 11 ◆ Channel wire components and accessories.

through the conductors from electrical interference or noise. Shields are usually constructed of aluminum foil bonded to a plastic film (*Figure 16*). If the wiring is not properly shielded, electrical noise may cause erratic or erroneous control signals, false indications, and improper operation of control devices.

2.8.2 Shield Grounding

A shield ground wire is a bare copper wire used to provide continuous contact with a specified grounding terminal. A shield ground wire allows connection of all the instruments within a loop to a common grounding system. In some electronic systems, the grounding wire is called a *drain wire*. Always refer to the loop diagram to determine whether or not the ground wire is to be terminated.

Usually, instrumentation is not grounded at both ends of the wire. This is to prevent unwanted ground loops in the system. If the ground is not to be connected at the end of the wire you are installing, do not remove the ground wire. Fold it back and tape it to the cable. This is called *floating the ground*. This is done in case the ground at the other end ever develops a problem. In that case, that ground can be removed and the wire taped to the cable can be untaped and installed.

2.8.3 Jackets

A plastic jacket covers and protects the components within the wire. Polyethylene (PE) and polyvinyl chloride (PVC) jackets are the most commonly used (*Figure 17* on page 461). Some jackets have a nylon rip cord that allows the jacket to be peeled back without the use of a knife or cable cutter. This eliminates nicking of the conductor insulation when preparing for termination.

Classroom

Discuss shields, ground wires, and jackets.

Audiovisual

Show Transparency 17 (Figure 16).

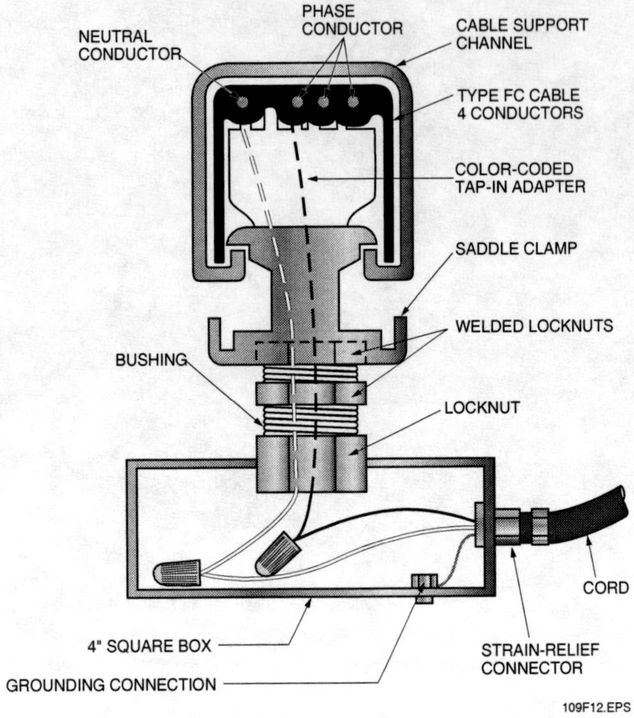

Figure 12 ◆ Type FC connection.

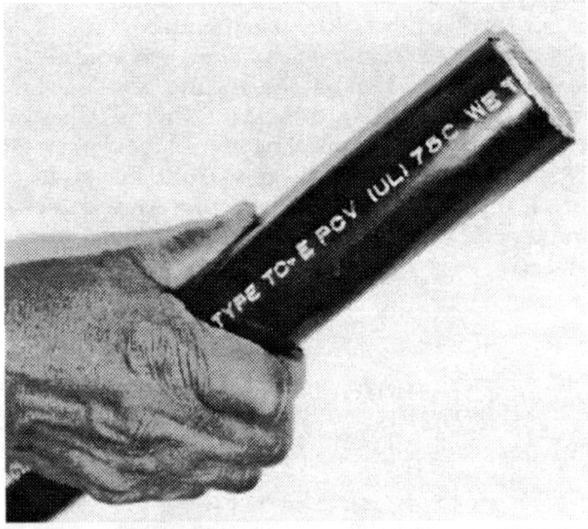

Figure 13 ◆ Type TC cable.

3.0.0 ◆ INSTALLING CONDUCTORS IN CONDUIT SYSTEMS

Conductors are installed in all types of conduit by pulling them through the conduit. This is done by using **fish tape,** pull lines, and pulling equipment.

3.1.0 Fish Tape

Fish tape can be made of flexible steel or nylon and is available in coils of 25 to 200 feet. It should be kept on a reel to avoid twisting. Fish tape has a hook or loop on one end to attach to the conductors to be pulled (*Figure 18* on page 461). Broken or damaged fish tape should not be used. To prevent electrical shock, fish tape should not be used near or in live circuits.

Fish tape is fed through the conduit from its reel. The tape usually enters at one outlet or junction box and is fed through to another outlet or junction box (*Figure 19* on page 462).

Classroom

Discuss types of fish tape and pass around examples for the trainees to examine.

Audiovisual

Show Transparencies 18 and 19 (Figures 18 and 19).

Instructor's Notes:

458 CHAPTER 11

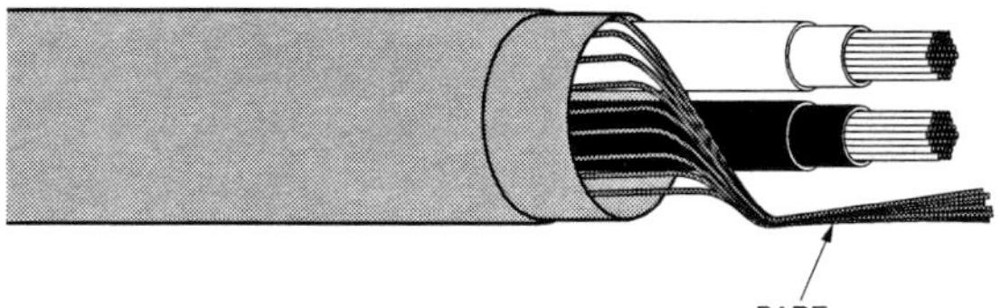

Figure 14 ◈ Type USE cable.

INSIDE TRACK

Type MC Cable

Metal clad cable (Type MC) is a type of cable that is widely used in both commercial and industrial environments. It is available in many configurations, with or without an outer jacket. Some of the special applications of MC cable include homerun cables, super neutrals, direct burial, fire alarm cable, and wiring in healthcare facilities.

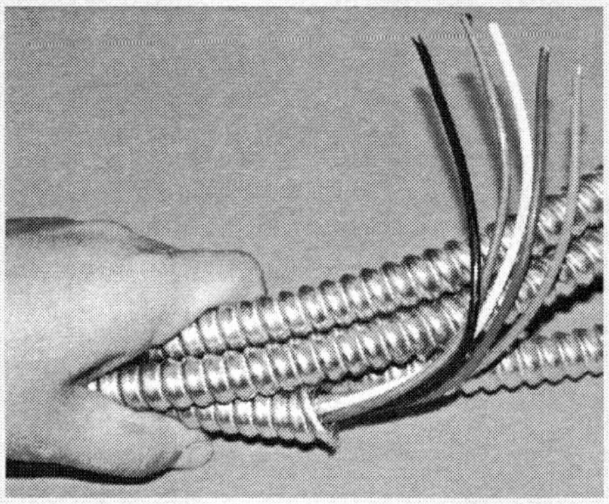

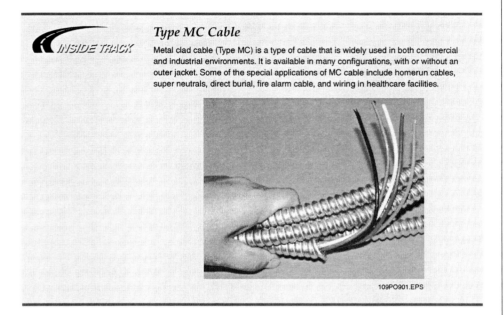

3.1.1 Power Conduit Fishing Systems

String lines can be installed by using different types of power systems. The power system is similar to an industrial vacuum cleaner and pulls a string or rope attached to a piston-like plug (sometimes called a **mouse**) through the conduit. Once the string emerges at the opposite end, either the conductor or a pull rope is then attached and pulled through the conduit, either manually or with power tools. See *Figure 20* on page 463.

The hose connection on these vacuum systems can also be reversed to push the mouse through the conduit as shown in *Figure 21* on page 463. In other words, the system can either suck or blow the mouse through the conduit, depending on which method is best in a given situation. In either case, a fish tape is then attached to the string for retrieving through the conduit.

3.1.2 Connecting Wire to a String Line

Once the string is installed in the conduit run, a fish tape is connected to it and pulled back through the conduit. Conductors are then attached to the hooked end of the fish tape or else

Classroom

Discuss power fishing systems.

Audiovisual

Show Transparencies 20 and 21 (Figures 20C and 21).

Demonstration

Demonstrate how to connect a wire to a string line.

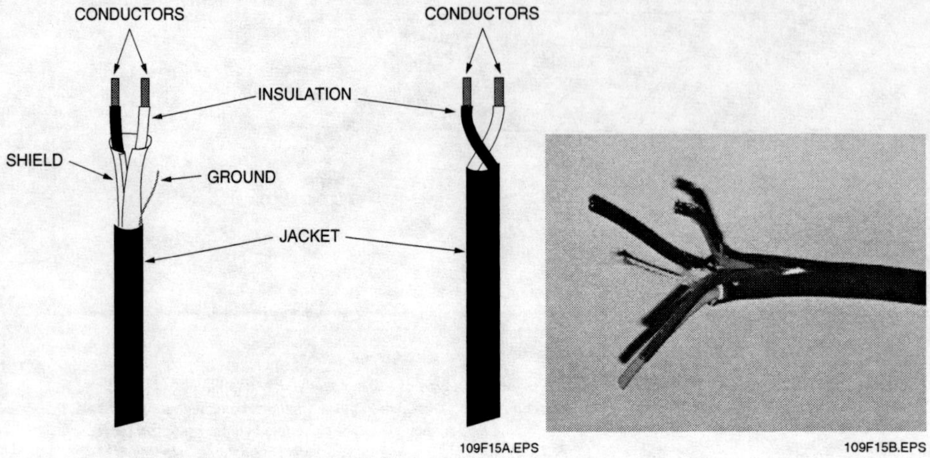

Figure 15 ◆ Instrumentation control wiring.

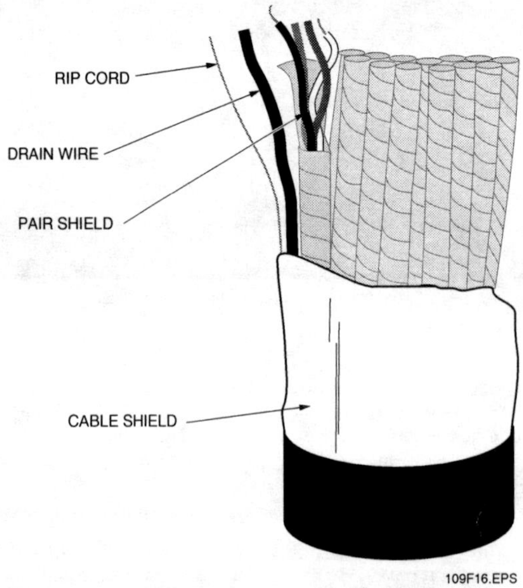

Figure 16 ◆ Multi-conductor instrumentation control wiring with shields.

Instructor's Notes:

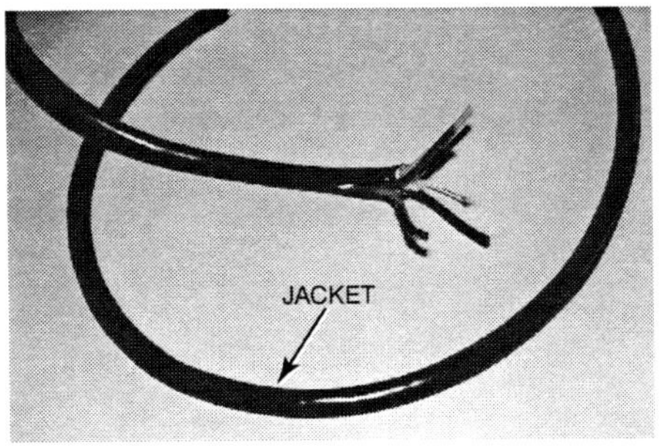

Figure 17 ◆ Wire jacket.

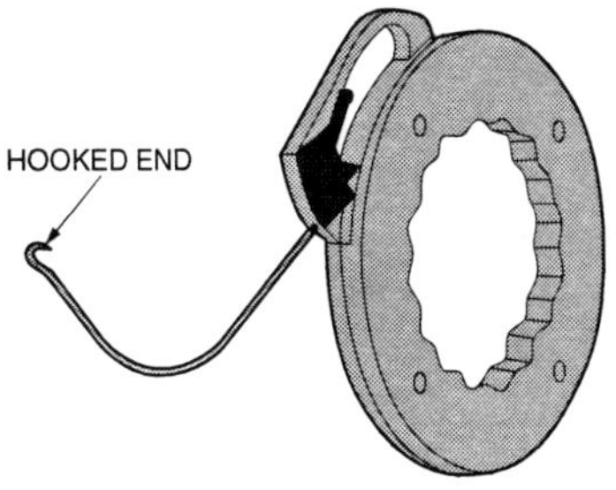

Figure 18 ◆ Fish tape.

INSIDE TRACK

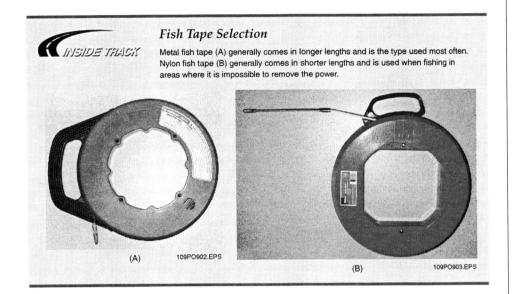

Fish Tape Selection

Metal fish tape (A) generally comes in longer lengths and is the type used most often. Nylon fish tape (B) generally comes in shorter lengths and is used when fishing in areas where it is impossible to remove the power.

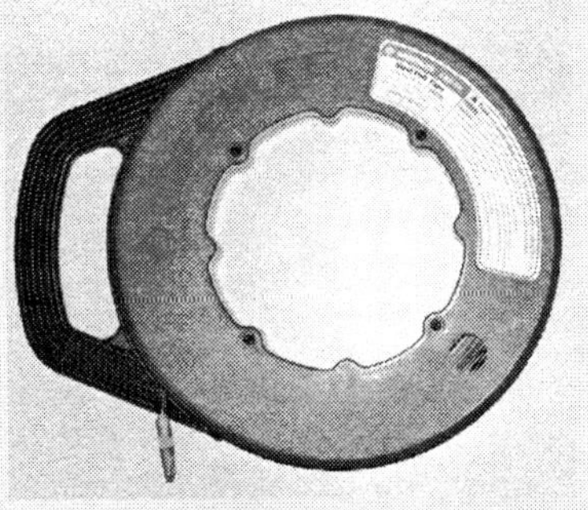

(A)

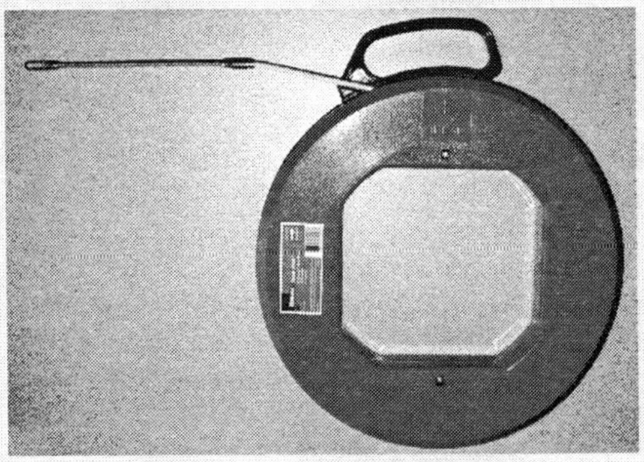

(B)

connected to a basket grip. In most cases, all required conductors are pulled at one time.

3.2.0 Wire Grips

Wire grips are used to attach the cable to the pull tape. One type of wire grip used is a basket grip (sometimes called *Chinese Fingers*). A basket grip is a steel mesh basket that slips over the end of a large wire or cable (*Figure 22* on page 464). The fish tape hooks onto the end and the pull on the fish tape tightens the basket over the conductor.

3.3.0 Pull Lines

After the tape has been inserted into the conduit run, you must determine if the pull on the wire will be easy or difficult. If the pull is going to be difficult because of bends in the conduit or the

Classroom

Discuss the purpose of wire grips and pass around examples for the trainees to examine.

Discuss when a pull line should be used.

CONDUCTORS 461

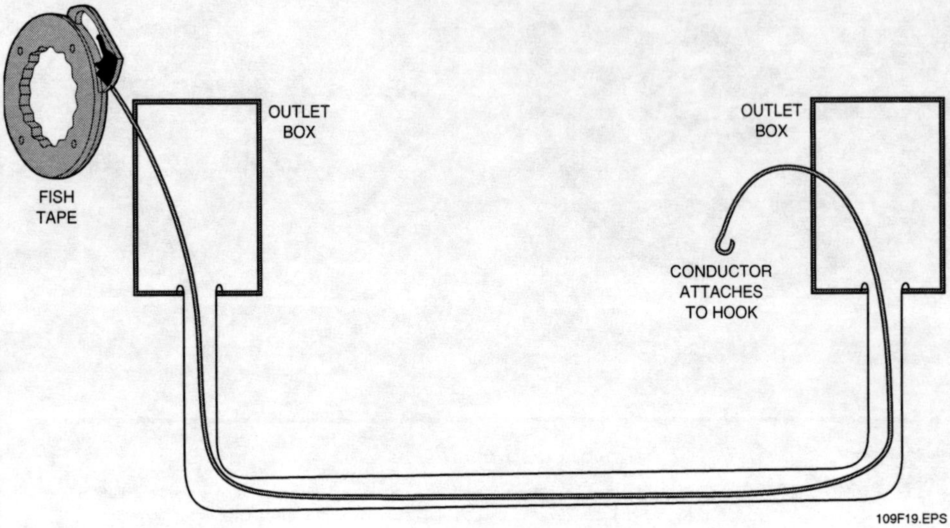

Figure 19 ◆ Fish tape installation.

size of the conductors, or if several conductors are to be pulled together, a pull line should be used.

A pull line is usually made of nylon or some other synthetic fiber. It is made with a factory-spliced eye for easy connection to fish tape or conductors.

Safety

Review the safety precautions associated with pulling cable.

> **WARNING!**
> When using pull lines, exercise extreme caution and *never* stand in a direct line with the pulling rope. If the rope breaks, the line will whip back with great force. This can result in serious injury or death.

3.4.0 Safety Precautions

The following are several important safety precautions that will help to reduce the chance of being injured while pulling cable.

- To avoid electrical shock, never use fish tape near or in live circuits. If wires must be pulled into boxes that contain live circuits, use rubber blankets over the exposed live circuitry.
- Read and understand both the operating and safety instructions for the pull system before pulling cable.
- When moving reels of cable, avoid back strain by using your legs to lift (rather than your back) and asking for help with heavy loads. Also, when manually pulling wire, spread your legs to maintain your balance and do not stretch.
- Select a rope that has a pulling load rating greater than the estimated forces required for the pull.
- Use only low-stretch rope such as multiplex and double-braided polyester for cable pulling. High-stretch ropes store energy much like a stretched rubber band. If there is a failure of the rope, pulling grip, conductors, or any other component in the pulling system, this potential energy will suddenly be unleashed. The whipping action of a rope can cause considerable damage, serious injury, or death.
- Inspect the rope thoroughly before use. Make sure there are no cuts or frays in the rope. Remember, the rope is only as strong as its weakest point.
- When designing the pull, keep the rope confined in conduit wherever possible. Should the rope break or any other part of the pulling system fail, releasing the stored energy in the rope,

Instructor's Notes:

462 CHAPTER 11

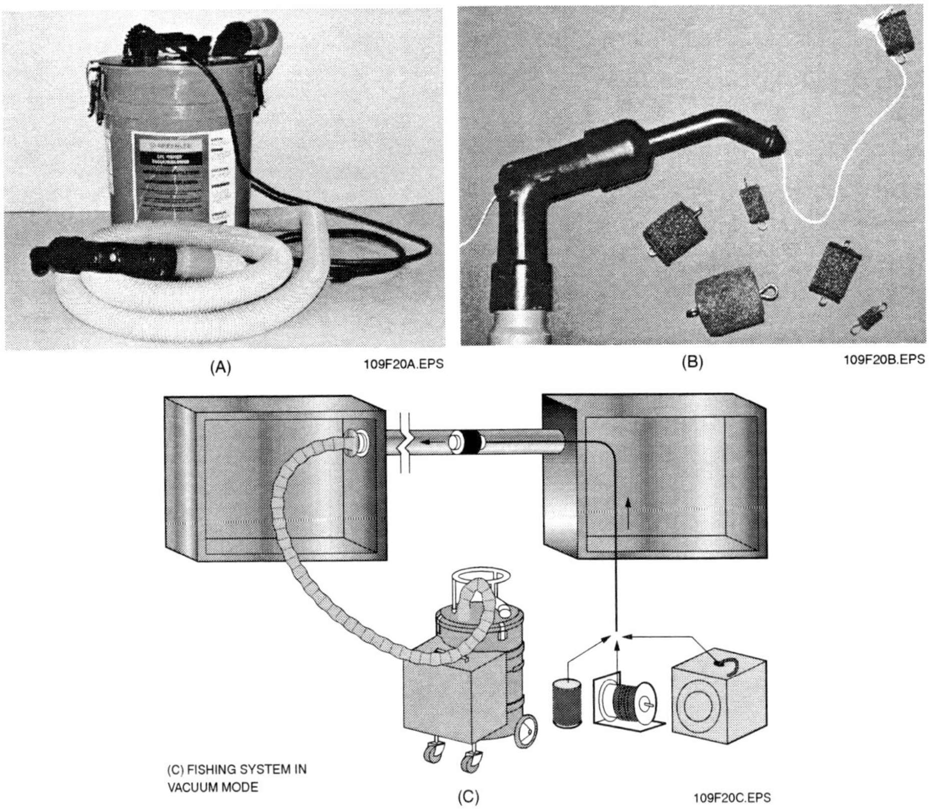

Figure 20 ◈ Power fishing system: (A) vacuum/blower unit; (B) foam plugs; (C) fishing system in vacuum mode.

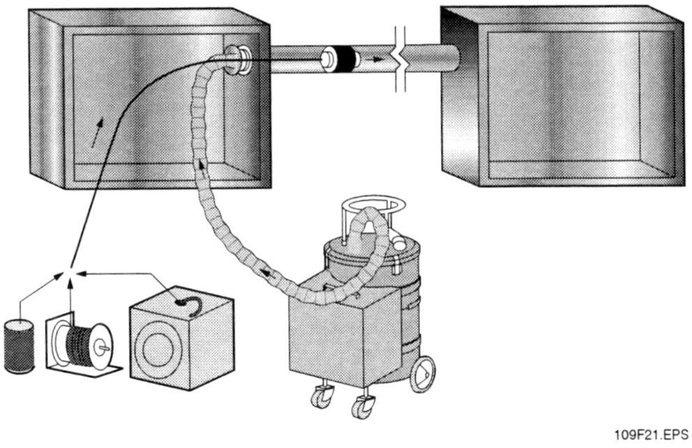

Figure 21 ◈ Power fishing system (blower mode).

463

Classroom

Explain how to straighten a bent fish tape using a nail slalom.

Audiovisual

Show Transparency 22.

Classroom

Discuss the applications of manual and electric wire pullers.

Audiovisual

Show Transparencies 23 through 25 (Figures 23 through 25).

Classroom

Show manual and electric power pullers and point out their components and features.

INSIDE TRACK

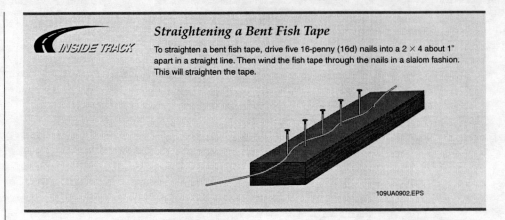

Straightening a Bent Fish Tape

To straighten a bent fish tape, drive five 16-penny (16d) nails into a 2 × 4 about 1" apart in a straight line. Then wind the fish tape through the nails in a slalom fashion. This will straighten the tape.

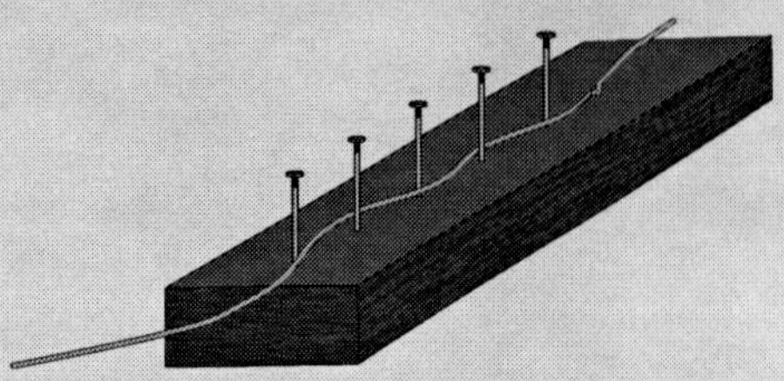

109UA0902.EPS

109F22.EPS

Figure 22 ◆ Basket grip.

the confinement in the conduit will work against the whipping action of the rope by playing out much of this energy within the conduit.

- Do not stand in a direct line with the pulling rope.
- Wrap up the pulling rope after use to prevent others from tripping over it.

3.5.0 Pulling Equipment

Many types of pulling equipment are available to help pull conductors through conduit. Pulling equipment can be operated both manually and electrically. A manually operated puller is used mainly for smaller pulling jobs where hand pulling is not possible or practical (*Figure 23*). It is also used in many locations where hand pulling would put an unnecessary strain on the conductors because of the angle of the pull involved.

Electrically driven power pullers are used where long runs, several bends, or large conductors are involved (*Figure 24*).

The main parts of a power puller are the electric motor, the chain or sprocket drive, the **capstan,** the sheave, and the pull line.

109F23.EPS

Figure 23 ◆ Manual wire puller.

The pull line is routed over the sheave to ensure a straight pull. The pull line is wrapped around the capstan two or three times to provide a good grip on the capstan. The capstan is driven by the electric motor and does the actual pulling. The pull line is unwound by hand at the same speed at which the capstan is pulling. This eliminates the need for a large spool on the puller to wind the pull line.

Attachments to power pullers, such as special application sheaves and extensions, are available for most pulling jobs (*Figure 25*). Follow the manufacturer's instructions for setup and operation of the puller.

Instructor's Notes:

464 CHAPTER 11

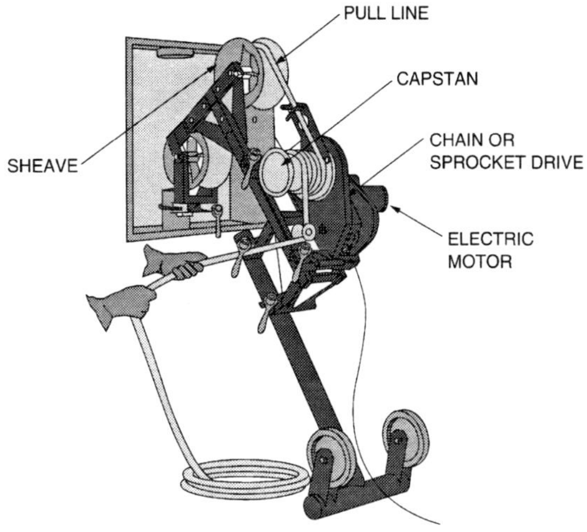

Figure 24 ◆ Power puller.

INSIDE TRACK

Replacing the Hook on a Metal Fish Tape

The hook on a fish tape can be replaced using a propane torch. IMPORTANT: Always work in an appropriate environment and wear the necessary protective equipment. Hold the fish tape securely in a pair of pliers, heat the end with a propane torch until it softens, then use a second set of pliers to form a hook in the fish tape. Allow it to air cool.

CAUTION

Before using power pullers, a qualified person must verify the amount of pull or tension that can be withstood by the conductors being pulled.

3.6.0 Feeding Conductors into Conduit

After the fish tape or pull line is attached to the conductors, they must be pulled back through the conduit. As the fish tape is pulled, the attached conductors must be properly fed into the conduit.

Usually, more than one conductor is fed into the conduit during a wire pull. It is important to keep the conductors straight and parallel, and free from kinks, bends, and crossovers. Conductors that are allowed to cross each other will form a bulge and make pulling difficult. This could also damage the conductors.

Spools and rolls of conductors must be set up so that they unwind easily, without kinks and bends.

When several conductors must be fed into the conduit at the same time, a reel cart is used (*Figure 26*). The reel cart will allow the spools to turn freely and help prevent the wires from tangling.

3.7.0 Conductor Lubrication

When conductors are fed into long runs of conduit or conduit with several bends, the wires are lubricated with a compound designed for wire lubrication.

Several types of formulated compounds designed for wire lubrication are available in either

Safety

Emphasize the importance of verifying allowable conductor tension during a pull.

Classroom

Discuss feeding conductors into conduit. Emphasize a slow and steady approach.

Discuss the use of reel carts.

Discuss conductor lubricants.

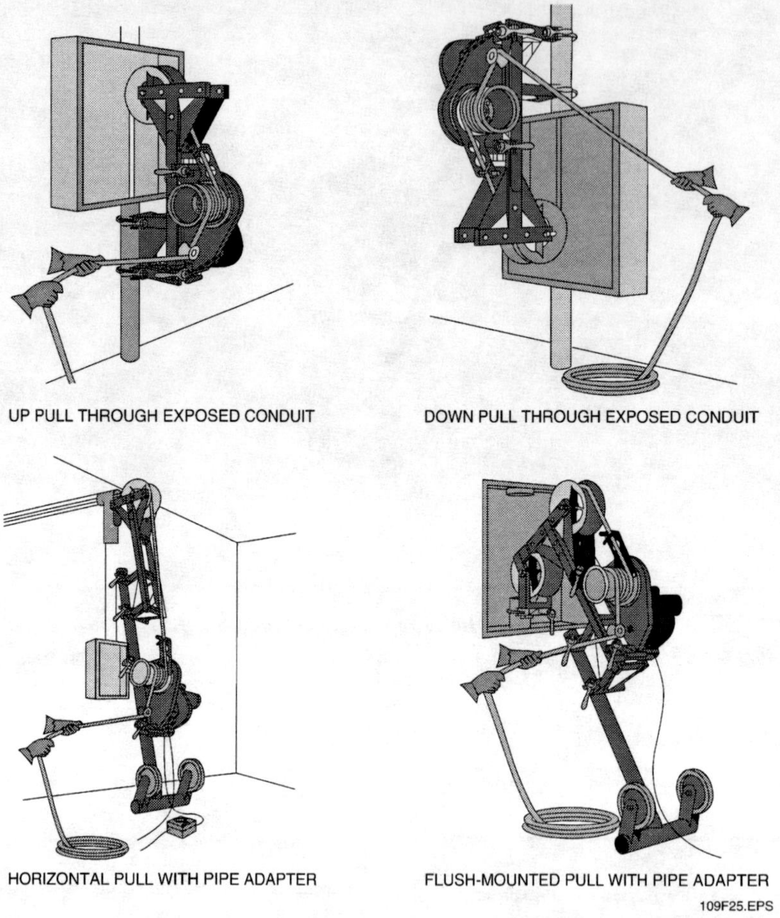

Figure 25 ◆ Power puller uses.

Classroom

Discuss conductor terminations.

dry powder, paste, or gel form. These compounds must be noncorrosive to the insulation material of the conductor and to the conduit itself. The compounds are applied by hand to the conductors as they are fed into the conduit. Avoid excess lubrication.

3.8.0 Conductor Termination

The amount of free conductor at each junction or outlet box must meet certain NEC specifications. For example, there must be sufficient free conductor so that bends or terminations inside the box, cabinet, or enclosure may be made to a radius as specified in the NEC. The NEC specifies a minimum of six inches for connections made to wiring devices or for splices. Where conductors pass through junction or pull boxes, enough slack should be provided for splices at a later date.

When a box is used as a pull box, the conductors are not necessarily spliced. They may merely

Instructor's Notes:

466 CHAPTER 11

enter the pull box via one conduit run and exit via another conduit run. The purpose of a pull box, as the name suggests, is to facilitate pulling conductors on long runs. A junction box, however, is not only used to facilitate pulling conductors through the raceway system, but it also provides an enclosure for splices in the conductors.

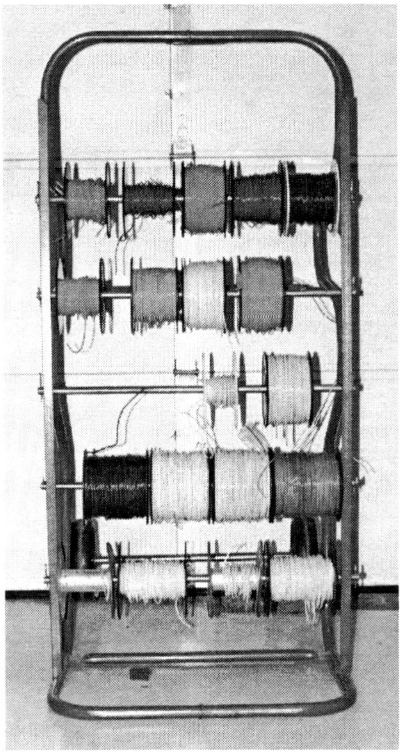

109F26.EPS

Figure 26 ◆ Reel cart.

INSIDE TRACK

Cutting Cable

Large cable is best cut using cable cutters.

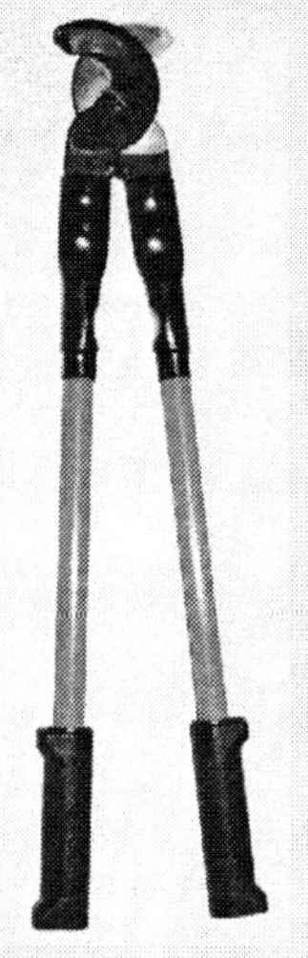

109PO904.EPS

Summary

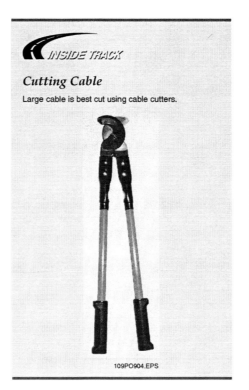

Knowing the different types of conductors, how they are rated, and what they are used for will help you when selecting conductors for a specific job. Pulling wires and cables through conduit systems is an important part of your job as an electrician. The more you learn about the concepts involved in pulling cable, the safer and more efficient you will be.

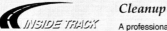

INSIDE TRACK

Cleanup

A professional always takes the time to clean up the work area, removing excess lubricant from the boxes, bushings, and conductors. Completing a job in a professional manner is not only good business, it is also a NEC requirement ***(NEC Section 110.12).***

Demonstration

Demonstrate how to pull, feed, lubricate, and terminate a single cable.

Laboratory

Have the trainees practice pulling, feeding, lubricating, and terminating a single cable. Note the proficiency of each trainee.

CONDUCTORS 467

Teaching Tip

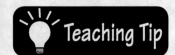

Discuss the "Think About It."

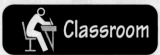

Classroom

Have the trainees complete the Review Questions and go over the answers prior to administering the Module Examination.

Examination

Administer the Module Examination. Record the results on Craft Training Report Form 200 and submit the results to the Training Program Sponsor.

Performance Testing

Administer the Performance Test and fill out Performance Profile Sheets for each trainee. If desired, trainee proficiency noted during laboratory sessions may be used to complete the Performance Test. Be sure to record the results on Craft Training Report Form 200 and submit the results to the Training Program Sponsor.

THINK ABOUT IT

Putting It All Together

Think about the design of conductor installations. How does the location of pull points affect the ease of the pull?

Review Questions

1. Where can the ampacity ratings of conductors be found in the NEC?
   a. *NEC Chapter 1*
   b. *NEC Articles 348 through 352*
   c. *NEC Tables 310.16 through 310.19*
   d. *NEC Chapter 9*

2. How many wires are contained in the three configurations of stranded cable?
   a. 7, 19, and 37
   b. 5, 7, and 9
   c. 3, 5, and 7
   d. 6, 18, and 36

3. The purpose of stranding is to _____ .
   a. stiffen the wire
   b. increase the conductivity of the wire
   c. decrease the conductivity of the wire
   d. increase the flexibility of the wire

4. The most common conductor material is _____ .
   a. silver
   b. aluminum
   c. copper
   d. platinum

5. All of the following help to determine how good a material will be for the construction of wire, *except* _____ .
   a. availability
   b. cost
   c. conductivity
   d. molecular weight

6. What letter must be included in the marking of a conductor if it is to be used in a wet, outdoor application?
   a. D
   b. O
   c. I
   d. W

7. What service conductor may be green or green with a yellow stripe?
   a. The grounding conductor of a multi-conductor cable
   b. The neutral conductor of a multi-conductor cable
   c. The ungrounded conductor of a multi-conductor cable
   d. The high leg of a four-wire, delta-connected secondary

8. What are the colors of insulation on the conductors for a three-conductor NM cable?
   a. One white, one red, and one black
   b. Two white and one red
   c. One white, one red, one black, and a grounding conductor
   d. One green, one white, one blue, and a grounding conductor

9. Type NMC cable is suitable for all of the following, *except* _____ .
   a. dry locations
   b. damp locations
   c. corrosive locations
   d. embedding in concrete

10. Type USE cable can be used for _____ .
    a. above-ground installation only
    b. underground installation within a special PVC pipeline
    c. underground installation including direct burial
    d. indoor applications only

Instructor's Notes:

468 CHAPTER 11

PROFILE IN SUCCESS

Matthew Fine, Elwood Electric

Matthew Fine twice competed in the National Craft Olympics during his apprentice training, winning first place in 1996. His story shows that youth is no barrier to advancement in the industry.

How did you choose a career in the electrical field?
My father was an electrician and I often helped him during the summer. Through that experience and the courses I had in high school, I became very interested in electrical control systems, especially motor controls.

What was your apprenticeship experience like?
The training I received during my four-year apprenticeship helped me to understand the principles behind the work I was doing on the job. Once you understand the principles, any task becomes easier. I should mention that I still think of myself as an apprentice because I am learning all the time.

The first time I competed in the Craft Olympics in 1995, I didn't place, but I did learn a lot about handling pressure. The next year I was able to apply what I learned and came away with a first-place finish.

Tell us about your present job.
I do residential and commercial installation and repair work, but I also work as a foreman, designer, estimator, and instructor. In a large company, there are different people for all of these jobs, but if you're going to succeed in a small company, you have to be able to wear more than one hat.

What factor contributed the most to your success?
My employer, Brad Elwood, gave me the opportunity to succeed. He has been my mentor in the business since high school. Without his help and direction, I would not be where I am today or have the attitude I have now.

What advice would you give to apprentice trainees?
Never pass up an opportunity to learn something new. The field changes constantly. If you don't keep up with it, you will be left behind. I myself plan to stay as far ahead as I can.

CONDUCTORS 469

GLOSSARY

Trade Terms Introduced in This Module

Ampacity: The current in amperes a conductor can carry continuously under the conditions of use without exceeding its temperature rating.

Capstan: The turning drum of the cable puller on which the rope is wrapped and pulled.

Fish tape: A hand device used to pull a wire through a conduit run.

Mouse: A cylinder of foam rubber that fits inside the conduit and is then propelled by compressed air or vacuumed through the conduit run, pulling a line or tape.

Wire grip: A device used to link pulling rope to cable during a pull.

Instructor's Notes:

Additional Resources

This module is intended to present thorough resources for task training. The following reference work is suggested for further study. This is optional material for continued education rather than for task training.

National Electrical Code Handbook, Latest Edition. Quincy, MA: National Fire Protection Association.

Answers to Review Questions

Answer	Section
1. c	2.1.0
2. a	2.3.2
3. d	2.3.2
4. c	2.4.0
5. d	2.4.0
6. d	2.5.2
7. a	2.5.3
8. c	2.5.3
9. d	2.7.2
10. c	2.7.10

Section 2.1.0 *Think About It—Overheating*

Insulation can become overheated as a result of excessive heat generated in the wire. When overheated, insulation becomes discolored. Eventually, it will melt or otherwise disintegrate and expose the enclosed wire. Factors which affect the ampacity of an insulated conductor are the physical size of the wire, the material the wire is made of, and the ambient temperature surrounding the wire.

Section 2.3.0 *Think About It—Wire Size*

The size of the wires used in a specific circuit depends on how much current the wires must conduct, how far they must conduct it, and in what type of environment or location the wires are to be placed. Use of too small a wire can result in overheating of both the wire and the load device(s) connected to the wire. This can cause poor operation of the load devices. If the overheating is excessive, it can severely damage both the wiring and load devices and may even cause a fire. Use of a wire size larger than needed only adds expense to an installation from which no benefit is derived.

Section 2.5.0 *Think About It—Conductor Insulation*

In addition to protection from shock, insulation on a wire provides corrosion protection and protection for the wires when pulling wires, and the insulation's color code helps in wire identification. Ground wires are often left bare. In addition, ***NEC Section 230.41*** allows the use of an uninsulated conductor in interior service-entrance cable in certain installations. ***NEC Section 230.30*** allows a grounded service-lateral conductor to be uninsulated under certain conditions.

Section 2.5.2 *Think About It—Insulation Types*

NEC Table 310.13 shows many cables with insulation suitable for use in wet locations. Two common types are THW and THWN.

Section 2.5.3 *Think About It—Color Coding Ungrounded Conductors*

Color coding of ungrounded conductors provides a method of identifying hot wires in a system and distinguishing between the different hot wires in multi-phase systems. Color coding also provides for standardization in wiring systems and aids when troubleshooting.

Section 3.8.0 *Think About It—Putting It All Together*

There is a direct relationship between the location and number of pull points provided in a conductor installation and how easy (or difficult) it will be to pull the cables. Using a generous number of pull points located in areas where there is adequate space to use pulling equipment and to maneuver while working will help to make the pulling process much easier.

CONTREN™ LEARNING SERIES — USER UPDATES

The NCCER makes every effort to keep these textbooks up-to-date and free of technical errors. We appreciate your help in this process. If you have an idea for improving this textbook, or if you find an error, a typographical mistake, or an inaccuracy in NCCER's Contren™ textbooks, please write us, using this form or a photocopy. Be sure to include the exact module number, page number, a detailed description, and the correction, if applicable. Your input will be brought to the attention of the Technical Review Committee. Thank you for your assistance.

Instructors – If you found that additional materials were necessary in order to teach this module effectively, please let us know so that we may include them in the Equipment/Materials list in the Instructor's Guide.

Write: Curriculum Revision and Development Department
National Center for Construction Education and Research
P.O. Box 141104, Gainesville, FL 32614-1104

Fax: 352-334-0932

E-mail: curriculum@nccer.org

Craft ______________________ Module Name ______________________

Copyright Date __________ Module Number __________ Page Number(s) __________

Description

____

____

____

(Optional) Correction

____

____

(Optional) Your Name and Address

____

____

Chapter 12

Conductor Terminations and Splices

26208-03

MODULE OVERVIEW

This course introduces the electrical trainee to the methods and procedures used when making conductor terminations and splices.

PREREQUISITES

Please refer to the Course Map in the Trainee Module. Prior to training with this module, it is recommended that the trainee shall have successfully completed the following modules:

Core Curriculum; Residential Electrical I, Chapters 1 through 11

OBJECTIVES

Upon completion of this module, the trainee will be able to:

1. Describe how to make a good conductor termination.
2. Prepare cable ends for terminations and splices.
3. Install lugs and connectors onto conductors.
4. Train cable at termination points.
5. Explain the role of the NEC in making cable terminations and splices.
6. Explain why mechanical stress should be avoided at cable termination points.
7. Describe the importance of using proper bolt torque when bolting lugs onto busbars.
8. Describe crimping techniques.
9. Select the proper lug or connector for the job.
10. Describe splicing techniques.
11. Explain how to use hand and power crimping tools.

PERFORMANCE TASKS

Under supervision of the instructor, the trainee should be able to:

1. Install heat-shrink insulators onto selected terminals.
2. Strip wires and make splices using wire nuts.
3. Terminate wires and cables using selected crimp-type and mechanical-type terminals and connectors.
4. Properly train cables using ratchet and hydraulic benders.
5. Tape selected types of wire splices and/or install a motor connection kit.

NCCER STANDARDIZED CRAFT TRAINING PROGRAM

The National Center for Construction Education and Research (NCCER) provides a standardized national program of accredited craft training. Key features of the program include instructor certification, competency-based training, and performance testing. The program provides trainees, instructors, and companies with a standard form of recognition through a National Craft Training Registry. The program is described in full in the *Guidelines for Accreditation*, published by the NCCER. For more information on standardized craft training, contact the NCCER by writing us at P.O. Box 141104, Gainesville, FL 32614-1104, calling 352-334-0911, or emailing info@nccer.org. More information can be found at our Web site, www.nccer.org.

HOW TO USE THIS ANNOTATED INSTRUCTOR'S GUIDE

Each page presents two sections of information. The larger section displays each page exactly as it appears in the Trainee Module. The narrow column ties suggested trainee and instructor actions to each page and provides icons to call your attention to material, safety, audiovisual, or testing requirements. The bottom of each page includes space for your notes.

Teaching Tip

If you see the Teaching Tip icon, that means there is a teaching tip associated with this section. Also refer to any suggested teaching tips at the end of the module.

SAFETY CONSIDERATIONS

Ensure that the trainees are equipped with appropriate personal protective equipment. Stress the importance of following the proper safety precautions and procedures used when making conductor terminations and splices.

PREPARATION

Before teaching this module, you should review the Module Outline, Objectives, Performance Tasks, and the Materials and Equipment List. Be sure to allow ample time to prepare your own training or lesson plan and gather all required equipment and materials.

MATERIALS AND EQUIPMENT LIST

Materials:

Transparencies

Markers/chalk

Copy of the latest edition of the *National Electrical Code*®

Assorted sizes and types of crimp connectors

Heat-shrink insulators

Assorted sizes and types of mechanical compression connectors

Assorted sizes and types of wire nuts

Assorted sizes of wire/cables and connectors

Heat-shrink and roll-on insulating cap motor connection kits

Assortment of electrical insulating tapes

Metal-clad (MC) cable

Type MC cable connectors

Module Examinations*

Performance Profile Sheets*

Equipment:

Overhead projector and screen

Whiteboard/chalkboard

Appropriate personal protective equipment

Wire strippers

Power cable strippers

Hand crimping tools and dies

Hydraulic crimping tools and dies

Torque wrenches

Heat gun for shrink insulators

Ratchet cable bender

Hydraulic cable bender

Propane torch

*Located in the Test Booklet packaged with this Annotated Instructor's Guide.

ADDITIONAL RESOURCES

This module is intended to present thorough resources for task training. The following reference works are suggested for both instructors and motivated trainees interested in further study. These are optional materials for continued education rather than for task training.

American Electricians' Handbook, Latest Edition. New York, NY: McGraw-Hill Publishing Company.

National Electrical Code Handbook, Latest Edition. Quincy, MA: National Fire Protection Association.

NOTES

The designations "National Electrical Code," "NE Code," and "NEC," where used in this document, refer to the *National Electrical Code*®, which is a registered trademark of the National Fire Protection Association, Quincy, MA. All National Electrical Code (NEC) references in this module refer to the 2002 edition of the NEC.

If you feel that additional math instruction would be helpful, Prentice Hall offers a basic math textbook entitled *Fundamentals of Electrical and Mechanical Mathematics.* It covers the basic math requirements for electrical trainees and may be ordered by contacting Prentice Hall Customer Service at 1-800-922-0579.

TEACHING TIME FOR THIS MODULE

An outline for use in developing your lesson plan is presented below. Note that each Roman numeral in the outline equates to one session of instruction. Each session has a suggested time period of 2½ hours. This includes 10 minutes at the beginning of each session for administrative tasks and one 10-minute break during the session. Approximately 7½ hours are suggested to cover *Conductor Terminations and Splices.* You will need to adjust the time required for hands-on activity and testing based on your class size and resources.

Topic	Planned Time
Session I. Introduction; Stripping and Cleaning Conductors; Wire Connections Under 600 Volts; Laboratory	
A. Introduction	____________
B. Stripping and Cleaning Conductors	____________
1. Stripping Small Wires	____________
2. Stripping Power Cables and Large Wires	____________
C. Wire Connections Under 600 Volts	____________
1. Aluminum Connections	____________
2. Heat-Shrink Insulators	____________
3. Wire Nuts	____________
D. Laboratory	____________
1. Under instructor supervision, have the trainees practice installing heat-shrink insulators onto shielded terminals.	____________
2. Under instructor supervision, have the trainees practice stripping wires and making connections using wire nuts.	____________
Session II. Guidelines for Installing Connectors; Laboratory; Bending Cable and Training Conductors; Laboratory; NEC Termination Requirements	
A. Guidelines for Installing Connectors	____________
1. Installing Crimp-Type Terminals and Connectors	____________
a. Crimping Tools	____________
b. General Compression (Crimp) Connector Installation Procedure	____________
2. Installing Mechanical-Type Terminals and Connectors	____________
3. Installing Specialized Cable Connectors	____________
B. Laboratory	____________
Under instructor supervision, have the trainees practice using selected crimp-on and mechanical connectors.	
C. Bending Cable and Training Conductors	____________
D. Laboratory	____________
Under instructor supervision, have the trainees practice using hydraulic and ratchet benders.	
E. NEC Termination Requirements	____________
1. Incoming Line Connections	____________
a. Main Disconnect	____________
b. Short Circuit Bracing	____________
c. Making Connections	____________
d. Making Grounding Conductor Connections	____________
2. Mounting Enclosures	____________

Session III. Taping Electrical Joints; Motor Connection Kits; Laboratory; Module Examination and Performance Testing

A. Taping Electrical Joints ____________

B. Motor Connection Kits ____________

C. Laboratory ____________

Under instructor supervision, have the trainees practice taping several types of wire splices and/or installing a motor connection kit.

D. Review ____________

E. Module Examination ____________

1. Trainees must score 70% or higher to receive recognition from the NCCER.
2. Record the testing results on Craft Training Report Form 200 and submit the results to the Training Program Sponsor.

F. Performance Testing ____________

1. Trainees must perform each task to the satisfaction of the instructor to receive recognition from the NCCER.
2. Record the testing results on Craft Training Report Form 200 and submit the results to the Training Program Sponsor.

Chapter 12

Module 26208-03

Conductor Terminations and Splices

Instructor's Notes:

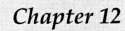

Course Map

This course map shows all of the modules in the first level of the Residential Electrical curriculum. The suggested training order begins at the bottom and proceeds up. Skill levels increase as you advance on the course map. The local Training Program Sponsor may adjust the training order.

RESIDENTIAL ELECTRICAL I

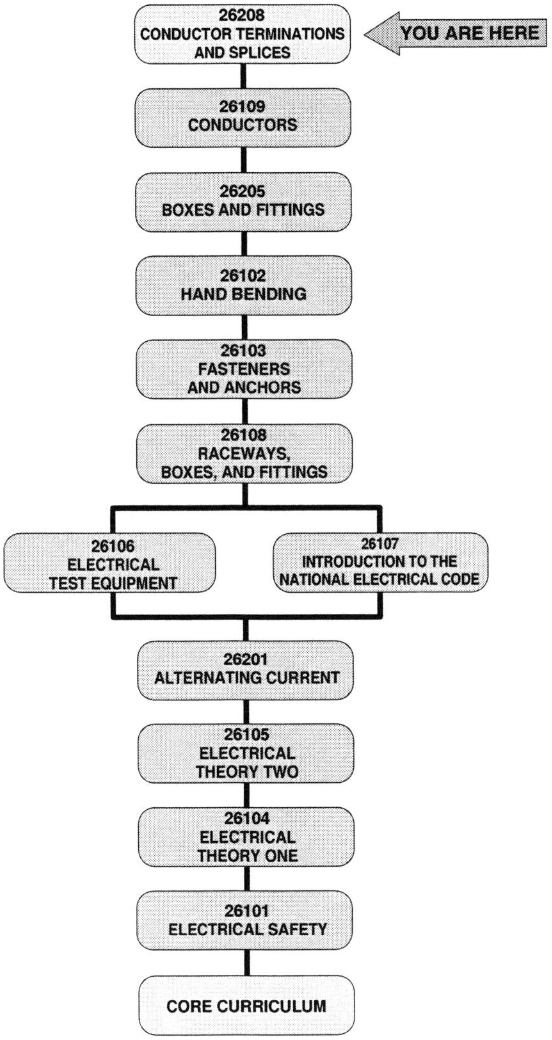

Homework

Assign reading of Module 26208.

MODULE 26208 CONTENTS

Figures

Tables

Instructor's Notes:

MODULE 26208

Conductor Terminations and Splices

Objectives

When you have completed this module, you will be able to do the following:

1. Describe how to make a good conductor termination.
2. Prepare cable ends for terminations and splices.
3. Install lugs and connectors onto conductors.
4. Train cable at termination points.
5. Explain the role of the NEC in making cable terminations and splices.
6. Explain why mechanical stress should be avoided at cable termination points.
7. Describe the importance of using proper bolt torque when bolting lugs onto busbars.
8. Describe crimping techniques.
9. Select the proper lug or connector for the job.
10. Describe splicing techniques.
11. Explain how to use hand and power crimping tools.

Prerequisites

Before you begin this module, it is recommended that you successfully complete the following: Core Curriculum; Electrical Level One; Residential Electrical I, Chapters 1 through 11.

Required Trainee Materials

1. Paper and pencil
2. Appropriate personal protective equipment
3. Copy of the latest edition of the *National Electrical Code*®

1.0.0 ◆ INTRODUCTION

Anyone involved with electrical systems of any type should have a good knowledge of wire **connectors** and splicing, as they are both necessary to make the numerous electrical joints required during the course of any electrical installation. A properly-made **splice** and **connection** will often last as long as the insulation on the wire itself, while a poorly-made connection will always be a source of trouble; that is, the joints will overheat under load and eventually fail with the potential for starting a fire.

The basic requirements for a good electrical connection include the following:

- It should be mechanically and electrically secure.
- It should be insulated as well as or better than the existing insulation on the conductors.
- These characteristics should last as long as the conductor is in service.

There are many different types of electrical joints, and the selection of the proper type for a given application will depend to a great extent on

Note: The designations "National Electrical Code," "NE Code," and "NEC," where used in this document, refer to the *National Electrical Code*®, which is a registered trademark of the National Fire Protection Association, Quincy, MA. *All National Electrical Code (NEC) references in this module refer to the 2002 edition of the NEC.*

Materials

Ensure that you have everything required to teach the course. Check the Materials and Equipment List at the front of this Instructor's Guide.

Classroom

Review the goals of the course.

Explain that terms shown in bold (blue) are defined in the Glossary at the back of this module.

Audiovisual

Show Transparency 1, Course Objectives.

Show Transparency 2, Performance Tasks.

Classroom

Explain the basic requirements of a good electrical connection.

Conductor Terminations and Splices

INSIDE TRACK

Poor electrical connections are responsible for a large percentage of equipment burnouts and fires. Many of these failures are a direct result of improper terminations, poor workmanship, and the use of improper splicing devices.

Classroom

Discuss the importance of proper wire stripping.

Emphasize the importance of using the proper tool to avoid damage to the conductors.

Explain that wire diameters differ depending on whether the conductor is solid or stranded.

how and where the splice or connection is used. Electrical joints are normally made with a solderless **pressure connector** or **lug** to save time.

2.0.0 ◆ STRIPPING AND CLEANING CONDUCTORS

Before any connection or splice can be made, the ends of the conductors must be properly stripped and cleaned. Stripping is the removal of insulation from the conductors at the end of the wire or at the location of the splice. Wires should only be stripped using the appropriate stripping tool. This will help to prevent cuts and nicks in the wire, which can reduce the conductor area as well as weaken the conductor.

Poorly stripped wire can result in nicks, scrapes, or burnishes. Any of these can lead to a stress concentration at the damaged cross section. Heat, rapid temperature change, mechanical vibration, and oscillatory motion can aggravate the damage, causing faults in the circuitry or even total failure.

Lost **strands** are a problem in splice or crimp-type **terminals,** while exposed strands might be a safety hazard.

Slight burnishes on conductors, as long as they had no sharp edges, were acceptable at one time. Now, however, most experts believe that under certain conditions, removing as little as 40 microinches of conductor plating from some wires can cause a failure.

Faulty stripping can pierce, scuff, or split the insulation. This can cause changes in dielectric strength and lower the wire's resistance to moisture and abrasion. Insulation particles often get trapped in solder and crimp joints. These form the basis for a defective **termination.** A variety of factors determine how precisely a wire can be stripped, including: wire size, insulation concentricity, adherence, and others.

It is a common mistake to believe that a certain gauge of stranded conductor has the same diameter as a solid conductor. This is a very important consideration in selecting the proper blades for strippers. *Table 1* shows the nominal sizes referenced for the different wire gauges.

Table 1 Dimensions of Common Wire Sizes

Size (AWG/kcmil)	Area (Circular Mils)	Overall Diameter in Inches: Solid	Overall Diameter in Inches: Stranded
18	1,620	0.040	0.046
16	2,580	0.051	0.058
14	4,130	0.064	0.073
12	6,530	0.081	0.092
10	10,380	0.102	0.116
8	16,510	0.128	0.146
6	26,240	—	0.184
4	41,740	—	0.232
3	52,620	—	0.260
2	66,360	—	0.292
1	83,690	—	0.332
1/0	105,600	—	0.373
2/0	133,100	—	0.419
3/0	167,800	—	0.470
4/0	211,600	—	0.528
250	—	—	0.575
300	—	—	0.630
350	—	—	0.681
400	—	—	0.728
500	—	—	0.813
600	—	—	0.893
700	—	—	0.964
750	—	—	0.998
800	—	—	1.03
900	—	—	1.09
1,000	—	—	1.15
1,250	—	—	1.29
1,500	—	—	1.41
1,750	—	—	1.52
2,000	—	—	1.63

To eliminate nicking, cutting, and fraying, wires should only be stripped using the appropriate stripping tool. The specific tool used depends on the size and type of wire being stripped.

2.1.0 Stripping Small Wires

There are many kinds of wire strippers available. *Figure 1* shows two common types of wire strip-

Instructor's Notes:

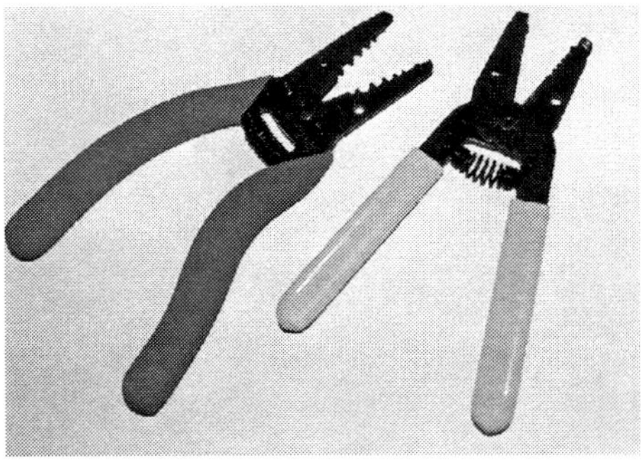

208F01.EPS

Figure 1 ◆ Wire strippers.

pers for small wires. Note that the one on the left is ergonomically designed with a curved soft foam handle. To use these tools, insert the wire into the proper size knife groove, then squeeze the tool handles. The tool cuts the wire insulation, allowing the wire to be easily removed without crushing its stripped end. The length of the strip is regulated by the amount of wire extending beyond the blades when the wire is inserted in the knife groove.

2.2.0 Stripping Power Cables and Large Wires

Figure 2 shows a heavy-duty stripper used to strip power cables with outside diameters ranging

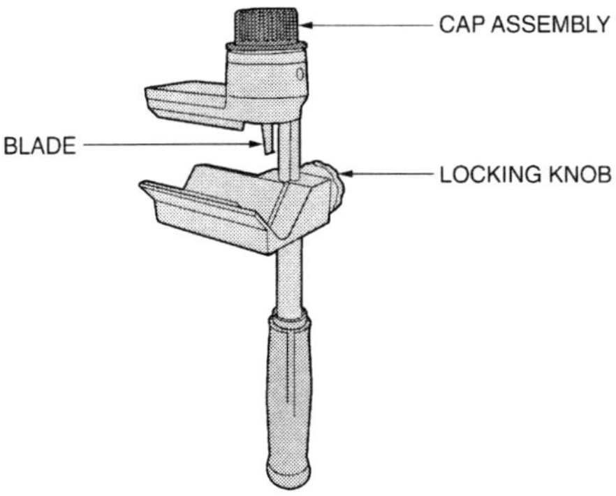

208F02.EPS

Figure 2 ◆ Cable stripper.

from ½" to 1¾". It can be used to strip insulation starting from the end of the cable, strip insulation from some point to the end of the cable, or to make a window cut. All stripping tools should be operated according to the manufacturer's instructions. The procedures for using the tool to strip insulation from the end of the cable and to make a window cut are described here.

To strip insulation from the end of the cable (*Figure 3*), proceed as follows:

Step 1 Loosen the locking knob to open the tool to the maximum position. Place the cable in the V-groove and close the tool firmly around the cable. Tighten the locking knob.

Step 2 Turn the cap assembly until the blade reaches the required depth.

> **CAUTION**
> Do not allow the blade to contact the conductor because damage to the conductor and/or the blade can result.

Step 3 Rotate the tool around the cable, advancing to the required strip length.

Step 4 Rotate the tool in the reverse direction to produce a square end cut (*Figure 3*).

Step 5 Loosen the locking knob to release the tool and remove it from the cable.

Step 6 Peel off the insulation.

To make a window cut (*Figure 3*), proceed as follows:

Step 1 With the tool opened to the maximum position, place the cable in the V-groove and close the tool firmly around the cable. Tighten the locking knob.

Step 2 Turn the cap assembly until the blade reaches the required depth.

Step 3 Rotate the tool to produce the first square cut.

Step 4 Rotate the tool in the reverse direction to cut the required window strip length.

Step 5 Rotate the tool in the original direction to produce the second square cut.

Step 6 Loosen the locking knob assembly to release the tool and remove it from the cable.

Step 7 Peel off the insulation.

Classroom

Discuss various wire strippers. Pass around examples for the trainees to examine.

Audiovisual

Show Transparency 3 (Figure 2).

Classroom

Discuss various types of cuts:

- **End terminations**
- **Window cuts**
- **Spiral cuts**
- **Square and longitudinal cuts**

Audiovisual

Show Transparency 4 (Figure 3).

Classroom

Discuss the use of slitting and ringing tools. Pass around an example for the trainees to examine.

Audiovisual

Show Transparency 5 (Figure 4).

Demonstration

Demonstrate how to strip various types of wires and cable.

Classroom

Discuss the NEC requirements for electrical connections.

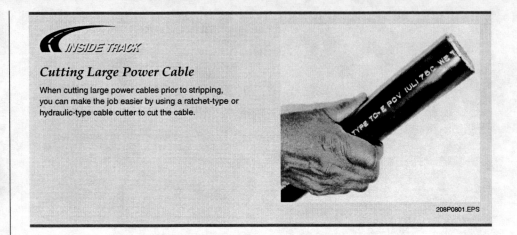

INSIDE TRACK

Cutting Large Power Cable

When cutting large power cables prior to stripping, you can make the job easier by using a ratchet-type or hydraulic-type cable cutter to cut the cable.

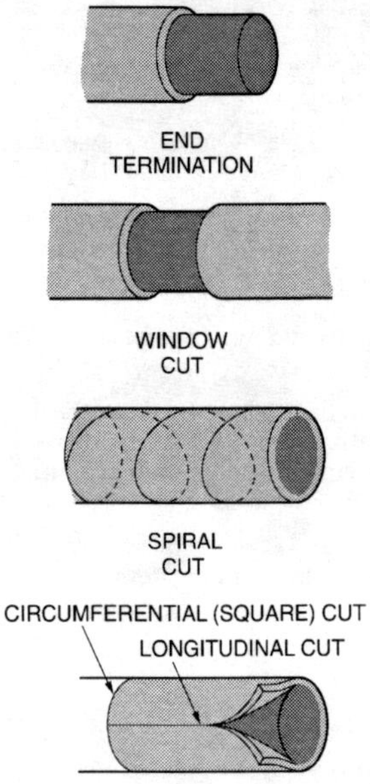

Figure 3 ◆ Types of cable stripping.

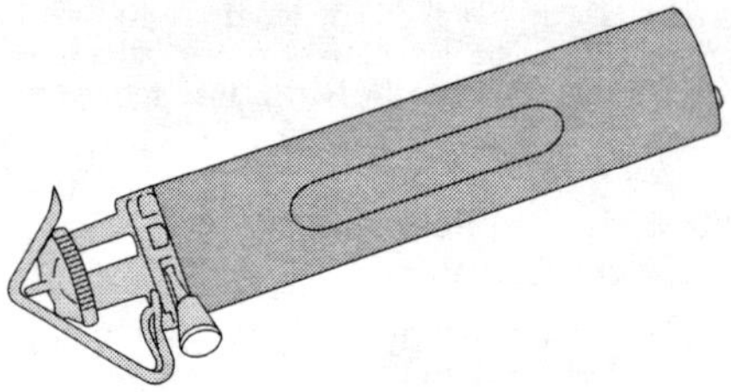

Figure 4 ◆ Round cable slitting and ringing tool.

Figure 4 shows a round cable slitting and ringing tool that can be used to strip single- or multiple-conductor cables. The tool can be used to cut around the cable (square cut) or slit the length of the cable jacket (longitudinal cut) for easy removal. The tool blade is adjustable to accommodate different jacket thicknesses.

3.0.0 ◆ WIRE CONNECTIONS UNDER 600 VOLTS

NEC Section 110.14 governs electrical connections, including terminations and splices. Wire connections are used to connect a wire or cable to such

Instructor's Notes:

INSIDE TRACK: Stripping Power and Large Cables

When using any bladed-type cable stripper, always be careful to keep your fingers away from the blade. Also, make sure to replace the blade whenever it becomes dull. Remember, a dull blade is more dangerous than a sharp one because you are more likely to apply undue pressure when using a dull blade, causing it to slip.

electrically operated devices as fan coil units, duct heaters, oil burners, motors, pumps, and control circuits of all types.

A variety of wire connectors for stranded wire are shown in *Figure 5*. These connectors are available in various sizes to accommodate wire sizes No. 22 AWG and larger. They can be installed with crimping tools having a single indenter or double indenter. The range is normally stamped on the tongue of each terminal.

Mechanical compression-type terminators are also available to accommodate wires from No. 8 AWG through 1,000 kcmil. One-hole lugs, two-hole lugs, split-bolt connectors, and other types are shown in *Figure 6*.

Crimp-type connectors made to accommodate wires smaller than No. 8 are normally made to accept at least two wire sizes and are often color coded. For example, one manufacturer's color code used for such connectors is red for No. 18 or No. 20 wire, blue for No. 16 or No. 14 wire, and yellow for No. 12 or No. 10 wire. Crimp-type connectors used for wire sizes No. 8 and larger, commonly called *lugs*, are made to accept one specific conductor size. Crimp-type **reducing connectors** are used to connect two different size wires. Mechanical compression-type connectors and lugs are made so that they accommodate a range of different wire sizes.

The parallel-tap connector with an insulated cover shown in *Figure 6* is one example of a pre-insulated, molded mechanical compression connector. There are several kinds available. They come in setscrew/pressure plate and insulation-piercing configurations made for use in a variety of feeder **tap** and splice applications. Because they are equipped with an insulating cover, the requirement for taping the joint is eliminated.

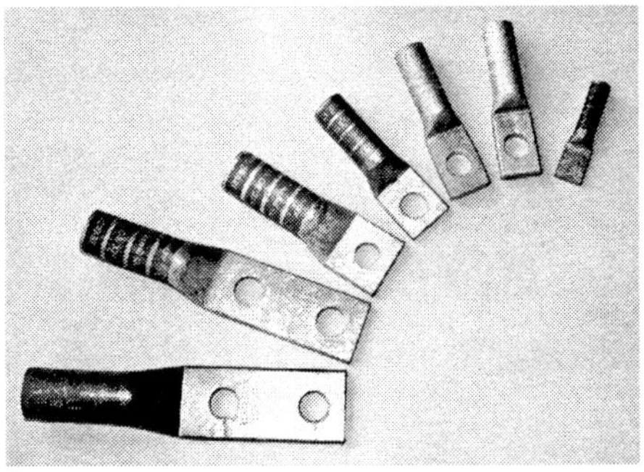

208F05.EPS

Figure 5 ◆ Crimp-on wire lugs.

3.1.0 Aluminum Connections

Aluminum has certain properties that are different from copper and must be understood if reliable connections are to be made. These properties are: cold flow, coefficient of thermal expansion, susceptibility to galvanic corrosion, and the formation of oxide film on the surface.

Because of the thermal expansion and cold flow of aluminum, standard copper connectors cannot be safely used on aluminum wire. Most manufacturers design their aluminum connectors with greater contact area to counteract these properties of aluminum. Tongues and barrels of all aluminum

THINK ABOUT IT: Mechanical Compression Connectors

What might be a use for a compression connector with a two-hole tongue?

Classroom

Discuss crimp-on and compression connectors and their applications. Pass around examples for the trainees to examine.

Audiovisual

Show Transparency 6 (Figure 6).

Classroom

Discuss the special requirements of aluminum connections.

Teaching Tip

Discuss the "Think About It." See the Teaching Tip at the end of this module.

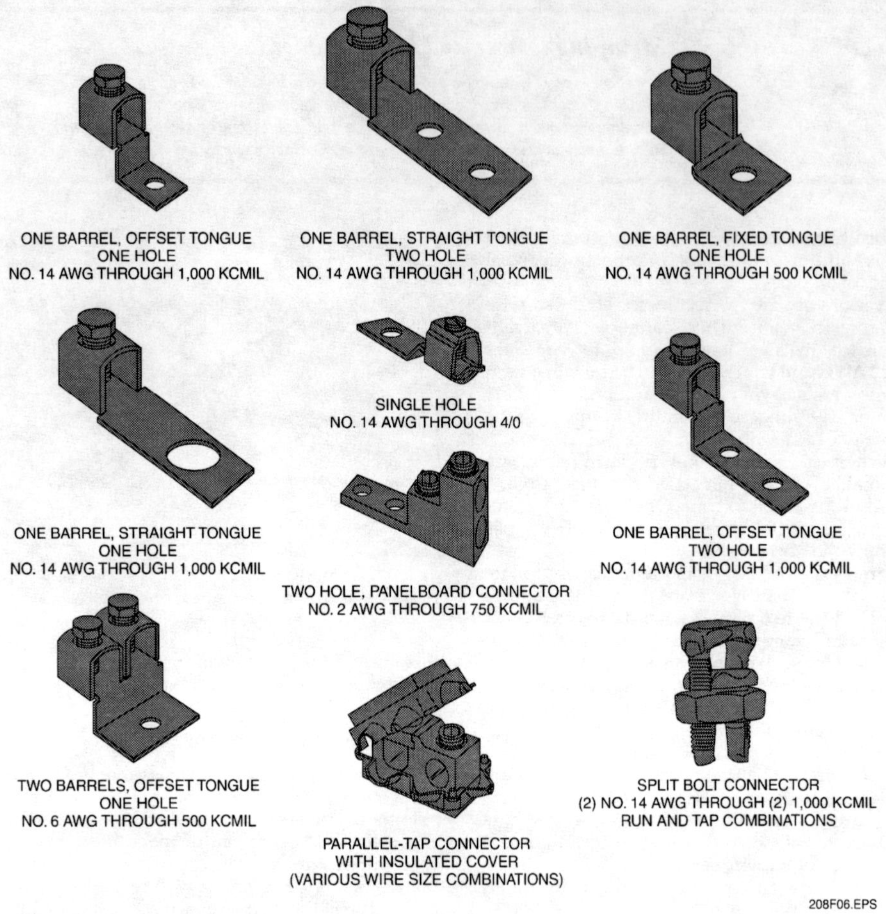

Figure 6 ◆ Various mechanical compression connectors.

connectors are larger or deeper than comparable copper connectors.

The electrolytic action between aluminum and copper can be controlled by plating the aluminum with a neutral metal (usually tin). The plating prevents electrolysis from taking place, and the joint remains tight. As an additional precaution, a joint-sealing compound should be used. Connectors should also be tin-plated and prefilled with an oxide-inhibiting compound.

The insulating aluminum oxide film must be removed or penetrated before a reliable aluminum joint can be made. Aluminum connectors are designed to bite through this film as they are applied to conductors. It is further recommended that the conductor be wire brushed and preferably coated with a joint compound to guarantee a reliable joint.

NEC Section 110.14 prohibits conductors made of dissimilar metals (copper and aluminum, copper and copper-clad aluminum, or aluminum and

Instructor's Notes:

486 CHAPTER 12

Tightening Compression Connector Screws and Bolts

INSIDE TRACK

Mechanical compression connectors must be tightened to a specified torque using a torque screwdriver or torque wrench. Overtightening can cut the wires or break the fitting, while undertightening may lead to loose connections, resulting in overheating and failure.

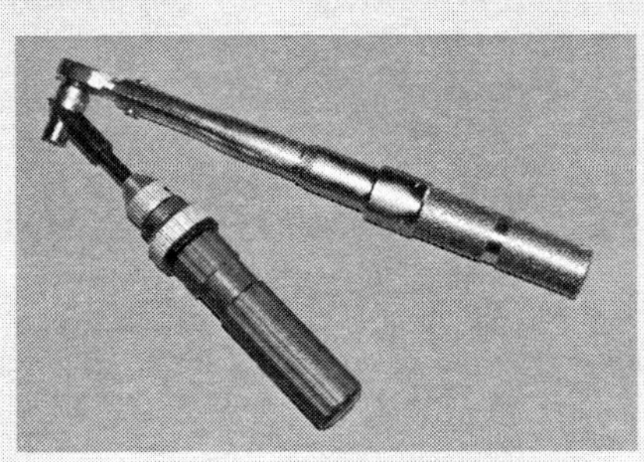

208P0802.EPS

CAUTION

Never use connectors designed strictly for use on copper conductors on aluminum conductors. Connectors listed for use on both metals will normally be marked AL-CU. All connectors must be applied and installed in the manner for which they are listed and labeled.

copper-clad aluminum) from being intermixed in a terminal or splicing connector unless the device is identified for the purpose and for the conditions under which it may be used. As a general rule:

- Connectors marked with only the wire size should only be used with copper conductors.
- Connectors marked with AL and the wire size should only be used with aluminum wire.
- Connectors marked with **AL-CU** and the wire size may be safely used with either copper or aluminum.

3.2.0 Heat-Shrink Insulators

Heat-shrink insulators for small connectors provide skintight insulation protection and are fast and easy to use. They are designed to slip over wires, taper pins, connectors, terminals, and splices. When heat is applied, the insulation becomes semi-rigid and will provide positive strain relief at the flex point of the conductor. A vaporproof band will seal and protect the conductor from abrasion, chemicals, dust, gasoline, oil, and moisture. Extreme temperatures, both hot and cold, will not affect the performance of these insulators. The source of heat can be any number of types, but most manufacturers of these insulators also produce a heat gun especially designed for use on heat-shrink insulators. It closely resembles and operates the same as a conventional hair dryer, as shown in *Figure 7*.

In general, a heat-shrink insulator may be thought of as tubing with a memory. After it is initially manufactured, it is heated and expanded to

Connection of Conductors Made of Dissimilar Materials

INSIDE TRACK

Unless specifically stated on the shipping carton or on the connector itself, wires made of copper, aluminum, or copper-clad aluminum may not be spliced together in the same connector.

Safety

Emphasize the importance of using the proper connectors with aluminum wire.

Classroom

Discuss the use of heat-shrink insulators and pass around examples for the trainees to examine.

Audiovisual

Show Transparency 7 (Figure 7).

Demonstration

Demonstrate the use of a heat gun to install a heat-shrink insulator.

Laboratory

Have the trainees practice installing heat-shrink insulators onto selected terminals. Note the proficiency of each trainee.

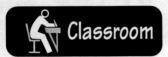

Classroom

Discuss the use of tubing selector guides.

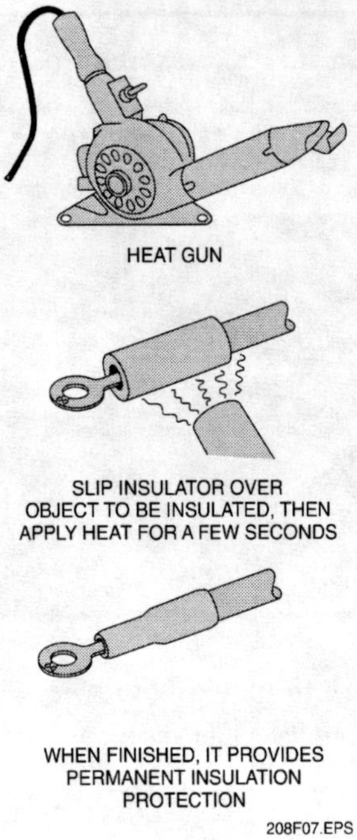

HEAT GUN
SLIP INSULATOR OVER OBJECT TO BE INSULATED, THEN APPLY HEAT FOR A FEW SECONDS
WHEN FINISHED, IT PROVIDES PERMANENT INSULATION PROTECTION
208F07.EPS

Figure 7 ◆ Method of installing heat-shrink insulators.

a predetermined diameter and then cooled. Upon application of heat through various methods, the tubing compound "remembers" its original size and shrinks to that smaller diameter. It is available in a range of sizes and is designed to shrink easily over any wire or device when heat is applied. This property enables it to conform to the contours of any object. The following describes some of the types currently available. A typical tubing selector guide appears in *Table 2.*

- *PVC*—This type is a general-purpose, economical tubing that is widely used in the electronics industry. The PVC compound is irradiated by being bombarded with high-velocity electrons. This results in a denser, cross-linked material with superior electrical and mechanical properties. It also ensures that the tubing will resist cracking and splitting.
- *Polyolefin*—Polyolefin tubing has a wide range of uses for wire bundling, harnessing, strain relief, and other applications where cables and components require additional insulation. It is irradiated, flame-retardant, flexible, and comes in a wide variety of colors.
- *Double wall*—This type is available and designed for outstanding protective characteristics. It is a semi-rigid tubing with an inner wall that melts and an outer wall that shrinks to conform to the melted area.

Table 2 Tubing Selector Guide

Type	Material	Temp. Range (°C)	Shrink Ratio	Max. Long. Shrinkage (%)	Tensile Strength (psi)	Colors	Dielectric Strength (V/mil)
Nonshrinkable	PVC	+105	—	—	2,700	White, red, clear, black	800
Shrinkable	PVC	−35 to +105	2:1	10	2,700	Clear, black	750
Nonshrinkable	Teflon®	−65 to +260	—	—	2,700	Clear	1,400
Shrinkable	Flexible polyolefin	−55 to +135	2:1	5	2,500	Black, white, red, yellow, blue, clear	1,300
Nonshrinkable	Teflon®	−65 to +260	—	—	7,500	Clear	1,400
Shrinkable	Polyolefin double wall	−55 to +110	6:1	5	2,500	Black	1,100
Shrinkable	Kynar®	−55 to +175	2:1	10	8,000	Clear	1,500
Shrinkable	Teflon®	+250	1.2:2	10	6,000	Clear	1,500
Shrinkable	Teflon®	+250	1½:1	10	6,000	Clear	1,500
Shrinkable	Neoprene	+120	2:1	10	1,500	Black	300

Instructor's Notes:

488 CHAPTER 12

- *Teflon®*—This type is considered by many users to be the best overall heat-shrink tubing—physically, electrically, and chemically. Its high-temperature rating of 250°C resists brittleness and loss of translucency from extended exposure to high heat and will not support combustion.
- *Neoprene*—Components that warrant extra protection from abrasion require a highly durable yet flexible tubing. Irradiated neoprene tubing offers this optimal coverage.
- *Kynar®*—Irradiated Kynar® is a thin-wall, semi-rigid tubing with outstanding resistance to abrasion. This transparent tubing enables easy inspection of components that are covered and retains its properties at its rated temperature.

Most tubing is available in a wide variety of colors and configurations. The manufacturer's tubing selector guide can help in the selection of the best tubing for any given application.

3.3.0 Wire Nuts

Ever since its invention in 1927, the **wire nut,** also known as the *Wirenut®* and the *solderless connector* (*Figure 8*), has been a favorite wire connector for use on residential and commercial branch circuit applications. Several varieties of wire nuts are available, but the following are the ones used most often:

- Those for use on wiring systems 300V and under
- Those for use on wiring systems 600V and under (1,000V in lighting fixtures and signs)

Most brands are UL listed for aluminum to copper in dry locations only; aluminum to aluminum only; and copper to copper only. The maximum temperature rating is 105°C (221°F).

Wire nuts are frequently used for all types of splices in residential and commercial applications and are considered to be the fastest connectors on the market for this type of work.

To use a wire nut, trim the bare conductors using the appropriate tool, and then screw on the wire nut. The wire nut draws the conductors and insulation into the shirt of the connector, which increases resistance to flashover. The internal spring is designed to tightly thread the conductors into the wire nut and then hold them with a positive grip. Some types of wire nuts have thin wings on each side of the connector to facilitate their installation. See *Figure 9.* Wire nuts are normally made in sizes to accommodate conductors as small as No. 22 AWG up to as large as No. 10 AWG, with practically any combination of those sizes in between.

Specially designed wire nuts are also made specifically for use in wet locations and/or direct burial applications. These wire nuts have a water repellent, non-hardening sealant inside the body that completely seals out moisture to protect the conductors against moisture, fungus, and corrosion. The sealant remains in a gel state and will not melt or run out of the wire nut body throughout the life of the connection. Unlike other types of

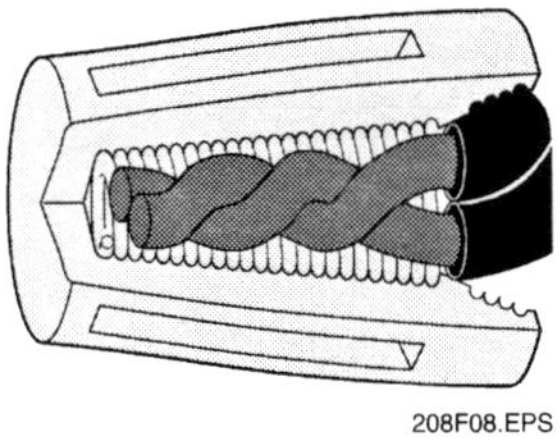

Figure 8 ◆ Typical wire nut showing interior arrangement.

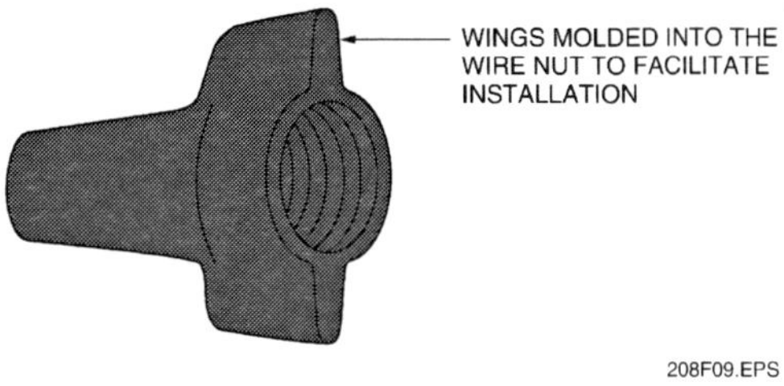

Figure 9 ◆ Some wire nuts have thin wings on each side to facilitate installation.

Classroom

Discuss the use of wire nuts.

Audiovisual

Show Transparencies 8 and 9 (Figures 8 and 9).

Demonstration

Demonstrate how to strip wires and make connections using wire nuts.

Laboratory

Have the trainees practice stripping wires and making connections using wire nuts. Note the proficiency of each trainee.

CONDUCTOR TERMINATIONS AND SPLICES 489

INSIDE TRACK

Insulated Spring Connectors

Insulated spring connectors, commonly called *wire nuts* or *Wirenuts®*, are solderless connectors made in various color-coded sizes that allow for splicing the hundreds of different solid or stranded wire combinations typically encountered in branch-circuit and fixture splicing applications.

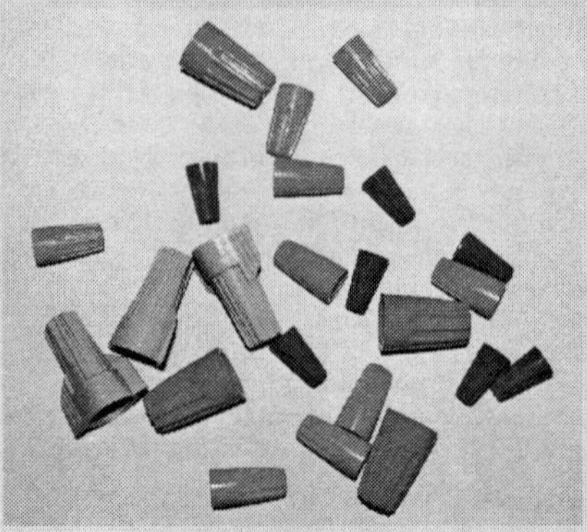

208P0803.EPS

wire nuts, this type can be used one time only. The wire nut can be backed off, eliminating the need to cut the wires for future or retrofit applications, but once removed, it must be discarded.

The general procedure for splicing wires with wire nuts is as follows:

Step 1 Select the proper size wire nut to accommodate the wires being spliced. Wire nut packages contain charts that list the allowable combinations of wires by size. Refer to the label on the wire nut box or container for this information (*Figure 10*).

Step 2 Strip the insulation from the ends of the wires to be spliced. The length of insulation stripped off is typically about ½"; however, it depends on the wire sizes and the wire nut being used. Follow the manufacturer's directions given on the wire nut package.

Step 3 Stick the ends of the wires into the wire nut and turn clockwise until tight. Note that some manufacturers of wire nuts require that the wires be pre-twisted before screwing on the nut. Also, some manufacturers recommend using a nut driver to tighten the wire nut. Always follow the manufacturer's instructions.

Classroom

Explain the importance of selecting the proper size connector for the job.

Audiovisual

Show Transparency 10 (Figure 10).

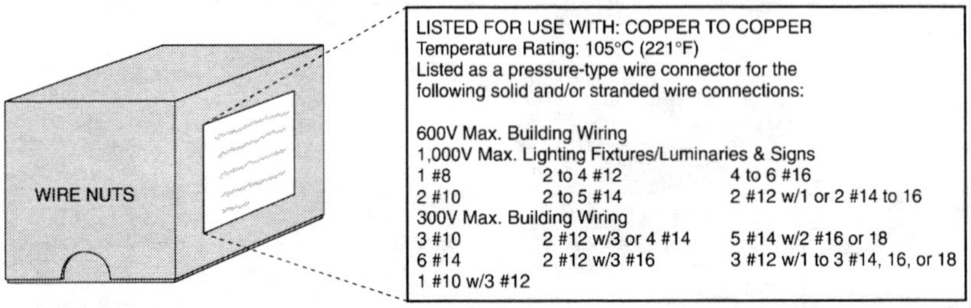
WIRE NUTS
LISTED FOR USE WITH: COPPER TO COPPER
Temperature Rating: 105°C (221°F)
Listed as a pressure-type wire connector for the following solid and/or stranded wire connections:
600V Max. Building Wiring
1,000V Max. Lighting Fixtures/Luminaries & Signs
1 #8 | 2 to 4 #12 | 4 to 6 #16
2 #10 | 2 to 5 #14 | 2 #12 w/1 or 2 #14 to 16
300V Max. Building Wiring
3 #10 | 2 #12 w/3 or 4 #14 | 5 #14 w/2 #16 or 18
6 #14 | 2 #12 w/3 #16 | 3 #12 w/1 to 3 #14, 16, or 18
1 #10 w/3 #12

208F10.EPS

Figure 10 ◆ Read package label to find allowable wire combinations.

Instructor's Notes:

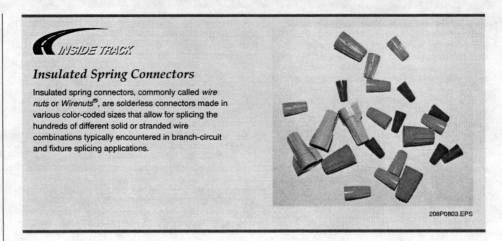

Inside Track: Making Box Connections

After making the connections within a box, tuck the wires neatly into the back of the box, as shown here. That way, when the painters and plasterers come along to finish the walls, the wires will not be covered in drywall compound or paint.

208P0804.EPS

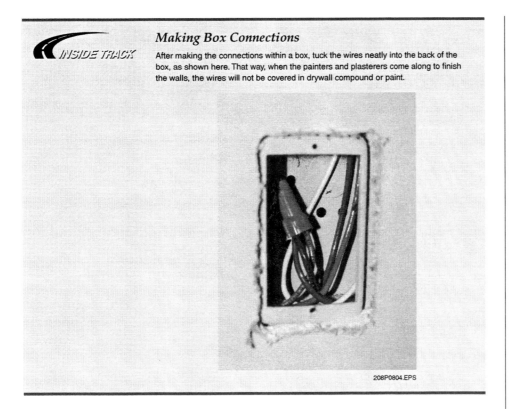

4.0.0 ◆ GUIDELINES FOR INSTALLING CONNECTORS

Guidelines for installing the different types of connectors covered in this module are given in the following sections.

4.1.0 Installing Crimp-Type Terminals and Connectors

The task of fastening a compression-type connector to a wire requires the use of the proper connector, crimping tool, and installation procedure.

4.1.1 Crimping Tools

Compression-type connectors and lugs must be attached to wires using the appropriate crimping tool. Since connectors, crimping tools, and crimping tool dies are designed as a unit for specific wire sizes, only the recommended tool and dies should be used. There are a wide variety of crimping tools available, from simple to complex. Most crimping tools are designed for use with either insulated connectors or non-insulated connectors. Quality tools used to crimp larger connectors and wires are typically made with hand-operated ratcheting mechanisms or hydraulic mechanisms that do not release until adequate crimping pressure has been applied. Some crimping tools require the use of related die sets. Some types are

Classroom

Discuss the tools used with compression-type connectors. Pass around examples for the trainees to examine.

Audiovisual

Show Transparencies 11 and 12 (Figures 11 and 12).

Demonstration

Demonstrate the use of crimping tools to make connections:
- **Hand-operated**
- **Hydraulic**

Safety

Stress the importance of reading the manufacturer's instructions when using power tools.

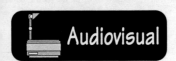

dieless. Many manufacturers color code both their connectors and matching crimping tool dies as an aid in selecting the proper die. They also mark their dies with die index numbers to identify the use of the proper connector/die combination.

Figure 11 shows one manufacturer's hand-operated tool. It is designed to crimp connectors to wires ranging in size from No. 8 AWG through No. 22 AWG stranded. This tool has interchangeable die sets to accommodate different connector sizes. It also has a full-cycle ratchet mechanism that provides for a complete, positive crimp. Once started, the ratchet mechanism does not allow the tool handles to be opened until after the full ratcheting cycle is completed, unless the user actuates the emergency release lever. This ensures a completed crimp every time.

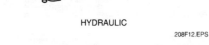

Figure 12 shows hand-operated and hydraulic crimping (compressor) tools typical of those used to crimp connectors for stranded wires ranging from No. 8 AWG to 750 kcmil. Both tools normally develop about 12 tons of compression force at 10,000 pounds per square inch (psi). Guidelines for the use of these tools are described here. Many other tools operate in a similar manner.

To use a hand-operated crimping tool, proceed as follows:

Step 1 Select the proper dies for use with the connector to be crimped. Do not operate the tool without the dies.

Step 2 Push the die release button on the C-head and slide one of the die halves into position until the retainer snaps. Insert the other die half in the piston body by pushing the die release button and sliding the die in until the retainer snaps.

Step 3 Place the tool C-head in position over the connector to be crimped. Pump the handle until compression is complete, as indicated by the dies touching at their flat surfaces nearest the throat of the C-head.

Step 4 Retract the ram and remove the connector after completion of the crimp. This is done by raising the pump handle slightly, rotating it clockwise until it stops, then pushing the handle down in a pumping motion until the pressure release snaps.

> **WARNING!**
> Always read and follow the manufacturer's instructions when using power tools.

To operate a hydraulic crimping tool, proceed as follows:

Step 1 Using a suitable hydraulic hose, connect the hydraulic pump to the crimping tool.

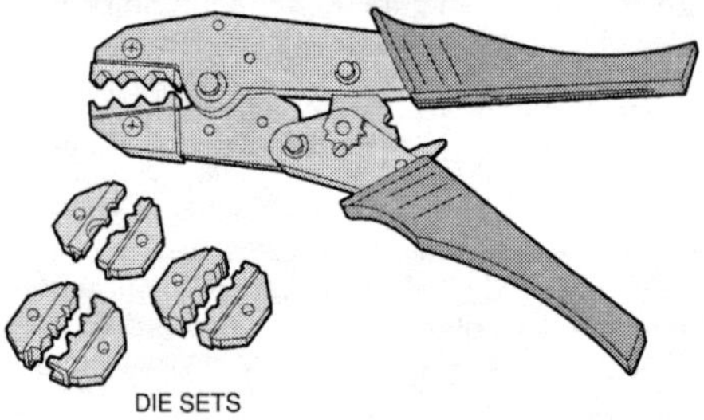

Figure 11 ◆ Crimping tool used to crimp small connectors.

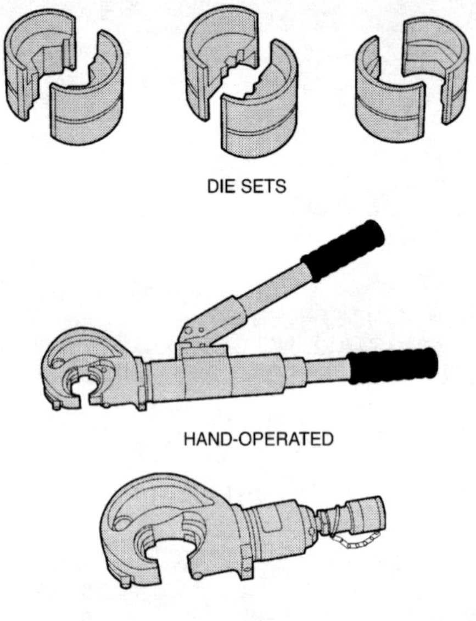

Figure 12 ◆ Crimping (compression) tools used to crimp large connectors.

Instructor's Notes:

492 CHAPTER 12

Step 2 Select the proper dies for use with the connector to be crimped. Do not operate the tool without the dies.

Step 3 Push the die release button on the C-head and slide one of the die halves into position until the retainer pin snaps. In a similar manner, install the other die half in the piston body.

Step 4 Place the tool C-head in position over the connector to be crimped. Operate the remote pump until compression is complete, as indicated by the dies touching on the frame side.

Step 5 Release the pressure at the hydraulic pump to retract the lower die half, then remove the connector from the tool.

4.1.2 General Compression (Crimp) Connector Installation Procedure

Step 1 Select a crimp-type connector of the proper size and of appropriate material for the wire size you are using. Copper connectors should be used with copper wires and aluminum connectors with aluminum wires. Dual-rated connectors may be used with both copper and aluminum wires.

Step 2 Using a suitable wire stripper, remove the insulation from the end of the wire, being careful not to nick the wire. Strip the insulation back far enough so that the bare conductor will go fully into the connector. Make sure not to strip off too much insulation; it should fit close to the connector when the wire is fully inserted into the connector.

Step 3 Clean the stripped portion of the wire. Use a wire brush for large wire sizes. Also clean the related unplated terminal pad and the surface to which the connector will be attached.

Step 4 Obtain the crimping tool and dies made for the type and size of connector to be crimped.

Step 5 Insert the stripped end of the wire completely into the connector. Position the crimping tool in place over the connector, then operate the tool to fully crimp the connector, as directed in the tool manufacturer's instructions. Make sure that the crimping tool jaws are fully closed, indicating that a full compression crimp has been made. Failure to make a secure crimp will create a weak joint.

Step 6 Using a bolt or screw and washers (if required), secure the crimped connector and attached wire to the correct terminal in the equipment. Tighten the terminal bolt and torque to the level specified by the equipment manufacturer. Too little or too much torque can adversely affect the performance of the connection.

Table 3 lists some torque values typical of those used for tightening common sizes of steel and aluminum terminal bolts.

4.2.0 Installing Mechanical-Type Terminals and Connectors

The procedure for installing mechanical-type connectors is basically the same as that described above for compression-type connectors, with the following exceptions. Before installing the mechanical connector on the wire, an oxide-inhibiting joint compound should be applied liberally to the conductor to prevent the formation of surface oxides once the connection is made (also apply the compound to any terminal pad). Following this, the connector is installed on the wire, then the connector's bolt/screw is tightened to the torque level specified by the connector manufacturer. Proper torque is important. Too much torque may sever the wires or break the connector; too little torque can cause overheating and failure.

4.3.0 Installing Specialized Cable Connectors

There are a wide variety of cables that require specially designed connectors, commonly called

Table 3 Recommended Tightening Torques for Various Bolt Sizes

Steel Hardware		Aluminum Hardware	
Bolt Size	Recommended Torque (Inch-Pounds)	Bolt Size	Recommended Torque (Inch-Pounds)
¼–20	80	½–13	300
5/16–18	180	5/8–11	480
3/8–16	240	¾–10	650
½–13	480	—	—
5/8–11	660	—	—
¾–10	1,900	—	—

Demonstration

Demonstrate how to use mechanical connectors.

Laboratory

Have the trainees practice terminating wires using selected crimp-on and mechanical connectors. Note the proficiency of each trainee.

Tightening Torques

INSIDE TRACK

Many terminations and types of equipment are marked with tightening torque information. For items of equipment, torque information is often marked on the equipment and/or is given in the manufacturer's installation instructions. When specific torque requirements are given, always follow the manufacturer's recommendations.

In cases where no tightening requirements are given, guidelines for tightening screw and bolt-type mechanical compression connectors can be found in the *National Electrical Code Handbook* comments pertaining to ***NEC Section 110.14.*** Other good sources of tightening information are manufacturers' catalogs and product literature.

Classroom

Discuss the use of Type MC cable and pass around an example for the trainees to examine.

Demonstration

Demonstrate how to terminate Type MC cable using a typical connector.

Audiovisual

Show Transparency 13 (Figure 13).

terminators, to secure them to equipment enclosures. Normally, all such specialized connectors are supplied with complete instructions for their installation. This section will introduce you to one type of specialized connector designed for use with metal-clad (Type MC) cable. Its construction and installation are typical of many specialized connectors.

Type MC cable is a factory assembly of one or more insulated circuit conductors with or without optical fiber members. It is enclosed in a metallic sheath of interlocking tape or a smooth or corrugated tube. Throughout the industry, there is increasing use of Type MC cable installed in trays and on racks instead of non-armored cable in conduit. At the appropriate locations, the cables are routed from the trays or racks, then along the structure to the various items of equipment. ***NEC Article 330*** governs the installation of Type MC cable. There are several types of connectors that can be used with Type MC cable. The specific connector used is determined by the size and type of cable and the application.

For the purpose of an example, the procedure for installing one manufacturer's weatherproof connector (*Figure 13*) designed for use with Type MC cable is given here:

Step 1 Select the correct connector size. This is normally done by comparing the physical dimensions of the cable to a cross-reference table given in the manufacturer's product literature and/or installation instructions.

Step 2 Strip back the jacket and armor of the cable as needed to meet equipment requirements. Expose the cable armoring further by stripping the cable jacket for a specified distance (L), as shown in *Figure 13.* This distance will be found in the manufacturer's installation instructions.

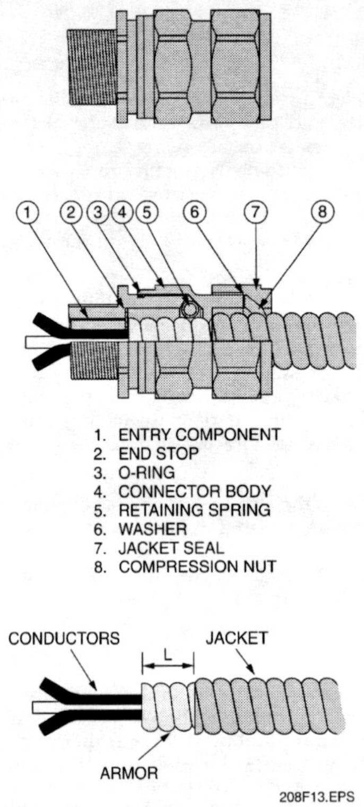

Figure 13 ◆ Weatherproof connector used with Type MC cable.

Instructor's Notes:

494 CHAPTER 12

Step 3 Make sure that the jacket seal and retaining spring are in their uncompressed state. If necessary, loosen the connector body and the compression nut. Note that it is not necessary to separate the connector parts.

Step 4 Screw the connector body into the equipment if it has a threaded entry, or secure it with a locknut if it has an unthreaded entry.

Step 5 Pass the cable through the connector until the armor makes contact with the end stop. If it is not possible for the insulated wires to pass through the end stop, then the end stop should be removed so that the wires can move past it and the armor can make contact with the integral end stop within the entry component.

Step 6 Tighten the connector body to compress the retaining spring and secure the armor. Normally, this is hand-tight plus one and a half full turns.

Step 7 Tighten the outer compression nut to form a seal on the cable jacket. Normally, this is hand-tight plus one full turn.

Step 8 If appropriate, terminate the individual wires contained in the cable using compression or mechanical-type connectors, as described above.

5.0.0 ◆ BENDING CABLE AND TRAINING CONDUCTORS

Training is the positioning of cable so that it is not under tension. Bending is the positioning of cable that is under tension. When installing cable or any large conductors, the object is to limit the tension so that the cable's physical and electrical characteristics are maintained for the expected service life. Training conductors, rather than bending them, also reduces the tension on lugs and connectors, extending their service life considerably.

All bends made in cable must comply with the NEC. Minimum bending radii are determined by the cable diameter and, in some instances, by the construction of the cable. For example, bends in Type MC cable must be made so that the cable is not damaged, and the radius of the curve of the inner edge of any bend shall not be less than the following:

- *Smooth sheath*—Ten times the external diameter of the metallic sheath for cable not more than ¾" in external diameter; twelve times the external diameter of the metallic sheath for cable more than ¾" but not more than 1½" in external diameter; and fifteen times the external diameter of the metallic sheath for cable more than 1½" in external diameter.
- *Interlocked-type armor or corrugated sheath*—Seven times the external diameter of the metal sheath.
- *Shielded conductors*—Twelve times the overall diameter of the individual conductors or seven times the overall diameter of the multiconductor cable, whichever is greater.

The two types of cable bending tools in common use are the ratchet bender and the hydraulic bender (*Figure 14*). The ratchet cable bender in *Figure 14(A)* bends 600V copper or aluminum conductors up to 500 kcmil, while the hydraulic bender in *Figure 14(B)* is designed for cables from 350 kcmil through 1,000 kcmil. In addition, the hydraulic bender is capable of one-shot bends up to 90° and automatically unloading the cable when the bend is finished. Either type simplifies and speeds cable installation.

Conductors at terminals or conductors entering or leaving cabinets or cutout boxes and the like must comply with certain NEC requirements, many of which are covered in ***NEC Article 312.*** The bending radii for various sizes of conductors that do not enter or leave an enclosure through the wall opposite its terminal are shown in *Table 4.* When using this table, the bending space at terminals must be measured in a straight line from the end of the lug or wire connector (in the direction

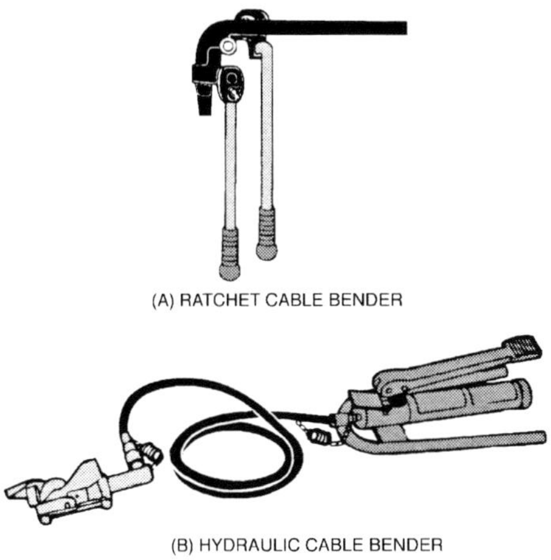

208F14.EPS

Figure 14 ◆ Types of cable bending tools.

Classroom

Explain the terms *training* and *bending.* Discuss cable bending tools and their applications.

Audiovisual

Show Transparency 14 (Figure 14).

Demonstration

Demonstrate the use of ratchet and hydraulic cable benders.

Laboratory

Have the trainees practice using ratchet and hydraulic benders. Note the proficiency of each trainee.

Classroom

Discuss minimum wire bending space and gutter width.

Audiovisual

Show Transparencies 15 and 16 (Figures 15 and 16).

Classroom

Discuss the NEC requirements for the bending space at terminals.

Table 4 Minimum Wire Bending Space and Gutter Width

AWG or Circular-Mil Size of Wire	Wires per Terminal 1	2	3	4	5
14–10	Not Specified	—	—	—	—
8–6	1½	—	—	—	—
4–3	2	—	—	—	—
2	2½	—	—	—	—
1	3	—	—	—	—
1/0–2/0	3½	—	—	—	—
3/0–4/0	4	5	—	—	—
250 kcmil	4½	6	8	—	—
300–350 kcmil	5	6	8	10	—
400–500 kcmil	6	8	10	12	—
600–700 kcmil	8	8	10	12	14
750–900 kcmil	8	10	12	14	16
1,000–1,250 kcmil	10	12	14	16	18
1,500–2,000 kcmil	12	—	—	—	—

that the wire leaves the terminal) to the wall, barrier, or obstruction, as shown in *Figure 15.*

An unshielded cable can tolerate a sharper bend than a shielded cable. This is especially true of cables having helical metal tapes which, when bent too sharply, can separate or buckle and cut into the insulation. The problem is compounded by the fact that most tapes are under jackets that conceal such damage. The shielding bedding tapes or extruded polymers may initially have sufficient conductivity and coverage to pass acceptance testing, but they often fail prematurely at the shield/insulation interface.

> **Note**
> Remember that cable offsets are bends.

When conductors enter or leave an enclosure through the wall opposite its terminals (*Figure 16*), ***NEC Table 312.6(B)*** applies. See *Table 5.* In using this table, the bending space at terminals must be measured in a straight line from the end of the lug or wire connector in a direction perpendicular to the enclosure wall. For removable and lay-in wire terminals intended for only one wire, the bending space in the table may be reduced by the number of inches shown in parentheses.

6.0.0 ◆ NEC TERMINATION REQUIREMENTS

There are several NEC requirements governing the termination of conductors as well as the installation of enclosures containing conductors. ***NEC Sections 110.14 and 312.6*** cover most installations and terminations. However, other sections, such as ***NEC Sections 300.4(F) and 430.10,*** will apply for specific applications.

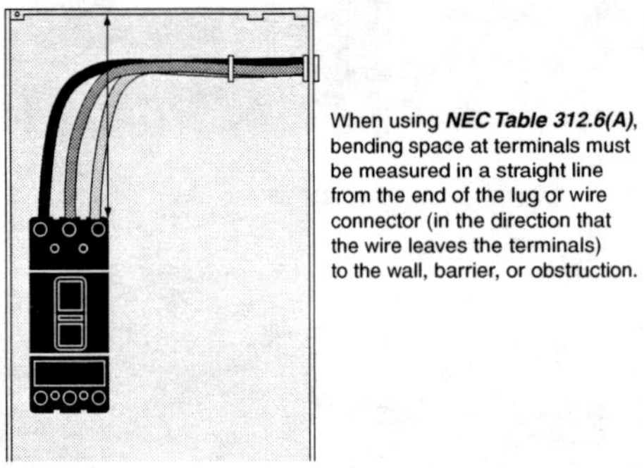
When using NEC Table 312.6(A), bending space at terminals must be measured in a straight line from the end of the lug or wire connector (in the direction that the wire leaves the terminals) to the wall, barrier, or obstruction.
208F15.EPS

Figure 15 ◆ Bending space at terminals is measured in a straight line.

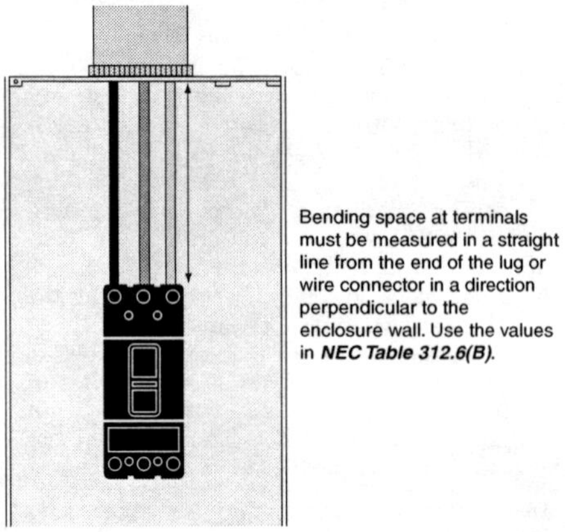
Bending space at terminals must be measured in a straight line from the end of the lug or wire connector in a direction perpendicular to the enclosure wall. Use the values in NEC Table 312.6(B).
208F16.EPS

Figure 16 ◆ Conductors entering an enclosure opposite the conductor terminals.

Instructor's Notes:

496 CHAPTER 12

Table 5 Minimum Wire Bending Space at Terminals

AWG or Circular-Mil Size of Wire	Wires per Terminal 1	2	3	4 or More
14–10	Not Specified	—	—	—
8	1½	—	—	—
6	2	—	—	—
4	3	—	—	—
3	3	—	—	—
2	3½	—	—	—
1	4½	—	—	—
1/0	5½	5½	7	—
2/0	6	6	7½	—
3/0	6½ (½)	6½ (½)	8	—
4/0	7 (1)	7½ (1½)	8½ (½)	—
250	8½ (2)	8½ (2)	9 (1)	10
300	10 (3)	10 (2)	11 (1)	12
350	12 (3)	12 (3)	13 (3)	14 (2)
400	13 (3)	13 (3)	14 (3)	15 (3)
500	14 (3)	14 (3)	15 (3)	16 (3)
600	15 (3)	16 (3)	18 (3)	19 (3)
700	16 (3)	18 (3)	20 (3)	22 (3)
750	17 (3)	19 (3)	22 (3)	24 (3)
800	18	20	22	24
900	19	22	24	24
1,000	20	—	—	—
1,250	22	—	—	—
1,500–2,000	24	—	—	—

6.1.0 Incoming Line Connections

In general, all ungrounded conductors in a motor control center (MCC) installation require some form of overcurrent protection to comply with *NEC Section 240.20.* Such overcurrent protection for the incoming lines to the MCC is usually in the form of fuses or a circuit breaker located at the transformer secondary that supplies the MCC. The conductors from the transformer secondary constitute the feeder to the MCC, and the rules of *NEC Section 240.21* apply. These rules allow the disconnect means and overcurrent protection to be located in the MCC, provided that the feeder taps from the transformer are sufficiently short and other requirements are met.

6.1.1 Main Disconnect

A circuit breaker or a circuit interrupter combined with fuses controlling the power to the entire MCC may provide the required overcurrent protection as described above or there may be a supplementary disconnect (isolation) means. *Figure 17* shows a main disconnect with stab load connectors.

When the MCC has a main disconnect, the incoming lines (feeders) are brought to the line terminals of the circuit breaker or circuit interrupter. The load side of the circuit breaker or the load side of the fuses associated with the circuit interrupter is usually connected to the MCC busbar distribution system. In cases where the main disconnect is rated at 400A or less, the load connection may be made with stab connections to vertical busbars that connect to the horizontal bus distribution system.

6.1.2 Short Circuit Bracing

All incoming lines to either incoming line lugs or main disconnects must be braced to withstand the mechanical force created by a high fault current. If the cables are not anchored sufficiently or the lugs are not tightened correctly, the connections become the weakest part of a panelboard or motor control center when a fault develops. In most cases, each incoming line compartment is equipped with a two-piece spreader bar located at a certain distance from the conduit entry. This spreader bar should be used along with appropriate lacing material to tie cables together where they can be bundled and to hold them apart where they must be separated. In other words, the incoming line cables should first be positioned and then anchored in place.

Manufacturers of electrical panelboards and motor control centers normally furnish detailed information on recommended methods of short circuit bracing; follow this information exactly.

6.1.3 Making Connections

Before beginning work on incoming line connections, refer to all drawings and specifications dealing with the project at hand. Details of terminations are usually furnished on larger installations.

WARNING!

All incoming line compartments present an obvious hazard when the door is opened or covers are removed with power on. When working in this area, the incoming feeder should be deenergized.

Discuss the NEC requirements for main disconnects.

Show Transparency 17 (Figure 17).

Discuss the NEC requirements for:

- **Short circuit bracing**
- **Grounding conductor connections**

CONDUCTOR TERMINATIONS AND SPLICES 497

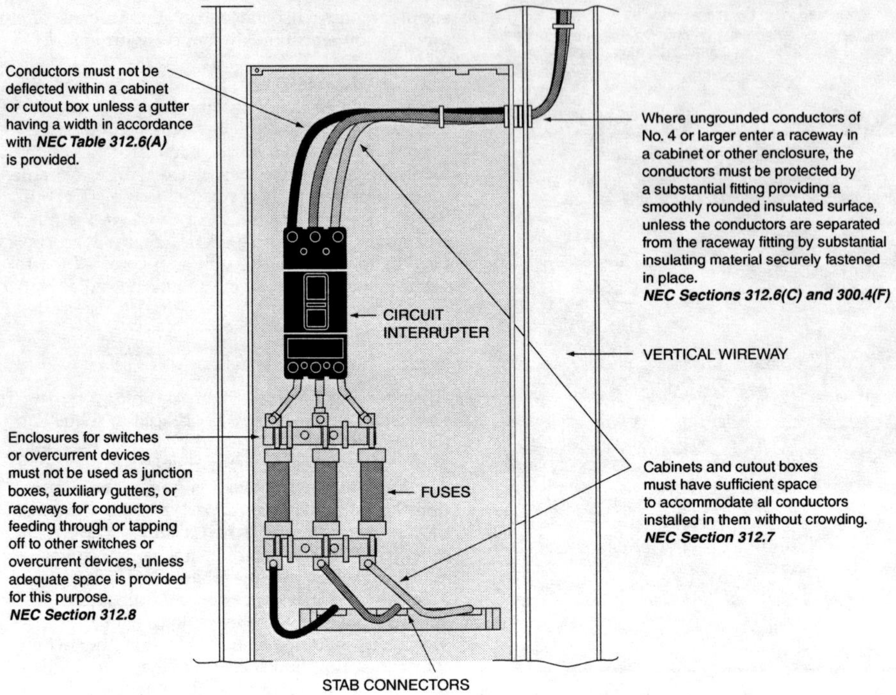

Conductors must not be deflected within a cabinet or cutout box unless a gutter having a width in accordance with NEC Table 312.6(A) is provided.
Where ungrounded conductors of No. 4 or larger enter a raceway in a cabinet or other enclosure, the conductors must be protected by a substantial fitting providing a smoothly rounded insulated surface, unless the conductors are separated from the raceway fitting by substantial insulating material securely fastened in place.
NEC Sections 312.6(C) and 300.4(F)
CIRCUIT INTERRUPTER
VERTICAL WIREWAY
Enclosures for switches or overcurrent devices must not be used as junction boxes, auxiliary gutters, or raceways for conductors feeding through or tapping off to other switches or overcurrent devices, unless adequate space is provided for this purpose.
NEC Section 312.8
FUSES
Cabinets and cutout boxes must have sufficient space to accommodate all conductors installed in them without crowding.
NEC Section 312.7
STAB CONNECTORS
208F17.EPS

Figure 17 ◈ Main disconnect with stab load connectors.

6.1.4 Making Grounding Conductor Connections

NEC Section 250.8 requires that grounding conductors be connected to boxes and enclosures using listed pressure conductors, clamps, or other listed means. When more than one equipment grounding conductor enters a box, all the conductors must be spliced or joined either within the box or to the box using devices suitable for this purpose. The arrangement of the grounding connections must be made so that the disconnection or removal of a receptacle, fixture, etc., fed from the box will not interfere with or interrupt the grounding continuity.

When connecting grounding conductors in metal outlet boxes or enclosures, *NEC Section 250.148* requires that a connection be made between the grounding conductors and the metal box by means of a grounding screw that shall not be used for any other purpose, or a listed grounding device.

Note

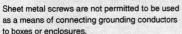

Sheet metal screws are not permitted to be used as a means of connecting grounding conductors to boxes or enclosures.

Another method approved by the NEC for connecting grounding conductors to a metal box is to use a listed grounding device such as a grounding clip. In nonmetallic boxes, the grounding conductors must be connected so that a connection can be made to any fitting or device in the box that requires grounding.

6.2.0 Mounting Enclosures

NEC Article 312 covers the installation and construction specifications of cabinets, cutout boxes, and meter socket enclosures. Part A begins with

Classroom

Discuss the NEC requirements for mounting enclosures.

Instructor's Notes:

Exothermic Welded Connections

INSIDE TRACK

Per **NEC Section 250.64(C)**, grounding electrode conductors may be spliced at any location by means of irreversible compression-type connectors listed for that purpose or by using the exothermic welding process.

In grounding systems, connections are often the weakest link, especially if they are subjected to high currents and corrosion. For this reason, exothermic welded connections are typically used in commercial and industrial structures to splice and/or connect the grounding electrode conductor system in a building. An exothermic welded connection produces a joint or connection that is better than compression-type connectors because of its welded molecular bond. It normally will not loosen, corrode, or increase in resistance over the lifetime of the installation.

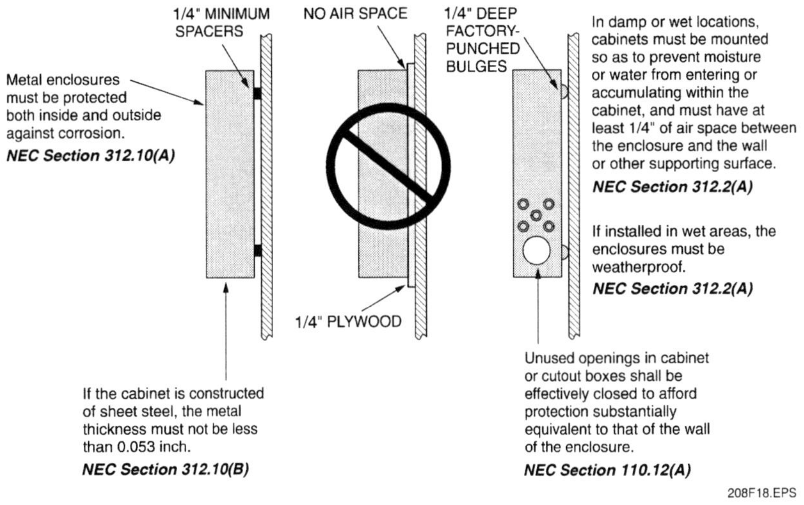

Figure 18 ◈ NEC requirements governing the mounting of enclosures in damp or wet locations.

the installation of such equipment in damp and wet locations. In general, enclosures mounted in damp or wet locations must have a minimum of 1/4" air space between the enclosure and the wall or other supporting surface. If the enclosure is mounted in a wet location, the enclosure must also be weatherproof. See *Figure 18.*

The installation of enclosures used in hazardous (classified) locations must conform to *NEC Articles 500 through 517* and will be covered later in your training.

The NEC requirements governing the position of enclosures in walls is covered in *NEC Section 312.3;* a summary of these requirements appears in *Figure 19.*

7.0.0 ◈ TAPING ELECTRICAL JOINTS

When it is not practical to protect a spliced joint by some other means, electrical tape may be used to insulate the joint. When tape is used, the joint should be taped carefully to provide the same quality of insulation over the splice as over the rest of the wires.

There are a wide variety of electrical **insulating tapes** made from nonconductive materials for use

Audiovisual

Show Transparencies 18 and 19 (Figures 18 and 19).

Classroom

Show various types of electrical tape and explain when each type is used.

CONDUCTOR TERMINATIONS AND SPLICES 499

Figure 19 ◆ NEC requirements governing the position of enclosures in walls.

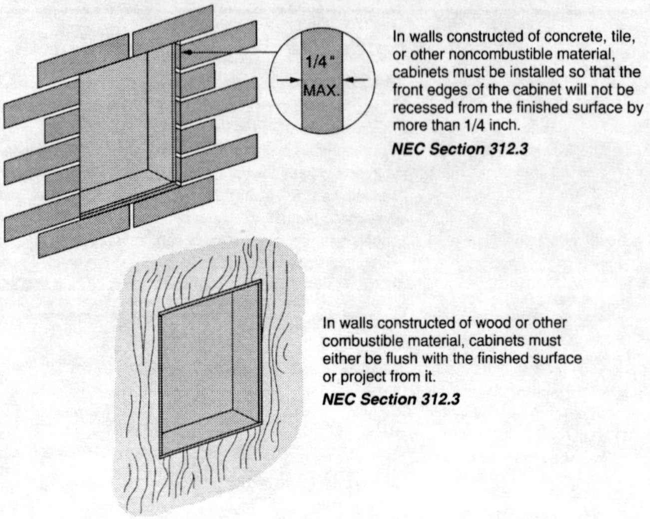
1/4"
MAX.
In walls constructed of concrete, tile, or other noncombustible material, cabinets must be installed so that the front edges of the cabinet will not be recessed from the finished surface by more than 1/4 inch.
NEC Section 312.3
In walls constructed of wood or other combustible material, cabinets must either be flush with the finished surface or project from it.
NEC Section 312.3
208F19.EPS

Demonstration

Demonstrate how to properly tape different types of splices and motor connections.

Audiovisual

Show Transparencies 20 and 21 (Figures 20 and 21B).

in specific applications. Some common types of electrical tape include vinyl plastic tape, linerless rubber tape, high-temperature silicone rubber tape, and glass cloth tape. Electrical tapes made of vinyl plastic are widely used as a primary insulation on joints made with thermoplastic-insulated wires. They are used for splices up to 600V and for fixture and wire splices up to 1,000V. Depending on the product, they are made for indoor use, outdoor use, or both. Linerless rubber splicing tape provides for a tight, void-free, moisture-resistant insulation without loss of electrical characteristics. It is typically used as a primary insulation with all solid dielectric cables through 70kV. Other applications include jacketing on high-voltage splices and terminals, moisture-sealing electrical connections, busbar insulations, and end sealing high-voltage cables. High-temperature silicone rubber tapes are used as a protective overwrap for terminating high-voltage cables. Glass cloth electrical tapes provide a heat-stable insulation for hot-spot applications such as furnace and oven controls, motor leads, and switches. They are also used to reinforce insulation where heavy loads cause high heat and breakdown of insulation, such as in motor control exciter feeds, etc. All-metal braid tapes are also available. These are used to continue electrostatic shielding across a splice. When taping a splice, begin by selecting the correct tape for the job. Always follow the tape manufacturer's recommendations.

A general procedure describing one method of taping a splice or joint, such as encountered when connecting motor lugs, is shown in *Figure 20.* A method for taping a split-bolt connector is shown in *Figure 21.* Prior to taping, you should make sure that the joined lugs are securely fastened together with the appropriate hardware. Pieces of a suitable filler tape (or putty) should be wrapped or molded around the lugs and attaching hardware so as to fill the voids and eliminate any sharp edges. It also helps to provide a smooth, even surface that will make taping easier.

> **Note**
>
> For all splices and joints where it is likely that the tape will have to be removed at some future date to perform work on the joint, an upside down (that is, adhesive side up) wrap of tape should be applied to the joint before applying the final layers of insulating tape to the joint in the usual manner. This will keep the area free from tape residue and facilitate the removal of the tape later on, if necessary.

Instructor's Notes:

500 CHAPTER 12

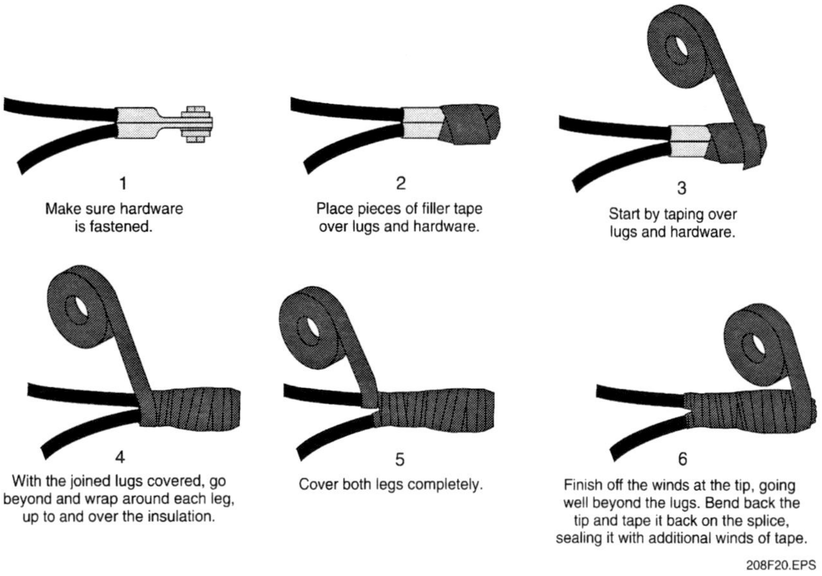
1
Make sure hardware is fastened.
2
Place pieces of filler tape over lugs and hardware.
3
Start by taping over lugs and hardware.
4
With the joined lugs covered, go beyond and wrap around each leg, up to and over the insulation.
5
Cover both legs completely.
6
Finish off the winds at the tip, going well beyond the lugs. Bend back the tip and tape it back on the splice, sealing it with additional winds of tape.
208F20.EPS

Figure 20 ◈ Typical method for taping motor lug connections.

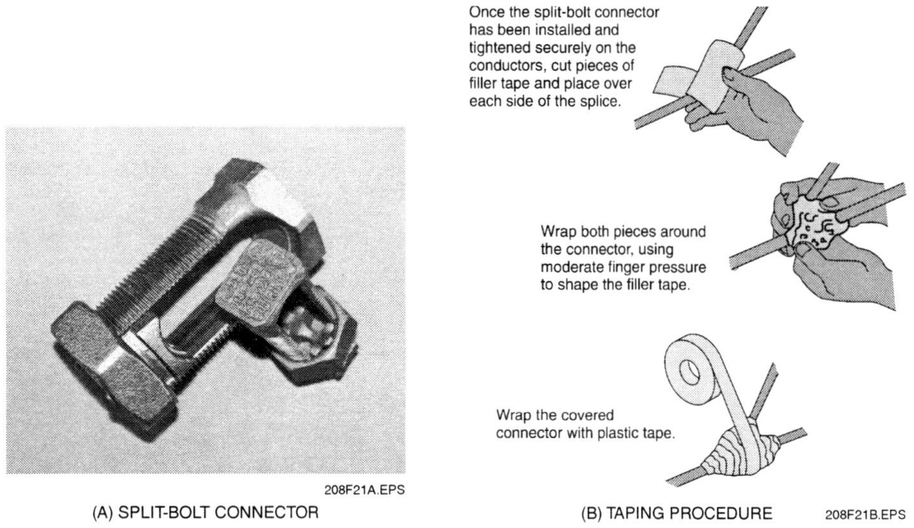
Once the split-bolt connector has been installed and tightened securely on the conductors, cut pieces of filler tape and place over each side of the splice.
Wrap both pieces around the connector, using moderate finger pressure to shape the filler tape.
Wrap the covered connector with plastic tape.
208F21A.EPS
(A) SPLIT-BOLT CONNECTOR
(B) TAPING PROCEDURE
208F21B.EPS

Figure 21 ◈ Typical method of taping a split-bolt connector.

CONDUCTOR TERMINATIONS AND SPLICES 501

Electrical Tape

INSIDE TRACK

Most non-electricians think of electrical tape as only the simple black vinyl variety found in nearly every home toolbox. Electrical tape actually comes in a wide range of colors to be used for labeling various conductors when making terminations. See photo (A). High-voltage electrical tape, as shown in photo (B), can be used to make connections on wires up to 1,000V, depending on the listing of the product being used. Always check the product label to be sure it matches the intended application.

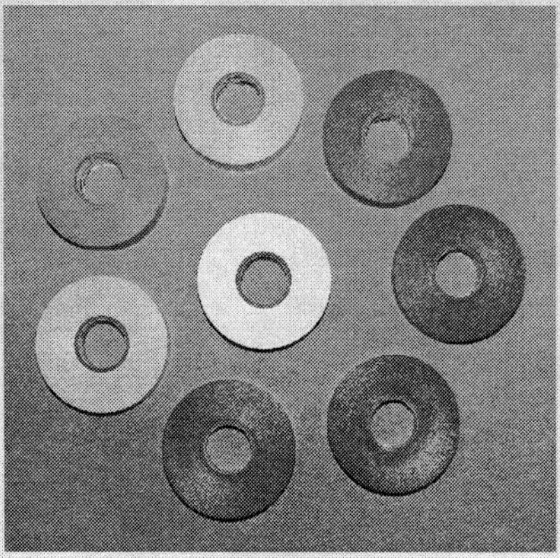

208P0805.EPS

(A)

208P0806.EPS

(B)

8.0.0 ◆ MOTOR CONNECTION KITS

Motor connection kits are available to insulate bolted splice connections, such as those in motor terminal boxes. These kits eliminate the need for taping and the use of filler tape or putty. To aid in joint reentry during rework, the insulator strips off easily, leaving a clean bolt area and thus eliminating the need to remove old tape and putty. Motor connection kits are available for use with stub (butt splice) connections (*Figure 22*) where there is insufficient room to make in-line connections.

They are also made to insulate in-line splice connections where space permits. These insulating kits incorporate a high-voltage mastic, which seals the splice against moisture, dirt, and other contamination. One type of motor connection kit insulator is heat-shrinkable. It installs easily using heat from a propane torch to shrink the insulator in a manner similar to that of heat-shrink tubing (*Figure 23*). Another type of kit used for insulating stub connections comes in the form of an elastomeric insulating cap that is cold-applied by rolling it over the stub splice (*Figure 23*). Always follow the kit manufacturer's recommendations when selecting a kit to use for a particular application. Typically, the kit is selected based on the size of the motor feeder cable.

Making Connections to Equipment Terminals

THINK ABOUT IT

Are there any requirements as to how many conductors can be connected to the terminals provided in electric equipment?

Demonstration

Show examples of heat-shrink and roll-on insulating cap motor connection kits. Demonstrate how to install these kits on stub and in-line motor connection splices.

Audiovisual

Show Transparency 22 (Figure 22).

Laboratory

Have the trainees practice taping several types of wire splices and/or installing a motor connection kit. Note the proficiency of each trainee.

Teaching Tip

Discuss the "Think About It." See the Teaching Tip at the end of this module.

Instructor's Notes:

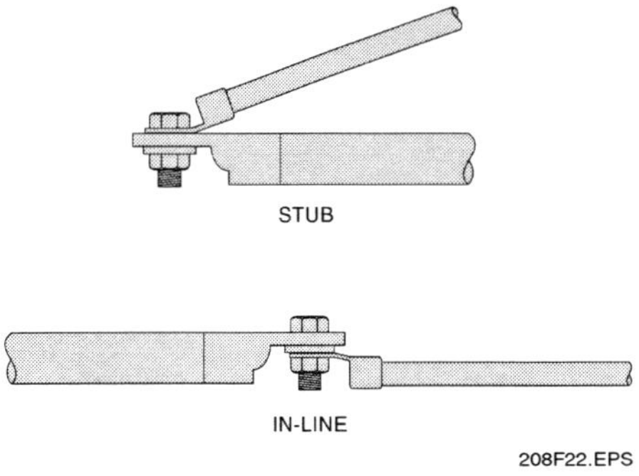

Figure 22 ◈ Stub and in-line splice connections.

THINK ABOUT IT

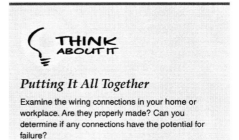

Putting It All Together

Examine the wiring connections in your home or workplace. Are they properly made? Can you determine if any connections have the potential for failure?

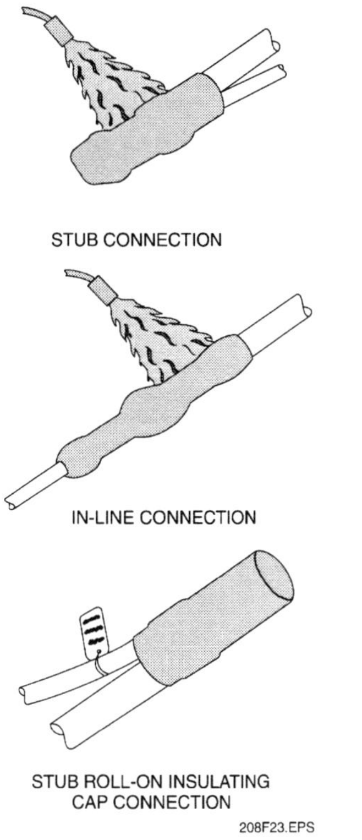

Figure 23 ◈ Motor connection kits installed on splices.

Teaching Tip

Discuss the "Think About It." Have the trainees report their findings to the rest of the class.

CONDUCTOR TERMINATIONS AND SPLICES

Classroom

Have the trainees complete the Review Questions, and go over the answers prior to administering the Module Examination.

Examination

Administer the Module Examination. Record the results on Craft Training Report Form 200 and submit the results to the Training Program Sponsor.

Performance Testing

Administer the Performance Test and fill out Performance Profile Sheets for each trainee. If desired, trainee proficiency noted during laboratory sessions may be used to complete the Performance Test. Be sure to record the results on Craft Training Report Form 200 and submit the results to the Training Program Sponsor.

Summary

Solderless connectors (wire nuts) are devices used to join wires without the need for solder. Such connectors are convenient and save much time on the job. Wire nuts are used to splice smaller conductors (No. 10 AWG and smaller) on residential and some commercial installations.

Crimp connectors or terminals are used for larger wire sizes. They are convenient for terminating conductors at terminal boards, control wiring terminals, and the like.

Lugs are provided for the larger wire sizes on panelboards and motor control centers. It is very important to tighten these lugs properly to provide a sound electrical connection as well as to provide short circuit bracing.

There are a variety of insulating tapes available to seal spliced joints. Always match the tape with the application.

Review Questions

1. How long should a good conductor splice last?
   a. 60 days.
   b. 6 years.
   c. As long as current flows.
   d. As long as the insulation on the wire itself.
2. Which of the following best describes the term *stripping* as it applies to conductor splicing?
   a. Removal of the insulation from conductors.
   b. Removal of the packing material from the carton.
   c. Removal of any excess strands of wire from a splice.
   d. Removal of the pulling ring from lead sheath conductors.
3. Which of the following is *not* a type of stripping?
   a. End termination.
   b. Window cut.
   c. Spiral cut.
   d. Indent cut.
4. What is the diameter, in inches, of 2,000 kcmil wire?
   a. 0.893
   b. 0.998
   c. 1.29
   d. 1.63
5. Which of the following best describes the purpose of a reducing connector?
   a. To temporarily connect a circuit.
   b. To join two different size conductors.
   c. To join conductors of the same size.
   d. To make a 90° bend in parallel conductors.
6. How long should heat be applied to a heat-shrink insulator in order for it to take the required shape?
   a. A few hours.
   b. A few minutes.
   c. A few seconds.
   d. A few days.
7. Which of the following sizes of wire would you normally use wire nuts on?
   a. No. 22 through No. 10 AWG.
   b. No. 8 through No. 2 AWG.
   c. No. 1 through 1/0 AWG.
   d. 4/0 through 250 kcmil.
8. Which of the following tools is best for positioning cable?
   a. Split-bolt connectors.
   b. A heat gun.
   c. A power connector indenter.
   d. Wire-bending tools (either ratchet or hydraulic).
9. Which of the following types of electrical tape is best for taping most joints for voltages up to 600V?
   a. High-temperature silicone rubber.
   b. Rubber.
   c. Glass cloth.
   d. Vinyl plastic.
10. Crimping tools include all of the following *except* _____ .
    a. hand-operated crimping tools
    b. hydraulic compression tools
    c. die sets
    d. pliers

Instructor's Notes:

504 CHAPTER 12

GLOSSARY

Trade Terms Introduced in This Module

AL-CU: An abbreviation for aluminum and copper, commonly marked on terminals, lugs, and other electrical connectors to indicate that the device is suitable for use with either aluminum conductors or copper conductors.

Connection: That part of a circuit that has negligible impedance and joins components or devices.

Connector: A device used to physically and electrically connect two or more conductors.

Insulating tape: Adhesive tape that has been manufactured from a nonconductive material and is used for covering wire joints and exposed parts.

Lug: A device for terminating a conductor to facilitate the mechanical connection.

Pressure connector: A connector applied using pressure to form a cold weld between the conductor and the connector.

Reducing connector: A connector used to join two different sized conductors.

Splice: The electrical and mechanical connection between two pieces of cable.

Strand: A group of wires, usually stranded or braided.

Tap: A splice connection of a wire to another wire. Also called a *tap-off.*

Terminal: A device used for connecting cables.

Termination: The connection of a cable.

Wire nut: A form of wire connector that is tightened or loosened by screwing the device clockwise or counterclockwise, respectively. Some wire nuts are provided with two thin, flat wings extending from opposite sides to facilitate tightening the connectors.

CONDUCTOR TERMINATIONS AND SPLICES 505

Additional Resources

This module is intended to present thorough resources for task training. The following reference works are suggested for further study. These are optional materials for continued education rather than for task training.

American Electricians' Handbook, Latest Edition. New York: McGraw-Hill.

National Electrical Code Handbook, Latest Edition. Quincy, MA: National Fire Protection Association.

Instructor's Notes:

Answers to Review Questions

Answer	Section
1. d	1.0.0
2. a	2.0.0
3. d	2.2.0/Fig. 3
4. d	2.0.0/Tab. 1
5. b	3.0.0
6. c	3.2.0/Fig. 7
7. a	3.3.0
8. d	5.0.0
9. d	7.0.0
10. d	4.1.1

Section 3.0.0 *Think About It—Mechanical Compression Connectors*

A compression connector with a two-hole tongue is ideal for bolting to busbars where two bolts prevent the connector from turning.

Section 8.0.0 *Think About It—Making Connections to Equipment Terminals*

The Underwriters Laboratory General Information Directory states that "product terminals (including wire connectors and terminal screws) are acceptable for connection of only one conductor, unless there is a marking or wiring diagram indicating the number of conductors which may be connected."

An example is a panelboard used as service equipment where the neutral bar is listed for single conductors per terminal. However, it may be permitted to have two equipment grounding conductors (not neutrals) under a single termination when marked on the panelboard label. Another example is that some circuit breaker manufacturers permit two conductors to be connected to a circuit breaker terminal.

CONTREN™ LEARNING SERIES — USER UPDATES

The NCCER makes every effort to keep these textbooks up-to-date and free of technical errors. We appreciate your help in this process. If you have an idea for improving this textbook, or if you find an error, a typographical mistake, or an inaccuracy in NCCER's Contren™ textbooks, please write us, using this form or a photocopy. Be sure to include the exact module number, page number, a detailed description, and the correction, if applicable. Your input will be brought to the attention of the Technical Review Committee. Thank you for your assistance.

Instructors – If you found that additional materials were necessary in order to teach this module effectively, please let us know so that we may include them in the Equipment/Materials list in the Instructor's Guide.

Write: Curriculum Revision and Development Department
National Center for Construction Education and Research
P.O. Box 141104, Gainesville, FL 32614-1104

Fax: 352-334-0932

E-mail: curriculum@nccer.org

Craft ______________________ Module Name ______________________

Copyright Date __________ Module Number __________ Page Number(s) __________

Description

____

____

____

____

(Optional) Correction

____

____

____

(Optional) Your Name and Address

____

____

____

Photo Credits

Chapter 1

Tim Ely, 101P0102.EPS
Mike Powers, 101P0101.EPS, 101P0103.EPS, 101P0104.EPS, 101P0105.EPS, 101P0106.EPS, 101P0107.EPS
Veronica Westfall, 101F02.EPS, 101F08.EPS, 101F09.EPS

Chapter 2

Veronica Westfall, 104F08B.EPS

Chapter 3

Walter Johnson, 105P0501.EPS

Chapter 4

Walter Johnson, 201P0101.EPS
Veronica Westfall, 201F13C.EPS

Chapter 5

Walter Johnson, 106F09.EPS, 106P0602.EPS
Veronica Westfall, 106P0601.EPS, 106P0603.EPS, 106P0604.EPS, 106P0605.EPS, 106P0606.EPS

Chapter 6

Tim Ely, 107P0703.EPS, 107P0704.EPS
Walter Johnson, 107P0701.EPS, 107P0702.EPS

Chapter 7

Tim Ely, 108P0805.EPS
Walter Johnson, 108P0801.EPS, 108F04.EPS, 108P0803.EPS, 108P0808.EPS
Veronica Westfall, 108P0802.EPS, 108F08.EPS, 108F12.EPS, 108F13A.EPS, 108F13B.EPS, 108F14.EPS, 108F15.EPS, 108P0804.EPS, 108F16.EPS, 108F17.EPS, 108F18.EPS, 108F19.EPS, 108F23.EPS, 108P0806.EPS, 108F28.EPS, 108P0807.EPS, 108F47.EPS, 108P0809.EPS

Chapter 8

Walter Johnson, 103F35.EPS, 103P0305.EPS
Veronica Westfall, 103P0301.EPS, 103P0302.EPS, 103P0303.EPS, 103P0304.EPS, 103P0306.EPS, 103P0307.EPS, 103F46B.EPS, 103P0308.EPS, 103P0309.EPS

Chapter 9

Veronica Westfall, 102F01.EPS, 102P0201.EPS, 102F03.EPS, 102F04.EPS, 102F05.EPS, 102P0202.EPS, 102P0203.EPS, 102F10B.EPS, 102F11B.EPS, 102F24B.EPS, 102F25.EPS, 102F26B.EPS, 102F27.EPS, 102P0204.EPS, 102P0205.EPS, 102P0206.EPS

Chapter 10

Walter Johnson, 205F25.EPS
Veronica Westfall, 205F01.EPS, 205F04.EPS, 205F06.EPS, 205F15.EPS, 205F16.EPS, 205F17.EPS, 205F18.EPS, 205F20C.EPS, 205P0501.EPS, 205P0502.EPS, 205P0503.EPS, 205P0504.EPS, 205F24.EPS, 205P0505.EPS, 205F30A.EPS, 205F30B.EPS, 205F31.EPS

Chapter 11

Walter Johnson, 109F08B.EPS, 109F09B.EPS
Veronica Westfall, 109F07B.EPS, 109F13.EPS, 109P0901.EPS, 109F15B.EPS, 109F17.EPS, 109P0902.EPS, 109P0903.EPS, 109F20A.EPS, 109F20B.EPS, 109F26.EPS, 109P0905.EPS

Chapter 12

Veronica Westfall, 208F01.EPS, 208P0801.EPS, 208F05.EPS, 208P0802.EPS, 208P0803.EPS, 208P0804.EPS, 208P0805.EPS, 208P0806.EPS, 208F21A.EPS

Index

NOTE: *f* refers to figures; *t* refers to tables.